Sixth Edition

CALCULUS
with Analytic Geometry

Student's Solutions Manual

Sixth Edition

CALCULUS

with Analytic Geometry

Dale Varberg
Hamline University

Edwin J. Purcell
University of Arizona

prepared by

Louis Guillou

Prentice Hall, Englewood Cliffs, New Jersey 07632

Editorial Production/ Supervision: *Benjamin D. Smith*
Supplement Acquisitions Editor: *Susan Black*
Acquisitions Editor: *Steven Conmy*
Manufacturing Buyers: *Paula Massenaro & Lori Bulwin*

©1992 by Prentice-Hall, Inc.
A Simon & Schuster Company
Englewood Cliffs, New Jersey 07632

Printed in the United States of America

10   9   8   7   6

ISBN 0-13-118035-5

Prentice-Hall International (UK) Limited, *London*
Prentice-Hall of Australia Pty. Limited, *Sydney*
Prentice-Hall Canada Inc. *Toronto*
Prentice-Hall Hispanoamericana, S.A., *Mexico*
Prentice-Hall of India Private Limited, *New Delhi*
Prentice-Hall of Japan, Inc., *Tokyo*
Simon & Schuster Asia Pte. Ltd., *Singapore*
Editora Prentice-Hall do Brasil, Ltda., *Rio de Janeiro*

TABLE OF CONTENTS

Sixth Edition

CALCULUS

with Analytic Geometry

Problem Set 1.1   The Real Number System

1. 10

3. $-4[3(-6 + 13) - 2(5 - 9)] = -4[3(7) - 2(-4)] = -4[21 + 8] = -116$

5. $-1/12 \approx -0.0833$ **7.** $1/24 \approx 0.0417$

9. $\dfrac{14}{33}\left[\dfrac{2}{3} - \dfrac{1}{7}\right]^2 = \dfrac{14}{33}\left[\dfrac{14}{21} - \dfrac{3}{21}\right]^2 = \dfrac{14}{33}\left[\dfrac{14 - 3}{21}\right]^2 = \dfrac{14}{33}\dfrac{11}{21}\dfrac{11}{21} = \dfrac{22}{189} \approx 0.1164$

11. $-5/16 \approx 0.3125$ **13.** $3/11 \approx 0.2727$

15. $(\sqrt{2} + \sqrt{3})(\sqrt{2} - \sqrt{3}) = \sqrt{4} - \sqrt{9} = 2 - 3 = -1$

17. -6 **19.** $36/49 \approx 0.7347$

21. $(2x-3)(2x+3) = 2x(2x+3) - 3(2x+3) = 4x^2 + 6x - 6x - 9 = 4x^2 - 9$

23. $6x^2 - 15x - 9$ **25.** $9t^4 - 6t^3 + 7t^2 - 2t + 1$

27. $\dfrac{x^2 - 4}{x - 2} = \dfrac{(x+2)(x-2)}{x-2} = x + 2 \;(x \neq 2)$

29. $\dfrac{(x-2)(x^2+2x+4)}{2(x-2)} = \dfrac{x^2+2x+4}{2} \;(x \neq 2)$

31. $\dfrac{18 - 4(x+3) + 6x}{x(x+3)} = \dfrac{2(x+3)}{x(x+3)} = \dfrac{2}{x} \;(x \neq -3)$

33. $\dfrac{x^2+x-6}{x^2-1} \cdot \dfrac{x^2+x-2}{x^2+5x+6} = \dfrac{(x+3)(x-2)}{(x+1)(x-1)}\dfrac{(x+2)(x-1)}{(x+3)(x+2)} = \dfrac{x-2}{x+1} \;(x \neq -3, -2, 1)$

35. (a) 0   (b) undefined   (c) 0   (d) undefined   (e) 1   (f) 0

---

37. (a) False since $(-20) - (-2) = -18$ which is not positive.

(b) True since $(1) - (-39) = 40$ which is positive.

(c) True since $(5/9) - (-3) = 32/9$ which is positive.

(d) True since $(-4) - (-16) = 12$ which is positive.

(e) True since $(34/39) - (6/7) = 4/273$ which is positive.

(f) False since $(-44/59) - (-5/7) = -13/413$ which is not positive.

39. $a < b \Rightarrow a+a < a+b$ and $a+b < b+b$ [To each side add a; b.]
     $\Rightarrow 2a < a+b < 2b$, from which the result immediately follows.

41. (a) $2^4 \cdot 3 \cdot 5$ (b) $2 \cdot 5 \cdot 31$ (c) $7 \cdot 17$ (d) $2^3 \cdot 3^3 \cdot 5^2$

43. Suppose $\sqrt{2} = p/q$ where p and q are natural numbers. Neither p nor
     q can equal 1. (If q = 1, then $\sqrt{2}$ = p, a natural number. If p = 1,
     then $\sqrt{2}$ would be less than 1; neither is true so $p \neq 1$ and $q \neq 1$.)
     $$\sqrt{2} = p/q \Rightarrow 2 = p^2/q^2 \Rightarrow p^2 = 2q^2.$$
     Therefore, p^2 has one more prime factor than q^2 (an extra factor of
     2). Then, either p^2 or q^2 has an odd number of prime factors (the
     other having an even number). This contradicts the result obtained in
     Problem 42. We conclude that $\sqrt{2}$ is not a rational number.

45. Let the two rational numbers be a/b and c/d where a,b,c,d are inte-
     gers and b,d are not zero. Then the sum is (ad+bc)/bd which is
     rational since (ad+bc) and bd are integers, and bd is not zero.

47. (a) Rational   (b) Rational   (c) Irrational

(d) Irrational   (e) Rational   (f) Irrational

49. Assume $\sqrt{m} = p/q$; then $m = p^2/q^2$. Consider the prime power decomposition
     of m, of p^2, and of q^2. Each prime factor of p^2 and each prime
     factor of q^2 appears an even number of times. Therefore, each prime
     factor of m occurs an even number of times, so m is a perfect square,
     contradicting the hypothesis of the problem.

51. Assume $\sqrt{2} - \sqrt{3} + \sqrt{6} = \frac{p}{q}$ for some integers p and q, $q \neq 0$. Then $\sqrt{2} +$
     $\sqrt{6} = \frac{p}{q} + \sqrt{3}$. Now square each side and solve for $\sqrt{3}$. $\sqrt{3} = \frac{p^2-5q^2}{2q(2q-p)}$, a
     rational number, contradicting result of Problem 44.

Problem Set 1.2 Decimals, Denseness, Calculators

1. 0.875

3.
$$
\begin{array}{r}
.15 \\
20\,\overline{)3.00} \\
2\ 0 \\
\hline
1\ 00 \\
1\ 00 \\
\hline
\end{array}
$$

5. 3.666 ...

7. 999x = 123, so x = 123/999.

9. Let x = 2.56 56 56···[Cycle has a 2-digit length so multiply each
        100x = 256.56 56 56 ···   side by 10^2.]
        Therefore, 99x = 254, so x = 254/99.

11. 9x = 1.8, so x = 18/90.

13. Each nonzero rational number which has a terminating decimal
        representation also has a nonterminating decimal representation which can
        be obtained from the terminating decimal representation by decreasing the
        last nonzero digit by one, and then following that by a nondeterminating
        string of 9's. [Also see the next problem.]

15. 0.000001 is a positive rational number less than 0.00001.
        0.00000101001000100001··· is an irrational number less than 0.00001.
            [Even though there is a pattern, it is not a repeating decimal.]

17. 3.1415910101000100001···is an irrational number between 3.14159 and π.

19. 0.9999···equals 1 so there is no number between 0.9999··· and 1.

21. Irrational. It is not a repeating decimal.

23. 14.071067812

25. 0.0449154472

27. 12.433227831

29. 78760.570071

31. 9.6221715736

33. $\pi r^2 h \approx 3(1.5)^2(17) \approx 115 \text{ in}^3$

35. $2\pi r \approx (6.3)(4000)(5000) \approx 126{,}000{,}000 \text{ ft.}$

37. $\pi r^2 h(12) \approx \pi (64)(270)(12) \approx 651{,}000 \text{ broad ft.}$

39. Using four-decimal-place accuracy:
        (a) 25.4828   (b) 9.1692   (c) 2046.9136

41. (a) -2   (b) -2   (c) $2.\overline{4} = 22/9$ (d) 1   (e) 3/2   (f) $\sqrt{2}$

3

Problem Set 1.3   Inequalities

1. (a)
   -4   1   x

(b)
   -4   1   x

(c)
   -4   1   x

(d)
   -4   1   x

(e)
   1   x

(f)
   -4   x

3. $4x - 7 < 3x + 5$.
   $x < 12$.
   Solution set: $(-\infty, 12)$.

Graph:
   12   x

5. $x \geq -5/3$.
   Solution set: $[-5/3, \infty)$

Graph:
   -5/3   x

7. $x > 2$.
   Solution set: $(2, \infty)$.

Graph:
   2   x

9. $-6 < 2x + 3 < -1$.
   $-9 < 2x < -4$.
   $-9/2 < x < -2$.
   Solution set: $(-9/2, -2)$.

Graph:
   -9/2   -2   x

11. $-2/5 \leq x < 3/5$
   Solution set: $[-2/5, 3/5)$.

Graph:
   -2/5   3/5   x

13. $2 + 3x < 5x + 1$ and $5x + 1 < 16$.
   $x > 1/2$ and $x < 3$.
   $1/2 < x < 3$.
   Solution set: $(1/2, 3)$

Graph:
   1/2   3   x

15. $x^2 + x - 12 < 0$.
   $(x + 4)(x - 3) < 0$.
   Split points: $-4, 3$.

Test points: $-5, 0, 4$.

Auxiliary axis for $(x+4)(x-3)$:

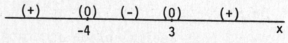
   (+)   (0)   (-)   (0)   (+)
   -4   3   x

Solution set: $(-4, 3)$.

Graph: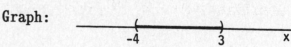
   -4   3   x

17. $(3x+1)(x-4) \leq 0$.
     Solution set: $[-1/3, 4]$.   Graph:

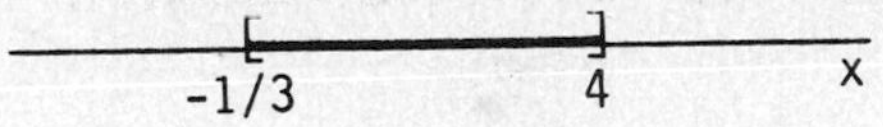

19. $(2x-1)(x+3) > 0$.
     Solution set: $(-\infty, -3) \cup (1/2, \infty)$.   Graph:

21. $\dfrac{x+5}{2x-1} \leq 0$
     Split points: $-5, 1/2$ Test points: $-6, 0, 1$

     Auxiliary axis for $\dfrac{x+5}{2x-1}$:

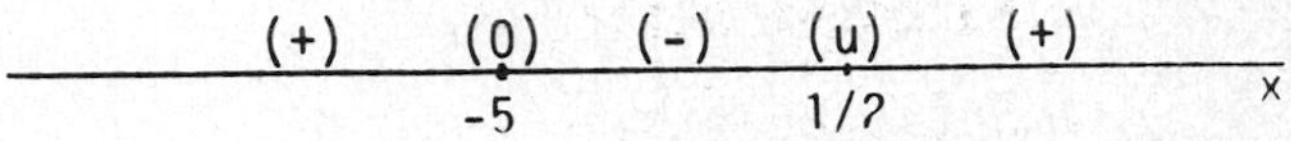

     Solution set: $[-5, 1/2)$ Graph:

23. $(1-5x)/x < 0$
     Solution set: $(-\infty, 0) \cup (1/5, \infty)$ Graph:

25. $3(4x-3)/(3x-2) \geq 0$
     Solution set: $(-\infty, 2/3) \cup [3/4, \infty)$ Graph:

27. $\dfrac{x-2}{x+4} < 2$

     $\dfrac{x-2}{x+4} - 2 < 0$

     $\dfrac{x-2 - 2(x+4)}{x+4} < 0$

     $\dfrac{-x-10}{x+4} < 0$

     Split points: $-10, -4$ Test points: $-11, -6, 0$

     Auxiliary axis for $\dfrac{-x-10}{x+4} < 0$:

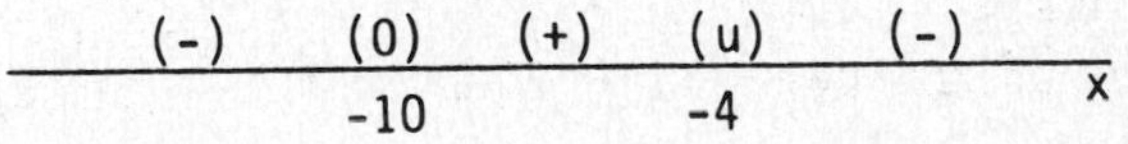

     Solution set: $(-\infty, -10) \cup (-4, \infty)$ Graph:

29. $(x+2)(2x-1)(3x+7) \geq 0$
Solution set: $[-7/3,-2] \cup [1/2,\infty)$ Graph:

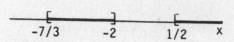

31. $(2x+3)(3x-1)^2(x-5) < 0$
Solution set: $(-3/2,1/3) \cup (1/3,5)$ Graph:

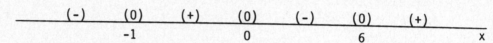

33. $x^3 - 5x^2 - 6x < 0$

$x(x^2 - 5x - 6) < 0$
$x(x+1)(x-6) < 0$

Split points: $-1,0,6$ Test points: $-2,-1/2,1,7$

Auxiliary axis for $x(x+1)(x-6)$:

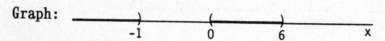

Solution set: $(-\infty,-1) \cup (0,6)$

Graph:

35. (a) $x > -2$ and $x < 1$
       $-2 < x < 1$
       Solution set: $(-2,1)$

(b) $x > -2$ and $x > -5/2$
    $x > -2$
    Solution set: $(-2,\infty)$

(c) $x > -2$ and $x < -5/2$
    Solution set: $\emptyset$

37. (a) $(x+1)(x^2+2x-7) - (x+1)(x-1) \geq 0;$ $(x+1)(x^2+x-6) \geq 0$
$(x+1)(x+3)(x-2) \geq 0$

Auxiliary axis for $(x+1)(x+3)(x-2)$:

$$\frac{(-)\quad(0)\quad(+)\quad(0)\quad(-)\quad(0)\quad(+)}{\quad\ -3\qquad\ -1\qquad\quad 2\qquad\ x}$$

Solution set: $[-3,-1]\cup[2,\infty)$

(b) $x^4-2x^2-8 \geq 0$ $(x^2-4)(x^2+2) \geq 0;$ $(x+2)(x-2)(x^2+2) \geq 0;$ $(x+2)(x-2) \geq 0$

Auxiliary axis for $(x+2)$ $(x-2)$:

$$\frac{(+)\qquad(0)\qquad(-)\qquad(0)\qquad(+)}{\qquad -2\qquad\qquad 2\qquad\qquad x}$$

Solution set: $(-\infty,-2]\cup[2,\infty)$

(c) $[(x^2+1)-2][(x^2+1)-5] < 0;$ $(x^2-1)(x^2-4) < 0$

$(x+1)(x-1)(x+2)(x-2) < 0$

Auxiliary axis for $(x+1)(x-1)(x+2)(x-2)$:

$$\frac{(+)\quad(0)\quad(-)\quad(0)\quad(+)\quad(0)\quad(-)\quad(0)\quad(+)}{\quad\ -2\qquad -1\qquad\quad 1\qquad\quad 2\qquad\ x}$$

Solution set: $(-2,-1)\cup(1,2)$

39. $\frac{1}{20} \leq \frac{1}{R_1} \leq \frac{1}{10}, \frac{1}{30} \leq \frac{1}{R_2} \leq \frac{1}{20},$ and $\frac{1}{40} \leq \frac{1}{R_3} \leq \frac{1}{30}$

Therefore, $\frac{1}{20} + \frac{1}{30} + \frac{1}{40} \leq \frac{1}{R_1} + \frac{1}{R_2} + \frac{1}{R_3} \leq \frac{1}{10} + \frac{1}{20} + \frac{1}{30}$

$$\frac{13}{120} \leq \frac{1}{R} \leq \frac{11}{60}; \qquad \frac{120}{13} \geq R \geq \frac{60}{11}; \qquad \frac{60}{11} \leq R \leq \frac{120}{13}$$

Problem Set 1.4   Absolute Values, Square Roots, Squares

1. $-5 < x < 3$
Solution set: $(-5,3)$

3. $|3x+4| < 8$
$-8 < 3x+4 < 8$
$-12 < 3x < 4$
$-4 < x < 4/3$ Solution set: $(-4, 4/3)$

5. $-12 \le x \le 24$
Solution set: $[-12,24]$

7. $x < 2$ or $x > 5$
Solution set: $(-\infty,2) \cup (5,\infty)$

9. $|4x+2| \ge 10$
$4x+2 \le -10$ or $4x+2 \ge 10$

$4x \le -12$ or $4x \ge 8$
$x \le -3$ or $x \ge 2$ Solution set: $(-\infty,-3] \cup [2,\infty)$

11. $(3x+5)/x < 0$ or $(x+5)/x > 0$
Solution set:
$(-\infty,-5) \cup (-5/3,0) \cup (0,\infty)$

13. $x = \dfrac{5\pm\sqrt{57}}{4} \approx -0.6, \; 3.1$
Solution set: $[-0.6,3.1]$
    (approximate)

15. $4x^2+x-2 > 0$

$x = \dfrac{-1 \pm \sqrt{33}}{8} \approx -0.8, \; 0.6$ if $4x^2+x-2 = 0$
Split points: $-0.8, \; 0.6$ Test points: $-1,0,1$

Auxiliary axis for $4x^2+x-2$: $\underset{\qquad -0.8 \qquad\qquad\quad 0.6 \qquad\qquad}{\dfrac{(+)\qquad (0)\qquad (-)\qquad (0)\qquad (+)}{\qquad\qquad\qquad\qquad\qquad\qquad\qquad x}}$

Solution set: $(-\infty,-0.8) \cup (0.6,\infty)$ (approximate)

17. $|x-3| < 0.5 \Rightarrow 5|x-3| < 2.5 \Rightarrow |5x-15| < 2.5$

19. $|x-2| < \epsilon/6 \Rightarrow 6|x-2| < \epsilon \Rightarrow |6x-12| < \epsilon$

21. $|x-5| < \delta \Rightarrow 3|x-5| < 3\delta$

$\Rightarrow |3(x-5)| < 3\delta$

$\Rightarrow |3x-15| < 3\delta$

$\Rightarrow |3x-15| < \epsilon$ if $\delta = \epsilon/3$

23. $|6x+36| < \epsilon \Leftrightarrow |x+6| < \epsilon/6$,   so let $\delta = \epsilon/6$

25. $\left|d - \dfrac{10}{\pi}\right| < \dfrac{.02}{\pi} \approx 0.0064$ in

27. $(x-2)^2 < 9(x+7)^2$
    $(4x+19)(2x+23) > 0$
    Solution set: $(-\infty, -11,5) \cup (-4.75, \infty)$

29. $2|2x-3| < |x+10|$

    $4(2x-3)^2 < (x+10)^2$ [Squaring each side]

    $4(4x^2-12x+9) < x^2+20x+100$

    $16x^2-48x+36 < x^2+20x+100$

    $15x^2-68x-64 < 0$

    $(3x-16)(5x+4) < 0$

    Split points: $-4/5, 16/3$ Test points: $-1, 0, 6$

    Auxiliary axis for $(3x-16)(5x+4)$:

    $$\frac{\overset{(+)}{} \overset{(0)}{\underset{-4/5}{}} \overset{(-)}{} \overset{(0)}{\underset{16/3}{}} \overset{(+)}{}}{ x}$$

    Solution set: $(-4/5, 16/3)$

31. (1) Multiply each side by nonnegative number $|x|$; by $|y|$.
    (2) Transitivity of "<" and definition of squaring.
    (3) $|a||b| = |ab|$ property and definition of "$|\ \ |$" [since x^2 and y^2 are nonnegative].
    (4) Same as for (3).
    (5) Subtract $|y|^2$ from each side.
    (6) Factor (Distributivity).
    (7) Divide each side by $(|x|+|y|)$ which is positive since $x^2 < y^2$.
    (8) Add $|y|$ to each side.

33. (a) $|a-b| = |a+(-b)| \le |a|+|-b| = |a|+|b|$.

    (b) $|a| = |(a-b)+b| \le |a-b|+|b|$, so $|a|-|b| \le |a-b|$.

    (c) $|a+b+c| = |(a+b)+c| \le |a+b|+|c| \le |a|+|b|+|c|$.

35. $\left|\dfrac{x-2}{x^2+9}\right| = \dfrac{|x-2|}{|x^2+9|} \le \dfrac{|x|+|-2|}{x^2+9} \le \dfrac{|x|+2}{9}$.

37. $|x^4 + x^3/2 + x^2/4 + x/8 + 1/16| \le x^4 + |x|^3/2 + x^2/4 + |x|/8 + 1/16$.
    $\le 1 + 1/2 + 1/4 + 1/8 + 1/16 < 2$.

39. $(a-1/a)^2 \ge 0 \Rightarrow a^2 - 2 + 1/a^2 \ge 0 \Rightarrow a^2 + 1/a^2 \ge 2$.

9

41. $0 < a < b \Rightarrow aa < ab$ and $ab < bb$

$\qquad\qquad \Rightarrow a^2 < ab < b^2$

$\qquad\qquad \Rightarrow a < \sqrt{ab} < b$ (by Problem 29)

43. $[(a+b)/2]^2 \geq ab$ (by Problems 41 and 31)

Problem Set 1.5   The Rectangular Coordinate System

1. $\sqrt{9+16} = 5$ $\qquad\qquad\qquad$ **3** $\sqrt{(2-4)^2 + (4-2)^2} = \sqrt{8} \approx 2.8284$

5. $\sqrt{(\pi - 1.232)^2 + (\sqrt{2} - 4.153)^2} \approx 3.339$

7. Let $A = (5,3)$, $B = (-2,4)$, $C = (10,8)$. $d(A,B) = d(A,C) = \sqrt{50}$

9. See the diagram following Case II.

(Case I) $(3,-1)$ and $(3,3)$ are diagonally opposite vertices. The diagonal is vertical, its midpoint is $(3,1)$, and its length is 4. The other pair of vertices is $(1,1)$ and $(5,1)$.

(Case II) $(3,-1)$ and $(3,3)$ are end points of a side. There are two possible pairs of other vertices: $(-1,-1)$ and $(-1,3)$, and $(7,-1)$ and $(7,3)$.

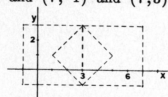

11. The midpoint of the segment joining $(-2,-2)$ and $(4,3)$ is $(1,1/2)$. The distance between this point and (-23) is $\sqrt{9 + 6.25} \approx 3.9051$.

13. $(x-1)^2 + (y+2)^2 = 36$

15. The radius is $d[(2,-1), (5,3)] = \sqrt{(5-2)^2 + (3+1)^2} = 5$
Equation of circle: $(x-2)^2 + (y+1)^2 = 25$

17. The center is $(1,5)$, the midpoint of segment AB, and the radius is $\sqrt{13}$, the distance between $(1,5)$ and $(3,8)$; equation is $(x-1)^2 + (y-5)^2 = 13$.

19. Solving $(3-1)^2 + (y+2)^2 = 36$, obtain $y = -2 \pm \sqrt{32} \approx -7.6569,\ 3.6569$.

21. $x^2+y^2+2x-10y+25 = 0$

$x^2+2x \qquad + y^2-10y \qquad = -25$

$(x^2+2x+1) + (y^2-10y+25) = -25+1+25$

$(x+1)^2 + (y-5)^2 = 1$

Center: $(-1,5)$ Radius: 1

23. $(x-6)^2 + y^2 = 1$; center is $(6,0)$; radius is 1.

25. $(x+1/2)^2 + (y-3/2)^2 = 9/4$; center is $(-1/2,3/2)$; radius is 3/2.

27. Length of each side is 4. Therefore, the radius of the inscribed circle is 2. Its center is $(4,1)$, the midpoint of the diagonals. Then the equation is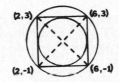

$(x-4)^2 + (y-1)^2 = 4.$

Length of a diameter is $\sqrt{16+16} = \sqrt{32}$. Therefore, the radius of the

circumscribed circle is $\dfrac{\sqrt{32}}{2} = \sqrt{8}$.

Its center is also $(4,1)$. The equation is

$(x-4)^2 + (y-1)^2 = 8.$

29. $d(A,C) = \sqrt{77837}$.

Cost by plane $= 4.82\sqrt{77837} = 1344.75$ dollars

Cost by truck $3.71(214 + 179) = 1458.03$ dollars.

31. Placing the triangle as indicated yields

$d(A,B) = \sqrt{a^2+b^2}$ and $M = (a/2, b/2)$. Then

$d(0,M) = \sqrt{(a/2)^2 + (b/2)^2} = (1/2)\sqrt{a^2+b^2}$

$d(A,M) = d(A,B)/2 = (1/2)\sqrt{a^2+b^2}$

$d(B,M) = d(A,B)/2 = (1/2)\sqrt{a^2+b^2}$

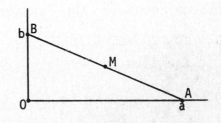

33. $x^2 + y^2 - 4x - 2y - 11 = 0$ $\qquad$ $x^2 + y^2 + 20x - 12y + 72 = 0$

$(x - 2)^2 + (y - 1)^2 = 16$ $\qquad$ $(x + 10)^2 + (y - 6)^2 = 64$

Center: (2,1) $\quad$ Radius: 4 $\qquad$ Center: (-10,6) $\quad$ Radius: 8

The circles do not intersect since the distance between the centers is

$\sqrt{(-10-2)^2+(6-1)^2}$ = 13 which is greater than 12, the sum of the radii.

35.

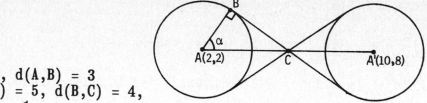

$d(A,A') = 10,\ d(A,B) = 3$

Thus, $d(A,C) = 5,\ d(B,C) = 4,$

$\alpha = \sin^{-1}(4/5)$

L = 2[circumference of circle - length of smaller arc] + 4[d(B,C)]

$= 2[2\pi(3) - 2(3)\sin^{-1}(0.8)] + 4[4] = 16 + 12\pi - 12\sin^{-1}(0.8) \approx$ 42.5716.

37. Let a,b,c denote the lengths of the respective sides of the triangle.

Semicircles: $Area(A) + Area(B) = \pi(a/2)^2 + \pi(b/2)^2$

$= \pi(a^2+b^2)/4 = \pi(c^2)/4 = \pi(c/2)^2 = Area(C)$

Equilateral Triangles: $Area(A) + Area(B) = (\sqrt{3}/4)a^2 + (\sqrt{3}/4)b^2$

$= (\sqrt{3}/4)(a^2+b^2) = (\sqrt{3}/4)c^2 = Area(C)$

General Theorem: $\quad$ If A,B,C are the areas of similar regions on respective sides, a,b,c (hypotenuse) of a right triangle, then A + B =C.

39. $2\pi(2) + d_1 + d_2 + d_3 = 4\pi + 8 + \sqrt{68} + 10 \approx 38.81$

Problem Set 1.6 $\quad$ The Straight Line

1. 5/2

3. $\dfrac{0 - 2}{3 - (-4)} = \dfrac{-2}{7}$

5. -5/3

7. 1.161/6.047 $\approx$ 0.1920

9. y - 3 = 4(x-2) $\qquad$ [point-slope form]

y - 3 = 4x - 8

4x - y - 5 = 0 $\qquad$ [general form]

11. $y = -2x+4$
     $2x+y-4 = 0$

13. Slope is 5/2.
     $y-3 = (5/2)(x-2)$
     $5x-2y-4 = 0$

15. $x = 2$ [The slope is undefined so the line is vertical.]

     $x-2 = 0$ [general form]

17. $y = (2/3)x - 4/3$
     Slope: 2/3 y-intercept: -4/3

19. $y = (-2/3)x + 2$
     Slope: -2/3 y-intercept: 2

21. $y = 2x+5$ has a slope of 2

     (a) Slope is 2 [Parallel lines have the same slope.]
          Equations: $y-(-3) = 2(x-3)$ [point-slope form]
                      $y+3 = 2x-6$
                      $y = 2x-9$ [slope-intercept form]
     (b) Slope is -1/2 [Perpendicular lines have slopes that are negative
                        reciprocals of each other.]
          Equations: $y-(-3) = (-1/2)(x-3)$ [point-slope form]
                      $y+3 = (-1/2)x + 3/2$
                      $y = (-1/2)x - 3/2$ [slope-intercept form]
          $2x+3y = 6$ [or $y = (-2/3)x + 2$] has a slope of -2/3
     (c) Slope is -2/3 [Parallel lines have the same slope.]
          Equations: $y-(-3) = (-2/3)(x-3)$ [point-slope form]
                      $y+3 = (-2/3)x + 2$
                      $y = (-2/3) x - 1$ [slope-intercept form]
     (d) Slope is 3/2 [Perpendicular lines have slopes that are negative
                        reciprocals of each other.]
          Equations: $y-(-3) = (3/2)(x-3)$ [point-slope form]
                      $y+3 = (3/2)x - (9/2)$
                      $y = (3/2)x - (15/2)$ [slope-intercept form]
          Slope of the line through $(-1,2)$ and $(3,-1)$ is $\frac{-1-2}{3-(-1)} = -3/4$
     (e) Slope is -3/4 [Parallel lines have the same slope.]
          Equations: $y-(-3) = (-3/4)(x-3)$ [point-slope form]
                      $y+3 = (-3/4)x + (9/4)$
                      $y = (-3/4)x - (3/4)$ [slope-intercept form]
          $x = 8$ is a vertical line and has no slope.
     (f) The line is vertical. An equation of the line is $x = 3$.
     (g) The line is horizontal. An equation of the line is $y = -3$.

23. Slope is 2; an equation is y+4 = 2(x-0), or y = 2x-4.

25. (3,8) is on the line, so (3,9) is above the line.

27. $\begin{bmatrix} 2x + 3y = 4 \\ -3x + y = 5 \end{bmatrix} \Leftrightarrow \begin{bmatrix} 2x + 3y = 4 \\ 9x - 3y = -15 \end{bmatrix} \Rightarrow [11x = -11] \Leftrightarrow [x = -1]$

Set x equal to -1 in -3x+y=5, and obtain that y equals 2.
Point of intersection: (-1,2)

2x + 3y = 4 [or y = (-2/3)x + (4/3)] has slope of -2/3.
Therefore, the slope of a line perpendicular to 2x + 3y = 4 is 3/2.

Equations: y - 2 = (3/2)(x - [-1]) [point-slope form]
          y - 2 = (3/2)x + (3/2)
          y = (3/2)x + (7/2)   [slope-intercept form]

29. The point of intersection is (3,1) and the slope of 3x -4y = 5 is 3/4.
Therefore, an equation of the line indicated is y-1 = (-4/3)(x-3) or y = (-4/3)x + 5.

31. $d = \dfrac{|(3)(-3) + (4)(2) + (-6)|}{\sqrt{(3)^2+(4)^2}} = \dfrac{7}{5} = 1.4$

33. Express 5y = 12x + 1 as 12x - 5y + 1 = 0. [general form]
Then the distance is $\dfrac{|(12)(-2) + (-5)(-1) + (1)|}{\sqrt{(12)^2 + (-5)^2}} = \dfrac{18}{13} \approx 1.3846$

35. (2,0) is a point on 3x+4y = 6. The distance from (2,0) to 3x+4y-12 = 0 is

$d = \dfrac{|(3)(2) + (4)(0) +(-12)|}{\sqrt{(3)^2 + (4)^2}} = \dfrac{6}{5} = 1.2$

37. Depreciation is $9,600 per year until its value is $0. V = 120,000 - 9,600t until V = 0.

39. Two points on the line are (0, 700 000) and (10, 820 000). The slope is
$\dfrac{820,000 - 700,000}{10 - 0} = 12,000$; N-intercept is 700,000.

Therefore, the formula is N = 12,000n + 700,000. For the year 2000, n = 40; so N = 12,000(40) + 700,000 = 1,180,000. That is, the prediction is that 1,180,000 cases of eggs will be produced in Matlin County in the year 2000.

41. (a) P = -2000 when x = 0. That is, the company loses $2000 if no items are sold.

(b) The slope is 450. It indicates that the profit increases by $450 for each item sold. [Note: it does not indicate that the profit per item is $450, due to the -2000 term.]

43. (1) If B = 0 and A ≠ 0, Ax+By+C = 0 is equivalent to x = -C/A whose graph is the vertical line which intersects the x-axis at -C/A.

(2) If B ≠ 0, Ax+By+C = 0 is equivalent to y = (-A/B)x -C/B whose graph is the line with slope -A/B and whose y-intercept is -C/B.

45. First, note that for each real number k, 2x - y + 4 + k(x + 3y - 6) = 0 is equivalent to (2 + k)x + (-1 + 3k)y + (4 - 6k) = 0, which is an equation of a line since there is no value of k that will make both (2 + k) and (-1 + 3k) equal to zero. [See Problem 43.]

Second, note that the lines intersect since their slopes are different. Finally, note that (x_0,y_0) is the point of intersection

$$\Rightarrow \ 2x_0 - y_0 + 4 = 0 \text{ and } x_0 + 3y_0 - 6 = 0$$

$$\Rightarrow \ (2x_0 - y_0 + 4) + k(x_0 + 3y_0 - 6) = 0 + k\cdot 0 = 0$$

$$\Rightarrow \ (x_0,y_0) \text{ is on the line } 2x-y+4 + k(x+3y-6) = 0$$

47. [See figure at right.] The midpoint of segment AC is (2,5) and its slope is 1/2, so an equation of the perpendicular bisector of segment AC is y-5 = -2(x-2).The midpoint of segment AB is (1,2) and its slope is -2, so an equation of the perpendicular bisector of segment AB is y-2 = (1/2)(x-1). The point of intersection of these perpendicular bisectors is (3,3).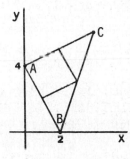

49. ax+by = 36 tangent to the circle at (a,b) (by Problem 48).
(12,0) is on ax+by = 36, so 12a = 36; a = 3.

(3,b) is on x^2+y^2 = 36, so $9+b^2$ = 36; b = ± 3√3.

Equations of the two tangent lines are 3x ± 3√3y = 36 or x ± √3y = 12.

51. m_{TU} = 0, so Tu∥OR.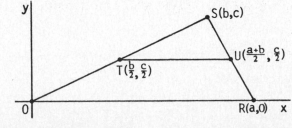

53. $y = 8$

Problem Set 1.7   Graphs of Equations

1. $y = -x^2 + 4$

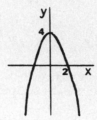

Symmetry: With respect to y-axis

x-intercepts: $-2, 2$ y-intercept: 4

3. $3x^2 + 4y = 0$

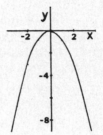

Symmetry: With respect to the y-axis
since $3(-x)^2 + 4y = 0$ is
equivalent to $3x^2 + 4y = 0$.

x-intercept: 0 y-intercept: 0

5. $x^2 + y^2 = 36$
The graph is the circle with center $(0,0)$
and radius 6.

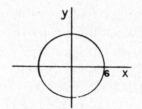

7. $4x^2 + 9y^2 = 36$

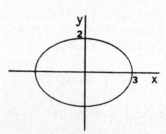

Symmetry: With respect to x-axis, y-axis,
and origin

x-intercepts: $-3, 3$ y-intercepts: $-2, 2$

9. $y = x^3 - 3x = x(x^2 - 3) = x(x+\sqrt{3})(x-\sqrt{3})$

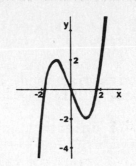

Symmetry: With respect to origin since
$(-y) = (-x)^3 - 3(-x)$ is equivalent
to $y = x^3 - 3x$.

x-intercepts: $-\sqrt{3}, 0, \sqrt{3}$ y-intercept: 0

11. $y = 1/(x^2 + 1)$

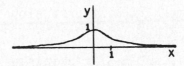

Symmetry: With respect to y-axis

x-intercepts: None   y-intercept: 1

13. $x^3 - y^2 = 0$

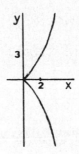

Symmetry: With respect to x-axis.

x-intercept: 0   y-intercept: 0

15. $y = (x-2)(x+1)(x+3)$

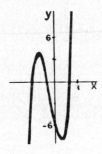

Symmetry: None of those discussed.

x-intercepts: $-3, -1, 2$ y-intercept: -6

17. $y = x^2(x-2)$

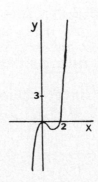

Symmetry: None of those discussed.

x-intercepts: $0, 2$ y-intercept: 0

19. $y = -x+1$ is a line with slope -1 and y-intercept 1.

$y = x^2+2x+1$ has none of the symmetries discussed, has x-intercept of -1 and y-intercept of 1.

Points of intersection are $(0,1)$ and $(-3,4)$

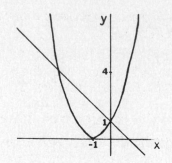

21. $y = -2x+1$ [a line with slope -2 and y-intercept 1]

$y = -x^2-x+3$

$-2x+1 = -x^2-x+3$ at points of intersection.

$x^2-x-2 = 0$
$(x-2)(x+1) = 0$
$x = 2$ or $x = -1$

If $x = 2$, $y = -3$; if $x = -1$, $y = 3$. Hence, intersection points are $(2,-3)$ and $(-1,3)$.

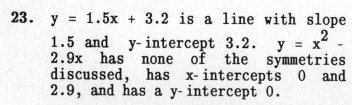

For $y = -x^2-x+3$:

Symmetry: None of those discussed

x-intercepts: Solve $x^2+x-3 = 0$

$$x = \frac{-(1) \pm \sqrt{(1)^2 - 4(1)(-3)}}{2(1)} = \frac{-1 \pm \sqrt{13}}{2}$$
$$x \approx -2.30, \ 1.30$$
y-intercept: 3

23. $y = 1.5x + 3.2$ is a line with slope 1.5 and y-intercept 3.2. $y = x^2 - 2.9x$ has none of the symmetries discussed, has x-intercepts 0 and 2.9, and has a y-intercept 0.

Points of intersection are approximately $(-0.64, 2.25)$ and $(5.04, 10.75)$.

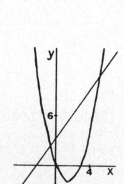

25. $y = 4x+3$ is a line with slope 4 and
     y-intercept 3.

     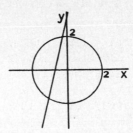

     $x^2+y^2 = 4$ is a circle with
     center $(0,0)$ and radius 2.

     Points of intersection are where

     $x = (-12 \pm \sqrt{59})/17$. Approximately,
     $(-1.16,-1.63)$ and $(-0.25, 1.98)$.

27. $y = 3x^4 - 2x + 1$
     $x = -1 \Rightarrow y = 3 + 2 + 1 = 6$ so one point is $(-1,6)$.
     $x = 1 \Rightarrow y = 3 - 2 + 1 = 2$ so the other point is $(1,2)$.

     The distance between $(-1,6)$ and $(1,2)$ is $\sqrt{4+16} \approx 4.4721$.

29. $y = 2^x+2^{-x}$ is symmetric with respect
     to the y-axis, has no x-intercepts,
     and has y-intercept 2.

     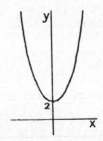

31. $y = (1 + x^{1.5})/x$

     See graph with
     Problem 32.

x	y	x	y	x	y
0.01	100.10	1.4	1.90	3.0	2.07
0.1	10.32	1.6	1.89	4.0	2.25
0.2	5.45	1.8	1.90	6.0	2.62
0.4	3.13	2.0	1.91	8.0	2.95
0.6	2.44	2.2	1.94	10.0	3.16
0.8	2.14	2.4	1.97	12.0	3.46
1.0	2.00	2.6	2.00	14.0	3.74
1.2	1.93	2.8	2.03	16.0	4.06

33. d > 0 means $ax^2+bx+c = 0$ has two real roots. Therefore, the graph of $y = ax^2+bx+c$ intersects the x-axis in two points.

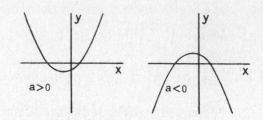

d = 0 means $ax^2+bx+c = 0$ has one real root. Therefore, the graph of $y = ax^2+bx+c$ intersects the x-axis in one point. That is, it is tangent to the x-axis.

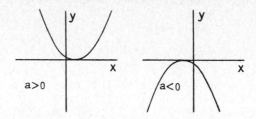

d < 0 means $ax^2+bx+c = 0$ has no real roots. Therefore, the graph of $y = ax^2+bx+c$ does not intersect the x-axis.

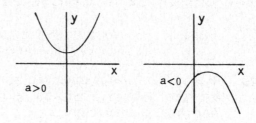

Problem Set 1.8   Chapter Review

True-False Quiz

1. False. C^{ex}: Let $p = \sqrt{2}$ and $q = 1$.

3. False. C^{ex}: $\sqrt{2}$ is irrational, but $\sqrt{2} - \sqrt{2} = 0$, which is rational.

5. False.   It is equal to 1.

7. False. C^{ex}: $(2*3)*2 = (2^3)^2 = 8^2 = 64$, but $2*(3*2) = 2^{(3^2)} = 2^9 = 512$.

9. True.   Assume $x \neq 0$. Then $\frac{|x|}{2}$ is positive and $\frac{|x|}{2} \leq |x|$. This contradicts that $|x|$ is less than every positive number, so our assumption that x is not equal to zero must have been false.

---

11. True. $a < b < 0 \Rightarrow 0 < -b < -a \Rightarrow 1/(b) > 1/(-a) \Rightarrow 1/b < 1/a$

13. True. $(a,b) \cap (c,d) \neq \emptyset$ $c < b \Rightarrow (c,b)$ is in (a,b) and in (c,d)

15. False. C^{ex}: $|-2| < |-7|$ but $-2 > -7$.

17. True. $|x+y| = -(x+y)$ [since $(x+y) < 0$]
    $= (-x) + (-y) = |x| + |y|$ [since $x < 0$ and $y < 0$]

19. True. Complete the squares and obtain the standard equation of a circle with center $(-a/2, -1/2)$ and square of radius $(a^2+1)/4$.

21. True. $ab > 0 \Rightarrow [a > 0$ and $b > 0]$ or $[a < 0$ and $b < 0]$
    $\Rightarrow (a,b)$ is in the 1st or in the 3rd quadrant.

23. True. $(y_2-y_1)^2$ must equal zero so $y_1 = y_2$.

25. False. It is not true for vertical lines.

27. False. If each line has a positive slope, the product of their slopes is positive, not -1, so they are not perpendicular.

29. False. The slopes of the lines are $-a$ and a, respectively. Their product is -1 only if $a = 1$ or $a = -1$.

Sample Test Problems

1. (a) 2, 6.25, 0.16   (b) 1, 9, 49   (c) 64, 8, 0.125

3. Let a/b and c/d be rational numbers. That is, a,b,c,d are integers and $bd \neq 0$. The average of a/b and c/d is

$$\frac{\frac{a}{b} + \frac{c}{d}}{2} = \frac{\left[\frac{a}{b} + \frac{c}{d}\right] \cdot bd}{2 \cdot bd} = \frac{ad + bc}{2bd}$$ which is a rational number since $(ad + bc)$ and $(2bd)$ are integers, and $2bd \neq 0$.

5. $1/2 = 0.5$ and $13/25 = 0.52$. $0.51010010001\cdots$ is an irrational number between them.

7. $x > -2$
   Solution set: $(-2, \infty)$ Graph:

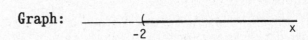

9. $2x^2 + 5x - 3 < 0$
$(2x-1)(x+3) < 0$
Split points: $-3, 1/2$ Test points: $-4, 0, 1$

Auxiliary axis for $(2x-1)(x+3)$: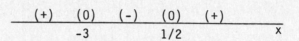

Solution set: $(-3, 1/2)$ Graph: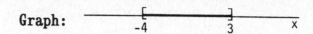

11. Solution set: $[-4, 3]$ Graph:

13. $(2x+1)/(x-1) \geq 0$
Solution set: $(-\infty, -1/2] \cup (1, \infty)$ Graph:

15. $\left| \dfrac{2x^2+3x+2}{x^2+2} \right| = \dfrac{|2x^2+3x+2|}{|x^2+2|} \leq \dfrac{2x^2 + 3|x| + 2}{x^2 + 2} \leq \dfrac{2(4) + 3(2) + 2}{2} = 8$

17. $(4,5)$ is the midpoint. $d[(3,-6), (4,5)] = \sqrt{122} \approx 11.0454$.

19. $(x-4)^2 + (y+3)^2 = 25$, so the center is $(4,-3)$ and the radius is 5.

21. (a) Slope: $\dfrac{3-1}{7-(-2)} = \dfrac{2}{9}$

Equations: $y-1 = (2/9)(x+2)$ [point-slope form]
$\qquad\qquad\quad y-1 = (2/9)x + (4/9)$
$\qquad\qquad\quad y \ \ = (2/9)x + (13/9)$ [slope-intercept form]

(b) The slope of $3x-2y = 5$ [or $y = (3/2)x - (5/2)$] is 3/2 so the slope of a line parallel to it is also 3/2.
Equations: $y-1 = (3/2)(x+2)$ [point-slope form]
$\qquad\qquad\quad y-1 = (3/2)x + 3$
$\qquad\qquad\quad y \ \ = (3/2)x + 4$ [slope-intercept form]

(c) The slope of $3x+4y = 9$ [or $y = (-3/4)x + 9/4$] is -3/4 so the slope of a line perpendicular to it is 4/3.
Equations: $y-1 = (4/3)(x+2)$ [point-slope form]
$\qquad\qquad\quad y-1 = (4/3)x + (8/3)$
$\qquad\qquad\quad y \ \ = (4/3)x + (11/3)$ [slope-intercept form]

(d) $y = 4$ is a horizontal line so each line perpendicular to it is a vertical line. The vertical line through $(-2,1)$ is $x = -2$.

(e) Slope: $\dfrac{1-3}{-2-0} = 1$ [since $(0,3)$ and $(-2,1)$ are on the line]. Then an equation of the line is $y = x+3$. [slope-intercept form]

23. $3y-4x = 6$ is a line with slope 4/3 and y-intercept 2.

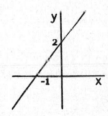

25. $y = 2x/(x^2+2)$

Symmetry: With respect to origin.

x-intercept: 0   y-intercept: 0

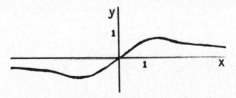

27. $y = x^2 - 2x + 4$
     $y - x = 4$ [or $y = x+4$]

Therefore, $x^2 - 2x + 4 = x+4$ at points of intersection.
$$x^2 - 3x = 0$$
$$x(x-3) = 0$$
$$x = 0 \quad \text{or} \quad x = 3$$

The corresponding values of y are $0 + 4 = 4$, and $3 + 4 = 7$, so the points of intersection are $(0,4)$ and $(3,7)$.

Problem Set 2.1   Functions and Their Graphs

1. (a) 0.   (b) 3.   (c) -1.   (d) k^2-1.   (e) 35.

   (f) -3/4   (g) $4t^2-1$.   (h) $9x^2-1$.   (i) $(1-x^2)/x^2$.

3. (a) $G(0) = \frac{1}{(0)-1} = -1$.   (b) $G(0.999) = \frac{1}{(0.999)-1} = -1000$.

   (c) $G(1.01) = \frac{1}{(1.01)-1} = 100$.   (d) $G(y^2) = \frac{1}{(y^2)-1} = \frac{1}{y^2-1}$.

   (e) $G(-x) = \frac{1}{(-x)-1} = \frac{-1}{x+1}$.   (f) $G(1/x^2) = \frac{1}{(1/x^2)-1} = \frac{x^2}{1-x^2}$.

5. (a) 1.1344.   (b) 138.4935.   (c) 530.6295.

7. (a) -3.2931.   (b) 1.1989.   (c) undefined.

9. (a) $x^2+y^2 = 4$.
   $y^2 = 4-x^2$.
   $y = \sqrt{4-x^2}$ or $y = -\sqrt{4-x^2}$.
   Therefore, $x^2+y^2 = 4$ does not determine a function with formula $y = f(x)$.

   (b) $xy+y+3x = 4$.
   $y(x+1) = 4-3x$.
   $y = \frac{4-3x}{x+1}$.
   Therefore, $xy+y+3x = 4$ does determine a function with formula $y = f(x) = \frac{4-3x}{x+1}$.

   (c) $x = \sqrt{3y+1}$.
   $x^2 = 3y+1, \ x \geq 0$.
   $y = \frac{x^2-1}{3}, \ x \geq 0$.

   Therefore, $x = \sqrt{3y+1}$ does determine a function with formula $y = f(x) = \frac{x^2-1}{3}$ and with domain $[0,\infty)$.

   (d) $3x = \frac{y}{y+1}$.
   $3x(y+1) = y$.
   $3xy + 3x = y$.
   $3xy-y = -3x$.
   $(3x-1)y = -3x$.
   $y = \frac{-3x}{3x-1}$.
   Thus, $3x = \frac{y}{y+1}$ does determine a function with formula $y = f(x) = \frac{-3x}{3x-1}$.

11. $\dfrac{f(a+h) - f(a)}{h} = \dfrac{[2(a+h)^2 - 1] - [2a^2 - 1]}{h} = \dfrac{h(4a+2h)}{h} = 4a+2h \quad [h \neq 0].$

13. $\dfrac{g(x+h)-g(x)}{h} = \dfrac{3/[(x+h)-2] - 3/[x-2]}{h} \quad \dfrac{-3h}{h(x+h-2)(x-2)} = \dfrac{-3}{(x+h-2)(x-2)}.$

15. (a) The radicand must be nonnegative.
$(2z+3 \geq 0)$ iff $(2\ z \geq -3)$ iff $(z \geq -1.5)$, so the natural domain is $[-1.5, \infty)$.

(b) The denominator must be nonzero.
$(4v-1 \neq 0)$ iff $(4v \neq 1)$ iff $(v \neq 1/4)$, so the natural domain is $\{v : v \neq 1/4\}$.

(c) The radicand must be nonnegative.
$(x^2-9 \geq 0)$ iff $(x^2 \geq 9)$ iff $(x \leq -3$ or $x \geq 3)$, so the natural domain is $(-\infty, -3] \cup [3,\infty)$.

(d) The radicand must be nonnegative.
$(625-y^4 \geq 0)$ iff $(625 \geq y^4)$ iff $(25 \geq y^2)$ iff $(-5 \leq y \leq 5)$, so the natural domain is $[-5, 5]$.

17. $f(-x) = -4 = f(x)$, so f is even.

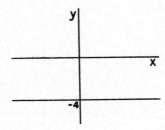

19. $F(-4) = -7$, but $F(4) = 9$, so F is neither even nor odd.

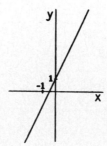

21. $g(x) = 3x^2 + 2x - 1$. $g(-1) = 0$, but
     $g(1) = 4$, so g is neither even nor
     odd.

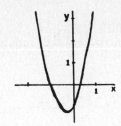

23. $g(-x) = -x/(x^2-1) = -g(x)$, so g is
     odd.

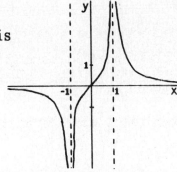

25. $f(-1)$ is undefined but $f(1) = 0$, so
     f is neither even nor odd.

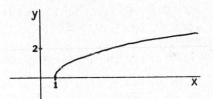

27. $f(x) = |2x|$.
     $f(-x) = |2(-x)| = |2x| = f(x)$,
     so f is even.

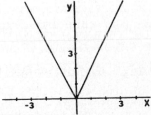

29. $g(-3) = -2$, but $g(3) = 1$, so g is
     neither even nor odd.

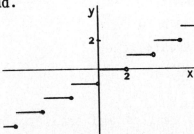

31. $g(-3) = 1$, but $g(3) = 8$, so g is neither even nor odd.

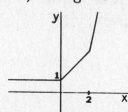

33. $T(x) = $ (direct cost for x refrigerators) + (overhead cost). $= 151x + 2200$, and $D_T = \{0,1,2,3,\cdots,100\}$.

$$u(x) = \frac{(\text{Total Cost})}{(\text{No. units})} = \frac{T(x)}{x} = \frac{151x + 2200}{x}, \text{ and } D_u = \{1,2,3,\cdots,100\}.$$

35. $E(x) = x - x^3$.

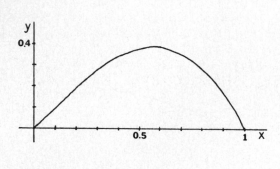

x	E(x)
0.0	0.000
0.1	0.099
0.2	0.192
0.3	0.273
0.4	0.336
0.5	0.375
0.6	0.386
0.7	0.357
0.8	0.288
0.9	0.171
1.0	0.000

For $x > 0$, x exceeds its cube by a maximum amount if x is near 0.6.

37. (a) $E(x) = 0.40x + 24$.   (b) Solve $120 = 0.40x + 24$ for x. $x = 240$ miles.

39. Length of circle (2 ends) $= \pi d$.
Length of each parallel side $= \frac{1 - \pi d}{2}$
Area = Area(rectangle) + Area(circle)
$= \frac{1 - \pi d}{2}$ (d) $+ \pi\left(\frac{d}{2}\right)^2 = \frac{d(2 - \pi d)}{4}$

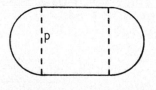

The semicircles could get smaller and smaller, but the largest they can get is when the length of the circle, πd, is 1 (when $d = 1/\pi$). Therefore, the natural domain is $(0, 1/\pi)$; or $(0, 1/\pi]$ if a circle is allowed.

41. $3/13 = 0.\overline{230769}$. Thus, range of f is $\{0,2,3,6,7,9\}$.

43. For t = 0: $f(x) = f(x+0) = f(x) + f(0) \Rightarrow f(0) = 0 = m \cdot 0$ for any m.

For t a natural number: $f(t) = f(1+1+ \cdots + 1)$
$$= f(1) + f(1) + \cdots + f(1) \text{ [t terms]}$$
$$= [f(1)]t$$

Thus, f(1) is a candidate for m.

For t = 1/p, p a natural number:
$$m = f(1) = f(1/p + 1/p + \cdots + 1/p) \text{ [p terms]}$$
$$= [f(1/p)]p$$

Therefore, $f(1/p) = m/p = m(1/p)$.

For t = q/p, q and p natural numbers:
$$f(q/p) = f(1/p + 1/p + \cdots + 1/p) \text{ [q terms]}$$

$$= [f(1/p)]q$$

$$= [(m/p)]q$$

$$= m(q/p)$$

For t a negative rational number: $0 = f(0) = f(-t+t) = f(-t) + f(t)$
Thus, $f(t) = -f(-t) = -[m(-t)] = mt$.

45. (a)   0.29945, 3.68518

(b) $(x^3 + 3*x - 5)/(x^2 + 4)$

-4	-4.0
-3	-3.15385
-2	-2.375
-1	-1.8
0	-1.25
1	- .2
2	1.125
3	2.38461
4	3.55

47. (a)   [-22,13]

(b)   [-1.07, 1.72] ∪ [4.35,5]

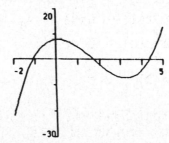

49. (a) $(1.33, 0)$, $(0, 2/3)$

(b) $(-\infty, \infty)$

(c) $x = -3, x = 2$

(d) $y = 0$

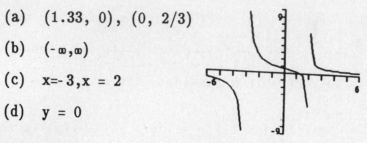

Problem Set 2.2 Operations on Functions

1. (a) $2 + \sqrt{5} \approx 4.2361$. (b) 0. (c) $2\sqrt{10}/3 \approx 2.1082$.

(d) Undefined. (e) 1.5 (f) 1.

3. $f(x) = x^3 + 2$, $D_f = \mathbf{R}$; and $g(x) = \dfrac{2}{x-1}$, $D_g = \{x: x \neq 1\}$.

(a) $(f+g)(x) = f(x) + g(x) = (x^3 + 2) + \dfrac{2}{x-1} = \dfrac{(x^3+2)(x-1) + 2}{x-1}$

$$= \dfrac{x^4 - x^3 + 2x}{x-1}, \quad D_{f+g} = D_f \cap D_g = \{x: x \neq 1\}.$$

(b) $\left(\dfrac{g}{f}\right)(x) = \dfrac{g(x)}{f(x)} = \dfrac{\frac{2}{x-1}}{x^3+2} = \dfrac{2}{(x-1)(x^3+2)}$.

$D_{f/g} = \{x: x \neq 1, -\sqrt[3]{2}\}$.

(c) $(f \circ g)(x) = f[g(x)] = f(\frac{2}{x-1}) = (\frac{2}{x-1})^3 + 2 = \dfrac{2x^3 - 6x^2 + 6x + 6}{(x-1)^3}$.

$D_{f \circ g} = \{x: x \neq 1\}$.

(d) $(g \circ f)(x) = g[f(x)] = g[x^3 + 2] = \dfrac{2}{(x^3+2) - 1} = \dfrac{2}{x^3+1}$.

$D_{g \circ f} = \{x: x \neq -1\}$.

5. $(f \circ g)(x) = \sqrt{|x|-4}$; $D_{f \circ g} = (-\infty, -4] \cup [4, \infty)$.

$(g \circ f)(x) = \sqrt{x-4}$; $D_{g \circ f} = [4, \infty)$.

7. 0.00001950.

9. Sequence of keys: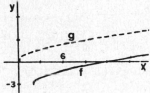

   Answer: Approximately 1.1134.

11. (a) Let $f(x) = x+7$, $g(x) = \sqrt{x}$. (b) Let $f(x) = x^2+x$, $g(x) = x^{15}$.

13. I. Let $h(x) = x^2+1$, $g(x) = \sqrt{x}$, $f(x) = \log x$. Then $p(x) = (f \circ g \circ h)(x)$.
    II. Let $h(x) = x^2$, $g(x) = \sqrt{x+1}$, $f(x) = \log x$. Then $p(x) = (f \circ g \circ h)(x)$.

15. Translate the graph of g two units
    right and three units down.

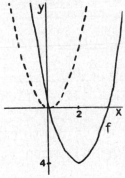

17. Let $a(x) = x^2$. Translate the graph
    of a two units right and four units
    down.

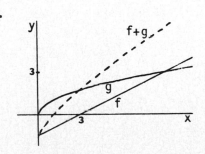

19.

21. $F(t) = \left\{ \begin{array}{l} \dfrac{t}{t} = 1, \ \ \text{if} \ t > 0 \\[3mm] \dfrac{-t}{t} = -1, \ \text{if} \ t < 0 \end{array} \right\}$.

Note that 0 is not in the domain of F.

23. (a) The sum of two even functions, f and g, is an even function.
Proof: $(f+g)(-x) = f(-x) + g(-x) = f(x) + g(x) = (f+g)(x)$.

(b) The sum of two odd functions, f and g, is an odd function.
Proof: $(f+g)(-x) = f(-x) + g(-x) = -f(x) + [-g(x)] = -(f+g)(x)$.

(c) The product of two even functions, f and g, is an even function.
Proof: $(fg)(-x) = f(-x)g(-x) = f(x)g(x) = (fg)(x)$.

(d) The product of two odd functions, f and g, is an even function.
Proof: $(fg)(-x) = f(-x)g(-x) = -f(x)[-g(x)] = f(x)g(x) = (fg)(x)$.

(e) The product of an even function f and an odd function g is odd.
Proof: $(fg)(-x) = f(-x)g(-x) = f(x)[-g(x)] = -f(x)g(x) = -(fg)(x)$.

25. (a) AF.   (b) PF.   (c) RF.   (d) PF.   (e) RF.   (f) AF.

27. (a) $P = \sqrt{29 - 3D + D^2} = \sqrt{29 - 3(2+\sqrt{t}) + (2+\sqrt{t})^2}$

$= \sqrt{29 - 6 - 3\sqrt{t} + 4 + 4\sqrt{t} + t} = \sqrt{27 + \sqrt{t} + t}$.

(b) $P(15) = \sqrt{27 + \sqrt{15} + 15} = \sqrt{42 + \sqrt{15}} \approx 6.7730$.

29. If t=0 at noon, $x(t) = \left\{ \begin{array}{l} 0 \ \ \ \ \ \ \ \ \ \ \ \ , \ \text{when} \ 0 \le t \le 1 \\ 300(t-1), \ \text{when} \ t > 1 \end{array} \right\}$.

$y(t) = 400t, \ \text{when} \ t \ge 0$.

$D(t) = \left\{ \begin{array}{l} 400t \ , \ \text{when} \ 0 \le t \le 1 \\ \sqrt{[400t]^2 + [300(t-1)]^2}, \ \text{when} \ t > 1 \end{array} \right\}$.

$= \left\{ \begin{array}{l} 400t \ \ \ \ \ \ \ \ \ \ \ \ \ \ \ , \ \text{when} \ 0 \le t \le 1 \\ 100\sqrt{25t^2 - 18t + 9}, \ \ \text{when} \ t > 1 \end{array} \right\}$.

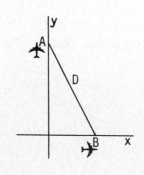

31. $f(f(x)) = f(\frac{ax+b}{cx-a}) = \dfrac{a\,\frac{ax+b}{cx-a} + b}{c\,\frac{ax+b}{cx-a} - a} = \dfrac{a(ax+b) + b(cx-a)}{c(ax+b) - a(cx-a)}$ (if $x \neq$ a/c)

$$= \frac{a^2x+ab+bcx-ab}{acx+bc-acx+a^2} = \frac{x(a^2+bc)}{a^2+bc} = x \text{ (if } a^2+bc \neq 0.)$$

33. (a) $f(\frac{1}{x}) = \dfrac{\frac{1}{x}}{\frac{1}{x} - 1} = \frac{1}{1-x}$ (if $x \neq 1$)

(b) $f(f(x)) = f(\frac{x}{x-1}) = \dfrac{\frac{x}{x-1}}{\frac{x}{x-1} - 1} = \frac{x}{x - (x-1)}$ (if $x \neq 1$) $= \frac{x}{1} = x$

(c) $f(\frac{1}{f(x)}) = f\left[\dfrac{1}{\frac{x}{x-1}}\right] = f(\frac{x-1}{x})$ (if $x \neq 1$)

$$= \dfrac{\frac{x-1}{x}}{\frac{x-1}{x} - 1} = \frac{x-1}{x-1 - x} \text{ (if } x \neq 0)$$

$$= \frac{x-1}{-1} = 1-x$$

35. (a)

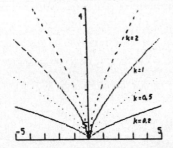

37.

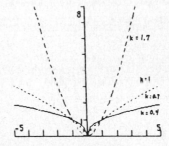

Problem Set 2.3   The Trigonometric Functions

1. (a) $4\pi/3$. (b) $-\pi/3$. (c) $-3\pi/4$. (d) 3π. (e) $10\pi/3$

(f) 4π. (g) $\pi/10$. (h) $\pi/8$. (i) $\pi/30$.

3. (a) $33.3^O = 33.3^O \times \dfrac{\pi \text{ rad}}{180^O} = \dfrac{33.3\pi}{180} \text{ rad} \approx 0.5812 \text{ rad.}$

(b) $471.5^O = 471.5^O \times \dfrac{\pi \text{ rad}}{180^O} = \dfrac{471.5\pi}{180} \text{ rad} \approx 8.2292 \text{ rad.}$

(c) $-391.4^O = -391.4^O \times \dfrac{\pi \text{ rad}}{180^O} = \dfrac{-391.4\pi}{180} \text{ rad} \approx -6.8312 \text{ rad.}$

(d) $14.9^O = 14.9^O \times \dfrac{\pi \text{ rad}}{180^O} = \dfrac{14.9\pi}{180} \text{ rad} \approx 0.2601 \text{ rad.}$

(e) $4.02^O = 4.02^O \times \dfrac{\pi \text{ rad}}{180^O} = \dfrac{4.02\pi}{180} \text{ rad} \approx 0.0702 \text{ rad.}$

(f) $-1.52^O = -1.52^O \times \dfrac{\pi \text{ rad}}{180^O} = \dfrac{-1.52\pi}{180} \text{ rad} \approx -0.0265 \text{ rad.}$

5. (a) 0.4368.   (b) 0.8996.   (c) 0.4855.

(d) -0.3532.   (e) 0.9355.   (f) -0.3775.

7. (a) 248.3004.   (b) 1.2828.

9. The following two triangles along with the triangle defintions of the trigonometric functions are helpful for evaluations involving $\pi/6$, $\pi/4$, and $\pi/3$.

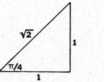 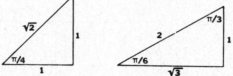

(a) $\tan(\pi/6) = \dfrac{1}{\sqrt{3}} \; \left[\dfrac{\text{OPP}}{\text{ADJ}}\right].$ (b) $\sec \pi = \dfrac{1}{\cos \pi} = \dfrac{1}{-1} = -1.$

(c) $\sec(3\pi/4) = -\sec(\pi/4) = -\dfrac{\sqrt{2}}{1} = -\sqrt{2}. \; \left[\dfrac{\text{HYP}}{\text{ADJ}}\right]$

(d) $\csc(\pi/2) = \dfrac{1}{\sin(\pi/2)} = \dfrac{1}{1} = 1.$

(e) $\cot(\pi/4) = \dfrac{1}{1} = 1. \; \left[\dfrac{\text{ADJ}}{\text{OPP}}\right]$

(f) $\tan(-\pi/4) = -\tan(\pi/4) = -\dfrac{1}{1} = -1. \left[\dfrac{\text{OPP}}{\text{ADJ}}\right]$

11. (a) $(1 + \sin z)(1 - \sin z) = (1 - \sin^2 z) = \cos^2 z = 1/\sec^2 z$.

   (b) $(\sec t - 1)(\sec t + 1) = \sec^2 t - 1 = \tan^2 t$.

   (c) $\sec t - \sin t \tan t = \dfrac{1}{\cos t} - \sin t \dfrac{\sin t}{\cos t} = \dfrac{1 - \sin^2 t}{\cos t} = \dfrac{\cos^2 t}{\cos t} = \cos t$,
       if $\cos t \neq 0$; i.e., t is not a multiple of $\pi/2$.

   (d) $\dfrac{\sec^2 t - 1}{\sec^2 t} = 1 - \dfrac{1}{\sec^2 t} = 1 - \cos^2 t = \sin^2 t$, if $\cos t \neq 0$.

   (e) $\cos t (\tan t + \cot t) = \sin t + \dfrac{\cos^2 t}{\sin t}$, if $\cos t \neq 0$.

   $= \dfrac{\sin^2 t + \cos^2 t}{\sin t} = \dfrac{1}{\sin t} = \csc t$.

13. (a) 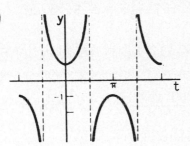   (b)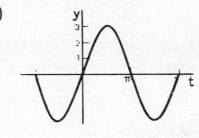

   (c) 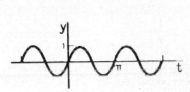   (d)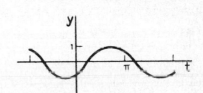

15. (a) 5.97 divided by $\pi/2$ is about 3.8. Thus, an angle of 5.07 radians
       can be obtained by rotating the positive x-axis counterclockwise
       through about 3.8 quadrants. Therefore, $P(x,y)$ is in the 4th
       quadrant, so $\cos(5.97)$ is positive.

   (b) 9.34 divided by $\pi/2$ is about 5.9. Thus, an angle of 9.34 radians
       can be obtained by rotating the positive x-axis counterclockwise
       through about 5.9 quadrants. Therefore, $P(x,y)$ is in the 2nd
       quadrant, so $\cos(9.34)$ is negative.

   (c) -.16.1 divided by $\pi/2$ is about -10.2. Thus, an angle of -16.1
       radians can be obtained by rotating the positive x-axis clockwise
       through about 10.2 quadrants. Therefore, $P(x,y)$ is in the 2nd
       quadrant, so $\cos(-16.1)$ is negative.

17. (a) Even, since $\sec(-t) = 1/\cos(-t) = 1/\cos t = \sec t$.
    (b) Odd, since $\csc(-t) = 1/\sin(-t) = 1/(-\sin t) = -\csc t$.
    (c) Even, since it is the product of two odd functions.
    (d) Odd, since it is the product of an odd and of an even function.
    (e) Even, since it is the product of two odd functions.

    (f) Neither, since $\sin(-\pi/4) + \cos(-\pi/4) = 0$, but $\sin(\pi/4) + \cos(\pi/4) = \sqrt{2}$.

19. $\tan(t+\pi) = \dfrac{\tan t + \tan \pi}{1 - \tan t \tan \pi} = \dfrac{\tan t + 0}{1 - (\tan t)(0)} = \tan t$.

21. Formula: $s = rt$. [s is length of arc; r is radius of circle; t is central angle measured in radians.]

    (a) $s = (2.5)(6) = 15$ cm.
    (b) $225^\circ = \dfrac{5\pi}{4}$ radians. Therefore, $s = (2.5)\dfrac{5\pi}{4} \approx 9.8175$ cm.

23. If $t = 1$ revolution $= 2\pi$ radians, $s = (2.5)(2\pi) = 5\pi$ ft. Now convert miles/hr to revolutions/min.

$$\frac{60 \text{ mi}}{1 \text{ hr}} \times \frac{5280 \text{ ft}}{1 \text{ mi}} \times \frac{1 \text{ hr}}{60 \text{ min}} \times \frac{1 \text{ rev}}{5\pi \text{ ft}} = \frac{316800 \text{ rev}}{300\pi \text{ min}} \approx 336.1352 \text{ rev/min.}$$

25. (a) $F(\pi/4) = \dfrac{50\mu}{\mu(\sqrt{2}/2) + \sqrt{2}/2} = \dfrac{100\mu}{\sqrt{2}(\mu+1)}$.

    (b) $F(0) = \dfrac{50\mu}{\mu(0) + 1} = 50\mu$.

    (c) $F(1) = \dfrac{50\mu}{\mu\sin(1) + \cos(1)} \approx \dfrac{50\mu}{0.8415\mu + 0.5403}$.

    (d) $F(\pi/2) = \dfrac{50\mu}{\mu(1) + 0} = 50$.

27. If m is the slope and a is the angle of inclination, $m = \tan a$. [From Problem 26].

    (a) $y = \sqrt{3}x - 7 \Rightarrow m m \sqrt{3} \Rightarrow \tan a = \sqrt{3} \Rightarrow a = \pi/3$.
    (b) $\sqrt{3}x + 3y = 6 \Rightarrow y = (-\sqrt{3}/3)x + 2 \Rightarrow m = -\sqrt{3}/3 \Rightarrow \tan a = -\sqrt{3}/3$
    $\Rightarrow a = 5\pi/6$. [Reference angle is $\pi/6$.]

29. Let θ be the angle from the first line to the second line.

    (a) $m_1 = 2$, $m_2 = 3$, $\tan\theta = \dfrac{3-2}{1 + 2\cdot3} = \dfrac{1}{7}$. $\theta \approx 0.1419$ radians.
    (b) $m_1 = 1/2$, $m_2 = -1$. $\tan\theta = \dfrac{-1 - 1/2}{1 + (1/2)(-1)} = -3$. $\theta \approx 1.8925$.
    (c) $m_1 = 1/3$, $m_2 = -2$. $\tan\theta = \dfrac{-2 - 1/3}{1 + (1/3)(-2)} = -7$. $\theta \approx 1.7127$.

31. $A = (1/2)(5)^2(2) = 25$ cm^2.

33. $(18 \text{ min})(\frac{100 \text{ rev}}{3 \text{ min}}) = 600$ rev

Length of groove $\approx$ (no. revolutions)(avg. length for one revolution)

$$= (600 \text{ rev}) \frac{2\pi(6) + 2\pi(3)}{2} \text{ in}$$

$$= 5400\pi \text{ in} \approx 16964.6 \text{ inn } [1413.72 \text{ ft}]$$

35. $s = r \sin(t/2)$; $h = r \cos(t/2)$
$A = \text{Area(triangle)} + \text{Area(semicircle)}$

$$= (1/2)(2s)(h) + (1/2)\pi s^2$$

$$= [r^2 \sin(t/2)\cos(t/2)] + (1/2)\pi [r \sin(t/2)]^2$$

$$= (1/2)r^2 \sin t + (1/2)\pi r^2 \sin^2(t/2)$$

$$= (r^2/2)[\sin t + \pi \sin^2(t/2)]$$

37. (a)   (b)

(c)   (d)

(e)   (f)

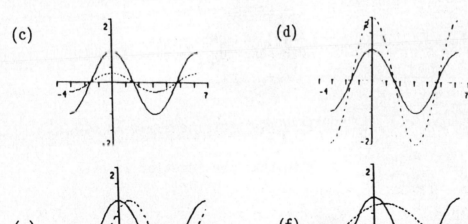

39.

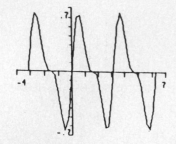

Problem Set 2.4   Introduction to Limits

1.   -2.

3. $(-2)^2 - 3(-2) + 1 = 11.$

5. $1/3.$

7. $\lim_{x \to 1} \dfrac{(x+4)(x-1)}{x-1} = \lim_{x \to 1} (x+4) = 5.$

9. $\lim_{x \to -3} \dfrac{2x^2 + 5x - 3}{x+3} = \lim_{x \to -3} \dfrac{(2x-1)(x+3)}{x+3} = \lim_{x \to -3} (2x-1) = 2(-3) - 1 = -7.$

11. $\lim_{x \to 9} \dfrac{(\sqrt{x}+3)(\sqrt{x}-3)}{\sqrt{x}-3} = \lim_{x \to 9} (\sqrt{x}+3) = 6.$

13. $\dfrac{(2)^2 + (2) - 6}{(2) + 2} = 0.$

15. $\lim_{t \to 2} \dfrac{t^2 - 5t + 6}{t^2 - t - 2} = \lim_{t \to 2} \dfrac{(t-2)(t-3)}{(t-2)(t+1)} = \lim_{t \to 2} \dfrac{t-3}{t+1} = \dfrac{(2)-3}{(2)+1} = -\dfrac{1}{3}.$

17.

x	$\dfrac{\tan x}{2x}$
0.5	0.5463
0.1	0.5017
0.01	0.50002

Note that the function is even.

It seems that the limit is 0.5.

19.

x	$\dfrac{x - \sin x}{x^3}$
0.5	0.1646
0.1	0.1666
0.01	0.1667
0.001	0.1667

Note that the function is even.

It seems that the limit is 0.1667 (1/6).

21. Note that $\dfrac{\sin t}{t^2}$ is an odd function

t	$\dfrac{\sin t}{t^2}$
0.5	1.9177
0.1	9.9833
0.01	99.9983

since it is the ratio of an odd
function and an even function.
Therefore, we only need to observe
what happens as x approaches zero
from one side.

It seems that $\lim\limits_{t \to 0} \dfrac{\sin t}{t^2}$ doesn't exist.

23.

x	$\dfrac{1+\cos x}{\sin 2x}$
3.1	-0.01041
3.14	-0.00040
3.141	-0.00015
3.142	0.00010
3.15	0.00210
3.2	0.01463

It seems that the limit is 0.

25.

x	$\dfrac{\sin 3x}{x-1}$
1.5	-1.9551
1.1	-1.5775
1.01	11.1361
1.001	138.1494

It seems that the limit does not exist.

27. (a) 2.   (b) 1.   (c) Doesn't exist.   (d) 2.5.

(e) 2.   (f) Doesn't exist.   (g) 2.   (h) 1

29. (a) 0.

(b) 2.

(c) Doesn't exist.

(d) 1.

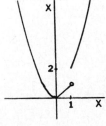

31. (a) 0.

(b) Doesn't exist.

(c) 1.

(d) 1/2.

33. $\displaystyle\lim_{x\to1^+} \frac{x^2-1}{|x-1|} = \lim_{x\to1^+} \frac{(x+1)(x-1)}{x-1} = \lim_{x\to1^+} (x+1) = 2.$

$\displaystyle\lim_{x\to1^-} \frac{x^2-1}{|x-1|} = \lim_{x\to1^-} \frac{(x+1)(x-1)}{-(x-1)} = \lim_{x\to1^-} \frac{(x+1)}{-1} = -2.$

Therefore, $\displaystyle\lim_{x\to1} \frac{x^2-1}{|x-1|}$ doesn't exist.

35. (a) Doesn't exist.   b) 0.

37. For $a = 0, 1, -1$: $\displaystyle\lim_{x\to0} f(x) = 0$; $\displaystyle\lim_{x\to1} f(x) = 1$; $\displaystyle\lim_{x\to-1} (x) = 1.$

39. (a) $\displaystyle\lim_{x\to1^+} \frac{|x-1|}{x-1} = \lim_{x\to1^+} \frac{x-1}{x-1} = \lim_{x\to1^+}(1) = 1$

$\displaystyle\lim_{x\to1^-} \frac{|x-1|}{x-1} = \lim_{x\to1^-} \frac{-(x-1)}{x-1} = \lim_{x\to1^-}(-1) = -1$

Therefore, $\displaystyle\lim_{x\to1} \frac{|x-1|}{x-1}$ doesn't exist.

(b) See line 2 of part (a).

(c) $\displaystyle\lim_{x\to1^-} \frac{x^2-|x-1|-1}{|x-1|} = \lim_{x\to1^-} \frac{x^2-[-(x-1)]-1}{-(x-1)} = \lim_{x\to1^-} \frac{x^2+x-2}{-(x-1)}$

$\displaystyle\qquad = \lim_{x\to1^-} \frac{(x+2)(x-1)}{-(x-1)} = \lim_{x\to1^-} [-(x+2)] = -(1+2) = -3$

(d) Does not exist since, for $x<1$, $\dfrac{1}{x-1} - \dfrac{1}{|x-1|} = \dfrac{1}{x-1} - \dfrac{1}{-(x-1)} = \dfrac{2}{x-1}.$

41. Does not exist   43. 0   45. 0.5

47. Does not exist   49. 6   51. -3

53. Experiment with various modifications of $f(x) = \sin(1/x)$

Problem Set 2.5   Rigorous Study of Limits

1. $\lim\limits_{t \to a} f(t) = M$ means that for each $\epsilon > 0$, there is a corresponding $\delta > 0$ such that $0 < |t-a| < \delta \Rightarrow |f(t)-M| < \epsilon$.

3. $\lim\limits_{z \to d} h(z) = P$ means that for each $\epsilon > 0$, there is a corresponding $\delta > 0$ such that $0 < |z-d| < \delta \Rightarrow |h(z)-P| < \epsilon$.

5. $\lim\limits_{x \to c} f(x) = L$ means that for each $\epsilon > 0$, there is a corresponding $\delta > 0$ such that $0 < c-x < \delta \Rightarrow |f(x)-L| < \epsilon$.

7. Let $\delta = \epsilon/5$. Then $0 < |x-3| < \delta \Rightarrow |5x-15| < \epsilon \Rightarrow |(5x-11)-4| < \epsilon$.

9. Note: In proving $\lim\limits_{x \to 5} \dfrac{x^2-25}{x-5} = 10$, we are not concerned with x being 5. Therefore, it will be sufficient to prove that $\lim\limits_{x \to 5} (x+5) = 10$ since

$\dfrac{x^2-25}{x-5} = \dfrac{(x-5)(x+5)}{x-5} = x+5$, if $x \neq 5$.

Proof: Let $\epsilon > 0$ be given. Choose $\delta = \epsilon$.
Then $0 < |x-5| < \delta \Rightarrow |x-5| < \epsilon$.
$\Rightarrow |(x-5) - 10| < \epsilon$.

Therefore, $\lim\limits_{x \to 5} (x+5) = 10$; hence, $\lim\limits_{x \to 5} \dfrac{x^2-25}{x-5} = 10$.

11. $(2x^2-x-3)/(x+1) = 2x-3$ for all $x \neq -1$, so just prove $\lim\limits_{x \to -1} (2x-3) = -5$.
Let $\delta = \epsilon/2$. Then $0 < |x+1| < \delta \Rightarrow |2x+2| < \epsilon \Rightarrow |(2x-3)-(-5)| < \epsilon$.

13. Let $\delta = \epsilon$. Then $0 < |x-2| < \delta \Rightarrow |x-2| < \epsilon \Rightarrow |x-2| < \dfrac{\epsilon}{\sqrt{2}}(\sqrt{2})$

$\Rightarrow |x-2| < \dfrac{\epsilon}{\sqrt{2}}(\sqrt{x}+\sqrt{2}) \Rightarrow \dfrac{\sqrt{2}|x-2|}{\sqrt{x}+\sqrt{2}} < \epsilon \Rightarrow \sqrt{2}\,|\sqrt{x}-\sqrt{2}| < \epsilon$

$\Rightarrow |\sqrt{2}x - 2| < \epsilon.$

15. $\dfrac{2x^3+3x^2-2x-3}{x^2-1} = \dfrac{(2x+3)(x^2-1)}{x^2-1} = 2x+3$, if $x \neq -1, 1$.

Proof: Let $\epsilon > 0$ be given. Choose $\delta = \min\{\epsilon/2, 2\}$ [to avoid $x = -1$].
Then $0 < |x-1| < \delta \Rightarrow |x-1| < \epsilon/2$

$\Rightarrow |2x-2| < \epsilon$

$\Rightarrow |(2x+3) - 5| < \epsilon.$

Therefore, $\lim\limits_{x\to 1} (2x+3) = 5$; hence, $\lim\limits_{x\to 1} \dfrac{2x^3+3x^2-2x-3}{x^2-1} = 5.$

17. Let $\delta = \min\{\epsilon/5, 1\}$. Then $0 < |x-3| < \delta \Rightarrow |x-3| < \epsilon/5$ and $|x-3| < 1$
$\Rightarrow |x-3| < \epsilon/5$ and $2 < x < 4 \Rightarrow |x-3| < \epsilon/5$ and $3 < x+1 < 5$
$\Rightarrow |x-3| < \epsilon/5$ and $1/5 < 1/|x+1| < 1/3 \Rightarrow |(x-3)| < \epsilon/|x+1|$
$\Rightarrow |(x-3)(x+1)| < \epsilon \Rightarrow |(x^2-2x) - 3| < \epsilon.$

19. Assume $L < M$. Let $\epsilon = (M-L)/2$. Then there is a $\delta > 0$ such that $0 < |x-c| < \delta \Rightarrow |f(x)-L| < \epsilon$ and $|f(x)-M| < \epsilon$. Then, $-\dfrac{M-L}{2} < f(x)-L < \dfrac{M-L}{2}$ and $-\dfrac{M-L}{2} < f(x)-M < \dfrac{M-L}{2}$. Thus, $\dfrac{3L-M}{2} < f(x) < \dfrac{M+L}{2}$ and $\dfrac{M+L}{2} < f(x) < \dfrac{3M-L}{2}$. The $f(x) < \dfrac{M+L}{2}$ and $\dfrac{M+L}{2} < f(x)$ parts constitute a contradiction. Obtain a similar contradiction assuming $M < L$. Therefore, $M = L$.

21. $0 \leq \sin^2(1/x) \leq 1 \Rightarrow 0 \leq x^4\sin^2(1/x) \leq x^4$, since $x^4 > 0$.
Therefore, $\lim\limits_{x\to 0} x^4 \sin^2(1/x) = 0$ [using Problems 20 and 18].

23. $\lim\limits_{x\to 0^+} |x| = \lim\limits_{x\to 0^+} x = 0;\ \lim\limits_{x\to 0^-} |x| = \lim\limits_{x\to 0^-} (-x) = 0;$ Thus, $\lim\limits_{x\to 0} |x| = 0.$

25. Let $\epsilon = 1$. Then there is a $\delta > 0$ such that
$$0 < |x-a| < \delta \Rightarrow |f(x)-L| < 1$$
$$\Rightarrow L-1 < f(x) < L+1$$
$$\Rightarrow |f(x)| < \max\{|L-1|, |L+1|\}$$

Thus, for all x in $(a-\delta, a+\delta)$, $|f(x)| \leq \max\{|f(a)|, |L-1|, |L+1|\}$
Let $c = a-\delta$, $d = a+\delta$, $M = \max\{|f(a)|, |L-1|, |L+1|\}$

27. (a) No.

(b) Yes. [The ϵ and δ symbols are merely interchanged.]

(c) Yes. [1/N and 1/M sort of play the roles of ϵ and δ, respectively. Because N and M are integers there is a little more to it than this; however, that can be easily taken care of since for each positive real number (ϵ and δ) there is a smaller positive number that is the reciprocal of some integer (N or M).]

(d) No.

29. (a) $g(x) = \dfrac{x^3 - x^2 - 2x - 4}{x^4 - 4x^3 + x^2 + x + 6}$ (b) No; $g(x)$ is unbounded on $[2,4]$ (c) 3

Problem Set 2.6   Limit Theorems

1. $7(3) - 4 = 17$. [using 5,3,1,2]

3. $\lim\limits_{x \to 2} [(x^2+1)(3x-1)] \overset{[6]}{=} \lim\limits_{x \to 2}(x^2+1) \cdot \lim\limits_{x \to 2}(3x-1)$

$\overset{[4,5]}{=} \lim\limits_{x \to 2}(x^2) + \lim\limits_{x \to 2}(1) \; \lim\limits_{x \to 2}(3x) - \lim\limits_{x \to 2}(1)$

$\overset{[8,1,3,1]}{=} \lim\limits_{x \to 2}(x)^2 + 1 \; 3\lim\limits_{x \to 2}(x) - 1$

$\overset{[2,2]}{=} [(2)^2 + 1][3(2) - 1] = 25.$

5. $\dfrac{2(4)}{3(4)^3 - 16} = 1/22$. [using 7,3,5,2,3,1,8,2]

7. $\sqrt{3(3)-5} = 2$.   [using 9,5,3,1,2]

9. $\displaystyle\lim_{t\to -2} (2t^3+15)^{13} \overset{[8]}{=} \left[\lim_{t\to -2} (2t^3+15)\right]^{13} \overset{[4]}{=} \left[\lim_{t\to -2} (2t^3) + \lim_{t\to -2} (15)\right]^{13}$

$\overset{[3,1]}{=} \left[2 \lim_{t\to -2} (t^3) + 15\right]^{13} \overset{[8]}{=} \left[2(\lim_{t\to -2} (t))^3 + 15\right]^{13}$

$\overset{[2]}{=} [2(-2)^3 + 15]^{13} = (-1)^{13} = -1.$

11. $\left[\dfrac{4(2)^3+8(2)}{(2)+4}\right]^{1/3} = 2$.   [using 9,7,4,4,3,3,2,1,8,2,2]

13. $\dfrac{(3)^4 - (3)^3 - 2(3)^2 + 1}{3(3)^2 - (5)(3) + 7} = \dfrac{37}{19} \approx 1.9474.$

15. $\displaystyle\lim_{x\to 4} \dfrac{x^2+2x-24}{x-4} = \lim_{x\to 4} \dfrac{(x+6)(x-4)}{x-4} = \lim_{x\to 4} (x+6) = 4+6 = 10.$

17. $\displaystyle\lim_{x\to -1} \dfrac{(x+6)(x+1)}{(x-5)(x+1)} = \lim_{x\to -1} \dfrac{x+6}{x-5} = \dfrac{-1+6}{-1-5} = \dfrac{5}{-6} \approx -0.8333.$

19. Doesn't exist since, as t approaches -1, the numerator approaches 1 and the denominator approaches 0, so the absolute value of the fraction is getting larger without bound.

21. $\displaystyle\lim_{y\to 1} \dfrac{(y-1)(y^2+2y-3)}{(y^2-2y+1)} = \lim_{y\to 1} \dfrac{(y-1)(y-1)(y+3)}{(y-1)^2} = \lim_{y\to 1} (y+3) = 1+3 = 4.$

23. $\sqrt{(3)^2+(-1)^2} = \sqrt{10} \approx 3.1623.$

25. $\sqrt[3]{(-1)}\,[(3)+3] = -6.$

27. $\lim\limits_{t \to a} [f(t) + (t-a)g(t)] = \lim\limits_{t \to a} f(t) + \lim\limits_{t \to a} (t-a) \lim\limits_{t \to a} g(t)$

$$= 3 + \left[\lim_{t \to a}(t) - \lim_{t \to a}(a) \right] (-1) = 3 + (a-a)(-1) = 3.$$

29. $\lim\limits_{x \to 2} \dfrac{5x^2 - 20}{x - 2} = \lim\limits_{x \to 2} \dfrac{5(x+2)(x-2)}{x-2} = \lim\limits_{x \to 2} 5(x+2) = 5(2+2) = 20.$

31. $\lim\limits_{x \to 2} \dfrac{1/x - 1/2}{x - 2} = \lim\limits_{x \to 2} \dfrac{2-x}{2x(x-2)} = \lim\limits_{x \to 2} \dfrac{-1}{2x} = \dfrac{-1}{2(2)} = \dfrac{-1}{4} = -0.25.$

33. Let $\epsilon > 0$ be given.

Choose δ_1 such that $0 < |x-c| < \delta_1 \Rightarrow |g(x) - M| < 1$

$\Rightarrow |g(x)| - |M| < 1$

$\Rightarrow |g(x)| < |M| + 1.$

Choose δ_2 such that $0 < |x-c| < \delta_2 \Rightarrow |f(x) - L| < \dfrac{\epsilon}{2(|M| + 1)}.$

Choose δ_3 such that $0 < |x-c| < \delta_3 \Rightarrow |g(x) - M| < \dfrac{\epsilon}{2(|L| + 1)}.$

Now let $\delta = \min \{\delta_1, \delta_2, \delta_3\}$ and use the hint.

Then $0 < |x-c| < \delta \Rightarrow |f(x)g(x) - LM| \le |g(x)||f(x)-L| + |L||g(x)-M|$

$\le (|M| + 1) \dfrac{\epsilon}{2(|M| + 1)} + |L| \dfrac{\epsilon}{2(|L| + 1)}$

$\le \dfrac{\epsilon}{2} + \dfrac{\epsilon}{2} = \epsilon.$

Therefore, $\lim\limits_{x \to c} f(x)g(x) = LM.$

35. $\lim\limits_{x \to c} f(x) = L \Leftrightarrow \lim\limits_{x \to c} f(x) - L = 0 \Leftrightarrow \lim\limits_{x \to c} f(x) - \lim\limits_{x \to c} L = 0$ [Thm A1]

$\Leftrightarrow \lim\limits_{x \to c} [f(x) - L].$ [Thm A5].

37. Choose δ such that $0 < |x-c| < \delta \Rightarrow |f(x) - L| < L/2$

$\Rightarrow -L/2 < f(x) - L < L/2 \Rightarrow L/2 < f(x) < 3L/2 \Rightarrow f(x) > 0.$

39. (a) Let $f(x) = \frac{1}{x}$; $g(x) = \frac{-1}{x}$. Then $f(x) + g(x) = 0$ [$x \neq 0$].

Thus, $\lim_{x \to 0} [f(x)+g(x)] = 0$, but $\lim_{x \to 0} f(x)$ and $\lim_{x \to 0} g(x)$ do not exist.

(b) Let $f(x) = \begin{Bmatrix} 1, & \text{if } x \text{ is rational} \\ 2, & \text{if } x \text{ is irrational} \end{Bmatrix}$; $g(x) = \begin{Bmatrix} 2, & \text{if } x \text{ is rational} \\ 1, & \text{if } x \text{ is irrational} \end{Bmatrix}$.

Then $\lim_{x \to 5} [f(x)g(x)] = 2$, but $\lim_{x \to 5} f(x)$ and $\lim_{x \to 5} g(x)$ do not exist.

41. $\sqrt{(1)-1} / [2(1)+1] = 0$

43. $\lim_{x \to 3^+} \dfrac{x-3}{\sqrt{(x+3)\ (x-3)}} = \lim_{x \to 3^+} \dfrac{\sqrt{x-3}}{\sqrt{x+3}} = \dfrac{\sqrt{3-3}}{\sqrt{3+3}} = 0.$

45. $\lim_{x \to 2^+} \dfrac{(x^2+1)\ x}{(3x-1)^2} = \lim_{x \to 2^+} \dfrac{(x^2+1) \cdot 2}{(3x-1)^2} = \dfrac{[(2)^2+1]2}{[3(2)-1]^2} = \dfrac{10}{25} = 0.4.$

47. $\lim_{x \to 0^-} \dfrac{x}{-x} \lim_{x \to 0^-} (-1) = -1.$

49. Assume $\lim_{x \to a} f(x) = L$.

Then $\lim_{x \to a} [f(x)g(x)] = \left[\lim_{x \to a} f(x)\right] \left[\lim_{x \to a} g(x)\right] = (L)(0) = 0 \neq 1.$

51. (a) $\lim_{x \to 0^+} \dfrac{\text{Perimeter } (\Delta NOP)}{\text{Perimeter } (\Delta MOP)}$

$= \lim_{x \to 0^+} \dfrac{1 + \sqrt{x^2+x} + \sqrt{x^2+(\sqrt{x}-1)^2}}{1 + \sqrt{x^2+x} + \sqrt{(x-1)^2+x}}$

$= \dfrac{1+0+1}{1+0+1} = 1$

(b) $\lim_{x \to 0^+} \dfrac{\text{Area}(\Delta NOP)}{\text{Area}(\Delta MOP)} = \lim_{x \to 0^+} \dfrac{(1/2)(1)(x)}{(1/2)\ (1)(\sqrt{x})} = \lim_{x \to 0^+} (\sqrt{x}) = 0$

Problem Set 2.7 Continuity of Functions

1. Is.

3. g is not continuous at 2 since g(2) is not defined.

5. Is not since h(2) is undefined.

7. Is not since left hand limit is 1 and right-hand limit is 2.

9. g is not continuous at 2 since g(2) is not defined.

11. Is.   13. Is.

15. $\lim\limits_{x\to 3} f(x) = \lim\limits_{x\to 3} \dfrac{x^2-9}{x-3} = \lim\limits_{x\to 3} \dfrac{(x+3)(x-3)}{x-3} = \lim\limits_{x\to 3} = 6$, so define f(3) = 6.

17. $\lim\limits_{t\to 1} \dfrac{(\sqrt{t}+1)(t-1)}{t-1} = \lim\limits_{t\to 1} (\sqrt{t}+1) = 2$, so define f(1) = 2.

19. $\lim\limits_{x\to -1} \dfrac{(x^2+3))(x+1)(x-1)}{x+1} = \lim\limits_{x\to 1} (x^2+3)(x-1) = -8$, so define $\theta(-1) = -8$.

21. f is discontinuous only at -2 and 3 since f is a rational function whose denominator is zero only at -2 and 3 [Theorem A].

23. Discontinuous nowhere.

25. Discontinuous everywhere on (-∞, -5.].

27. F is continuous on **R** since $1 + x^2$ is always positive [Theorems A,C].

29. Discontinuous nowhere.

31. f is discontinuous at each integer.

33. There are infinitely many. The following is one that satisfies all the conditions.

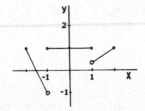

35. $f(x) = x^3+3x-2$ is continuous on [0,1], f(0) = -2, f(1) = 2, and 0 is a number between f(0) and f(1). Therefore, (by the Intermediate Value Theorem) there must be a number c between 0 and 1 such that f(c) = 0. Such a number c is a root of the equation.

37. f is continuous at c

    $\Leftrightarrow$ [for each $\epsilon > 0$, there is a $\delta > 0$ such that
    $0 < |x-c| < \delta \Rightarrow |f(x)-f(x)| < \epsilon$]

    $\Leftrightarrow$ [for each $\epsilon > 0$, there is a $\delta > 0$ such that
    $0 < |t-0| < \delta \Rightarrow |f(t+c)-f(c)| < \epsilon$] (letting $x = t+c$)

    $\Leftrightarrow \lim_{t \to 0} f(t+c) = f(c)$.

39. [Intuitive] If f is continuous on
[0,1] and all the f(x) values lie
between 0 and 1, then the graph of f
has to touch the line y=x at an end
point or cross it somewhere in
between. And f(x) = x at such a
point of intersection.

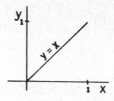

    [Formal] I. If f(0) =0 or f(1) = 1, then 0 or 1 is a fixed point. (This
            is the end point possibility.)

           II. If f(0) > 0 and f(1) < 1, define g(x) = x - f(x). Then
               g is continuous on [0,1]; g(0) = 0 - f(0) < 0; g(1) = 1-f(1)
               > 0; so 0 is between g(0) and g(1). Therefore, by the
               Intermediate Value Theorem, there is at least one number c
               between 0 and 1 such that g(c) = 0.

Thus, c - f(c) = 0 [since g(c) = c - f(c)].
Therefore, f(c) = c, so c is a fixed point of f.

41. Let f be the function with domain [0,1] defined by: if x is the
coordinate of a point of the stretched string, then f(x) is the
coordinate of that point of the relaxed string. Then f satisfies the
hypotheses of Problem 39, so f(c) = c for some point in [0,1].

43. Let f be the function defined on $[0,\pi]$
by: $f(\theta)$ = Temp(P) - Temp(Q), where θ is
the angle formed with the positive side
of the x-axis.

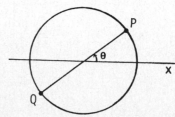

If f(0) = 0, we are done.

If f(0) $\neq$ 0, then f(0) and $f(\pi)$ have opposite signs.

Therefore, by the Intermediate Value Theorem, there is some number in
$(0,\pi)$ such that f(w) = 0. That is, the temperatures at the ends of the
diameter are equal if the diameter forms an angle of w with the
positive side of the x-axis.

45. Let f be the function with domain $[0, \pi/2]$ which is defined by:
$f(\theta)$ = (length of sides of rectangle whose angle of inclination is θ)
- (length of sides of rectangle whose angle of inclination is $\theta + \frac{\pi}{2}$).

Note: (1) f is continuous on $[0, \pi/2]$.
   (2) f(0) and f($\pi/2$) have opposite signs; say one is k and the other is -k.
   (3) 0 is between -k and k.

Therefore, by the Intermediate Value Theorem, there is a number c between 0 and $\pi/2$ such that f(c) = 0. That is, the rectangle corresponding to the angle c is such that the lengths of all of its sides are equal.

47. (a) $f(0) = f(0+0) = f(0) + f(0) \Rightarrow f(0) = 0$.

Now let a be any real number other than 0.

Then $\lim_{x \to a} f(x) = \lim_{h \to 0} f(a+h) = \lim_{h \to 0} [f(a) + f(h)]$

$= \lim_{h \to 0} f(a) + \lim_{h \to 0} f(h) = f(a) + \lim_{h \to 0} f(h) = f(a) + 0 = $
f(a), so f is continuous at each real number, a.

(b) Thus, $f(x) = mx$ on the reals by Problem 43 of Section 2.1 and Problems 46 and 47(a) of this section, since we know that $g(x) = mx$ satisfies $g(x+y) = g(x) + g(y)$ on the reals.

49. (a) $\left[-\frac{3}{4}, \frac{3}{4} \right]$, $\left\{ -\frac{3}{4}, 0, \frac{3}{4} \right\}$, (b) at x = 0 (c) $-\frac{3}{4}, 0, \frac{3}{4}$

Problem Set 2.8 Chapter Review

True-False Quiz

1. True. $y = f(x) = x^2/(3-x)$.

3. True. Auxiliary axis for $\frac{x}{4-x}$.

         (-) (0) (+) (u) (-)
         ———————————————————— x
           0   4

   The radicand must be nonnegative. Thus, the natural domain of f is $[0,4)$.

5. True. $(f+g)(-x) = f(-x)+g(-x) = f(x)+g(x) = (f+g)(x)$.

7. True. It is the ratio of an odd function and an even function.

9. False. C^{ex}: Consider $f(x) = 6$. The natural domain is $\mathbb{R}$ which contains at least two numbers, but the range is 6.

11. True. $f \circ g(x) = f(x^3) = x^6$; $g \circ f(x) = g(x^2) = x^6$.

13. True. $f([a-h] + h) = f(a) = 0$.

15. False. C^{ex}: $\pi/2$ is not in the natural domain of the tan function.

17. False. C^{ex}: Same as for next problem.

19. True.   Has the same graph as $y = x+5$ except for the hole.

21. False. C^{ex}: The function defined in Problem 18 will do.

23. True.   f is continuous at 2. Use the limit definition of continuity with $= 1.001f(2) - f(2) = .001f(2)$.

25. True.   Use the Squeeze Theorem.

27. False C^{ex}: Let $f(x) = \begin{cases} x, & \text{if } x \neq 0 \\ 1, & \text{if } x = 0 \end{cases}$ and let $g(x) = -x$.

              Then $f(x) \neq g(x)$ for all x, but $\lim_{x \to 0} f(x) = \lim_{x \to 0} g(x) = 0$.

29. True. $|f(x)-b| < \epsilon \Rightarrow ||f(x)| - |b|| < \epsilon$.

Sample Test Problems

1. (a) $- 1/2 = -0.5$.   (b) 4.   (c) Does not exist.

    (d) $1/t - 1/(t-1)$.   (e) $t/(t+1) - t$.

3. (a) The denominator must be nonzero. $[x^2 - 1 \neq 0]$ iff $[x \neq -1, 1]$. Therefore, the natural domain is $\{x : x \neq -1, 1\}$.

    (b) The radicand must be nonnegative.

    $[4 - x^2 \geq 0]$ iff $[4 \geq x^2]$ iff $[2 \geq |x|]$ iff $[-2 \leq x \leq 2]$, so the natural domain is $[-2, 2]$.

    (c) The denominator must be nonzero. $[|2x+3| \neq 0]$ iff $[x \neq -3/2]$. Therefore, the natural domain is $\{x : x \neq -3/2\}$.

5. (a)   (b)   (c)

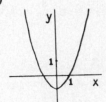   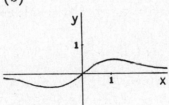   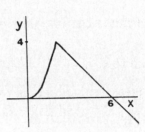

7. $V(x) = (32 - 2x)(24 - 2x)x$; $D_V = (0, 12)$.

9. (a) $y = (1/4)x^2$.

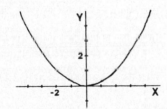

    (b) $y = (1/4)(x+2)^2$. Translate the graph in (a) 2 units left.

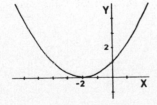

(c) $y = -1 + (1/4)(x+2)^2$. Translate the graph in (b) 1 unit down.

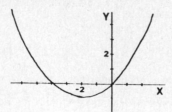

11. Let $k(x) = \sin x$; $h(x) = x^2$; $g(x) = 1+x$; $f(x) = \sqrt{x}$.

13. (a) $-\sin t = -0.8$.   (b) $-\sqrt{1-\sin^2 t} = -\sqrt{1-(0.8)^2} = -0.6$.

     (c) $2\sin t \cos t = 2(.8)(-6) = -9.6$.

     (d) $\sin t / \cos t = 0.8/(-0.6) = -4/3$.

     (e) $\sin t = 0.8$.   (f) $-\sin t = -0.8$.

15. The circumference of the circle is $2\pi \cdot 9$ inches $= 18\pi$ inches. Now transform "revolutions per minute" to "inches per second."

$$\frac{20 \text{ rev}}{1 \text{ min}} \times \frac{18\pi \text{ in}}{1 \text{ rev}} \times \frac{1 \text{ min}}{60 \text{ sec}} = \frac{360\pi \text{ in}}{60 \text{ sec}} = \frac{6\pi \text{ in}}{1 \text{ sec}}.$$

In 1 second the fly travels $6\pi \approx 18.8496$ inches.

17. $\displaystyle\lim_{u \to 1} \frac{(u+1)(u-1)}{u-1} = \lim_{u \to 1} (u+1) = 2$.

19. $\displaystyle\lim_{x \to 2} \frac{x-2}{x(x+2)(x-2)} = \lim_{x \to 2} \frac{1}{x(x+2)} = \frac{1}{8} = 0.125$.

21. $\displaystyle\lim_{x \to 0} \frac{\tan x}{\sin 2x} = \lim_{x \to 0} \frac{(\sin x)/(\cos x)}{2 \sin x \cos x} = \lim_{x \to 0} \frac{1}{2 \cos^2 x} = \frac{1}{2(1)^2} = 0.5$.

23. $\displaystyle\lim_{x \to 4} \frac{(\sqrt{x}+2)(x-4)}{x-4} = \lim_{x \to 4} (\sqrt{x}+2) = 4$.

25. See Problem 27 of Problem Set 2.6.

27. $\lim\limits_{t\to 2^-}([t]-t) = \lim\limits_{t\to 2^-}[t] - \lim\limits_{t\to 2^-}(t) = 1 - 2 = -1.$

29. (a) -1 [since $f(-1)$ is undefined] and 1 [since $\lim\limits_{x\to 1} f(x)$ doesn't exist].

　　(b) Define $f(-1) = -1$ [since $\lim\limits_{x\to -1} f(x) = -1$].

31. (a) $2(3)-4(-2) = 14.$

　　(b) $-2\lim\limits_{x\to 3}\dfrac{(x+3)(x-3)}{x-3} = -2\lim\limits_{x\to 3}(x+3) = -2(6) = -12.$

　　(c) $g(3) = \lim\limits_{x\to 3} g(x) = -2.$　　　　(d) $g\left[\lim\limits_{x\to 3} f(x)\right] = g(3) = -2.$

　　(e) $\sqrt{(3)^2-8(-2)} = 5.$　　　　　　(f) $\dfrac{|(-2)-(-2)|}{(3)} = 0.$

33. The only possible points of discontinuity are 0 and 1.
　　At $x = 0$: $f(0) = -1$; $\lim\limits_{x\to 0^+} f(x) = \lim\limits_{x\to 0^+}(ax+b) = b$; $\lim\limits_{x\to 0^-} f(x) = \lim\limits_{x\to 0^-}(-1) = -1.$
　　　　　Let $b = -1$. Then f is continuous at 0.
　　At $x=1$: $f(1) = 1$; $\lim\limits_{x\to 1^-} f(x) = \lim\limits_{x\to 1^-}(ax-1) = a-1$; $\lim\limits_{x\to 1^+} f(x) = \lim\limits_{x\to 1^+}(1) = 1.$
　　　　　Let $a-1 = 1$ (i.e. $a = 2$). Then f is continuous at 1.
　　Conclusion: $a=2$, $b=-1.$

Problem Set 3.1   Two Problems with One Theme

1. $6/1.5 = 4$.

3. The tangent line seems to go through points (1,8) and (4,0). Therefore, the slope is about -8/3.

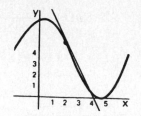

5. The tangent line seems to go through (0,1.5) and (2.5,6.5). Therefore, the slope is about $5/(2.5) = 2$.

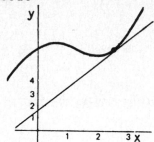

7. (d) $m_{sec} = \dfrac{[4-(3.01)^2] - [-5]}{3.01-3} = -6.01$.

   (e) $m_{tan} = \lim_{h\to0} \dfrac{[4-(3+h)^2] - [-5]}{h} = \lim_{h\to0} \dfrac{h(-6-h)}{h} = \lim_{h\to0} (-6-h) = -6$.

9. Let $f(x) = x^2 - 3x + 2$. Then the slope of the tangent line at x=c is

$$m_{tan} = \lim_{h \to 0} \frac{f(c+h) - f(c)}{h} = \lim_{h \to 0} \frac{[(c+h)^2 - 3(c+h) + 2] - [c^2 - 3c + 2]}{h}$$

$$= \lim_{h \to 0} \frac{c^2 + 2ch + h^2 - 3c - 3h + 2 - c^2 + 3c - 2}{h} = \lim_{h \to 0} \frac{h(2c + h - 3)}{h}$$

$$= \lim_{h \to 0} (2c + h - 3) = 2c - 3. \text{See table below for slopes at x=c.}$$

c	-2	1.5	2	5
2c-3	-7	0	1	7

11. $m_{tan} = \lim_{h \to 0} \frac{f(1+h) - f(1)}{h} = \lim_{h \to 0} \frac{-1}{2(h+2)} = \frac{-1}{4}$.

Equation of the tangent at (1, 1/2) is
y - (1/2) = (-1/4)(x - 1) or y = (-1/4)x + (3/4).

13. (a) 256 - 144 = 112 ft.   (b) 112/1 = 112 ft/sec.

(c) $[16(3.02)^2 - 16(3)^2]/(3.02 - 3) = 96.32$ ft/sec.

(d) $\lim_{h \to 0} \frac{16(3+h)^2 - 16(3)^2}{h} = \lim_{h \to 0} (96 + 16h) = 96$ ft/sec.

15. (a) Let $s(t) = \sqrt{t}$. The instantaneous velocity at time t=c (c>0) is

$$v(c) = \lim_{h \to 0} \frac{s(c+h) - s(c)}{h} = \lim_{h \to 0} \frac{\sqrt{c+h} - \sqrt{c}}{h}$$

$$= \lim_{h \to 0} \frac{(\sqrt{c+h} - \sqrt{c})(\sqrt{c+h} + \sqrt{c})}{h \quad (\sqrt{c+h} + \sqrt{c})} = \lim_{h \to 0} \frac{c+h - c}{h(\sqrt{c+h} + \sqrt{c})}$$

$$= \lim_{h \to 0} \frac{1}{\sqrt{c+h} + \sqrt{c}} = \frac{1}{2\sqrt{c}} \text{ [ft/sec].}$$

(b) Solve $\frac{1}{2\sqrt{c}} = \frac{1}{6}$ for c and obtain c = 9.

It will reach a velocity of 1/6 feet per second in 9 seconds.

17. (a) $[(1/2)(2.01)^2+1] - [3] = 0.02005$ grams.

(b) $0.02005/0.01 = 2.005$ grams/hour.

(c) $\lim\limits_{h\to0} \dfrac{[(1/2)(2+h)^2+1] - [3]}{h} = \lim\limits_{h\to0} (2 + h/2) = 2$ grams/hour.

19. (a) $(5^3-3^3)/2 = 49$ gm/cm.

(b) $\lim\limits_{h\to0} \dfrac{(3+h)^3-3^3}{h} = \lim\limits_{h\to0} (27+9h+h^2) = 27$ gm/cm

21. The rate of growth when t = 10 [weeks] is

$$\lim\limits_{h\to0} \frac{W(10+h) - W(10)}{h} = \lim\limits_{h\to0} \frac{[0.2(10+h)^2 - 0.09(10+h)] - [0.2(10)^2 - 0.09(10)]}{h}$$

$$= \lim\limits_{h\to0} \frac{20 + 4h + 0.2h^2 - 0.9 - 0.09h - 20 + 0.9}{h}$$

$$= \lim\limits_{h\to0} \frac{h(4 + 0.2h - 0.09)}{h} = \lim\limits_{h\to0} (3.91 + 0.2h)$$

$$= 3.91 \text{ grams/week.}$$

23. For day: $\dfrac{100 - 800}{24 - 0} \approx -29.17$, so they average about 29,170 gal/hr.

At 8:00 a.m.: Draw a tangent line at the point where the time is 8 and determine its slope.

$$\frac{100 - 800}{12 - 6} \approx -116.7$$

Therefore, the water was being used at the rate of about 116,700 gal/hr at 8:00 a.m. (Answers will vary depending on the tangent line that is drawn.)

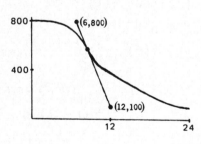

---

25. If t denotes the number of days after the start of the oil spill, then the radius of the spill is r = 2t, and the area is $A = \pi r^2 = \pi(2t)^2$, or $A(t) = 4\pi t^2$. Therefore, when t = 3 the rate of change of area with respect to time is $\lim\limits_{h \to 0} \dfrac{A(3+h) - A(3)}{h} = \lim\limits_{h \to 0} \dfrac{4\pi(3+h)^2 - 36\pi}{h}$

$$= \lim\limits_{h \to 0} (24\pi + 4\pi h) = 24\pi \approx 75.3982 \text{ km}^2/\text{day}.$$

27.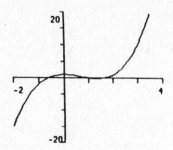

29. 2.81833 ft/s

Problem Set 3.2   The Derivative

1. $\lim\limits_{h \to 0} \dfrac{[(3+h)^2 - (3+h)] - [3^2 - 3]}{h} = \lim\limits_{h \to 0} (5+h) = 5.$

3. $f(x) = x^3 + 2x^2.$

$f'(-1) = \lim\limits_{h \to 0} \dfrac{f(-1+h) - f(-1)}{h} = \lim\limits_{h \to 0} \dfrac{[(-1+h)^3 + 2(-1+h)^2] - [(-1)^3 + 2(-1)^2]}{h}$

$= \lim\limits_{h \to 0} \dfrac{[(-1+3h-3h^2+h^3) + 2(1-2h+h^2)] - [1+2]}{h}$

$= \lim\limits_{h \to 0} \dfrac{-h-h^2+h^3}{h} = \lim\limits_{h \to 0} \dfrac{h(-1-h+h^2)}{h} = \lim\limits_{h \to 0} (-1-h+h^2) = -1.$

5. $\lim\limits_{h \to 0} \dfrac{[5(x+h)-4] - [5x-4]}{h} = \lim\limits_{h \to 0} (5) = 5.$

7. $\lim\limits_{h \to 0} \dfrac{[8(x+h)^2-1] - [8x^2-1]}{h} = \lim\limits_{h \to 0} (16x+8h) = 16x.$

9. $f(x) = ax^2 + bx + c$.

$$f'(x) = \lim_{h \to 0} \frac{f(x+h) - f(x)}{h} = \lim_{h \to 0} \frac{[a(x+h)^2 + b(x+h) + c] - [ax^2 + bx + c]}{h}$$

$$= \lim_{h \to 0} \frac{[ax^2 + 2axh + ah^2 + bx + bh + c] - [ax^2 + bx + c]}{h}$$

$$= \lim_{h \to 0} \frac{h(2ax + ah + b)}{h} = \lim_{h \to 0} (2ax + ah + b) = 2ax + b.$$

11. $\displaystyle\lim_{h \to 0} \frac{[(x+h)^3 - 2(x+h)] - [x^3 - 2x]}{h} = \lim_{h \to 0} (3x^2 + 3xh + h^2 - 2) = 3x^2 - 2.$

13. $\displaystyle\lim_{h \to 0} \frac{3/[5(x+h)] - 3/5x}{h} = \lim_{h \to 0} \frac{-3}{5x(x+h)} = \frac{-3}{5x^2}.$

15. $F(x) = \dfrac{6}{x^2 + 1}.$

$$F'(x) = \lim_{h \to 0} \frac{F(x+h) - F(x)}{h} = \lim_{h \to 0} \frac{\dfrac{6}{(x+h)^2 + 1} - \dfrac{6}{x^2 + 1}}{h}$$

$$= \lim_{h \to 0} \frac{\dfrac{6}{x^2 + 2xh + h^2 + 1} - \dfrac{6}{x^2 + 1}}{h} \cdot \frac{(x^2 + 2xh + h^2 + 1)(x^2 + 1)}{(x^2 + 2xh + h^2 + 1)(x^2 + 1)}$$

$$= \lim_{h \to 0} \frac{6(x^2 + 1) - 6(x^2 + 2xh + h^2 + 1)}{h(x^2 + 2xh + h^2 + 1)(x^2 + 1)}$$

$$= \lim_{h \to 0} \frac{h(-12x - 6h)}{h(x^2 + 2xh + h^2 + 1)(x^2 + 1)} = \lim_{h \to 0} \frac{-12x - 6h}{(x^2 + 2xh + h^2 + 1)(x^2 + 1)} = \frac{-12x}{(x^2 + 1)^2}.$$

17. $\displaystyle\lim_{h \to 0} \frac{[2(x+h) - 1]/[(x+h) - 4] - (2x - 1)/(x - 4)}{h} = \lim_{h \to 0} \frac{-7}{(x+h-4)(x-4)} = \frac{-7}{(x-4)^2}.$

19. $\displaystyle\lim_{h \to 0} \frac{\sqrt{3(x+h)} - \sqrt{3x}}{h} = \lim_{h \to 0} \frac{3}{\sqrt{3(x+h)} + \sqrt{3x}} = \frac{3}{2\sqrt{3x}}.$

21. $H(x) = \dfrac{3}{\sqrt{x-2}}$.

$$H'(x) = \lim_{h\to 0} \frac{H(x+h) - H(x)}{h} = \lim_{h\to 0} \frac{\dfrac{3}{\sqrt{(x+h)-2}} - \dfrac{3}{\sqrt{x-2}}}{h}$$

$$= \lim_{h\to 0} \frac{\dfrac{3}{\sqrt{(x+h)-2}} - \dfrac{3}{\sqrt{x-2}}}{h} \frac{\sqrt{x+h-2}\ \sqrt{x-2}}{\sqrt{x+h-2}\ \sqrt{x-2}}$$

$$= \lim_{h\to 0} \frac{3\sqrt{x-2} - 3\sqrt{x+h-2}}{h\sqrt{x+h-2}\ \sqrt{x-2}} = 3 \lim_{h\to 0} \frac{(\sqrt{x-2} - \sqrt{x+h-2})}{h\sqrt{x+h-2}\sqrt{x-2}} \frac{(\sqrt{x-2} + \sqrt{x+h-2})}{(\sqrt{x-2} + \sqrt{x+h-2})}$$

$$= 3 \lim_{h\to 0} \frac{(x-2) - (x+h-2)}{h\sqrt{x+h-2}\sqrt{x-2}(\sqrt{x-2} + \sqrt{x+h-2})}$$

$$= 3 \lim_{h\to 0} \frac{-h}{h\sqrt{x+h-2}\sqrt{x-2}(\sqrt{x-2} + \sqrt{x+h-2})}$$

$$= 3 \lim_{h\to 0} \frac{-1}{\sqrt{x+h-2}\ \sqrt{x-2}(\sqrt{x-2} + \sqrt{x+h-2})} = \frac{-3}{\sqrt{x-2}\sqrt{x-2}(\sqrt{x-2} + \sqrt{x-2})}$$

$$= \frac{-3}{(x-2)2\sqrt{x-2}} = \frac{-3}{2(x-2)^{3/2}}.$$

23. $\displaystyle\lim_{t\to x} \frac{(t^2-3t) - (x^2-3x)}{t-x} = \lim_{t\to x} \frac{(t+x)(t-x) - 3(t-x)}{t-x} = \lim_{t\to x} [(t+x)-3] = 2x-3.$

25. $\displaystyle\lim_{t\to x} \frac{t/(t-5) - x/(x-5)}{t-x} = \lim_{t\to x} \frac{-5(t-x)}{(t-x)(t-5)(x-5)} = \lim_{t\to x} \frac{-5}{(t-5)(x-5)} = \frac{-5}{(x-5)^2}.$

27. Derivative of $f(x) = 2x^3$ at $x = 5$.

29. $f'(2)$ for $f(x) = x^2$.   31. $f'(x)$ for $f(t) = t^2$.

33. Derivative of $f(x) = 2/x$ at $x = t$.

35. $f'(x)$ for $f(x) = \cos x$.

37. $f'(0) \approx -1/2$; $f'(2) \approx 1$; $f'(5) \approx 1$; $f'(7) \approx -1$.

39. Use the values obtained in Problem 37 as well as those in the following table obtaining by estimating the slope of tangents.

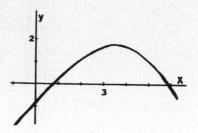

x	$f'(x)$
0.7	0.0 (horizontal tangent)
5.8	0.0 (horizontal tangent)
-1.0	-2.0
3.4	1.7 (steepest tangent line)

41. (a) $f(2) \approx 2.6$; $f'(2) \approx 0.8$; $f(0.5) \approx 1.8$; $f'(0.5) \approx -1.2$.

(b) $[f(2.5)-f(0.5)]/[2.5-0.5] \approx (2.8-1.8)/2 = 1/2$.

(c) At 5.

(d) At 3 and at 5.

(e) At 1, at 3, and at 5.

(f) At 0.

(g) At about -0.7 and 1.8, and everywhere on the interval $(5,7)$.

43. f is differentiable everywhere except possibly where x is 2.

$$f'(2) = \lim_{h \to 0} \frac{f(2+h) - f(2)}{h}.$$

Consider $\displaystyle\lim_{h \to 0^+} \frac{f(2+h) - f(2)}{h} = \lim_{h \to 0^+} \frac{(2+h)^2 - 4}{h} = \lim_{h \to 0^+} (4+h) = 4.$

$\displaystyle\lim_{h \to 0^-} \frac{f(2+h) - f(2)}{h} = \lim_{h \to 0^-} \frac{[m(2+h) + b] - 4}{h} = \lim_{h \to 0^-} \left[m + \frac{2m+b-4}{h}\right]$

This left-hand limit exists only if $2m+b-4 = 0$, in which case its value is m.

Thus, in order for the two-sided limit to exist, it is necessary that $2m+b-4 = 0$ and $m = 4$; or equivalently, $m = 4$ and $b = -4$.

45. Let $z = x_0$.

(a) $f'(-z) = \lim\limits_{h \to 0} \dfrac{f(-z+h) - f(-z)}{h} = \lim\limits_{h \to 0} \dfrac{-f(z-h) + f(z)}{h}$ [since f is odd]

$= \lim\limits_{h \to 0} \dfrac{f(z-h) - f(z)}{-h}$

> Let $t = -h$.
> Then $t \to 0$ as $h \to 0$.

$= \lim\limits_{t \to 0} \dfrac{f(z+t) - f(z)}{t} = f'(z) = m.$

(b) $f'(-z) = \lim\limits_{h \to 0} \dfrac{f(-z+h) - f(-z)}{h} = \lim\limits_{h \to 0} \dfrac{f(z-h) - f(z)}{h}$ [since f is even]

$= -\lim\limits_{h \to 0} \dfrac{f(z-h) - f(z)}{-h}$

> Let $t = -h$.
> Then $t \to 0$ as $h \to 0$.

$= -\lim\limits_{t \to 0} \dfrac{f(z+t) - f(z)}{t} = -f'(z) = -m.$

47. (a) $(0, 2.67)$

(b) $(0, 2.67)$

(c) $f'(x) < 0 \Rightarrow f(x)$ decreasing

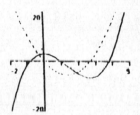

Problem Set 3.3 Rules for Finding Derivatives

1. $6x^2$.

3. $y = \pi x^2 \Rightarrow Dy = \pi \cdot 2x^1 = 2\pi x.$

5. $9x^{-4}$.

7. $8x^{-5}$.

9. $y = \dfrac{3}{5x^5} = \dfrac{3}{5}x^{-5} \Rightarrow Dy = \dfrac{3}{5}(-5)x^{-6} = \dfrac{-3}{x^6}.$

11. $-3x^2 + 2$.

13. $-4x^3 + 6x - 6.$

61

15. $y = 5x^6 - 3x^5 + 11x - 9 \Rightarrow Dy = 30x^5 - 15x^4 + 11.$

17. $-15x^{-6} - 6x^{-4}.$ **19.** $-2x^{-2} + 2x^{-3}.$

21. $y = \dfrac{1}{2x} + 2x = \dfrac{1}{2}x^{-1} + 2x \Rightarrow Dy = \dfrac{1}{2}(-1)x^{-2} + 2 = \dfrac{-1}{2x^2} + 2 = \dfrac{4x^2 - 1}{2x^2}.$

23. $3x^2 + 1.$ **25.** $8x + 4.$

27. $y = (x^2 + 2)(x^3 + 1) = x^5 + 2x^3 + x^2 + 2 \Rightarrow Dy = 5x^4 + 6x^2 + 2x.$

29. $5x^4 + 42x^2 + 2x - 51.$ **31.** $60x^3 - 30x^2 - 32x + 14.$

33. $y = \dfrac{1}{3x^2 + 1} \Rightarrow Dy = \dfrac{(3x^2 + 1)\,D(1) - (1)\,D(3x^2 + 1)}{(3x^2 + 1)^2} = \dfrac{(3x^2 + 1)(0) - (6x)}{(3x^2 + 1)^2}$

$\qquad\qquad = \dfrac{-6x}{(3x^2 + 1)^2}.$

35. $\dfrac{(4x^2 - 3x + 9)(0) - 1(8x - 3)}{(4x^2 - 3x + 9)^2} = \dfrac{-8x + 3}{(4x^2 - 3x + 9)^2}.$

37. $\dfrac{(x+1)(1) - (x-1)(1)}{(x+1)^2} = \dfrac{2}{(x+1)^2}.$

39. $y = \dfrac{2x^2 - 1}{3x + 5} \Rightarrow Dy = \dfrac{(3x + 5)\,D(2x^2 - 1) - (2x^2 - 1)\,D(3x + 5)}{(3x + 5)^2}$

$\qquad\qquad = \dfrac{(3x + 5)(4x) - (2x^2 - 1)(3)}{(3x + 5)^2} = \dfrac{6x^2 + 20x + 3}{(3x + 5)^2}.$

41. $\dfrac{(2x + 1)(4x - 3) - (2x^2 - 3x + 1)(2)}{(2x + 1)^2} = \dfrac{4x^2 + 4x - 5}{(2x + 1)^2}.$

43. $\dfrac{(x^2 + 1)(2x - 1) - (x^2 - x + 1)(2x)}{(x^2 + 1)^2} = \dfrac{x^2 - 1}{(x^2 + 1)^2}.$

45. (a) $(f \cdot g)'(0) = f(0)\, g'(0) + g(0) f'(0) = (4)(5) + (-3)(-1) = 23.$

   (b) $(f+g)'(0) = f'(0) + g'(0) = (-1) + (5) = 4.$

   (c) $(f/g)'(0) = \dfrac{g(0) f'(0) - f(0)\, g'(0)}{[g(0)]^2} = \dfrac{(-3)(-1) - (4)(5)}{(-3)^2} = \dfrac{-17}{9} - 1.8889.$

47. $D[f(x)f(x)] = f(x)f'(x) + f(x)f'(x) = 2f(x)f'(x).$

49. $Dy = 6x - 6$. Slope at $x=1$ is 0, so the line is horizontal. An equation of the horizontal line through $(1,-2)$ is $y = -2$.

51. $y = x^3 - x^2 \rightarrow Dy = 3x^2 - 2x.$

   Horizontal lines have slope zero so set $3x^2 - 2x = 0$ and solve for x.
   $x(3x - 2) = 0$ iff $x = 0$ or $x = 2/3.$

   If $x = 0$, $y = 0$; if $x = 2/3$, $y = -4/27$. Therefore, the points at which the tangents are horizontal are $(0,0)$ and $(2/3, -4/27).$

53. (a) $v(t) = -32t + 40$, so $v(2) = -24$ ft/sec.

   (b) $v(t) = 0$ if $t = 40/32 = 1.25$ sec.

55. (i) (x_0, y_0) is on the curve so (*) $y_0 = 4x_0 - x_0^2.$

   (ii) $Dy = 4 - 2x$ so the slope of the tangent at (x_0, y_0) is $4 - 2x_0$.
   The slope is also $(y_0 - 5)/(x_0 - 2)$, so (**) $(y_0 - 5)/(x_0 - 2) = 4 - 2x_0.$

   Solving (*) and (**) simultaneously obtain the points $(3,3)$ and $(1,3)$. The slope of the line through $(2,5)$ and $(3,3)$ is -2, so the equation of the line is $y - 3 = -2(x - 3)$ or $y = -2x + 9.$

   The slope of the line through $(2,5)$ and $(1,3)$ is 2, so the equation of the line is $y - 3 = 2(x - 1)$ or $y = 2x + 1.$

57. We will find the first point (a,b) on the graph of $y = 7 - x^2$ at which the tangent line passes through (4,0).

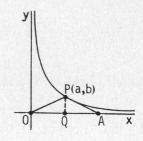

Dy = -2x, so the slope of the tangent at (a,b) is -2a. On the other hand, the slope of the line through (a,b) and (4,0) is (b-0)/(a-4).

Thus, -2a = b/(a-4); equivalently, $b = -2a^2 + 8a$ (since a ≠ 4). Since (a,b) is on the graph of $y = 7 - x^2$, it is also true that $b = 7 - a^2$.

Therefore, $-2a^2 + 8a = 7 - a^2$; $a^2 - 8a + 7 = 0$; (a-1)(a-7) = 0; a = 1 or a = 7. Thus, the insects will first see each other when a = 1 and b = 6; i.e., when the fly is at the point (1,6).

$$d[(1,6),(4,0)] = \sqrt{(4-1)^2 + (0-6)^2} = \sqrt{45} \approx 6.7082 \text{ units.}$$

59. If r denotes the radius of the watermelon and t denotes the time after the watermelon began to form, then r = 2t. Let V_R denote the volume of the rind.

$$V_R = (4/3)\pi r^3 - (4/3)\pi(0.9r)^3 = (1.084\pi/3)r^3 = (8.672\pi/3)t^3, \text{ and the}$$

derivative of V_R with respect to t is $D(V_R) = 8.672\pi t^2$.

Thus, at the end of the fifth week the volume of the rind is growing at the rate of $8.672\pi(5)^2 = 216.8\pi \approx 681.1$ cc/week.

Problem Set 3.4 Derivatives of Sines and Cosines

1. 3cosx + 5sinx.

3. $y = \cot x = \dfrac{\cos x}{\sin x} \Rightarrow Dy = \dfrac{(\sin x) D(\cos x) - (\cos x) D(\sin x)}{(\sin x)^2}$

$$= \frac{(\sin x)(-\sin x) - (\cos x)(\cos x)}{(\sin x)^2} = \frac{-\sin^2 x - \cos^2 x}{\sin^2 x}$$

$$= \frac{-(\sin^2 x + \cos^2 x)}{\sin^2 x} = \frac{-1}{\sin^2 x} = -\csc^2 x.$$

CALCULUS Chapter 3

5. $\dfrac{(\sin x)(0)-(1)(\cos x)}{\sin^2 x}=\dfrac{-\cos x}{\sin^2 x}-\csc x\cot x.$

7. $\dfrac{(\sin x+\cos x)(\cos x)-(\sin x)(\cos x-\sin x)}{(\sin x+\cos x)^2}=\dfrac{1}{1+2\sin x\cos x}=\dfrac{1}{1+\sin 2x}.$

9. $y=\sin^2 x+\cos^2 x=1 \Rightarrow Dy=D(1)=0.$

11. $\dfrac{x(-\sin x)-(\cos x)(1)}{x^2}=\dfrac{x\sin x+\cos x}{-x^2}.$

13. Let $y=f(x)=\sin x.$ Then $f'(x)=\cos x.$
Therefore, $f'(1)=\cos(1)\approx 0.5403$ is the slope of the tangent at $x=1.$
$f(1)=\sin(1)\approx 0.8415$ is the y-coordinate of the point where x=1.

Tangent: $y-\sin(1)=[\cos(1)](x-1)$ or $y=[\cos(1)]x+[\sin(1)-\cos(1)].$
$y=0.5403x+0.3012$ (approximately).

15. The rate at which the seat is moving horizontally is the derivative of the horizontal coordinate, 30 cos 2t [See Example 4.]

$D(30\cos 2t)=30\,D(\cos 2t)$
$=30\,D(1-2\sin^2 t)$
$=30[D(1)-2\,D(\sin^2 t)]$
$=30[0-2\sin 2t]$ [Use the result from Problem 6.]
$=-60\sin 2t.$

The derivative evaluated at $t=\pi/4$ is $-60\sin(\pi/2)=-60,$ so the seat is moving horizontally to the left at 60 feet per second when t is $\pi/4$ seconds.

17. $\lim\limits_{x\to 0}\dfrac{\sin 2x}{3x}=\lim\limits_{x\to 0}\dfrac{2\sin 2x}{(3)2x}=\dfrac{2}{3}\lim\limits_{x\to 0}\dfrac{\sin 2x}{2x}=\dfrac{2}{3}(1)=\dfrac{2}{3}.$

19. $\lim\limits_{x\to 0}\dfrac{\sin 2x}{\tan x}=\lim\limits_{x\to 0}\dfrac{2\sin x\cos x}{\sin x/\cos x}=\lim\limits_{x\to 0}2\cos^2 x=2(1)^2=2.$

65

21. $\displaystyle\lim_{x\to 0}\frac{\tan x - \sin x}{x \cos x} = \lim_{x\to 0}\frac{\frac{\sin x}{\cos x} - \sin x}{x \cos x} = \lim_{x\to 0}\frac{\sin x - \sin x \cos x}{x \cos^2 x}$

$\displaystyle\qquad = \lim_{x\to 0}\frac{(\sin x)(1 - \cos x)}{x \cos^2 x} = \lim_{x\to 0}\frac{\sin x}{x}\frac{1 - \cos x}{\cos^2 x}$

$\displaystyle\qquad = \lim_{x\to 0}\frac{\sin x}{x}\lim_{x\to 0}\frac{1 - \cos x}{\cos^2 x} = (1)\frac{1 - 1}{1^2} = 0.$

23. The point of intersection occurs when $\sin x = \cos x$ (at $x = \pi/4$). The derivatives of the functions are $\sqrt{2}\cos x$ and $-\sqrt{2}\sin x$. The values of the derivatives where $x = \pi/4$ are 1 and -1, respectively, so the tangents are perpendicular.

25. $\displaystyle\lim_{h\to 0}\frac{\sin(x+h)^2 - \sin x^2}{h} = \lim_{h\to 0}\frac{\sin\left[x^2 + (2xh + h^2)\right] - \sin x^2}{h}$

$\displaystyle\quad = \lim_{h\to 0}\frac{\sin x^2\cos(2xh + h^2) + \cos x^2\sin(2xh + x^2) - \sin x^2}{h}$

$\displaystyle\quad = \lim_{h\to 0}\left[(\sin x^2)\frac{\cos(2xh + h^2) - 1}{h} + (\cos x^2)\frac{\sin(2xh + h^2)}{h}\right]$

$\displaystyle\quad = (\sin x^2)\lim_{h\to 0}(2x + h)\frac{\cos\left[h(2x+h)\right] - 1}{h(2x+h)} + (\cos x^2)\lim_{h\to 0}(2x+h)\frac{\sin\left[h(2x+h)\right]}{h(2x+h)}$

$\displaystyle\quad = (\sin x^2)(2x)(0) + (\cos x^2)(2x)(1) = 2x\cos x^2.$

27. $\sin x = \sin 2x \Rightarrow \sin x = 2\sin x\cos x$

$\qquad\qquad \Rightarrow \sin x(1 - 2\cos x) = 0$

$\qquad\qquad \Rightarrow \sin x = 0$ or $\cos x = 1/2$

The smallest positive value of x that satisfies this is $x_0 = \pi/3$.

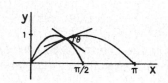

$y = \sin x \Rightarrow Dy = \cos x$

$\qquad \Rightarrow$ slope of graph of $y = \sin x$

$\qquad\qquad$ at $x_0 = \pi/3$ is $1/2$ [since $\cos(\pi/3) = 1/2$]

$y = \sin 2x = 2\sin x\cos x \Rightarrow Dy = (2\sin x)(-\sin x) + (\cos x)(2\cos x)$

$\qquad\qquad\qquad = 2(\cos^2 x - \sin^2 x) = 2\cos 2x$

$\qquad\qquad\qquad \Rightarrow$ slope of graph of $y = \sin 2x$ at $x_0 = \pi/3$ is -1

$\qquad\qquad\qquad\qquad$ [since $2\cos(2\pi/3) = -1$]

Using the result of Problem 28, Section 2.3, the angle θ from the graph of $y = \sin 2x$ to the graph of $y = \sin x$ satisfies

$\tan\theta = \dfrac{(1/2) - (-1)}{1 + (-1)(1/2)} = \dfrac{3/2}{1/2} = 3$, so $\theta \approx 1.2490$ radians [71.57°].

29. (a) Any guess is acceptable (but some will be more accurate).

(b) $\dfrac{D}{E} = \dfrac{(1/2)(1-\cos t)(\sin t)}{(1/2)(1)(t) - (1/2)(\cos t)(\sin t)} = \dfrac{(1-\cos t)(\sin t)}{t - \sin t \cos t}$

(c) For t = 0.01, D/E ≈ 0.7499965 on one calculator, so 0.75 seems to be a good estimate for the limit of D/E as $t \to 0^+$.

31. Compare your computer generated answers to the answers for problems 3.4.11 - 16.

33. (a)

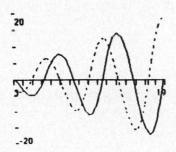

(b) 6, 5

(c) f'(x) = 0 can have more than n-1 solutions

(d) About 25

Problem 3.5   The Chain Rule

1. $15(2-9x)^{14}(-9) = -135(2-9x)^{14}$.

3. $y = (5x^2+2x-8)^5$.

$D_x y = 5(5x^2+2x-8)^4 \ D_x(5x^2+2x-8)$

$= 5(5x^2+2x-8)^4(10x+2) = 10(5x+1)(5x^2+2x-8)^4$.

5. $9(x^3-3x^2+11x)^8(3x^2-6x+11)$.   **7.** $-3(3x^4+x-8)^{-4}(12x^3+1)$.

67

9. $y = (3x^4+x-8)^{-9}$.

$D_x y = -9(3x^4+x-8)^{-10} D_x(3x^4+x-8) = -9(3x^4+x-8)^{-10}(12x^3+1)$

$= \dfrac{-9(12x^3+1)}{(3x^4+x-8)^{10}}$.

11. $[\cos(3x^2+11x)](6x+11)$.   **13.** $3\sin^2 x \cos x$.

15. $y = \left[\dfrac{x^2-1}{x+4}\right]^4$.

$D_x y = 4\left[\dfrac{x^2-1}{x+4}\right]^3 D_x\left[\dfrac{x^2-1}{x+4}\right] = 4\left[\dfrac{x^2-1}{x+4}\right]^3 \dfrac{(x+4)(2x)-(x^2-1)(1)}{(x+4)^2}$

$= 4\left[\dfrac{x^2-1}{x+4}\right]^3 \dfrac{2x^2+8x-x^2+1}{(x+4)^2} = \dfrac{4(x^2-1)^3(x^2+8x+1)}{(x+4)^5} = \dfrac{4(x-1)^3(x+1)^3(x^2+8x+1)}{(x+4)^5}$.

17. $\dfrac{17}{(2x+5)^2} \cos\left[\dfrac{3x-1}{2x+5}\right]$. [See Problem 16 for derivative of $(3x-1)/(2x+5)$.]

19. $(4x-7)^2(2) + (2x+3)[2(4x-7)(4)] = 2(4x-7)(12x+5)$.

21. $y = (2x-1)^3(x^2-3)^2$.

$D_x y = (2x-1)^3 D_x[(x^2-3)^2] + (x^2-3)^2 D_x[(2x-1)^3]$ [Product Rule]

$= (2x-1)^3[2(x^2-3)D_x(x^2-3)] + (x^2-3)^2[3(2x-1)^2 D_x(2x-1)]$ [Chain Rule]

$= (2x-1)^3 2(x^2-3)(2x) + (x^2-3)^2 3(2x-1)^2(2)$

$= 2(2x-1)^2(x^2-3)[2x(2x-1) + 3(x^2-3)]$ [Factoring out common factor]

$= 2(2x-1)^2(x^2-3)(7x^2-2x-9) = 2(2x-1)^2(x^2-3)(7x-9)(x+1)$.

23. $\dfrac{(3x-4)[2(x+1)(1)] - (x+1)^2(3)}{(3x-4)^2} = \dfrac{(x+1)(3x-11)}{(3x-4)^2}$.

25. $\dfrac{(2x^2-5)[2(3x^2+2)(6x)] - (3x^2+2)^2(4x)}{(2x^2-5)^2} = \dfrac{4x(3x^2+2)(3x^2-17)}{(2x^2-5)^2}$.

27. $D_t \left[\dfrac{3t-2}{t+5}\right]^3 = 3\left[\dfrac{3t-2}{t+5}\right]^2 D_t \left[\dfrac{3t-2}{t+5}\right] = 3\left[\dfrac{3t-2}{t+5}\right]^2 \dfrac{(t+5)(3)-(3t-2)(1)}{(t+5)^2}$

$= 3\left[\dfrac{3t-2}{t+5}\right]^2 \dfrac{3t+15-3t+2}{(t+5)^2} = 3\left[\dfrac{3t-2}{t+5}\right]^2 \dfrac{17}{(t+5)^2} = \dfrac{51(3t-2)^2}{(t+5)^4}.$

29. $\dfrac{(t+5)\left[3(3t-2)^2(3) - (3t-2)^3(1)\right]}{(t+5)^2} = \dfrac{(3t-2)^2(6t+47)}{(t+5)^2}.$

31. $3\sin^2\theta \cos\theta.$

33. $D_x \left[\dfrac{\sin x}{\cos 2x}\right]^3 = 3 \left[\dfrac{\sin x}{\cos 2x}\right]^2 D_x \left[\dfrac{\sin x}{\cos 2x}\right]$

$= \dfrac{3\sin^2 x}{\cos^2 2x} \dfrac{(\cos 2x)(\cos x) - (\sin x)\left[(-\sin 2x)(2)\right]}{(\cos 2x)^2}$

$= \dfrac{3\sin^2 x(\cos 2x \cos x + 2\sin 2x \sin x)}{\cos^4 2x}.$

35. $f'(x) = 3\left[\dfrac{x^2+1}{x+2}\right]^2 \dfrac{(x+2)(2x)-(x^2+1)(1)}{(x+2)^2}$; $f'(3) = 3\left[\dfrac{10}{5}\right]^2 \dfrac{(5)(6)-(10)}{(5)^2} = 9.6$.

37. $F'(t) = (2t+3) \cos(t^2+3t+1)$; $F'(1) = 5 \cos 5 \approx 1.4183.$

39. $D_x\left[\sin^4(x^2+3x)\right] = D_x\left[\sin(x^2+3x)\right]^4$

$= 4\left[\sin(x^2+3x)\right]^3 D_x\left[\sin(x^2+3x)\right]$

$= 4\left[\sin(x^2+3x)\right]^3 \left[\cos(x^2+3x)\right] D_x(x^2+3x)$

$= 4\left[\sin(x^2+3x)\right]^3 \left[\cos(x^2+3x) (2x+3)\right.$

$= 4(2x+3) \sin^3(x^2+3x) \cos(x^2+3x).$

[or think of it as] $y = u^4$, $u = \sin v$, $v = x^2+3x.$
Then $D_x y = D_u y \ D_v u \ D_x v$

$= (4u^3)(\cos v)(2x+3)$

$= 4(\sin^3 v)(\cos v)(2x+3)$

$= 4(2x+3) \sin^3(x^2+3x) \cos(x^2+3x).$

41. $3[\sin^2(\cos t)][\cos(\cos t)](-\sin t) = -3(\sin t)(\sin^2 \cos t)(\cos \cos t)$.

43. $[4\cos^3(\sin\theta^2)][-\sin(\sin\theta^2)][\cos\theta^2][2\theta]$

$$= -8\theta(\sin \sin\theta^2)(\cos^3 \sin\theta^2)(\cos\theta^2).$$

45. $D_x(\sin[\cos(\sin 2x)]) = \cos[\cos(\sin 2x)] \ D[\cos(\sin 2x)]$

$$= \cos[\cos(\sin 2x)] \quad [-\sin(\sin 2x)] \ D(\sin 2x)$$

$$= \cos[\cos(\sin 2x)] \quad [-\sin(\sin 2x)] \ (\cos 2x) \quad D(2x)$$

$$= \cos[\cos(\sin 2x)] \quad [-\sin(\sin 2x)] \ (\cos 2x) \quad (2)$$

$$- 2 \cos[\cos(\sin 2x)] \ \sin(\sin 2x) \quad (\cos 2x).$$

47. $D_x y = (x^2+1)^3 \ 2(x^4+1)(4x^3) + (x^4+1)^2 \ 3(x^2+1)^2(2x)$.

Slope at x=1: $(2)^3 \ 2(2)(4) + (2)^2 \ 3(2)^2(2) = 224$.
Equation of tangent: $(y-32) = 224(x-1)$; $y = 224x-192$.

49. (a) 4 rev/sec = 8π rad/sec. Since $\theta = 0$
when t=0, the angle of revolution is $\theta =$
$(8\pi$ rad/sec$)$ $(t$ sec$) = 8\pi t$ rad.
$P = (10\cos 8\pi t, 10\sin 8\pi t)$.

(b) $y(t) = 10\sin 8\pi t$.
$y'(t) = 10(\cos 8\pi t)(8\pi) = 80\pi\cos 8\pi t$.
$y'(1) = 80\pi$. P is rising at $80\pi \approx 251.33$ cm/sec.

51. (a) First convert minutes to seconds and revolutions to radians.
$$60 \text{ rev/min} = \frac{60 \text{ rev}}{1 \text{ min}} \times \frac{1 \text{ min}}{60 \text{ sec}} \times \frac{2\pi \text{ rad}}{1 \text{ rev}} = \frac{2\pi \text{ rad}}{1 \text{ sec}} = 2\pi \text{ rad/sec}.$$
Let the angle of revolution be $\beta = (2\pi$ rad/sec$)(t$ sec$) = 2\pi t$ rad.
Then $P = (\cos\beta, \sin\beta) = (\cos 2\pi t, \sin 2\pi t)$.

(b) $y_Q = y_P + L$ [See figure at right.]

$$= y_P + \sqrt{25 - x_P^2}.$$

$$y_Q(t) = \sin 2\pi t + \sqrt{25 - \cos^2 2\pi t}.$$

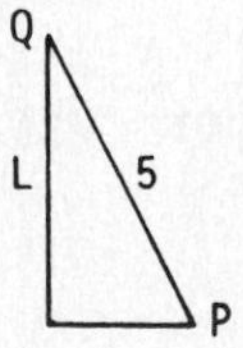

70

(c) The velocity of Q is then $v_Q(t)$

$$= y_Q'(t) = (\cos 2\pi t)(2\pi) + \frac{1}{2\sqrt{25 - \cos^2 2\pi t}}(-2\cos 2\pi t)(-\sin 2\pi t)(2\pi)$$

$$= 2\pi \cos 2\pi t + \frac{2\pi \sin 2\pi t \cos 2\pi t}{\sqrt{25 - \cos^2 2\pi t}}.$$

53. (a) $[|x^2-1|/(x^2-1)](2x) = 2x|x^2-1|/(x^2-1).$

(b) $(|\sin x|/\sin x)(\cos x) = (|\sin x|\cos x)/\sin x.$

55. (a) $f'(x) = 2\sin 2x + (\cos^2 2x)(-2\sin 2x) = 2\sin 2x (1-\cos^2 2x)$

$\qquad = 2\sin 2x \sin^2 2x = 2\sin^3 2x$

(b) $f'(x) = \frac{3}{8} - \frac{3}{8}\cos 4x - \frac{1}{8}[(\sin^3 2x)(-2\sin 2x)+(\cos 2x)(3\sin^2 2x\ 2\cos 2x)]$

$\qquad = \frac{3}{8} - \frac{3}{8}(1-2\sin^2 2x) + \frac{1}{4}\sin^4 2x - \frac{3}{4}\sin^2 2x \cos^2 2x$

$\qquad = \frac{3}{8} - \frac{3}{8} + \frac{3}{4}\sin^2 2x + \frac{1}{4}\sin^4 2x - \frac{3}{4}\sin^2 2x (1-\sin^2 2x)$

$\qquad = \frac{3}{4}\sin^2 2x + \frac{1}{4}\sin^4 2x - \frac{3}{4}\sin^2 2x + \frac{3}{4}\sin^4 2x = \sin^4 2x.$

57. $f(f(f(f(x)))) = (f \circ f \circ f \circ f)(x)$

$(f \circ f \circ f \circ f)'(x) = f'[(f \circ f \circ f)(x)]\ f'[(f \circ f)(x)]\ f'[f(x)]f'(x)$

$(f \circ f \circ f \circ f)'(0) = f'[(f \circ f \circ f)(0)]\ f'[(f \circ f)(0)]\ f'[f(0)]f'(0)$

$\qquad = f'(0)\ f'(0)\ f'(0)\ f'(0) = (2)(2)(2)(2) = 16.$

59. (a) odd \qquad\qquad\qquad (b) even

(c) 0.7 \qquad\qquad\qquad\quad (d) 1

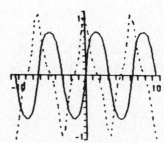

Problem Set 3.6 Leibniz Notation

1. $\Delta y = y_2 - y_1 = 3.25 - 3 = 0.25.$

3. $y = \dfrac{3}{x+1}.$ $y_1 = \dfrac{3}{3.34}$ and $y_2 = \dfrac{3}{3.31}$ so $\Delta y = y_2 - y_1 = \dfrac{3}{3.31} - \dfrac{3}{3.34} \approx 0.00814.$

5. $\dfrac{\Delta y}{\Delta x} = \dfrac{[(x+\Delta x)^2 - 3(x+\Delta x)] - [x^2 - 3x]}{\Delta x} = \dfrac{\Delta x(2x+\Delta x - 3)}{\Delta x} = 2x+\Delta x - 3.$ $[\Delta x \neq 0]$

   $\dfrac{dy}{dx} = \lim_{\Delta x \to 0} (2x+\Delta x - 3) = 2x - 3.$

7. $\dfrac{dy}{dx} = \dfrac{(x+\Delta x)/[(x+\Delta x)+1] - x/(x+1)}{\Delta x} = \dfrac{(x+\Delta x)(x+1) - x(x+\Delta x+1)}{\Delta x(x+\Delta x+1)(x+1)}$

   $= \dfrac{\Delta x}{\Delta x(x+\Delta x+1)(x+1)} = \dfrac{1}{(x+\Delta x+1)(x+1)}$ $[\Delta x \neq 0]$

   $\dfrac{dy}{dx} = \lim_{\Delta x \to 0} \dfrac{1}{(x+\Delta x+1)(x+1)} = \dfrac{1}{(x+1)^2}.$

9. $y = u^3$ and $u = x^2 + 3x.$
   $\dfrac{dy}{dx} = \dfrac{dy}{du}\dfrac{du}{dx} = (3u^2)(2x+3) = 3(x^2+3x)^2(2x+3).$

11. $\dfrac{dy}{dx} = [\cos(x^2)](2x) = 2x\cos(x^2).$

13. $\dfrac{dy}{dx} = 4\left[\dfrac{x^2+1}{\cos x}\right]^3 \dfrac{(\cos x)(2x) - (x^2+1)(-\sin x)}{\cos^2 x} = \dfrac{4(x^2+1)^3(2x\cos x + x^2\sin x + \sin x)}{\cos^5 x}.$

15. $y = \cos(x^2)\, \sin^2 x = [\cos(x^2)](\sin x)^2.$
   $\dfrac{dy}{dx} = [\cos(x^2)]\, \dfrac{d}{dx}[(\sin x)^2] + (\sin x)^2\, \dfrac{d}{dx}[\cos(x^2)]$

   $= [\cos(x^2)]\, 2(\sin x)\, \dfrac{d}{dx}(\sin x) + (\sin x)^2\, [-\sin(x^2)]\, \dfrac{d}{dx}(x^2)$

   $= [\cos(x^2)]\, 2(\sin x)(\cos x) + (\sin x)^2\, [-\sin(x^2)](2x)$

   $= 2\sin x\, [\cos(x^2)\, \cos x - x\, \sin(x^2)\, \sin x].$

17. $[4\sin^3(x^2+3)][\cos(x^2+3)]2x = 8x\, \sin^3(x^2+3)\, \cos(x^2+3).$

19. $2\cos\left[\dfrac{x^2+2}{x^2-2}\right]\left[-\sin\left[\dfrac{x^2+2}{x^2-2}\right]\right]\dfrac{(x^2-2)(2x)-(x^2+2)(2x)}{(x^2-2)^2} = \dfrac{8x}{(x^2-2)^2}\sin\dfrac{2(x^2+2)}{x^2-2}.$

21. $\dfrac{d}{dt}(\sin^3 t + \cos^3 t) = 3\sin^2 t\,\cos t + 3\cos^2 t\,(-\sin t)$

$\qquad\qquad\qquad\qquad\qquad = 3\sin t\,\cos t\,(\sin t - \cos t).$

23. $\pi[2(r+3)(1)] - 3\pi[(r)[2(r+2)(1)] + (r+2)^2(1)] = -\pi(9r^2+22r+6).$

25. $f'(x) = 4(x+x^{-1})^3(1-x^{-2}); \quad f'(2) = 4(2+1/2)^3(1-1/4) = 375/8 = 46.875.$

27. (a) $(f+g)'(3) = f'(3) + g'(3) = (-1) + (-4) = -5.$

(b) $(f\cdot g)'(3) = f'(3)g(3) + f(3)g'(3) = (-1)(3) + (2)(-4) = -11.$

(c) $(f/g)'(3) = \dfrac{g(3)f'(3) - f(3)g'(3)}{[g(3)]^2} = \dfrac{(3)(-1) - (2)(-4)}{(3)^2} = \dfrac{5}{9} \approx 0.5556.$

(d) $(f\circ g)'(3) = f'[g(3)]\cdot g'(3) = f'(3)\cdot(-4) = (-1)(-4) = 4.$

29. (a) $(f+g)'(4) = f'(4) + g'(4) \approx 0+1 = 1.$

(b) $(f\circ g)'(6) = f'(g(6))\,g'(6) \approx f'(2)\,(-1) \approx (1)(1) = -1.$

31. Let the length of an edge, the volume, and the surface area be denoted by x, V, and S, respectively. It is given that dx/dt = 16 cm/min.

(a) $V=x^3, \dfrac{dV}{dt} = 3x^2\dfrac{dx}{dt} = 3x^2(16) = 48x^2.$ At x=20; $\dfrac{dV}{dt} = 19200$ cm^3/min.

(b) $S=6x^2, \dfrac{dS}{dt} = 12x\dfrac{dx}{dt} = 12x(16) = 192x.$ At x=15; $\dfrac{dS}{dt} = 2880$ cm^2/min.

33. $y = x^2 \cos^2(x^2)$

Slope: $\frac{dy}{dx} = (x^2)[2\cos(x^2)][-\sin(x^2)](2x) + [\cos^2(x^2)](2x)$

$\frac{dy}{dx}\bigg|_{x=\sqrt{\pi}} = 0 + [1](2\sqrt{\pi}) = 2\sqrt{\pi}$

Point: $(\sqrt{\pi}, \pi)$

Equation of Tangent: $y - \pi = 2\sqrt{\pi}(x - \sqrt{\pi})$; $y = 2\sqrt{\pi}x - \pi$

To find where the tangent crosses the x-axis, let $y=0$ and solve for x.
$x = \pi/2\sqrt{\pi} = \sqrt{\pi}/2 \approx 0.8862$.

35. See write-up of Problem 57, Section 5.

$g'(x) = (f \circ f \circ f \circ f)'(x) = f'[(f \circ f \circ f)(x)]\ f'[(f \circ f)(x)]\ f'[f(x)]\ f'(x)$

$g'(x_1) = f'(x_2)\ f'(x_1)\ f'(x_2)\ f'(x_1) = [f'(x_1)\ f'(x_2)]^2$

$g'(x_2) = f'(x_1)\ f'(x_2)\ f'(x_1)\ f'(x_2) = [f'(x_1)\ f'(x_2)]^2$

Thus, $g'(x_1) = g'(x_2)$.

37. t = # minutes after 12:00

$\theta = (\frac{11}{12}t)(\frac{2\pi}{60})$ radians $= \frac{11\pi t}{360}$ radians

$s = [6^2 + 8^2 - 2(6)(8)\cos\frac{11\pi t}{360}]^{1/2}$

$= [100 - 96\cos\frac{11\pi t}{360}]^{1/2}$

$\frac{ds}{dt} = \frac{1}{2}[100 - 96\cos\frac{11\pi t}{360}]^{-1/2}[96\sin\frac{11\pi t}{360}]\frac{11\pi}{360}$

At t=20, $\frac{ds}{dt} = \frac{48}{\sqrt{100 - 96\cos\frac{11\pi}{18}}}(\sin\frac{11\pi}{18})(\frac{11\pi}{360}) = 0.375675$ in/min

Problem Set 3.7 High Order Derivatives

1. $dy/dx = 3x^2+6x-2$. $d^2y/dx^2 = 6x+6$. $d^3y/dx^3 = 6$.

3. $y = (2x+5)^4$.

$$\frac{dy}{dx} = 4(2x+5)^3(2) = 8(2x+5)^3.$$

$$\frac{d^2y}{dx^2} = 24(2x+5)^2(2) = 48(2x+5)^2.$$

$$\frac{d^3y}{dx^3} = 96(2x+5)^1(2) = 192(2x+5).$$

5. $dy/dx = (\cos 3x)(3) = 3\cos 3x$. $d^2y/dx^2 = 3(-\sin 3x)(3) = -9\sin 3x$.
    $d^3y/dx^3 = -9(\cos 3x)(3) = -27\cos 3x$.

7. $dy/dx = -(x-3)^{-2}$. $d^2y/dx^2 = 2(x-3)^{-3}$. $d^3y/dx^3 = -6(x-3)^{-4}$.

9. $f(x) = 2x^3-7$. $f'(x) = 6x^2$. $f''(x) = 12x$. $f''(2) = 12(2) = 24$.

11. $f'(t) = -t^{-2}$. $f''(t) = 2t^{-3} = 2/t^3$. $f''(2) = 1/4$.

13. $f'(x) = x\, 3(x^2+1)^2(2x) + (x^2+1)^3 = (x^2+1)^2(7x^2+1)$.
     $f''(x) = (x^2+1)^2(14x) + (7x^2+1)\, 2(x^2+1)\, 2x$. $f''(2) = 1860$.

15. [Before starting this problem, review the double-angle identity for the
     sine function on page 54 of the text. It is a useful identity for
     simplifying the work of derivative-taking in this and other problems.]

$f(x) = \sin^2(\pi x) = (\sin \pi x)^2$.
$f'(x) = 2(\sin \pi x)(\cos \pi x)(\pi) = \pi \sin 2\pi x$. [double-angle identity]
$f''(x) = \pi(\cos 2\pi x)(2\pi) = 2\pi^2 \cos 2\pi x$. $f''(2) = 2\pi^2 \cos 4\pi = 2\pi^2 \approx 19.7392$.

17. (Use mathematical induction) $D^1(x^1) = 1 = 1!$
     $D^{k+1}(x^{k+1}) = D^k[D(x^{k+1}) = D^k[(k+1)x^k] = (k+1)k! = (k+1)!$

19. (a) 0.   (b) 0.   (c) 0.

21. $f(x) = x^3 + 3x^2 - 45x - 6$.

$f'(x) = 3x^2 + 6x - 45 = 3(x^2 + 2x - 15) = 3(x+5)(x-3)$.

$\quad f'(x) = 0 \quad$ iff $\quad x = -5$ or $x = 3$.

$f''(x) = 6x + 6$.

$\quad f''(-5) = 6(-5) + 6 = -24. \quad f''(3) = 6(3) + 6 = 24$.

23. (a) $s(t) = 2t(6-t)$. $\quad v(t) = 4(3-t)$. $\quad a(t) = -4$.
(b) Object is moving to the right when $t < 3$.
(c) Object is moving to the left when $t > 3$.
(d) Acceleration is negative for all t.
(e)

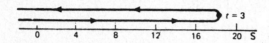

25. (a) $s(t) = t(t^2 - 9t + 24)$. $\quad v(t) = 3(t-2)(t-4)$. $\quad a(t) = 6(t-3)$.

(b) When t is in $(-\infty, 2) \cup (4, \infty)$.

(c) When t is in $(2, 4)$.

(d) When t is in $(-\infty, 3)$.

(e)

27. $s(t) = t^2 + \dfrac{16}{t} = t^2 + 16t^{-1}$, $t > 0$.

(a) $v(t) = s'(t) = 2t - 16t^{-2} = 2t - \dfrac{16}{t^2} = \dfrac{2t^3 - 16}{t^2} = \dfrac{2(t-2)(t^2 + 2t + 4)}{t^2}$.

Auxiliary axis for $v(t)$:

$a(t)$ $v'(t) = 2 + 32t^{-3} = 2 + \dfrac{32}{t^3} = \dfrac{2t^3 + 32}{t^3} = \dfrac{2(t^3 + 16)}{t^3}$.

Auxiliary axis for $a(t)$:

   (b)   The object is moving to the right (values of s increasing as t increases) when v is positive; i.e., when t is greater than 2.

   (c)   It is moving left (values of s decreasing as t increases) when v is negative; i.e., when t is between 0 and 2.

   (d)   The acceleration is never negative.

   (e)

29. $v(t) = 2t^3 - 15t^2 + 24t$. $a(t) = 6(t-1)(t-4)$.

      $a(t) = 1$ iff $t=1$ or $t=4$. $v(1) = 11$. $v(4) = -16$.

31.   (a) $v_1 = 4 - 6t$. $v_2 = 2t - 2$. $v_1 - v_2$ iff $t=3/4$.
      (b) $|v_1| = |v_2|$ iff $t=3/4$ or $t=1/2$.
      (c) $s_1 = s_2$ iff $t=0$ or $t=3/2$.

33.   Equations of Motion: $s(t) = 16t^2 + 48t + 256$, $t \geq 0$ ft.
                             $v(t) = s'(t) = -32t + 48$ ft/sec.

   (a)   Initial velocity is v when t=0. $v(0) = -32(0) + 48 = 48$ ft/sec.

   (b)   Maximum height is reached when v=0. $v(t)=0$ iff $t = \frac{48}{32} = 1.5$ sec.

   (c)   Maximum height is $s(1.5) = -16(1.5)^2 + 48(1.5) + 256 = 292$ ft.

   (d)   Hits ground when s=0. $s(t)=0$ iff $-16t^2 + 48t + 256 = 0$.

      Using the quadratic formula obtain $t = \frac{3 \pm \sqrt{73}}{2} = -2.77, 5.77$ sec.

      The former value is outside the domain. Therefore, the object hits the ground in about 5.77 seconds.

   (e) $v\left[\frac{3 + \sqrt{73}}{2}\right] = -32\left[\frac{3 + \sqrt{73}}{2}\right] + 48 \approx -136.70$ ft/sec.

      Therefore, it hits the ground with a speed of about 136.70 ft/sec.

35. $s(t) = v_0 t - 16t^2$; $v(t) = v_0 - 32t$.

$5280 = v_0 t - 16t^2$ and $0 = v_0 - 32t$ (at maximum height).
Solving simultaneously for t and v_0,

$t = \sqrt{330}$, $v_0 = 32\sqrt{330} \approx 581.3089$ ft/sec. (about 396 mph).

37. In this problem we will use the fact that if $f(x) = |x|$, $f'(x) = \dfrac{|x|}{x}$.
[See Problem 52 of Section 3.5.]

$s(t) = t^3 - 3t^2 - 24t - 6$.

$v(t) = s'(t) = 3t^2 - 6t - 24 = 3(t-4)(t+2)$.

Auxiliary axis for $v(t)$:

(+)	(0)	(-)	(0)	(+)
	-2		4	t

$a(t) = v'(t) = 6t - 6 = 6(t-1)$.

Auxiliary axis for $a(t)$:

(-)	(0)	(+)
	1	t

Let $r(t) = |v(t)|$ denote the speed. The rate of change of the speed with respect to time is then

$$r'(t) = \frac{|v(t)|}{v(t)} v'(t) = \frac{|v(t)|}{v(t)} a(t).$$

Use the equation for $r'(t)$ and the auxiliary axes for $v(t)$ and $a(t)$ to develop the table.

Interval	$v(t)$	$a(t)$	$r'(t)$
$(-\infty, -2)$	+	-	-
$(-2, -1)$	-	-	+
$(1, 4)$	-	+	-
$(4, \infty)$	+	+	+

The point is slowing down (speed is decreasing) when $r'(t)$ is negative; i.e., when t is less than -2 or between 1 and 4.

39. Let s(t) be the distance the car travels in t units of time.
(a) $s'(t) = ks$, $k \geq 0$.

(b) $s''(t) > 0$.

(c) I didn't say $s''(t) < 0$; I said $s^{(3)}(t) < 0$.

(d) $s''(t) = $ 10mph/min.

(e) $s''(t)$ is less than, but close to, 0.

(f) $s'(t) = k$, $k \geq 0$.

41. (a) $C'(t) > 0$ continues, and even $C''(t) > 0$. [C(t) is cost at time t.]

(b) $V'(t) < 0$ and $V''(t) > 0$. [V(t) is amount of oil at time t.]

(c) $P'(t) > 0$ and $P''(t) < 0$. [P(t) is population at time t.]

(d) $S'(t) > 0$ and $S''(t) = k > 0$. [S(t) is distance traveled in time t.]

(e) $\theta'(t) > 0$ and $\theta''(t) > 0$. [θ(t) is angle at time t.]

(f) $P'(t) > 0$ and $P''(t) < 0$. [P(t) is profit at time t.]

(g) $A'(t) < 0$ and $A''(t) > 0$. [A(t) is assets at time t.]

43. $D(uv) = uDv + vDu$
$D^2(uv) = uD^2v + 2DuDv + (D^2u)v$
$D^3(uv) = uD^3v + 3DuD^2v + 3D^2uDv + (D^3u)v$
$D^n(uv) = uD^nv + nDuD^{n-1}v + \dfrac{n(n-1)}{2} D^2uD^{n-2}v + \cdots + nD^{n-1}uDv + (D^nu)v$

45. (a) -1.28258

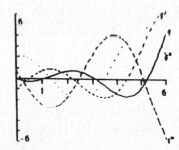

Problem Set 3.8 Implicit Differentiation

1. $D(x^2 - y^2) = D(9)$. $2x - 2yDy = 0$. $Dy = x/y$.

3. $xy = 4$. $D(xy) = D(4)$. $x(Dy) + y(1) = 0$.
Therefore, $Dy = \dfrac{-y}{x}$.

5. $D(xy^2-x+16) = D(0)$. (x) $(2yDy)+y^2-1 = 0$. $Dy = (1-y^2)/(2xy)$.

7. $D(4x^3+11xy^2-2y^3) = D(0)$. $12x^2+(11x)(2yDy)+11y^2-6y^2Dy = 0$.
$Dy = (12x^2+11y^2)/[2y(3y-11x)]$.

9. $6x-\sqrt{2xy}+xy^3 = y^2$. $D(6x-\sqrt{2xy}+xy^3) = D(y^2)$.

$6 - \dfrac{1}{2\sqrt{2xy}} D(2xy) + [(x) D(y^3) + (y^3)(1)] = 2yDy$.

$6 - \dfrac{[(2x)(Dy) + (y)(2)]}{2\sqrt{2xy}} + (x)(3y^2Dy) + y^3 = 2yDy$.

$12\sqrt{2xy} - 2x\,Dy - 2y + 6xy^2\sqrt{2xy}\,Dy + 2y^3\sqrt{2xy} = 4y\sqrt{2xy}\,Dy$.

$(-2x + 6xy^2\sqrt{2xy} - 4y\sqrt{2xy})\,Dy = -12\sqrt{2xy} + 2y - 2y^3\sqrt{2xy}$.

Therefore, $Dy = \dfrac{-6\sqrt{2xy} + y - y^3\sqrt{2xy}}{-x + 3xy^2\sqrt{2xy} - 2y\sqrt{2xy}}$.

11. $D(xy+siny) = D(x^2)$. $xDy+y + (cosy)Dy = 2x$. $Dy = (2x-y)/(x+cosy)$.

13. $D(x^3y+y^3x) = D(10)$. $x^3Dy+3x^2y+y^3+3xy^2Dy = 0$.
Slope at $(1,2)$: $Dy+6+8+12Dy = 0$, so $Dy = -14/13$.
Tangent at $(1,2)$: $y-2 = (-14/13)(x-1)$, or $y = (-14/13)x + (40/13)$.

15. $sin(xy) = y$. $D[sin(xy)] = D[y]$. $[cos(xy)] D(xy) = Dy$.

$[cos(xy)][x\,Dy + y(1)] = Dy$. $[cos(xy)][x\,Dy + y] = Dy$.

At $(\pi/2,1)$: $[cos(\pi/2)][(\pi/2)Dy + (1)] = Dy$.

$[0][(\pi/2)Dy + 1] = Dy$, so $Dy = 0$.

Therefore, at $(\pi/2,1)$, the slope of the tangent line is zero, so the tangent line is horizontal. Its equation is then $y = 1$.

17. $D(x^{2/3}-y^{2/3}-2y) = D(2) \cdot (2/3)x^{-1/3} - (2/3)y^{-1/3}Dy - 2Dy = 0$.
Slope at $(1,-1)$: $(2/3) + (2/3)Dy - 2\,Dy$, so $Dy = 1/2$.
Tangent at $(1,-1)$: $y+1 = (1/2)(x-1)$, or $y = (1/2)x - (3/2)$.

19. $5x^{2/3} + 1/2\sqrt{x}$.

21. $y = x^{1/3} + x^{-1/3}$.

$$\frac{dy}{dx} = (1/3)x^{-2/3} + (-1/3)x^{-4/3} = \frac{1}{3\sqrt[3]{x^2}} - \frac{1}{3\sqrt[3]{x^4}} = \frac{\sqrt[3]{x^2} - 1}{3\sqrt[3]{x^4}}.$$

23. $(1/4)(3x^2 - 4x)^{-3/4}(6x - 4) = (1/2)(3x - 2)/(3x^2 - 4x)^{3/4}$.

25. $(-2/3)(x^3 + 2x)^{-5/3}(3x^2 + 2) = -2(3x^2 + 2)/3(x^3 + 2x)^{5/3}$.

27. $y = \sqrt{x^2 + \sin x}$.

$$\frac{dy}{dx} = \frac{1}{2\sqrt{x^2 + \sin x}}(2x + \cos x) = \frac{2x + \cos x}{2\sqrt{x^2 + \sin x}}.$$

29. $(-1/3)(x^2 \sin x)^{-4/3}[x^2 \cos x + (\sin x)2x]$
$$= (-x/3)(x \cos x + 2\sin x)(x^2 \sin x)^{-4/3}.$$

31. $(1/4)[1 + \cos(x^2 + 2x)]^{-3/4}[-\sin(x^2 + 2x)](2x + 2)$
$$= (-1/2)(x+1)\sin(x^2 + 2x)[1 + \cos(x^2 + 2x)]^{-3/4}.$$

33. $s^2 t + t^3 = 1$.

$$\frac{d}{dt}(s^2 t + t^3) = \frac{d}{dt}(1). \qquad\qquad \frac{d}{ds}(s^2 t + t^3) = \frac{d}{dt}(1).$$

$$[(s^2)(1) + t(2s\frac{ds}{dt})] + 3t^2 = 0. \quad [(s^2)\frac{dt}{ds} + (t)(2s)] + 3t^2\frac{dt}{ds} = 0.$$

$$2st\frac{ds}{dt} = -3t^2 - s^2. \qquad\qquad (s^2 + 3t^2)\frac{dt}{ds} = -2st.$$

Therefore, $\dfrac{ds}{dt} = \dfrac{-3t^2 - s^2}{2st} = \dfrac{s^2 + 3t^2}{-2st}.$ Therefore, $\dfrac{dt}{ds} = \dfrac{-2st}{s^2 + 3t^2}.$

Notice that $\dfrac{ds}{dt} = \dfrac{1}{dt/ds}.$

35. $(x-2)^2 + y^2 = 1$.
$2(x-2)+2yy' = 0$. $y' = (2-x)/y$, so tangent has slope $(2-x)/y$. Line through origin has form $y = mx$, so tangent has slope y/x.

Solving $(2-x)/y = y/x$ and $y^2 = -x^2+4x-3$ (equation of circle) simultaneously, obtain $x = 3/2$, $y = \pm \sqrt{3}/2$. Thus, slope of tangent is $\pm\sqrt{3}/3$.

Therefore, equation of tangents are $y = (\pm \sqrt{3}/3)x$.

37. $y' = (-y)/(x+3y^2)$.
$xy''+2[(-y)/(x+3y^2)]+3y^2y''+6y[(-y)/(x+3y^2)]^2 = 0$. $y'' = (2xy)/(x+3y^2)^3$.

39. $2x^2y-4y^3 = 4$.

First Derivative: $D[2x^2y-4y^3] = D[4]$.
$$[(2x^2)(Dy) + (y)(4x)] - [(12y^2)(Dy)] = 0.$$
$$2x^2Dy + 4xy - 12y^2Dy = 0. \quad (*)$$
Second Derivative: $D[2x^2Dy + 4xy - 12y^2Dy] = D[0]$.
$$[(2x^2)(D^2y)+(Dy)(4x)]+[(4x)(Dy)+(y)(4)]-[(12y^2)(D^2y)+(Dy)(24y)(Dy)] = 0.$$
$$2x^2D^2y + 4x\,Dy + 4x\,Dy + 4y - 12y^2D^2y - 24y(Dy)^2 = 0.$$
$$(2x^2-12y^2)D^2y + 8x\,Dy - 24y(Dy)^2 + 4y = 0. \quad (**)$$
Now use $(*)$ at $(2,1)$: $2(4)Dy + 4(2)(1) - 12(1)Dy = 0 \Rightarrow Dy = 2$.
Now use $(**)$ at $(2,1)$ and the fact that $Dy = 2$ at $(2,1)$:
$$[2(4)-12(1)]D^2y + 8(2)(2) - 24(1)(4) + (4) = 0 \Rightarrow D^2y = -15.$$

41. $3x^2+3y^2y' = 3xy'+3y$.
At $(3/2,3/2)$: $3(9/4)+3(9/4)y' = 3(3/2)y'+3(3/2)$, so $y' = -1$. Normal lines has slope 1 and equation $y-3/2 = 1(x-3/2)$, or $y = x$, which passes through the origin.

43. For $2x^2+y^2 = 6$, $y' = -2x/y$. For $y^2 = 4x$, $y' = 2/y$.
At points of intersection, $x = y^2/4$ (from second equation).

Thus, slope of tangent to $2x^2+y^2 = 6$ at points of intersection must be $-2x/y = -2(y^2/4)/y = -y/2$.

We already have slope of tangent to $y^2 = 4x$ is $2/y$.

$(-y/2)(2/y) = -1$. [Note that $y \neq 0$ at points of intersection.]

45. Determine the point of intersection in the first quadrant:

$$\left[\begin{matrix} x^2-xy+2y^2 = 28 \\ y = 2x \end{matrix}\right] \Rightarrow x^2 - x(2x) + 2(2x)^2 = 28$$

$$\Rightarrow x^2 = 4$$

$$\Rightarrow x = 2 \quad \text{(in the first quadrant)}.$$

Therefore, the point of intersection is (2,4). Next determine the derivatives:

(*) y = 2x.   (**) $x^2-xy+2y^2 = 28$.

  Dy = 2. $D[x^2-xy+2y^2] = D[28]$.

                                               $2x - [xDy + y(1)] + 4y\,Dy = 0$.

                                               $(-x+4y)Dy + (2x-y) = 0$.

                                               $Dy = \dfrac{2x-y}{-x+4y}$.

Therefore, the slopes at (2,4) are 2(*) and 0(**).

Then use the formula in Problem 44 with $m_1 = 2$ and $m_2 = 0$:

$$\tan\theta\ \frac{(0) - (2)}{1 + (0)(2)} = -2, \text{ so } \theta \approx 2.0344 \text{ radians.} \quad [\text{About } 116.6°]$$

47. $x^2-xy+y^2 = 16$

$2x - (xy'+y) + 2yy' = 0; \ y' = \dfrac{y-2x}{2y-x}$

Points: (-4,0)   (4,0)

Slopes: $y'(-4,0) = \dfrac{0+8}{0+4} = 2$ $y'(4,0) = \dfrac{0-8}{0-4} = 2$

Equations: y-0 = 2(x+4)   y-0 = 2(x-4)

            y = 2x+8   y = 2x-8

49. h = 13/3

Problem Set 3.9 Related Rates

1. $V = x^3$, and dx/dt = 3.

$dV/dt = 3x^2(dx/dt) = 3x^2(3) = 9x^2$.

When x=10, dV/dt = 900 in^3/sec.

3. Step 1: Let t denote time elapsed after the plane is overhead [hr].
            Let x and y be as indicated in Figure 7 of the text [mi].

    Step 2: We are given that $\frac{dx}{dt}$ = 240 mph, the plane's speed.

            We are to find $\frac{dy}{dt}$ when t = 1/120 hr.

    Step 3: $1+x^2 = y^2$.   [Pythagorean Theorem]

    Step 4: $\frac{d}{dt}(1+x^2) = \frac{d}{dt}(y^2)$.

            $2x \frac{dx}{dt} = 2y \frac{dy}{dt}$ or $x \frac{dx}{dt} = y \frac{dy}{dt}$.

    Step 5: When t = 1/120, x = 2 (from hint) and
                        $y = \sqrt{1+4} = \sqrt{5}$ (from equation in Step 3).
            $(2)(240) = (\sqrt{5}) \frac{dy}{dx}$ (substituting into equation in Step 4).
            Therefore, $\frac{dy}{dt} = \frac{480}{\sqrt{5}} \approx 214.6625$.

    Conclusion: The distance of the plane from the observer is increasing at
                about 214.66 miles per hour 30 seconds after the plane goes
                overhead.

5. $x^2+y^2 = s^2$, dx/dt = 400, dy/dt = 500.
    $2x(dx/dt)+2y(dy/dt) = 2s(ds/dt)$ or $400x+500y = s(ds/dt)$.
    When t = 1, x = 600, y = 500, and $s = \sqrt{610000}$,
    so ds/dt = $(240000 + 25000)/\sqrt{610000} \approx 627.38$ mph.

7. $x^2+y^2 = 400$ and dx/dt = 2.
    $2x(dx/dt) + 2y(dy/dt) = 0$, so dy/dt = $-2x/y$.
    When x=4, $y = \sqrt{400-16} = \sqrt{384}$, so dy/dt = $-8/\sqrt{384} \approx 0.4082$ ft/sec.

9. Step 1:   Let t denote the time elapsed [sec].
              Let r and h be as indiated in Figure 8 [ft].
              Let V denote the volume of the conical pile [ft^3].

    Step 2:   We are given dV/dt = 16 ft^3/sec since the volume of the pile
              is increasing at the same rate as the sand pouring from the
              pipe. We are to find dh/dt when h=4 feet.

    Step 3: $V = (1/3)\pi r^2 h$ and $h = (1/4)(2r) = (1/2)r$,
              so $V = (1/3)\pi(2h)^2 h = (4\pi/3)h^3$.

    Step 4: $\frac{dV}{dt} = (4\pi/3)(3h^2)\frac{dh}{dt} = 4\pi h^2 \frac{dh}{dt}$.

    Step 5:   When h=4, $(16) = 4\pi(4)^2 \frac{dh}{dt}$, so $\frac{dh}{dt} = \frac{1}{4\pi} \approx 0.07958$.

    Conclusion: When the pile is 4 feet high, the altitude is increasing at
                 about 0.07958 ft/sec [almost 1 in/sec].

11. $V = (1/2)h\ell(20)$ [if $0 \le h \le 5$] and $\ell/h = 40/5 = 8$, so $V = 80h^2$.
     dV/dt = 40 and $dV/dt = 160h(dh/dt)$, so dh/dt = 1/4h.
     When h=3, dh/dt = 1/12 $\approx$ 0.08333 ft/min.

13. $A = \pi r^2$ and dr/dt = 0.02.
     $dA/dt = 2\pi r(dr/dt) = 0.04\pi r$.

     When r=8.1, $dA/dt = 0.04\pi(8.1) \approx 1.0179$ in^2/sec.

15. Step 1:   Let t denote time elapsed [min].
              Let x[km] and β[rad] be as indicated
              in the figure to the right.

    Step 2:   We are given $\frac{d\beta}{dt} = 4\pi$ rad/min.

              We are to find $\frac{dx}{dt}$ when x = 1/2 km.

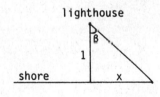

    Step 3: $\tan\beta = x/1$, so $x = \tan\beta$.

    Step 4: $\frac{dx}{dt} = \sec^2\beta \frac{d\beta}{dt} = 4\pi\sec^2\beta$.

Step 5: When x=1/2, $\sec\beta = \sqrt{1+\tan^2\beta} = \sqrt{1+x^2} = \sqrt{1+(1/4)} = \frac{\sqrt{5}}{2}$, so

$$\frac{dx}{dt} = 4\pi(\sqrt{5}/2)^2 = 5\pi \approx 15.7080.$$

Conclusion: The beam is moving along the shoreline at about 15.7080 kilometers per minute when it is 1/2 kilometer past the point opposite the lighthouse.

17. $6/s = 30/y$, $y = x+s$, and $\tan\theta = s/6$,
    so $y = 5s$, $x = y-s = 5s-s = 4s$, and $s = 6\tan\theta$.

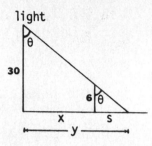

Therefore, $dy/dt = 5(ds/dt)$, $dx/dt = 4(ds/dt)$,
and $ds/dt = 6\sec^2\theta(d\theta/dt)$,
so $d\theta/dt = (1/6)\cos^2\theta(ds/dt)$.

$dx/dt = 2$, so $ds/dt = 1/2$, $dy/dt = 5(1/2) = 5/2$.
When s=6, $\theta=\pi/4$,
so $d\theta/dt = (1/6)[\cos^2(\pi/4)](1/2)$
$= 1/24 \approx 0.0417$ rad/sec.

19. $s^2 = x^2+y^2+(100)^2$, $dx/dt = 66$, and $dy/dt = 88$.

$2s(ds/dt) = 2x(dx/dt)+2y(dy/dt) = 2x(66)+2y(88)$, so $ds/dt = (66x+88y)/s$.
When t=10, x=660, y=880 and $s=100\sqrt{122}$.
Therefore, $ds/dt = [66(660)+88(880)]/(100\sqrt{122})$
≈ 109.55 ft/sec. [about 75 mph]

21. Step 1: Let t denote time elapsed [hr].
            Let h [ft] be as indicated in Figure 11 in the text.
            Let V denote the volume [cubic ft].

Step 2: We are given $\frac{dV}{dt} = -2$ ft^3/hr since the volume is decreasing at 2 cubic feet per hour.

We are to find $\frac{dh}{dt}$ when h = 3 ft.

Step 3: $V = \pi h^2[8-(h/3)] = 8\pi h^2 - (1/3)\pi h^3$.

Step 4: $\dfrac{dV}{dt} = (16\pi h - \pi h^2)\,\dfrac{dh}{dt}$.

Step 5: When $h=3$, $-2 = [16\pi(3) - \pi(9)]\,\dfrac{dh}{dt}$, so $\dfrac{dh}{dt} = \dfrac{-2}{39\pi} \approx -0.01632$.

Conclusion: When the water is 3 feet deep, the water level is decreasing at about 0.01632 feet per hour [almost 0.2 inches per hour].

23. $V = kP^{-1}$; $dV/dt = -kP^{-2}(dP/dt) = -k(k/V)^{-2}(dP/dt) = -k^{-1}V^2(dP/dt)$.
When $t=6.5$, $V=300$, $P=60$, and $dP/dt \approx -80/2 = -40$.
Therefore, $dV/dt \approx -(18000)^{-1}(300)^2(-40) = 200$ in^3/min.

25. Given: $dx/dt = 2$ ft/sec; so $d^2x/dt^2 = 0$.

Determine: (a) dy/dt when $a = 60°$

   (b) d^2y/dt^2 when $a = 60°$

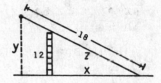

Notes: $z^2 = x^2 + 144$, and $\dfrac{y}{12} = \dfrac{18}{z}$, or $y = \dfrac{216}{z}$.

(1) $2z\,\dfrac{dz}{dt} = 2x\,\dfrac{dx}{dt}$ and $\dfrac{dy}{dt} = \dfrac{-216}{z^2}\,\dfrac{dz}{dt}$,

so $\dfrac{dz}{dt} = \dfrac{x}{z}\,\dfrac{dx}{dt}$, and then $\dfrac{dy}{dt} = \dfrac{-216}{z^2}\,\dfrac{x}{z}\,\dfrac{dx}{dt} = -216xz^{-3}\,\dfrac{dx}{dt}$.

(2) $\dfrac{d^2y}{dt^2} = \left(-216\,\dfrac{dx}{dt}\right)(z^{-3})\left(\dfrac{dx}{dt}\right) + (-216x)\left(-3z^{-4}\,\dfrac{dz}{dt}\right)\left(\dfrac{dx}{dt}\right)$

   $+ (-216x)(z^{-3})(0)$

$= \left(-216\,\dfrac{dx}{dt}\right)(z^{-3})\left(\dfrac{dx}{dt}\right) + (-216x)\left(-3z^{-4}\,\dfrac{x}{z}\,\dfrac{dx}{dt}\right)\left(\dfrac{dx}{dt}\right)$

$= -216z^{-5}\left(\dfrac{dx}{dt}\right)^2 (z^2 - 3x^2)$

(3) When $a = 60°$, $x^2 = 48$, $z^2 = 192$.

(a) $\left.\dfrac{dy}{dt}\right|_{a=60°} = \dfrac{-216(4\sqrt{3})}{(8\sqrt{3})^3}\,(2) = -1.125,$

so the vertical velocity of the top is -1.125 ft/sec when $a = 60°$.

(b) $\left.\dfrac{d^2y}{dt^2}\right|_{a=60°} = \dfrac{-216}{(8\sqrt{3})^5}\,(2)^2\,[192-3(48)] = \dfrac{-3\sqrt{3}}{64} \approx -0.0812,$

so the vertical acceleration of the top is about -0.0812 ft/sec^2.

27. (a) Surface area is $S = 4\pi r^2$, so it is given that $\dfrac{dV}{dt} = -4\pi r^2 k$ $(k > 0)$.

On the other hand, since volume is $V = \dfrac{4}{3}\pi r^3$, $\dfrac{dV}{dt} = 4\pi r^2 \dfrac{dr}{dt}$.

Thus, $-4\pi r^2 k = 4\pi r^2 \dfrac{dr}{dt}$, so $\dfrac{dr}{dt} = -k$.

That is, the radius decreases at a constant rate.

(b) If $V_0 = \dfrac{4}{3}\pi r_0^3$, then $\dfrac{8}{27}V_0 = (\dfrac{4}{3}\pi r_0^3)\dfrac{8}{27} = \dfrac{4}{3}\pi(\dfrac{2}{3}r_0)^3$, so when the volume is 8/27 of its original, the radius is 2/3 of its original.

From $\dfrac{dr}{dt} = -k$ [part (a)], we obtain $r = -kt + r_0$ $(r = r_0$ if $t = 0)$.

Then since $r = \dfrac{2}{3}r_0$ when $t = 1$ hour, $(\dfrac{2}{3}r_0) = -k(1) + r_0$, from which it follows that $k = \dfrac{1}{3}r_0$.

Therefore, $r = -(\dfrac{1}{3}r_0)t + r_0$; so when $r = 0$, $t = 3$ hours.

29. Let y denote the distance from the girl to the base of the light post. Then $dy/dt = 16/3$ when $x > 30$ and $dy/dt = 80/17$ when $x < 30$.

Problem Set 3.10 Differentials and Approximations

1. $dy = (4x-3)dx.$

3. $y = (3+2x^2)^{-4}.$ $dy = -4(3+2x^3)^{-5}(0+6x^2)dx = -24x^2(3+2x^3)^{-5}dx.$

5. $dy = (1/2)(4x^5+2x^4-5)^{-1/2}(20x^4+8x^3)dx = [2x^3(5x+2)(4x^5+2x^4-5)^{-1/2}]dx.$

7. $ds = (2/5)(t^2-3)^{-3/5}(2t)dt = [(4t/5)(t^2-3)^{-3/5}]dt.$

9. $y = f(x) = x^3.$
    $dy = 3x^2dx.$
    (a) When $x = 0.5$ and $dx = 1$,
         $dy = 3(0.25)(1) = 0.75.$

    (b) When $x = -1$ and $dx = 0.75$,
         $dy = 3(1)(0.75) = 2.25.$

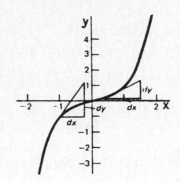

11. (a) $\Delta y = (1.5)^3 - (0.5)^3 = 3.25.$ (b) $\Delta y = (-0.25)^3 - (-1)^3 \approx 0.9844.$

13. $dy = 2xdx,$ $\Delta y = [(x+dx)^2 - 3] - [x^2-3] = 2xdx + dx^2.$

    (a) $dy = 2;$ $\Delta y = 2.25.$ (b) $dy = -0.72;$ $\Delta y \approx -0.7056.$

15. Let $y = f(x) = \sqrt{x};$ $f(x+dx) \approx f(x) + dy = \sqrt{x} + \dfrac{1}{2\sqrt{x}}\,dx.$

    402 is near 400 and $\sqrt{400}$ is 20, so let $x = 400$ and $dx = 2.$

    Then, $\sqrt{402} = f(402) = f(400 + 2) \approx \sqrt{400} + \dfrac{1}{2\sqrt{400}}(2) = 20 + 0.05 = 20.05.$

    Onc calculator gives 20.04993766.

17. If $y = x^{1/3}$, then $dy = (1/3)x^{-2/3}dx.$ $f(x+dx) \approx f(x) + dy.$
    When $x = 27$ and $dx = -0.09$, $dy = -1/300$, so

    $(26.91)^{1/3} \approx (27)^{1/3} - 1/300 = 2.99666\cdots$ Calculator: 2.996663.

19. $V = (4/3)\pi r^3,$ so $dV = 4\pi r^2dr.$ When $r = 5$ and $dr = 0.125,$
    $dV \approx 39.2699$ cubic centimeters.

21. Let $V = f(r) = \dfrac{4}{3}\pi r^3.$ Then $dV = 4\pi r^2dr.$

    When $r = 6$ ft and $dr = -0.3$ in $= \dfrac{-0.3 \text{ in}}{12 \text{ in/ft}} = \dfrac{-1}{40}$ ft.

    $dV = 4\pi(6)^2\left[\dfrac{-1}{40}\right] = -3.6\pi$ ft$^3.$

    Thus, volume of interior $\approx f(6) + dV = \dfrac{4}{3}\pi(6)^3 + (-3.6\pi) \approx 893.4690$ ft$^3.$

23. $c = 2\pi r$, $dc = 2\pi dr$. When $r = 4000$ and $dr = 2$, $dc = 4\pi \approx 12.5664$ ft.

25. If k is the diameter, $V = (4/3)\pi(k/2)^3 = \pi k^3/6$, $dV = (\frac{\pi k^2}{2})dk$.
When $k = 20$ and $dk = 0.1$, $V = 4000\pi/3$ and $dV = 20\pi$, so the volume is
$4000\pi/3 \pm 20\pi \approx 4188.79 \pm 62.83$ cm^3.

27. Let c denote the length of the third side, and β denote the angle
between the sides of length 151 centimeters. The Law of Cosines give

$$c = [(151)^2 + (151)^2 - 2(151)(151) \cos\beta]^{1/2} = 151\sqrt{2}\sqrt{1-\cos\beta}.$$

$$dc = 151\sqrt{2}\,\frac{1}{2\sqrt{1-\cos\beta}}\,(\sin\beta)d\beta = \frac{151\sqrt{2}\,\sin\beta}{2\sqrt{1-\cos\beta}}\,d\beta.$$

For $\beta = 0.53$ radians, and $d\beta = \pm0.005$ radians,

$$c = 151\sqrt{2}\sqrt{1-\cos(0.53)} \approx 79.097,\text{ and}$$

$$dc = \frac{151\sqrt{2}\,\sin(0.53)}{2\sqrt{1-\cos(0.53)}}\,(\pm0.005) \approx \pm 0.729.$$

Therefore, $c = 79.097 \pm 0.729$ centimeters.

29. $y = 3x^2-2x+11$, $dy = (6x-2)dx$, $d^2y/dx^2 = 6$. Let $M = 6$. When $x = 2$ and
$dx = 0.001$, $dy = 0.01$, so $|\Delta y - dy| \leq (1/2)(6)(0.001)^2 = 0.000003$.

31. $V = 100\pi r^2 + (4/3)\pi r^3$
$dV = (200\pi r + 4\pi r^2)dr$

If $r = 10$ cm and $dr = 0.1$ cm,

then $dV = 240\pi$ cm$^3 \approx 0.2$ gallons.

33. 9.47%

Problem Set 3.11   Chapter Review

True-False Quiz

1. False.   The tangent to $y = x^3$ at the origin crosses the curve.

3. True.   For example, if the velocity increases from -4 to -2, the speed would change from $|-4|$ to $|-2|$, which is a decrease.

   For a particular function, consider $s(t) = (t-4)^2$ when t changes from 2 to 3.

5. False.   C^{ex}: If $f(x) = x^2$ and $g(x) = x^2+1$, $f(2) \neq g(2)$, but for all x $f'(x) = g'(x) = 2x$.

7. True.   See Theorem A, Section 3.2

9. False.   C^{ex}: $D(x^2 \cdot x^3) = D(x^5) = 5x^4$,

   but $D(x^2) \cdot D(x^3) = (2x)(3x^2) = 6x^3$.

   $5x^4 \neq 6x^3$ if, for example, x is 1.

11. True. $f(x) = x^3 g(x) \Rightarrow f'(x) = x^3 g'(x) + 3x^2 g(x) = x^2 [xg'(x) + 3g(x)]$.

13. False. $D^2 y = f(x) \, g''(x) + 2f'(x)g'(x) + f''(x)g(x)$.

15. True.   By the derivative rules stated in Theorems A-F of Section 3.3.

17. True. $h'(c) = f(c)g'(c) + g(c)f'(c) = 0 + 0 = 0$.

19. True. $D^2 [kf(x)] = k \, D^2 [f(x)]$, and $D^2 [f(x) \cdot g(x)] = D^2 [f(x)] + D^2 [g(x)]$.

21. True. $(f \circ g)'(2) = f'[g(2)] \cdot g'(2) = f'[2] \cdot (2) = (2)(2) = 4$.

23. False. $V = (4/3)\pi r^3$, $dV/dt = 4\pi r^2 (dr/dt) = 12\pi r^2$ (since $dr/dt = 3$). This equals 27 only when $r = 3/(2\pi^{1/2})$.

25. True: $\lim_{x \to 0} \frac{\tan x}{3x} = \lim_{x \to 0} \left[\frac{\sin x}{x} \frac{1}{3\cos x} \right] = (1) \frac{1}{3(1)} = \frac{1}{3}$.

27. True. $V = \frac{4}{3}\pi r^3 \Rightarrow \frac{dV}{dt} = 4\pi r^2 \frac{dr}{dt}$.

Set $\frac{dV}{dt} = 3$; obtain $\frac{dr}{dt} = \frac{3}{4\pi r^2} > 0$, so the radius is increasing.

However, $\frac{dr}{dt}$ is positive but approaching zero as r increases.

29. True. $V = (4/3)\pi r^3$, $dV = 4\pi r^2 dr = Sdr = S\Delta r$.

Sample Test Problems

1. (a) $\frac{[(x+h)^2 - 5(x+h)] - [x^2 - 5x]}{h} = \frac{h(2x+h-5)}{h} = 2x+h-5 \to 2x-5$ (as $h\to 0$).

    (b) $\frac{1/(x+h-3) - 1/(x-3)}{h} = \frac{(x-3)-(x+h-3)}{h(x-3)(x+h-3)} = \frac{-1}{(x-3)(x+h-3)} \to \frac{-1}{(x-3)^2}$ ($h\to 0$).

    (c) $\frac{[9-(x+h)]^{1/2} - (9-x)^{1/2}}{h} = \frac{(9-x-h) - (9-x)}{h[(9-x-h)^{1/2} + (9-x)^{1/2}]} = \frac{-1}{(9-x-h)^{1/2} + (9-x)^{1/2}}$

    $\to (-1/2)(9-x)^{-1/2}$ (as $h \to 0$).

3. (a) Derivative of $f(x) = 3x^2$ at $x=2$.

    (b) Derivative of $f(x) = \tan x$ at $x=\pi/4$.

    (c) Derivative of $f(p) = 3/p$ at $p=x$.

5. $3x^2 - 6x - 2x^{-3}$.

7. $D(3x+2)^{2/3} = 2(3x+2)^{-1/3}$. $D^2(3x+2)^{2/3} = -2(3x+2)^{-4/3}$.

9. $(-1/2)(x^2+4)^{-3/2}(2x) = -x(x^2+4)^{-3/2}$.

11. $3[\cos^2 5x][-\sin 5x](5) = -15\sin 5x \cos^2 5x$.

13. $f'(x) = (x^2-1)^2(9x^2-4) + (3x^3-4x)2(x^2-1)(2x)$. $f'(2) = 672$.

15. $2x-y+2 = 0$, or equivalently, $y = 2x+2$ in slope-intercept form. The slope of this line is 2, so the slope of each line perpendicular to it is $-1/2$.

The equation of the given curve is $y = (x-2)^2$. Its derivative is $Dy=2(x-2)$.

Set $2(x-2) = -1/2$ and solve for x; obtain $x = 7/4$. Then substitute $x = 7/4$ into the equation of the curve; obtain $y = 1/16$. The point is $(7/4, 1/16)$.

17. $V = (4/3)\pi r^3$, $dV = 4\pi r^2 dr$. When r=5 and dr = 0.1, $dV = 10\pi \approx 31.42$ m^3.

19. $V = (1/2)bh(12) = 6bh$, and $b/h = 6/4$,

so $V = 6(3h/2)h = 9h^2$.

$dV/dt = 9$ and $dV/dt = 18h(dh/dt)$,

so $dh/dt = 1/2h$.

When h=3, $dh/dt = 1/6$.

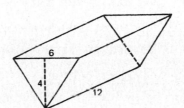

21. $s(t) = t^3-6t^2+9t$.

$v(t) = 3t^2-12t+9 = 3(t-1)(t-3)$.

Auxiliary axis for v(t):

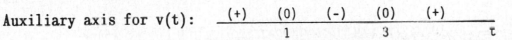

$a(t) = 6t-12 = 6(t-2)$.

Auxiliary axis for a(t):

(a) The object is moving to the left when the vleocity is negative; that is, when t is between 1 and 3.

(b) The velocity is zero when t=1 and when t=3. $a(1) = 6(-1) = -6$ and $a(3) = 6(1) = 6$.

(c) The acceleration is positive when t is greater than 2.

23. (a) $3x^2 + 3y^2y' = (x^3)(3y^2y') + (y^3)(3x^2)$. $\quad y' = (x^2y^3 - x^2)/(y^2 - x^3y^2)$.

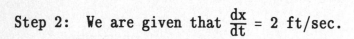

(b) $(x)[\cos(xy)](xy'+y) + \sin(xy) = 2x$. $\quad y' = \dfrac{2x - \sin(xy) - xy\cos(xy)}{x^2\cos(xy)}$.

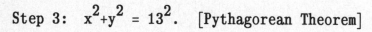

25. $dy = [\pi\cos(\pi x) + 2x]dx$.
When $x=2$ and $dx=0.01$, $dy = [\pi+4](0.01) \approx 0.0714$.

27. Step 1: Let t denote the time elapsed. [sec]
Let x and y be as indicated in the
figure at the right. [ft]

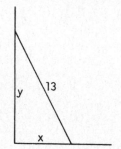

Step 2: We are given that $\dfrac{dx}{dt} = 2$ ft/sec.

We are to find $\dfrac{dy}{dt}$ when $y = 5$ ft.

Step 3: $x^2+y^2 = 13^2$. [Pythagorean Theorem]

Step 4: $\dfrac{d}{dt}(x^2+y^2) = \dfrac{d}{dt}(169)$.

$2x\dfrac{dx}{dt} + 2y\dfrac{dy}{dt} = 0$. (*).

Step 5: When $y=5$, $x = \sqrt{13^2 - 5^2} = 12$.

Substituting the values of x, y, and $\dfrac{dx}{dt}$ into (*),

$2(12)(2) + 2(5)\dfrac{dy}{dt} = 0$, so $\dfrac{dy}{dt} = \dfrac{-24}{5} = -4.8$.

Conclusion: The top of ladder is moving down the wall at 4.8 feet per
second when it is 5 feet above the ground.

29. $\dfrac{|\sin x|}{\sin x}(\cos x) = \dfrac{|\sin x|\cos x}{\sin x}$.

Problem Set 4.1 Maxima and Minima

1. $f'(x) = -2x+4$. Critical points are 0,2,3. $f(0) = -1$, $f(2) = 3$, and $f(3) = 2$, so the maximum value is 3 and the minimum value is -1.

3. $G(x) = \frac{1}{5}(2x^3+3x^2-12x)$, $I = [-3,3]$.

   G is continuous on $[-3,3]$ so the Max-Min Existence Theorem applies.

   $G'(x) = \frac{1}{5}(6x^2+6x-12) = \frac{6}{5}(x^2+x-2) = \frac{6}{5}(x+2)(x-1)$.
   $G'(x) = 0$ iff $x = -2,1$; $G'(x)$ is defined everywhere on $(-3,3)$.

   Critical Points: -3,3 (end points); -2,1 (stationary points).

x	G(x)		
-3	1.8		
-2	4		
1	-1.4	←	Minimum value is -1.4; it occurs at x=1.
3	9	←	Maximum value is 9; it occurs at x=3.

5. See Problem 6 and observe that the end points are not in the interval. Critical points arc -1 and 1. As x approaches the left end point $f(x)$ approaches 2.125, and as x approaches the right end point $f(x)$ approaches 19. Therefore, the minimum is -1 and there is no maximum, but the function gets arbitrarily close to 19.

7. $g'(x) = -2x(1+x^2)^{-2}$. Critical points are -2,0,1. $g(-2) = 0.2$, $g(0) = 1$, and $g(1) = 0.5$, so the maximum is 1 and the minimum is 0.2.

9. $f(x) = \dfrac{x}{x^2+2}$, $I = [-1,4]$.

f is continuous on $[-1,4]$ so the Max-Min Existence Theorem applies.

$f'x = \dfrac{(x^2+2)(1) - (x)(2x)}{(x^2+2)^2} = \dfrac{2-x^2}{(x^2+2)^2} = \dfrac{(\sqrt{2}+x)(\sqrt{2}-x)}{(x^2+2)^2}$.

$f'(x) = 0$ iff $x = -\sqrt{2}$ [not in $(-1,4)$], $\sqrt{2}$; $f'(x)$ is defined everywhere on $(-1,4)$.

Critical Points: $-1,4$ (end points); $\sqrt{2}$ (stationary point).

x	f(x)	
-1	$-1/3 \approx -0.3333$	← Minimum value is $-1/3$; it occurs at x=-1.
$\sqrt{2}$	$\sqrt{2}/4 \approx 0.3536$	← Maximum value is $\sqrt{2}/4$; it occurs at x=$\sqrt{2}$.
4	$2/9 \approx 0.2222$	

11. $f'(x) = |x-2|/(x-2)$. Critical points are $1,2,5$. $f(1) = 1$, $f(2) = 0$, and $f(5) = 3$, so the maximum is 3 and the minimum is 0.

13. $g'(x) = (2/5)x^{-3/5}$. Critical points are $-1,0,32$. $g(-1) = 1$, $g(0) = 0$, and $g(32) = 4$, so the maximum is 4 and the minimum is 0.

15. $h(x) = x^{2/5}$, $I = (-1,32)$.

Note that I is not a closed interval so the Max-Min Existence Theorem does not apply. However, since h could be defined and would be continuous on $[-1,32]$ we will consider h on that interval and then eliminate the end points from consideration in the final conclusion about the extreme values.

In Problem 13 it was found that on $I = [-1,32]$ $g(x) = x^{2/5}$ has a maximum value of 4 which occurs at x=32 (an end point), and has a minimum value of 0 which occurs at x=0 (a singular point).

Therefore, h has no maximum value on $(-1,32)$ [but gets arbitrarily close to 4 as x approaches 32], and has a minimum value of 0 at x=0.

17. Maximize $P(x) = x(10-x)$ on $[0,10]$. $P'(x) = 10-2x$. Critical points are $0,5,10$. $P(0) = P(10) = 0$, and $P(5) = 25$. If the interpretation is that the numbers need not be distinct, then the numbers are 5 and 5. If the interpretation requires distinct numbers, then there are no such numbers.

19. A = xy and 2x+2y = 200, so A(x) = x(100-x) on [0,100].

A'(x) = 100-2x. Critical points are 0, 50 and 100.

A(0) = A(100) = 0, and A(50) = 2500, so the area is maximum if the dimensions are 50 ft by 50 ft.

21. Let x denote the length (inches) of the edges of the squares to be cut out. Let V denote the volume (cubic inches) of the resulting box.

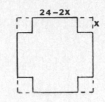

V = $(24-2x)^2$x = $4(x^3-24x^2+144x)$, where x is greater than 0 and less than 12.

That is, we wish to maximize the function
V(x) = $4(x^3-24x^2+144x)$ on I = (0,12). We will do so on [0,12].

V is continuous on [0,12] so the **Max-Min Existence Theorem** applies.
V'(x) = $4(3x^2-48x+144)$ = $12(x^2-16x+48)$ = 12(x-4)(x-12).

V'(x) = 0 iff 4, 12 [not in (0,12)]; defined everywhere on (0,12).

Critical Points: 0, 12 (end points); 4 (stationary point).

x	V(x)
0	0
4	1024
12	0

← Maximum value is 1024; it occurs at x=4.

Conclusion: The volume of the largest such box is 1024 cubic inches.

23. A = xy and 2x+y = 80, so A(x) = x(80-2x) on [0,40]. A'(x) = 80-4x. Critical points are 0,20,40. A(0) = A(40) = 0, A(20) = 800. The dimensions should be 20 ft (perpendicular to barn) by 40 ft.

25. A = xy and 2x+y+(y-100) = 180, so A(x) = x(140-x) on [0,40].

A'(x) = 140-2x. Critical points are 0,40. A(0) = 0, and A(40) = 4,000, so the dimensions should be 40 ft (perpendicular to barn) by 100 ft.

27. Let x and y (units) be as indicated in Figure 15 (in 1st quadrant).
     Let A denote the area (square units) of the rectangle.
     Then (1) $y = 12 - x^2$,
          (2) $A = (width)(height) = (2x)(y) = 2xy$.
     Therefore, $A = 2x(12 - x^2) = 24x - 2x^3$, where x is not negative or greater
     than $\sqrt{12}$.
     We wish to maximize the function $A(x) = 24x - 2x^3$ on $I = [0, \sqrt{12}]$.
     A is continuous on $[0, \sqrt{12}]$, so the Max-Min Existence Theorem applies.

     $A'(x) = 24 - 6x^2 = -6(x^2 - 4) = -6(x+2)(x-2)$.
     $A'(x) = 0$ iff $x = -2$ [not in $(0, \sqrt{12})$], 2; defined everywhere on I.
     Critical Points: 0, $\sqrt{12}$ (end points); 2 (stationary point).

x	A(x)
0	0
2	32
$\sqrt{12}$	0

     2 | 32 ← Maximum value of A is 32; it occurs at x=2.

     Conclusion: The dimensions of the rectangle of maximum area are 4 units
                 (wide) by 8 units (high).

29. The figure at the right is a cross-section of
     the gutter. If A is its area, then

     $A = (1/2)h(3+a)$, $h = 3\sin\theta$,
     $a = 3 + 2(3\cos\theta)$,
     so $A(\theta) = (1/2)(3\sin\theta)[3 + (3 + 6\cos\theta)]$
               $= (9/2)(2\sin\theta + \sin 2\theta)$.
     $A'(\theta) = 9(\cos\theta + \cos 2\theta)$

     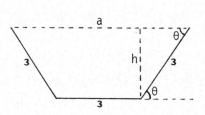

               $= 9(2\cos^2\theta + \cos\theta - 1)$
               $= 9(2\cos\theta - 1)(\cos\theta + 1)$.
     Critical points are $0, \pi/3, \pi/2$.

     $A(0) = 0$, $A(\pi/3) = 27\sqrt{3}/4 \approx 11.69$, $A(\pi/2) = 9$, so θ should be $\pi/3$.

31. If C is total cost (cents), t is time (hrs) of the run, and x is speed
     (mph) of the truck, $t = 400/x$, and $C(x) = 400(25 + x/4) + 1200(400/x)$ on
     $[40, 55]$. $C'(x) = 100(x + 40\sqrt{3})(x - 40\sqrt{3})/x^2$. Critical points are 40 and 55.
     $C(40) = 26000$, $C(55) = 24227$. The most economical speed is 55 mph.

33. Let s be the square of the distance between
(0,4) and (x,y) on $y = x^2/4$ for x in $[0,2\sqrt{3}]$.
$s = x^2 + (y-4)^2$
$s(y) = 4y + (y-4)^2 = y^2 - 4y + 16$ for y on $[0,3]$.

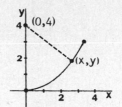

$s'(y) = 2y - 4$; $s'(y) = 0$ if $y = 2$.

Auxiliary axis for $s'(y)$:

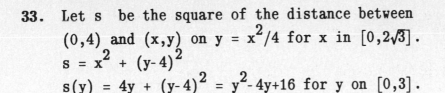

y	s(y)
0	16
2	12
3	13

Maximum value of s is 16; it occurs at y=0, x=0.
Minimum value of s is 12; it occurs at y=2, x=$2\sqrt{2}$.

Therefore, P is $(2\sqrt{2},2)$; Q is $(0,0)$.

35. Let V denote the volume of the box.   V = xyz
$$2x + 2y = 8$$
$$2x + z = 5$$

Therefore, $V(x) = x(4-x)(5-2x) = 2x^3 - 13x^2 + 20x$, x in $[0,2.5]$.

$V'(x) = 6x^2 - 26x + 20 = 2(3x-10)(x-1)$.
$V'(x) = 0$ if x=1, which yields the maximum value of V. [x=0 obviously
yields the minimum value of V=0.]
Thus, V is maximum if x=1, y=3, and z=3.

37. (a) -1,0,1.57079, 4.71239, 5; max: f(1.57079) = 3.5708; min f(4.71239) =
- 2.7123

   (b) -1,0,1.57079, 3.44998, 4.31239, 5; max: g(1.57079) = 3.5708; min:
g(3.44998) = 0.

Problem Set 4.2 Monotonicity and Concavity

1. $f'(x) = 2(x-2)$. f is decreasing on $(-\infty,2]$; increasing on $[2,\infty)$.

3. $F(x) = x^3 - 1$ is continuous on $\mathbf{R}$.
   $F'(x) = 3x^2$.

   Auxiliary axis for $F'(x)$:

   Therefore, F is increasing on $(\infty, 0]$ and on $[0, \infty)$; and so, on $\mathbf{R}$.

5. $g'(t) = 4(t+1)(t^2 - t + 1)$. g is dec. on $(-\infty, -1]$; inc. on $[-1, \infty)$.

7. $f'(x) = 10x^2(x-3)^2$. f is increasing on $\mathbf{R}$.

9. $H(t) = \sin^2 2t$, t in $[0, \pi]$. H is continuous on $[0, \pi]$.
   $H'(t) = 2(\sin 2t)(\cos 2t)(2) = 4 \sin 2t \cos 2t = 2\sin 4t$ (Double-angle id.)
     Set $H'(t) = 0$: $\sin 4t = 0$.

$$4t = 0, \pm\pi, \pm 2\pi, \pm 3\pi, \pm 4\pi, \cdots.$$

$$t = 0, \frac{\pm\pi}{4}, \frac{\pm\pi}{2}, \frac{\pm 3\pi}{4}, \pm\pi \cdots.$$

   The only values in $(0, \pi)$ are $\frac{\pi}{4}$, $\frac{\pi}{2}$, and $\frac{3\pi}{4}$.

   Auxiliary axis for $H'(t)$:

   Therefore, H is increasing on $[0, \pi/4]$ and on $[\pi/2, 3\pi/4]$;

   decreasing on $[\pi/4, \pi/2]$ and on $[3\pi/4, \pi]$.

11. $f'(x) = 2(x-3)$; $f''(x) = 2$. f is concave up on $\mathbf{R}$. No inflection point.

13. $F'(x) = 3x^2 - 12$; $F''(x) = 6x$. F is concave down on $(-\infty, 0)$; concave up on $(0, \infty)$. $(0,0)$ is the only inflection point.

15. $g(x) = 3x^2 - x^{-2}$.

$g'(x) = 6x + 2x^{-3}$.

$g''(x) = 6 - 6x^{-4} = \dfrac{6x^4 - 6}{x^4} = \dfrac{6(x^2+1)(x+1)(x-1)}{x^4}$.

Auxiliary axis for $g''(x)$:

(+)	(0)	(−)	(u)	(−)	(0)	(+)
	−1		0		1	x

Therefore, g is concave up on $(-\infty,-1)$ and on $(1,\infty)$;
          concave down on $(-1,0)$ and on $(0,1)$.

Inflection points occur at x=−1 since g is concave up to the left of −1 and concave down to the right of −1, and at x=1 since g is concave down to the left of 1 and concave up to the right of 1. The inflection points are (−1,2) and (1,2), since g(−1) = 2 and g(1) = 2.

17. $g'(x) = 12x^5 + 60x^3 + 180x + 120$; $g''(x) = 60(x^4 + 3x^2 + 3)$, which is never zero. g is concave up on **R**. There are no inflection points.

19. $f'(x) = 3(x+1)(x-1)$.
        f is dec. on $[-1,1]$;
        inc. on $(-\infty,-1]$ and on $[1,\infty)$.

$f''(x) = 6x$.
        f is concave down on $(-\infty,0)$;
        concave up on $(0,\infty)$.

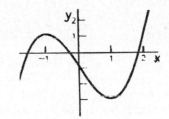

Some points are (−1,1), (1,−3), (2,1) and (0,−1), an inflection point.

21. $g(x) = 3x^4 - 4x^3 + 2$, $D_g = $ **R**.

$g'(x) = 12x^3 - 12x^2 = 12x^2(x-1)$.

$g''(x) = 36x^2 - 24x = 12x(3x-2)$.

g'

(−)	(0)	(−)	(0)	(+)
	0		1	x

g"

(+)	(0)	(−)	(0)	(+)
	0		2/3	x

Use the information shown on the two auxiliary axes to develop the following table summarizing the "activity" of g. Then plot at least x-intercepts, y-intercepts and points at which g'(x) or g''(x) is 0.

Inc = increasing; Dec = decreasing; CU = concave up; CD = concave down.

Intervals and Split Points	Activity of g	x	g(x)
$(-\infty,0)$	Dec and CU	0	2
at 0	Levels off	2/3	1.41
$(0,2/3)$	Dec and CD	1	1
at 2/3	Inflection	−1	9
$(2/3,1)$	Dec and CU	2	18
at 1	Levels off		
$(1,\infty)$	Inc and cu		

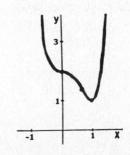

23. $G'(x) = 15x^2(x+1)(x-1)$.
 G is dec. on $[-1,1]$;
 inc. on $(-\infty,-1]$ and on $[1,\infty)$.

$G''(x) = 30x(2x^2-1)$.
 G is concave up on $(-0.71,0)$ and on
 $(0.71,\infty)$; concave down on $(-\infty,-0.71)$
 and on $(0,0.71)$.

Some points are $(-1,3),(0,1),(-1.41,-1.70)$,

$(1.41,3.70)$ and inflection points

$(-0.71,2.24)$ and $(0.71,-0.24)$.

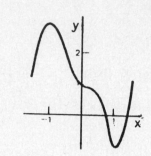

25. $f'(x) = (1/2)(\cos x)(\sin x)^{-1/2}$.
 f is inc. on $[0,\pi/2]$; dec. on $[\pi/2,\pi]$.
$f''(x) = -[1+\sin^2 x]/4(\sin x)^{3/2}$.
 f is concave down on $[0,\pi]$.

Some points are $(0,0)$, $(\pi/2,1)$, $(\pi,0)$.

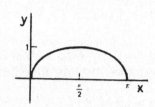

27. $f(x) = x^{2/3}(1-x) = x^{2/3}-x^{5/3}, D_f = \mathbb{R}$.

$f'(x) = (2/3)x^{-1/3} - (5/3)x^{2/3} = \dfrac{5x-2}{-3\sqrt[3]{x}}$.

$f''(x) = (-2/9)x^{-4/3} - (10/9)x^{-1/3} = \dfrac{2+10x}{-9^3\sqrt{x^4}}$.

f'	(-)	(u)	(+)	(0)	(-)		f"	(+)	(0)	(-)	(u)	(-)
		0		0.4	x				-0.2		0	x

Intervals and Split Points	Activity of f	x	f(x)
$(-\infty,-0.2)$	Dec and CU	0	0
at -0.2	Inflection	-0.2	0.41
$(-0.2,0)$	Dec and CD	0.4	0.33
at 0	Cusp	1	0
$(0,0.4)$	Inc and CD	-1	2
at 0.4	Levels off	2	-1.59
$(0.4,\infty)$	Dec and CD	-2	4.76
		3	-4.16

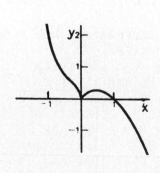

29. 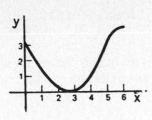   31.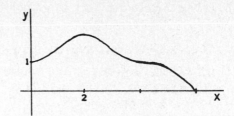

33. $f(x) = ax^2 + bx + c$, $a \neq 0$.
     $f'(x) = 2ax + b$.
     $f''(x) = 2a$.

     $a > 0 \Rightarrow f$ is concave up on $\mathbf{R}$; $a < 0 \Rightarrow f$ is concave down on $\mathbf{R}$. Hence, no change of concavity, so no inflection point for any quadratic function.

35. Assume that there are points a and b in I such that $a < b$ and $f'(a)$ and $f'(b)$ have opposite signs. Then since f' is continuous on the interval $[a,b]$, there is a point c in (a,b) such that $f'(c) = 0$ [by the Intermediate Value Theorem]. This contradicts the hypothesis that $f'(x) \neq 0$ at all interior points of I. Therefore, $f'(x)$ is positive on the interior of I or $f'(x)$ is negative on the interior of I. Hence, f is increasing on I or f is decreasing on I.

37. (a) If $f(x) = x^2$, then $f'(x) = 2x > 0$ on $(0,\infty)$, so f is increasing on $(0,\infty)$. Hence $0 < x < y \Rightarrow f(x) < f(y) \Rightarrow x^2 < y^2$.

     (b) If $g(x) = \sqrt{x}$, then $f'(x) = 1/2\sqrt{x} > 0$ on $(0,\infty)$, so f is increasing on $(0,\infty)$. Hence $0 < x < y \Rightarrow g(x) < g(y) \Rightarrow \sqrt{x} < \sqrt{y}$.

     (c) If $h(x) = 1/x$, then $f'(x) = -1/x^2 < 0$ on $(0,\infty)$, so f is decreasing on $(0,\infty)$. Hence $0 < x < y \Rightarrow h(x) > h(y) \Rightarrow 1/x > 1/y$.

39. $f(x) = ax^{1/2} + bx^{-1/2}$; $f'(x) = (1/2)ax^{-1/2} - (1/2)bx^{-3/2}$

     $f''(x) = (-1/4)ax^{-3/2} + (3/4)bx^{-5/2} = (-1/4)x^{-5/2}(ax - 3b)$

     (1) (4,13) on graph of $f \Rightarrow 13 = a(2) + b/2$ or $4a + b = 26$.

     (2) $f''(x) = 0$ at $(4,13) \Rightarrow a(4) - 3b = 0$ or $4a - 3b = 0$.

     Solve $4a + b = 26$ and $4a - 3b = 0$ simultaneously: $a = 39/8$, $b = 13/2$.

41. (a) $(f+g)'(x) = f'(x) + g'(x) > 0$, so no additional conditions needed.

(b) $(fg)'(x) = f'(x)g(x) + f(x)g'(x) > 0$ if we also have $f(x) > 0$ and $g(x) > 0$. [There are some less simple conditions that would also work].

(c) $(f \circ g)'(x) = f'(g(x))g'(x) > 0$, so no additional conditions needed.

43. (a), (d), (e)

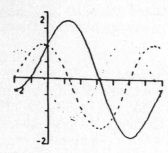

(b) (1.3, 5)

(c) (-0.3, 3.2)

45. (-0.71, 0.68)

Problem Set 4.3 Local Maxima and Minima

1. $f'(x) = 3x(x-2)$; $f''(x) = 6(x-1)$. Critical points are 0 and 2. Using either test, obtain a local maximum at 0 and a local minimum at 2.

3. $f(x) = \frac{x}{2} - \sin x$ on $(0, 2\pi)$. f is continuous on $(0, 2\pi)$.

   $f'(x) = \frac{1}{2} - \cos x.$

   ```
        (-)      (0)     (+)     (0)     (-)
   ├──────────┼───────┼───────┼───────┤─── x
   0         π/3             5π/3           2π
   ```

   $f''(x) = \sin x.$
   Critical points: $\pi/3$, $5\pi/3$ (stationary points).

   1st Derivative Test: Use the auxiliary axis for f' to conclude that $\pi/3$ gives a local minimum; $5\pi/3$ gives a local maximum.

   2nd Derivative Test: $f''(\pi/3) = \sin(\pi/3) = \sqrt{3}/2 > 0$ so $\pi/3$ gives a local minimum; $f''(5\pi/3) = \sin(5\pi/3) = -\sqrt{3}/2 < 0$ so $5\pi/3$ gives a local maximum.

5. $g'(x) = x^3$; $g''(x) = 3x^2$. Critical point is 0. Use 1st Derivative test and obtain a local minimum at 0.

7. $f'(x) = (3/4)(x+2)(x-2)$; $f''(x) = 3x/2$. Critical points are -2 and 2. $f''(-2) = -3 < 0$ so -2 gives a local max. of $f(-2) = 3$. $f''(2) = 3 > 0$ so 2 gives a local minimum of $f(2) = -5$.

9. $h(x) = x^4 + 2x^3$. h is continuous on $\mathbf{R}$.

   $h'(x) = 4x^3 + 6x^2 - 2x^2(2x+3).$

   ```
           (-)      (0)     (+)     (0)     (+)
   ─────────────┼───────────────┼─────────── x
              -3/2             0
   ```

   Therefore, critical points are 0, -3/2 (stationary). Use the auxiliary axis for h' to conclude by the First Derivative Test that -3/2 gives a local minimum value of $h(-3/2) = -27/16$; 0 does *not* give a local extremum.

11. $g'(t) = (-2/3)(t-1)^{-1/3}$. Critical point is 1. $g'(x)$ is positive to the left of 1 and negative to the right of 1, so 1 gives a local maximum value of $g(1) = 2$.

13. $f'(x) = x^{-2}(x+1)(x-1)$. Critical points are -1 and 1. Use an auxiliary axis for f' and conclude that -1 gives a local max. value of $f(-1) = -2$ and 1 gives a local minimum value of $f(1) = 2$.

15. $f(t) = \dfrac{\sin t}{2 + \cos t}$, $0 < t < 2\pi$. f is continuous on $(0, 2\pi)$.

$$f'(t) = \frac{(2 + \cos t)(\cos t) - (\sin t)(-\sin t)}{(2+\cos t)^2} = \frac{2\cos t + \cos^2 t + \sin^2 t}{(2 + \cos t)^2}$$

$$= \frac{2\cos t + 1}{(2+\cos t)^2}.$$

Use the auxiliary axis for f′ to conclude by the First Derivative Test that the stationary point $2\pi/3$ gives a local maximum value of $f(2\pi/3) = \sqrt{3}/3$; the stationary point $4\pi/3$ gives a local minimum value of $f(4\pi/3) = -\sqrt{3}/3$.

17. See work in Problem 18. The global maximum value is $F(1) = 3$; the global minimum value is $F(9) = -9$.

19. $f'(x) = \dfrac{-64\cos x}{\sin^2 x} + \dfrac{27\sin x}{\cos^2 x} = \dfrac{27\sin^3 x - 64\cos^3 x}{\sin^2 x \cos^2 x}$ which equals zero if

$3\sin x = 4\cos x$ or $\tan x = 4/3$ or $x = \tan^{-1}(4/3) \approx 0.9273$. Use an auxiliary for f′ and conclude that there is no global maximum but there is a global minimum of $f(\tan^{-1}[4/3]) = 64/(4/5) + 27/(3/5) = 125$.

21. $g(x) = x^2 + \dfrac{16x^2}{(8-x)^2}$, $x > 8$. g is continuous on $(8, \infty)$.

$$g'(x) = 2x + \frac{(8-x)^2(32x) - (16x^2)2(8-x)(-1)}{(8-x)^4}$$

$$= 2x + \frac{(8-x)(32x) - (16x^2)2(-1)}{(8-x)^3} = 2x + \frac{256x}{(8-x)^3}$$

$$= 2x\left[1 + \frac{128}{(8-x)^3}\right] = 0 \text{ if } x=0 \text{ [not on } (8,\infty)] \text{ or } x = 8 + \sqrt[3]{128} \approx 13.0397.$$

Therefore, the global minimum value is $g(8 + \sqrt[3]{128}) \approx 277.1477$.

23. $f'(x)$ auxiliary axis: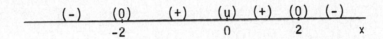

Local maximum where x is -2; local minimum where x is 3.

25. [Based on where the graph of f' is above or below the x-axis]

$f'(x)$ auxiliary axis:

$$\frac{\quad(+)\quad(0)\quad(-)\quad(0)\quad(+)\quad(u)\quad(-)\quad(0)\quad(-)\quad}{\qquad\quad -3\qquad\quad -1\qquad\quad 0\qquad\quad 2\qquad\quad x}$$

[Based on where the graph of f' is increasing or decreasing]

$f''(x)$ auxiliary axis:

$$\frac{\quad(-)\quad(0)\quad(+)\quad(u)\quad(+)\quad(0)\quad(-)\quad}{\qquad\quad -2\qquad\qquad\quad 0\qquad\quad 2\qquad\quad x}$$

(a) From the $f'(x)$ auxiliary axis, we observe that f is increasing on $(-\infty,-3)$ and $[-1,0]$; decreasing on $[-3,-1]$ and on $[0,\infty)$.

(b) From the $f''(x)$ auxiliary axis, we observe that f is concave up on $(-2,0)$ and $(0,2)$; concave down on $(-\infty,-2)$ and on $(2,\infty)$.

(c) From the $f'(x)$ auxiliary axis, we observe that f attains local maxima where x is -3 and where x is 0 [noting that f is continuous at 0 even though $f'(0)$ doesn't exist]; local minima where x is -1.

(d) From the $f''(x)$ auxiliary axis, we observe that f has inflection points where x is -2 and where x is 2.

f could have the following graph:

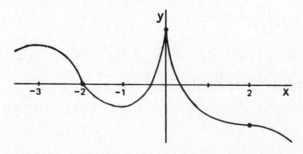

27. Critical points of a function.
    $f(x) = x^5 - 5x^3 + 4$

x	f(x)	
-2	12	End point
-1.73205	14.3923	Local max
1.73205	-6.3923	Local min
2.5	23.5312	End point

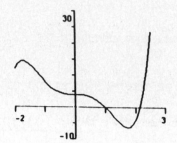

Absolute max = 23.5312
Absolute min = -6.3923

29. Local max at x=1, local min at x=2.2.

Problem Set 4.4 More Max-Min Problems

1. Let a (a>0) and b be the two numbers. Minimize $S = a^2 + b^2$ if ab = -12. Then $S(a) = a^2 + 144a^{-2}$ on $(0,\infty)$. $S'(a) = 2a^{-3}(a^4 - 144)$. Critical point is $2\sqrt{3}$. Use an auxiliary axis and conclude that S is minimum at the critical point. Therefore, the two numbers are $2\sqrt{3}$ and $-2\sqrt{3}$.

3. Step 1: Let d denote the distance between (x,y) and (10,0). Let $s = d^2$.

   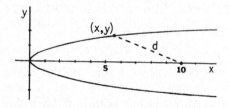

   Step 2: Note that if s is minimal, then d is minimal, since $d = \sqrt{s}$ and the square root function is an increasing function.

   Step 3: $x = 2y^2$ for points on the curve, so s can be expressed as a function of y only by $s(y) = (2y^2 - 10)^2 + y^2$.

   Step 4: y can be any real number. However, s is an even function, so we can restrict our search to $[0,\infty)$ and then use the symmetry.

   Step 5: $s'(y) = 2(2y^2 - 10)(4y) + 2y = 16y^3 - 78y = 2y(8y^2 - 39)$
           $= 2y(\sqrt{8}y + \sqrt{39})(\sqrt{8}y - \sqrt{39})$.

	(-)	(0)	(+)
0		$\sqrt{39/8}$	y

Step 6: The global minimum of s is where $y = \sqrt{39/8} \approx 2.21$, and
$$x = 2(\sqrt{39/8})^2 = 9.75.$$

Conclusion: The points on $x = 2y^2$ that are closest to (10,0) are approximately (9.75,2.21) and (9.75,-2.21).

5. Minimize $L = 3y+4x$ if $900 = xy$. Then $L(x) = 3(900/x) + 4x$ on $(0,\infty)$.
    $L'(x) = 4x^{-2}(x^2-675)$. Critical point is $15\sqrt{3}$, at which at L is minimum.
    Should use $15\sqrt{3} \approx 26'$ for x; use $20\sqrt{3} \approx 34'8''$ for y.

7. $L = 4y+6x$ and $900 = xy$, so $L(x) = 3600/x + 6x$ and $L'(x) = 6x^{-2}(x^2-600)$.
    Critical point is $10\sqrt{6}$ at which L is minimum. The minimum amount of fence
    is needed if x is $10\sqrt{6} \approx 24'6''$ and y is $15\sqrt{6} \approx 36'9''$.

9. Step 1:   Let x and y (ft) be as in the
              figure to the right.
              Let k (a constant) be the cost
              per square foot of base.
              Let C denote the total cost.

    Step 2:   Minimize C = (cost of base)+(cost of 4 sides)+(cost of top)
              $= k(x^2)+k(4xy)+2k(x^2) = 3kx^2+4kxy$.

    Step 3: $12000 = x^2y$. Therefore, C can be expressed as a function of x
              only by $C(x) = 3kx^2 + 4kx(12000/x^2) = 3kx^2 + \dfrac{48000k}{x}$.

    Step 4: $x > 0$.

    Step 5: $C'(x) = 6kx - 48000kx^{-2} = \dfrac{6kx^3-48000k}{x^2} = \dfrac{6k(x-20)(x^2+20x+400)}{x^2}$.

    $$\begin{array}{c c c c}
    & (-) & (0) & (+) \\
    \hline
    0 & & 20 & x
    \end{array}$$

    Step 6: C is minimum when $x = 20$; $y = 12000/(20)^2 = 30$.

    Conclusion: The most economical dimensions for the cistern are a base
                that is 20' × 20' and a depth of 30'.

11. Maximize $S = k(2x)(2y)^2 = 8kxy^2$ if $9x^2+8y^2 = 72$.
    Then $S(x) = 9k(8x-x^3)$ on $(0,\sqrt{8})$.
    $S'(x) = 9k(8-3x^2)$. Critical point is $\sqrt{8/3}$, at
    which S is maximum. The dimensions should be
    $2\sqrt{8/3} \approx 3.27$ wide by $2\sqrt{6} \approx 4.90$ deep.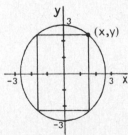

---

13. Same as Problem 15 except $T(x) = (1/3)(x^2+4)^{1/2} + (1/4)(10-x)$.

$T'(x) = (x/3)(x^2+4)^{-1/2} - 1/4$. Critical point is $6/\sqrt{7}$, at which T is minimum. Should land the boat $6/\sqrt{7} \approx 2.27$ miles down the shore from P.

15. Step 1: Let x denote the distance (mi) indicated.
Let T_b and T_w denote the times (hr) by boat and, walking, respectively.

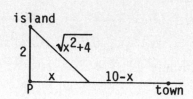

Let $T = T_b + T_w$.

Step 2: Minimize $T = T_b + T_w$.

Step 3: $T(x) = (\sqrt{x^2+4})/20 + (10-x)/4$.   Recall: time $= \dfrac{\text{distance}}{\text{velocity}}$.

Step 4: $0 \le x \le 10$.

Step 5: $T'(x) = \dfrac{x - 5\sqrt{x^2+4}}{20\sqrt{x^2+4}}$ which is never 0 since $5\sqrt{x^2+4} > x$.

Hence, there are no stationary points. There are no singular points since x^2+4 is never zero. Therefore, the minimum must occur at an end point. The correct end point is obviously at $x=10$ (boat all the way) rather than $x=0$.

Conclusion: Take the boat right to the town.

17. See figure at right. Let the positions of the ships be A and B, and let t denote the time in hours after 7:00 a.m. Minimize S, the square of the distance between the ships.

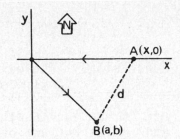

$x = 60 - 20t$.

$a = (30t)\cos(-\pi/4) = 15\sqrt{2}t$.

$b = (30t)\sin(-\pi/4) = -15\sqrt{2}t$.

$S = (x-a)^2 + b^2$.

Therefore, $S(t) = (60 - 20t - 15\sqrt{2}t)^2 + (-15\sqrt{2}t)^2$ for ≥ 0.

$S'(t) = -200(12 + 9\sqrt{2} - 13t - 6\sqrt{2}t)$. Critical point is $(12+9\sqrt{2})/(13+6\sqrt{2})$ which is approximately equal to 1.1509, and S is minimum there. Thus, the ships are closest about 1.1509 hrs after 7:00 a.m., at 8:09:03 a.m.

19. Maximize $V = \pi R^2 h$. $R^2 + (h/2)^2 = r^2$, so

$V(h) = \pi(r^2 h - h^3/4)$ on $(0, 2r)$.

$V'(h) = (\pi/4)(4r^2 - 3h^2)$. Critical point is $2r/\sqrt{3}$, at which V has maximum value of

$\pi(2r^2/3)(2r/\sqrt{3}) = (4\pi/3\sqrt{3})r^3 \approx 2.4184 r^3$.

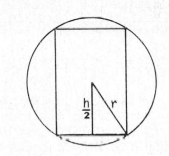

21. Step 1: Let r (a constant) denote the radius of the sphere. Let x denote the radius of the base of the cylinder. Let 2y denote the altitude of the cylinder. Let S denote the curved surface area.

Step 2: Maximize $S = 2\pi(\text{radius of base})(\text{altitude}) = 2\pi(x)(2y) = 4\pi xy$.

Step 3: $x^2 + y^2 = r^2$, so $y = \sqrt{r^2 - x^2}$.

Then express S as a function of x only by $S(x) = 4\pi x\sqrt{r^2 - x^2}$.

Step 4: $0 < x < r$.

Step 5: Maximizing S is equivalent to maximizing $f(x) = x\sqrt{r^2 - x^2}$ (since 4π is positive). This is the same function we maximized in Problem 18, except that we have x and r (instead of X and a).

Step 6:   Thus, (see Problem 18) S is maximum if $x = r/\sqrt{2}$;

$$y = \sqrt{r^2 - (r/\sqrt{2})^2} = r/\sqrt{2}.$$

Conclusion: The maximum surface area is obtained if the altitude of the cylinder is the same as the diameter of its base.

23.   Let x be the length of one side of the triangle, and S be the sum of the areas. Then
$s(x) = (\sqrt{3}/4)x^2 + [(100-3x)/4]^2$ on $[0,100/3]$.

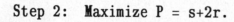
3x   100-3x

$S'(x) = [(4\sqrt{3}+9)/8]x + [-75/2]$.   The Critical points are 0, 100/3, and $300/(4\sqrt{3}+9)$.   Evaluate S at the critical points and conclude:

(a)   S is minimum if $900/(4\sqrt{3}+9) \approx 56.50$ cm is used for the triangle.

(b)   S is maximum if all of the wire is used for the square.

25.   Minimize $C = (1)(2\pi rh) + (2)(2\pi r^2)$, given that the volume is $V = \pi r^2 h + (2/3)\pi r^3$.   Therefore, $C(r) = 2Vr^{-1} + (8\pi/3)r^2$ for $r > 0$. $C'(r) = -2Vr^{-2} + (16\pi/3)r$.   A stationary point is $(3V/8\pi)^{1/3}$, and C is minimum there. The corresponding altitude is twice that.

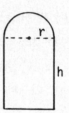

27.   Step 1:   Let s be as indicated.
Let P denote the perimeter.

Step 2:   Maximize $P = s+2r$.

Step 3: $s = r\theta$ and $A = \frac{1}{2}\theta r^2$.   [Formulas for arc length and area]
Then P can be expressed as a function of r by
$$P(r) = r\theta + 2r = r(2Ar^{-2}) + 2r = 2r + 2Ar^{-1}.$$

Step 4: $0 < r$.

Step 5: $P'(r) = 2 - 2Ar^{-2} = \dfrac{-2(A-r^2)}{r^2} = 0$ iff $r = \sqrt{A}$.

Step 6: $P''(r) = 4Ar^{-3} = \dfrac{4A}{r^3} > 0$, where $P'(r) = 0$.

Therefore, P is minimum where $r = \sqrt{A}$.
Conclusion: $r = \sqrt{A}$; $\theta = 2A/r^2 = 2$ radians (about 114.6°).

29. Minimize $T(x) = \dfrac{\sqrt{a^2+x^2}}{c_1} + \dfrac{\sqrt{b^2+(d-x)^2}}{c_2}$ on $(0,d)$.

$$T'(x) = \frac{x}{c_1\sqrt{a^2+x^2}} - \frac{d-x}{c_2\sqrt{b^2+(d-x)^2}} = \frac{1}{c_1}\sin\theta_1 - \frac{1}{c_2}\cdot\sin\theta_2 = \frac{\sin\theta_1}{c_1} - \frac{\sin\theta_2}{c_2}.$$

31. Maximize x.
$u+v = 27$,
so $x\sec a + 8\csc a = 27$, a in $(0,\pi/2)$

Thus, $x(a) = \dfrac{27-8\csc a}{\sec a} = 27\cos a - 8\cot a$

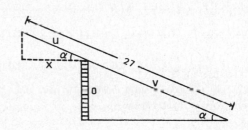

$x'(a) = -27\sin a + 8\csc^2 a = \dfrac{27\sin^3 a + 8}{\sin^2 a}$

$x'(a) = 0$ if $\sin a = 2/3$

Auxiliary axis for $x'(a)$:

```
    (+)         (0)         (-)
|————————————|———————————|——————————| α
0           sin⁻¹(2/3)              π/2
```

Thus, x is maximum if $\sin a = 2/3$.
$\sin a = 2/3 \Rightarrow 2/3 = 8/v \Rightarrow v = 12 \Rightarrow u = 15$.

$\sin a = 2/3 \Rightarrow \cos a = \sqrt{5}/3 \Rightarrow x/15 - \sqrt{5}/3 \Rightarrow x - 5\sqrt{5} \approx 11/18$.

The maximum horizontal overhang is about 11.18 ft [11 ft 2.16 in.]

33. (a) Let M denote the area of triangle A.
$M = \frac{1}{2}(a-x)y$

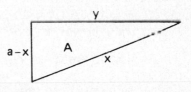

$y^2 = x^2 - (a-x)^2 = 2ax - a^2$

Thus, $M(x) = \frac{1}{2}(a-x)\sqrt{2ax-a^2}$, x in $[a/2, a]$

[Note that determining $M'(x)$ maybe a bit involved. We can avoid it by noting that $M(x)$ will be maximum for the same value of x for which $N(x) = [2M(x)]^2$ is maximum.]

$N(x) = (x-a)^2(2ax-a^2)$

$N'(x) = 2(x-a)(2ax-a^2) + (x-a)^2(2a) = 2a(x-a)(3x-2a)$

Auxiliary axis for $N'(x)$:
```
        (+)     (0)     (-)
    |————————|———————|————————| x
   a/2      2a/3        a
```

Thus, the area of A is maximum if x is 2a/3.

113

(b) Let **M** denote the area of triangle B.

$$M = \frac{1}{2} x \sqrt{z^2 - x^2}$$

$$\frac{\sqrt{z^2 - x^2}}{a} = \frac{x}{\sqrt{2ax - a^2}}$$

$$\sqrt{z^2 - x^2} = \frac{ax}{\sqrt{2ax - a^2}}$$

Thus, $M(x) = \dfrac{1}{2} x \dfrac{ax}{\sqrt{2ax - a^2}} = \dfrac{ax^2}{2\sqrt{2ax - a^2}}$, x in $[a/2, a]$.

[We'll get a bit fancier this time. Note that $M(x)$ will be minimum for the same value of x for which $N(x) = (1/a)[2M(x)]^2 = x^4/(2x - a)$ is minimum.]

$$N'(x) = \frac{2x^3(3x - 2a)}{(2x - a)^2}$$ [Use quotient rule, simplify, and factor.]

Auxiliary axis for $N'(x)$:

$$\underset{a/2 \qquad\quad 3a/4 \qquad\quad a}{\left[\underline{\quad\quad \overset{(-)}{\quad} \quad \overset{(0)}{\quad} \quad \overset{(+)}{\quad}\quad}\right]x}$$

Thus, the area of B is minimum if x is $2a/3$.

(c) $\sqrt{z^2 - R^2} = \dfrac{ax}{\sqrt{2ax - a^2}}$ [from (b)]

Solve for z^2; obtain $z^2 = \dfrac{2x^3}{2x - a}$

[Note that z will be minimum for the same value of x that z^2 is minimum, so we will minimize $N(x) = \dfrac{2x^3}{2x - a}$.]

$$N'(x) = \frac{2x^2(4x - 3a)}{(2x - a)^2}$$

Auxiliary axis for $N'(x)$:

Thus, the length z is minimum if x is $3a/4$.

35. (a) $L'=3$, $L=4$, $\theta=\pi/2$ (b) $L'=5$, $L=12$, $\theta=\pi/2$

  (c) L' is maximized when $\theta=\pi/2$ in which case $h=\sqrt{m^2-h^2}$ and $L'=h$.

  (d) Note that since θ increases at a constant rate with respect to time, L increases most rapidly at the time when $L'(\theta)$ is largest. Now

$$L''(\theta) = \frac{L(\theta)hm\cos\theta - hm\sin\theta\, L'(\theta)}{L^2(\theta)}.\quad \text{Thus}\quad L''(\theta)=0 \Rightarrow L(\theta)\cos\theta = L'(\theta)\sin(\theta)$$

$$\Rightarrow L(\theta)\cos\theta = \frac{hm\sin^1\theta}{L(\theta)} \Rightarrow L^2(\theta)\cos\theta = hm(1-\cos^2\theta)$$

$$\Rightarrow (h^2+m^2-2hm\cos\theta)\cos\theta = hm - hm\cos^2\theta$$

$$\Rightarrow hm\cos^2\theta - (h^2+m^2)\cos\theta + hm = 0 \Rightarrow \cos\theta - \frac{h}{m}$$

$$\Rightarrow \varphi = \pi/2 \Rightarrow \sin\theta = \frac{L}{m} \Rightarrow L = \sqrt{h^2+m^2-2mh\frac{h}{m}} = \sqrt{m^2-h^2}$$

$$\Rightarrow L' = \frac{hm\sin\theta}{L} = \frac{hm\frac{L}{m}}{L} = h$$

$f(x) = 15^*\sin(x)^*(34-30^*\cos(x))^{-.5}$

x	f(x)	
0	0	End point
.9723	3	Local max
3.14159		End point

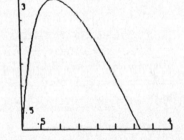

Absolute max = 3
Absolute min = 0

$f(x) = 65^*\sin(x)^*(194-130^*\cos(x))^{-.5}$

x	f(x)	
0	0	End point
1.176	5	Local max
3.14159	0	End point

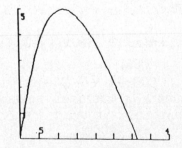

Absolute max = 5
Absolute min = 0

Problem Set 4.5   Economic Applications

1. $C(x) = 8000 + 110x$.

3. $P(x) = R(x) - C(x) = [300x-(x^2/2)]- [8000+110x] = -8000+190x-(x^2/2)$.

5. $C(1800)/1800 \approx \$3.56$ is the average cost.

    $C'(x) = 3.25-0.0004x$; $C'(1800) \approx \$2.53$
        is the marginal cost at $x = 1800$.

7. Let $A(x) = C(x)/x = (80000 - 400x + x^2)/40000$ be the average cost.
    $A'(x) = (-400+2x)40000$. Stationary point is 200, at which the average
    cost is minimum.

9. (a) $R(x) = xp(x) = 20x+4x^2-(1/3)x^3$ is the revenue function.
         $R'(x) = 20+8x-x^2$
                $= (10-x)(2+x)$ is the marginal revenue function.

    (b)

    R'   (+)   (0)   (-)
    |_____
    0   10   x

        R is increasing on $[0,10]$.

    (c) $R''(x) = 8-2x = 2(4-x)$ [This is the 1st derivative of R'.]
         $R''(x) = 0$ iff $x=4$
         $R^{(3)}(x) = -2$ [This is the 2nd derivative of R'.]

        Therefore $R^{(3)}(4)$ is negative. Make use of the Second Derivative
        Derivative Test and conclude that the marginal revenue is maximum
        when $x=4$.

11. $R(x) = xp(x) = 800x/(x+3) - 3x$. $R'(x)= 2400(x+3)^{-2} - 3$. Stationary
    point is $\sqrt{800} - 3 \approx 25.28$, at which R is maximum. $R(25) = 639.29$ and
    $R(26) = 639.24$ so 25 units yields a maximum revenue of 639.29. The
    corresponding marginal revenue is $R'(25) = 0.06$.

13. $p(x) = 6 - \frac{x-4000}{250} (0.15) = 8.4 - 0.0006x$ for $x \geq 4000$.

    $R(x) = xp(x) = 8.4x - 0.0006x^2$.
    $R'(x) = 8.4 - 0.0012x = -0.0012(x-7000)$. R is maximum for $x = 7000$.
    A price of $p(7000) = \$4.20$ will maximize revenue.

15. Fixed cost = 6000.

Variable cost = cost of material + cost of labor

$$= \begin{cases} 1.00x + 0.40x, & \text{if } 0 \le x \le 4500 \\ 1.00x + 0.40(4500) + 0.60(x-4500), & \text{if } x > 4500 \end{cases}.$$

(a) Total cost = Fixed cost + Variable cost

$$= \begin{cases} 6000 + 1.40x & \text{if } 0 \le x \le 4500 \\ 5100 + 1.60x, & \text{if } x > 4500 \end{cases}.$$

(b) $p(x) = \begin{cases} 7.00 & \text{if } 0 \le x \le 4000 \\ 7.00 - (0.10)\dfrac{x-4000}{100} = 11.00 - 0.001x, & \text{if } x > 4000 \end{cases}.$

(c) $P(x) = R(x) - C(x) = xp(x) - C(x)$

$$= \begin{cases} 7.00x - [6000 + 1.40x] & \text{if } 0 \le x \le 4000 \\ x[11.00 - 0.001x] - [6000 + 1.40x], & \text{if } 4000 < x \le 4500 \\ x[11.00 - 0.001x] - [5100 + 1.60x], & \text{if } x > 4500 \end{cases}$$

$$= \begin{cases} 5.60x - 6000, & \text{if } 0 \le x \le 4000 \\ -6000 + 9.60x - 0.001x^2, & \text{if } 4000 < x \le 4500 \\ -5100 + 9.40x - 0.001x^2, & \text{if } x > 4500 \end{cases}.$$

$$P'(x) = \begin{cases} 5.60, & \text{if } 0 < x < 4000 \\ 9.60 - 0.002x, & \text{if } 4000 < x < 4500 \\ 9.40 - 0.002x, & \text{if } x > 4500 \end{cases}.$$

$P'(x)$ is undefined for $x = 0$, 4000, and 4500.

$P'(x) = 0$ for $x - 4700$.

P is continuous on $[0,\infty)$, and the auxiliary axis for P' is

```
         (+)      (u)      (+)      (u)      (+)      (0)      (-)
    +-----+--------+--------+--------+--------+--------+----------  x
    0    4000             4500             4700
```

P is maximum if 4700 units/month are produced.

17. $P(x) = \left\{ \begin{array}{l} -200 + 6x + 0.009x^2, \ 0 \le x \le 300 \\ (10x - 0.001x^2) - (800 + 3x - 0.01x^2), \ 300 < x \le 450 \end{array} \right\}.$

$P'(x) = \left\{ \begin{array}{l} 6 + 0.018x, \ 0 < x < 300 \\ 7 + 0.018x, \ 300 < x < 450 \end{array} \right\}.$

Critical points are 0, 300, and 450. Maximum profit of \$4172.50 is obtained for x = 450 units.

19. Using formula derived in Problem 18,

$$\frac{dR}{dm} = \frac{5m^3 + 130m}{(m^2+13)^{3/2}} [(10x - 0.1x^2) + x(10 - 0.2x)].$$

When m = 6, x = 180/7, so dR/dm ≈ \$1713.14 dollars/employee.

21. Let T denote the total annual cost associated with inventory [dollars]. If x denotes the economic order quantity (EOQ), we will need to place an order N/x times per year.

T = (N/x)[ordering cost per order] + (x/2) [holding cost per item]

$T(x) = (N/x)(F+Bx) + (x/2)(A) = FNx^{-1} + NB + (A/2)x$, x in (0,N]

$$T'(x) = -FNx^{-2} + A/2 = \frac{Ax^2 - 2FN}{2x^2} = 0 \ \text{if} \ x = \sqrt{2FN/A}$$

$T''(x) = 2FNx^{-3} > 0$ for x positive, and $\sqrt{2FN/A}$ is the only stationary point, so the minimum value of T occurs there.

Problem Set 4.6 Limits at Infinity, Infinite Limits

1. $\frac{3-2x}{x+5} = \frac{3/x - 2}{1 + 5/x} \to -2$ (as $x \to \infty$).

3. $\lim\limits_{x \to -\infty} \frac{2x^2 - x + 5}{5x^2 + 6x - 1} = \lim\limits_{x \to -\infty} \frac{2 - (1/x) + (5/x^2)}{5 + (6/x) - (1/x^2)} = \frac{2 - 0 + 0}{5 + 0 - 0} = \frac{2}{5}$.

5. $\frac{2x^3 - 3x^2 + 1}{5x^3 - 4x + 7} = \frac{2 - 3/x + 1/x^3}{5 - 4/x^2 + 7/x^3} \to \frac{2}{5}$.

7. $\dfrac{3x^3-4x+1}{x^4-1} = \dfrac{3/x - 4/x^3 + 1/x^4}{1 - 1/x^4} \to 0 \ (\text{as } x \to -\infty).$

9. $\lim_{x\to\infty} \sqrt[3]{\dfrac{1+8x^2}{x^2+4}} = \lim_{x\to\infty} \sqrt[3]{\dfrac{(1/x^2) + 8}{1 - (4/x^2)}} = \sqrt[3]{\dfrac{0 + 8}{1 + 0}} = 2.$

11. $\dfrac{2x+1}{(x^2+1)^{1/2}} = \dfrac{2 + 1/x}{(1 + 1/x^2)^{1/2}} \to 2 \quad (\text{as } x \to \infty)$

13. $(\sqrt{2x^2+3} - \sqrt{2x^2-5}) = \dfrac{(2x^2+3) - (2x^2-5)}{\sqrt{2x^2+3} + \sqrt{2x^2-5}} = \dfrac{8}{\sqrt{2x^2+3} + \sqrt{2x^2-5}}$

$$= \dfrac{8/x}{\sqrt{2 + 3/x^2} + \sqrt{2 - 5/x^2}} \to 0 \quad (\text{as } x \to \infty).$$

15. $\lim\limits_{y\to -\infty} \dfrac{9y^3+1}{y^2-2y+2} = \lim\limits_{y\to -\infty} \dfrac{9y+(1/y^2)}{1-(2/y)+(2/y^2)} = -\infty.$

17. $-\infty.$

19. $+\infty$

21. $\lim\limits_{x\to 1/5^-} \dfrac{4x+1}{2x-3} = -\infty$ since $\left[\dfrac{\to 7}{\to 0^-}\right].$

23. $\lim\limits_{x\to 3} \dfrac{2x}{(x-3)^2} = +\infty.$

25. $\lim\limits_{x\to 3^-} \dfrac{(x-3)(x+2)}{x-3} = \lim\limits_{x\to 3^-} (x+2) = 5.$

27. $\lim\limits_{x\to 0^+} \dfrac{[x]}{x} = \lim\limits_{x\to 0^+} \dfrac{0}{x}$ [since $[x] = 0$ if x is near and on the right of 0]

$$= \lim\limits_{x\to 0^+} (0) = 0.$$

29. $\lim\limits_{x\to 0^-} \dfrac{-x}{x} = \lim\limits_{x\to 0^-} (-1) = -1.$

31. $-\infty.$

33. $f(x) = \dfrac{3}{x+1}$.

Vertical asymptotes: Denominator is 0 when x = -1 so we suspect that the line x=-1 is a vertical asymptote.

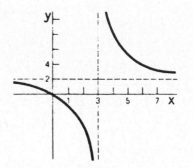

It is since $\displaystyle\lim_{x\to-1^-}\dfrac{3}{x+1} = -\infty$ $\left[\dfrac{\to 3}{\to 0^-}\right]$

and $\displaystyle\lim_{x\to-1^+}\dfrac{3}{x+1} = +\infty$ $\left[\dfrac{\to 3}{\to 0^+}\right]$.

Horizontal asymptotes: The horizontal line y=0 since

$$\lim_{x\to-\infty}\dfrac{3}{x+1} = 0 \quad\text{and}\quad \lim_{x\to\infty}\dfrac{3}{x+1} = 0.$$

35. $\displaystyle\lim_{x\to 3^-}\dfrac{2x}{x-3} = -\infty$ and $\displaystyle\lim_{x\to 3^+}\dfrac{2x}{x-3} = +\infty$.

Therefore, x=3 is a vertical asymptote.

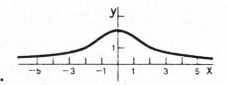

$$\lim_{x\to\pm\infty}\dfrac{2x}{x-3} = \lim_{x\to\pm\infty}\dfrac{2}{1-3/x} = 2.$$

Therefore, y=2 is a horizontal asymptote.

37. No vertical asymptote.

$$\lim_{x\to\pm\infty}\dfrac{14}{2x^2+7} = \lim_{x\to\pm\infty}\dfrac{14/x^2}{2+7/x^2} = 0.$$

Therefore, y=0 is a horizontal asymptote.

39. $\dfrac{2x^4+3x^3-2x-4}{x^3-1} = 2x+3 - \dfrac{1}{x^3-1}$, so y = 2x+3 is an oblique asymptote since

$$\lim_{x\to\pm\infty}\left[\dfrac{2x^4+3x^3-2x-4}{x^3-1}\right] - (2x+3) = \lim_{x\to\pm\infty}\left[\left[2x+3 - \dfrac{1}{x^3-1}\right] - (2x+3)\right]$$

$$= \lim_{x\to\pm\infty}\dfrac{-1}{x^3-1} = 0.$$

41. (a) Let f be defined on (c,b) for some number a.
Then $\lim\limits_{x\to c^+} f(x) = -\infty$ iff for each negative number **M**, there
is a corresponding number $\delta > 0$ such that
$0 < x - c < \delta \Rightarrow f(x) > $ **M**.

(b) Let f be defined on (a,c) for some number a.
Then $\lim\limits_{x\to c^-} f(x) = -\infty$ iff for each negative number **M**, there
is a corresponding number $\delta > 0$ such that
$0 < c - x < \delta \Rightarrow f(x) < $ **M**.

43. Let $\epsilon > 0$ be given. Then there are numbers **M** and **N** such that
$x > $ **M** $\Rightarrow |f(x) - A| < \epsilon/2$ and $x > $ **N** $\Rightarrow |g(x) - B| < \epsilon/2$.

Let **M** $= \max\{$**M,N**$\}$. Then $x > $ **M** $|f(x) - A| < \epsilon/2$ and $|g(x) - B| < \epsilon/2$

$$\Rightarrow |[f(x) + g(x)] - [A+B] = |[f(x) - A] + [g(x) - B]|$$

$$\leq |f(x) - A| + |g(x) - B| < \epsilon/2 + \epsilon/2 = \epsilon.$$

45. (a) Doesn't exist. (As $x \to \infty$, sinx continues to take on each value in
the interval $[-1,1]$.)

(b) $\lim\limits_{x\to\infty} \sin(1/x) = \lim\limits_{t\to 0^+} \sin t = 0$ [Letting t = 1/x]

(c) $\lim\limits_{x\to\infty} x\sin(1/x) = \lim\limits_{x\to\infty} \dfrac{\sin(1/x)}{1/x} = \lim\limits_{t\to 0^+} \dfrac{\sin t}{t} = 1$ [Letting t = 1/x]

(d) $\lim\limits_{x\to\infty} x^{3/2} \sin(1/x) = \lim\limits_{x\to\infty} (x^{1/2}) [x\sin(1/x)] = \infty$

[First factor $\to \infty$; second factor $\to 1$.]

(e) $\lim\limits_{x\to\infty} x^{-1/2} \sin x = 0$

$[-x^{-1/2} \leq x^{-1/2} \sin x \leq x^{-1/2}$; outside terms $\to 0$ as $x\to\infty$.]

(f) $\lim\limits_{x\to\infty} \sin(\frac{\pi}{6} + \frac{1}{x}) = \sin[\lim\limits_{x\to\infty} (\frac{\pi}{6} + \frac{1}{x})] = \sin(\frac{\pi}{6}) = \frac{1}{2}$

(g) Doesn't exist. [Same reason as for (a)]

(h) $\lim\limits_{x\to\infty} [\sin(x + \frac{1}{x}) - \sin x] = \lim\limits_{x\to\infty} [\sin x \cos(\frac{1}{x}) - \cos x \sin(\frac{1}{x}) - \sin x]$

$= \lim\limits_{x\to\infty} \sin x [\cos(\frac{1}{x}) - 1] - \cos x \sin(\frac{1}{x}) = 0$

[Use Squeeze Theorem on each term as we did in (e); each term has one factor that approaches 0 and another factor that is bounded between -1 and 1.]

47. 1.5   **49.** -1.06066   **51.** 1   **53.** ∞   **55.** -1   **57.** -∞   **59.** ∞

Problem Set 4.7 Sophisticated Graphing

1. $f(x) = x(x+2)(x-2)$; domain is **R**; intercepts are at $(0,0)$, $(2,0)$, and $(-2,0)$; symmetry with respect to the origin.

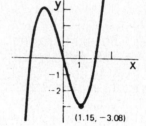

$f'(x) = 3x^2 - 4$; decreasing on (approximately) $[-1.15, 1.15]$; increasing on $[-\infty, -1.15]$ and on $[1.15, \infty)$; local maximum at $(-1.15, 3.08)$;

local minimum at $(1.15, -3.08)$.

$f''(a) = 6x$; concave down on $(-\infty, 0)$; concave up on $(0, \infty)$; $(0,0)$ is an inflection point.

3. $F(x) = \frac{1}{20} (x^4 - 18x^2 + 20)$.
    Domain: **R**.
    Symmetry: With respect to the y-axis. [F is an even function.]

    Intercepts: Set $F(x)$ equal to zero and use the quadratic equation to
        solve for x^2, then obtain the square roots.

    $$x^2 = \frac{18 \pm \sqrt{324 - 80}}{2} = 9 \pm \sqrt{61}, \text{ so } x = \pm\sqrt{9 + \sqrt{61}}, \pm\sqrt{9 - \sqrt{61}}.$$

    That is, x-intercepts are approximately ±4.10, ±1.09. y-intercept is 1.

    Monotonicity: $F'(x) = \frac{1}{20} (4x^3 - 36x) = \frac{1}{5} x(x+3)(x-3).$

Dec (-)	(0)	Inc (+)	(0)	Dec (-)	(0)	Inc (+)
	-3		0		3	x

Concavity: $F''(x) = \frac{1}{20}(12x^2 - 36) = \frac{3}{5}(x + \sqrt{3})(x - \sqrt{3})$.

CU (+)	(0)	CD (−)	(0)	CU (+)

$-\sqrt{3}$ $\sqrt{3}$
-1.73 1.73

Asymptotes: None.

Summary	
At 0	Levels off
(0,1.73)	Dec and CD
At 1.73	Inflection
(1.73,3)	Dec and CU
At 3	Levels off
(3,∞)	Inc and CU

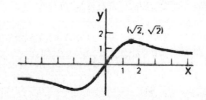

x	F(x)	
0	1	(local max)
1.73	−1.25	(inflection point)
3	−3.05	(global min)
5	9.75	

Use symmetry to obtain the graph for $x < 0$.

5. $g(x) = 4x(x^2+2)^{-1}$; domain is $\mathbb{R}$; intercept is at $(0,0)$; symmetry with respect to origin.

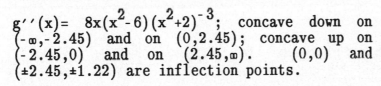

$g'(x) = 4(2-x^2)(x^2+2)^{-2}$; decreasing on $(-\infty,-1.41]$ and on $[1.41,\infty)$; increasing on $[-1.41,1.41]$; global minimum at $(-1.41,-1.41)$; global maximum at $(1.41,1.41)$.

$g''(x) = 8x(x^2-6)(x^2+2)^{-3}$; concave down on $(-\infty,-2.45)$ and on $(0,2.45)$; concave up on $(-2.45,0)$ and on $(2.45,\infty)$. $(0,0)$ and $(\pm2.45,\pm1.22)$ are inflection points.

$g(x) \to 0$ as $x \pm \infty$, so $y = 0$ is a horizontal asymptote.

7. $h(x) = x/(x+1)$; domain is $\{x:x \neq -1\}$;
    intercept at $(0,0)$.

    $h'(x) = (x+2)^{-2}$; increasing on $(-\infty,-1)$ and
    on $(-1,\infty)$.

    $h''(x) = -2(x+1)^{-3}$; concave up on $(-\infty,-1)$;
    concave down on $(-1,-\infty)$.

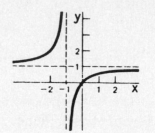

    $h(x) \to \infty$ as $x \to -1^-$, and $h(x) \to -\infty$ as $x \to -1^+$,
    so $x=-1$ is a vertical asymptote;
    $h(x) \to 1$ as $x \to \pm\infty$, so $y=1$ is a horizontal
    asymptote.

9. $f(x)\ \dfrac{x^2}{x^2-4} = \dfrac{x^2}{(x+2)(x-2)}$.

    Domain: $\{x:x \neq -2,2\}$.
    Symmetry: With respect to the y-axis. [f is an even function.]
    Intercepts: x-intercept is 0; y-intercept is 0.

    Montonicity: $f'(x) = \dfrac{-8x}{(x^2-4)^2} = \dfrac{-8x}{(x+2)^2(x-2)^2}$.

Inc (+)	(u)	Inc (+)	(0)	Dec (-)	(u)	Dec (-)	
	-2		0		2		x

    Concavity: $f''(x) = \dfrac{8(3x^2+4)}{(x^2-4)^3} = \dfrac{8(3x^2+4)}{(x+2)^3(x-2)^3}$.

CU (+)	(u)	CD (-)	(u)	CU (+)	
	-2		2		x

    Asymptotes: $x=-2$ (vertical) since $\displaystyle\lim_{x \to -2^-} f(x) = \infty\ \left[\dfrac{\to 4}{\to 0^+}\right]$.

    and $\displaystyle\lim_{x \to -2^+} f(x) = -\infty\ \left[\dfrac{\to 4}{\to 0^-}\right]$.

x=2 (vertical) since $\lim\limits_{x \to -2^-} f(x) = -\infty$ $\left[\dfrac{\to 4}{\to 0^-}\right]$.

and $\lim\limits_{x \to -2^+} f(x) = \infty$ $\left[\dfrac{\to 4}{\to 0^+}\right]$.

y=1 (horizontal) since $\lim\limits_{x\to\pm\infty} f(x) = \lim\limits_{x\to\pm\infty} \dfrac{1}{1-(4/x^2)} = 1.$

Summary

(0,2)	Dec and CD
At 2	Vertical Asymptote
(2,∞)	Dec and CU

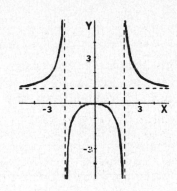

x	f(x)
0	0 (local max)
1	-1/3
3	1.8

Use symmetry to obtain the graph for x < 0.

11. $g(x) = 2x(x+3)^{1/2}$; domain is $[-3,\infty)$; intercepts are at (0,0) and (-3,0).
$g'(x) = 3(x+2)(x+3)^{-1/2}$; decreasing on $[-3,-2]$; increasing on $[-2,\infty)$; global minimum at (-2,-4).
$g''(x) = (3/2)(x+4)(x+3)^{-3/2}$; concave up on $(-3,\infty)$.

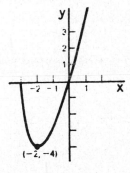

13. $H(x) = \left\{ \begin{array}{ll} -x^2, & x < 0 \\ x^2, & x \geq 0 \end{array} \right\}$.

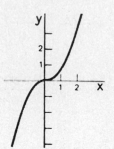

15. $f(x) = |\sin x|$. The graph is the same as that of $g(x) = \sin x$, except that all portions of the graph of g that are below the x-axis (where $y < 0$) are reflected with respect to the x-axis.

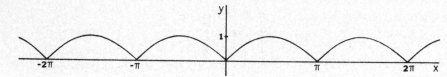

17.

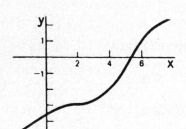

19.

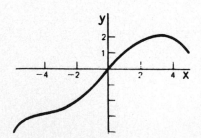

21. Auxiliary axis for f'(x): [Based on where graph of f' is above or below
the the x-axis.]

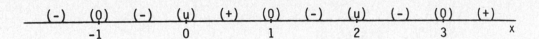

    (-)   (0)   (-)   (ψ)   (+)   (0)   (-)   (ψ)   (-)   (0)   (+)
          -1   0   1   2   3   x

Auxiliary axis for f''(x): [Based on where graph of f' is increasing or
decreasing.]

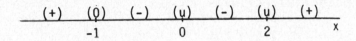

       (+)   (0)   (-)   (ψ)   (-)   (ψ)   (+)
             -1   0   2   x

Summary: f(0) = 0   and f(2) = 0.

f is increasing on [0,1] and on [3,∞).
f is decreasing on (-∞,0] and on [1,3].
Relative maximum where x is 1; minima where x is 0 and 3.
Slopes of tangents to left of x-0 and on both sides of x = 2
approach -∞.
Slopes of tangents to right of x=0
approach ∞.

f is concave up on (-∞,-1), and on (2,∞).
f is concave down on (-1,0), and on (0,2).
Inflection points where   x is -1 and 2.

A possible graph for f (No. 20): A possible graph for f (No. 21):

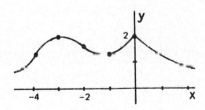
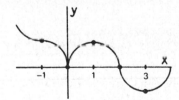

23. (a) $f(x) = x * \sqrt{(x * x - 6 * x + 40)}$

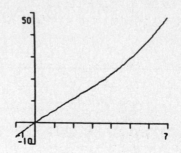

x	f(x)	
-1	-6.85565	End point
7	47.9896	End point

Absolute max point: (7, 47.99)

Absolute min point: (-1, -6.86)

Inflection point: (2.2, 12.4)

(b) $f(x) = \sqrt{(abs(x)) * (x * x - 6 * x + 40)}$

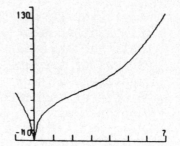

Bad derivative near 0

Absolute max point: (7, 124.35)

Absolute min point: (0,0)

Inflection point: (2.3, 47.8)

(c) $f(x) = \sqrt{(x * x - 6 * x + 40)/(x - 2)}$

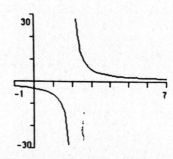

Bad value near 2

Absolute max point: None

Absolute min point: None

Inflection point: None

(d) $f(x) = \sin((x\overset{*}{}x - 6\overset{*}{}x + 40)/6)$

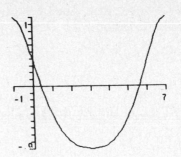

x	f(x)	
-1	.99979	End point
3	-.89858	Local min
7	.99979	End point

Absolute max point: (7, 1.00)
Absolute min point: (3, -0.90)
Inflection point: (0.05, 0.33), (5.9, 0.3)

Problem Set 4.8   The Mean Value Theorem

1. MVT applies; c=0.

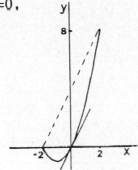

3. The Mean Value Theorem applies since
    $g(x) = x^3/3$ is continuous on $[-2,2]$
    and $g'(x) = x^2$ exists on $(-2,2)$.
    Therefore, for some c in $(-2,2)$

    $$g'(c) = \frac{g(2) - g(-2)}{2 - (-2)} = \frac{8/3 - (-8/3)}{4} = \frac{4}{3}.$$

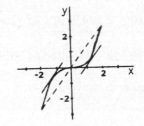

Therefore $c^2 = 4/3$ or $c = \pm 2/\sqrt{3}$. That is, the tangent lines at x = $\pm 2/\sqrt{3} \approx \pm 1.15$ are parallel to the secant line connecting the points (-2,-8/3) and (2,8/3).

5. MVT does not apply since F is
    discontinuous at 3.

7. MVT applies; c = 16/27.   9. MVT does not apply since ∅ is
                                              discontinuous at 0.

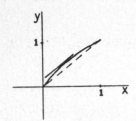

11. Assume f is continuous on $[0,2]$ (i.e., no time gaps) and differentiable
    on $(0,2)$ (i.e., no head-on battle with a semi). The MVT states that the
    instantaneous velocity at some time in $(0,2)$ equals the averge velocity
    on $[0,2]$, which is $112/2 = 56$ mph.

13. If $f \geq (a) = f(b)$, the concluding equation of the MVT reduces to $f'(x) = 0$,
    the concluding equation of Rolle's Theorem.

15. f is increasing on $(a,x_0]$ and on $[x_0,b)$ by the monotonicity theorem.

    Case I: If x_1 and x_2 are both in $(a,x_0]$ or both in $[x_0,b)$, then
    $$f(x_1) < f(x_2) \text{ since } f \text{ is increasing}$$
    on both intervals.

    Case II: x_1 is in (a,x_0) and x_2 is in (x_0,b)
    $$\Rightarrow \ x_1 < x_0 \text{ and } x_0 < x_2$$
    $$\Rightarrow \ f(x_1) < f(x_0) \text{ and } f(x_0) < f(x_2)$$
    $$\Rightarrow \ f(x_1) < f(x_2).$$

    Therefore, for all x in (a,b), $x_1 < x_2 \Rightarrow f(x_1) < f(x_2)$, so f is
    increasing on (a,b).

17. Let $G(x) = 0$. Then $F'(x) = G'(x) \Rightarrow F(x) = G(x) + C = C$ for some C.

19. Let $G(x) = Dx$. Then $G'(x) = F'(x) = D$, $F(x) = G(x) + C = Dx + C$ for some C.

21. By the Intermediate Value Theorem, $f(x) = 0$ has *at least* one solution between a and b.

Assume that $f(x)$ 0 has two distinct solutions, α and β, $\alpha < \beta$.

Then f satisfies the conditions of Rolle's Theorem, so there must be a point c between α and β such that $f'(c) = 0$. But c is also between a and b, which contradicts that $f'(x) \neq 0$ on (a,b).

Hence, the assumption that there are two distinct solutions must be incorrect.

23. Let a and b be successive distinct zeros of f'. That is, $f'(a) = 0$. $f'(b) = 0$, and $f'(x) \neq 0$ on (a,b). Assume x and y are distinct zeros of f in (a,b). That is, $f(x) = f(y) = 0$. Then (by Rolle's Theorem) there is a number c in (a,b) such that $f'(c) = 0$, which contradicts that a and b are successive zeros of f'.

25. Assume $x_1 < x_2$. By the MVT, applied to $[x_1, x_2]$, there is a number c between x_1 and x_2 such that $\dfrac{f(x_2) - f(x_1)}{x_2 - x_1} = f'(c)$.

Therefore, $\dfrac{|f(x_2) - f(x_1)|}{x_2 - x_1} = |f'(c)| \leq M$, so $|f(x_2) - f(x_1)| \leq M|x_2 - x_1|$.

27. (a)   (b)

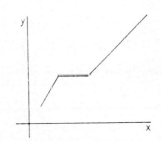
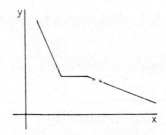

29. $(d/dx)[f^2(x)] = 2f(x)f'(x) \geq 0$ on I. Now use the result obtained in Problem 28.

31. $f(x) = \sqrt{x}$ is continuous on $[x, x+2]$ and differentiable on $(x, x+2)$ for all $x \geq 0$. Applying the MVT to the interval $[x, x+2]$, there is a number c in $(x, x+2)$ such that $[f(x+2) - f(x)]/[(x+2) - x] = f'(c)$. That is, $[\sqrt{x+2} - \sqrt{x}]/2 = 1/2\sqrt{c}$. Thus, $\sqrt{x+2} - \sqrt{x} = 1/\sqrt{c}$. As $x \to \infty$, $c \to \infty$ and $1/2\sqrt{c} \to 0$, so $(\sqrt{x+2} - \sqrt{x}) \to 0$.

33. Let $x_1(t)$ and $x_2(t)$ be the distances the two horses have traveled in time t, for t in $[0,T]$, where T is the time it took them to run the course.

Let $f(t) = x_1(t) - x_2(t)$.

Then (1) f is continuous on $[0,T]$
     (2) f is differentiable of $(0,T)$,
     (3) $f(0) = f(T) = 0$.

Therefore, there is a number c in $(0,T)$ such that $f'(c) = \frac{f(T)-f(0)}{T-0} = 0$.
Therefore, $x_1'(c) - x_2'(c) = 0$; $x_1'(c) = x_2'(c)$ [speeds are equal.]

35. An equation of the tangent line at c is $y = f(c) + f'(c)(x-c)$.
Let x be any number such that $x \neq c$. Then there is some number $\bar{x}$ between x and c such that $f'(\bar{x}) = \frac{f(x)-f(c)}{x-c}$ or $f(x) = f(c) + f'(\bar{x})(x-c)$.
Now, f concave up $\Rightarrow$ f' is increasing
$$\Rightarrow [x < \bar{x} < c \Rightarrow f'(x) < f'(\bar{x}) < f'(c)]$$
$$\Rightarrow \text{and } [x > \bar{x} > c \Rightarrow f'(x) > f'(\bar{x}) > f'(c)]$$
$$\Rightarrow [f'(\bar{x}) - f'(c)](x-c) > 0$$
$$\Rightarrow f'(\bar{x})(x-c) > f'(c)(x-c)$$
$$\Rightarrow f(c) + f'(\bar{x})(x-c) > f(c) + f'(c)(x-c)$$
$$\Rightarrow f(x) > f(c) + f'(c)(x-c)$$

Problem Set 4.9   Chapter Review

True-False Quiz

1. True.   By the Min-Max Existence Theorem.

3. True.   For example, the cosine function has a critical point at each multiple of π.

5. True. $f''(x) = 90x^4 + 48x^2 + 4 > 0$ on $\mathbb{R}$.

7. True.   Use Monotonicity Theorem.

9.   True.   See Problem 33, Section 4.2.

11. False.   See discussion on pages 184-185 of the text.

13. True. $\lim\limits_{x\to\pm\infty} \dfrac{x^2+1}{1-x^2} = -1$

15. True.   f is continuous on $[0,2]$ and $f'(x) = \dfrac{1}{2\sqrt{x}}$ exists on $(0,2)$.

17. True.   See Problem 17, Section 4.8.

19. True.   At each integer multiple of π.

21. True.   If a,b,c are the x-intercepts, with $a < b < c$, then by Rolle's Theorem, there are at least two points, one in (a,b) and one in (b,c), where there are horizontal tangent lines.

23. False. C^{ex}: $f(x) = x$ and $g(x) = x^3$ are increasing on $\mathbb{R}$, but $(fg)(x) = x^4$ is not increasing on $\mathbb{R}$.

25. True.   See Problem 28 of Section 4.8 for the "if" part. For the "only if" part consider that $[f(b)-f(a)]/(b-a) \geq 0$ with all the conditions given will imply $f'(x) \geq 0$.

27. False. C^{ex}: For the function graphed, f is concave up on $\mathbb{R}$, but f has y=0 as a horizontal asymptote. Choose such a function where $f''(x) > 0$ everywhere.

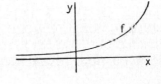

29. True.   If f had two local maxima on an open interval, then it would also have to have a local minimum on the interval. However, $f'(x)$ is quadratic so there are at most two critical points on the open interval.

Sample Test Problems

1. $f'(x) = -2x^{-3}$. Critical points are $-2,-1/2$. $f(-2) = 1/4$ is the minimum value; $f(-1/2) = 4$ is the maximum value.

3. $f(x) = 3x^4 - 4x^3$; $f'(x) = 12x^3 - 12x^2 = 12x^2(x-1)$.

Critical Points: -2 and 3 (end points); 0 and 1 (stationary points).

x	f(x)
-2	80
0	0
1	-1 ← Minimum value is -1.
3	135 ← Maximum value is 135.

5. $f'(x) = 10x^3(x-2)$. Critical points are -1,0,2,3. $f(-1) = 0$, $f(0) = 7$, $f(2) = -9$ (the minimum value), $f(3) = 88$ (the maximum value).

7. $f'(x) = 3(x+1)(x-1)$; $f''(x) = 6x$. f is increasing on $(-\infty,-1]$ and on $[1,\infty)$, and is concave down on $(-\infty,0)$.

9. $f(x) = x^4 - 4x^5$.
$f'(x) = 4x^3 - 20x^4 = 4x^3(1-5x)$.

Therefore, f is increasing on $[0,0.2]$.
$f''(x) = 12x^2 - 80x^3 = 4x^2(3-20x)$.

Therefore, f is concave down on $(0.15, \infty)$.

11. $f(x) = x^2(x-4)$; $f'(x) = x(3x - 8)$ f is increasing on $[-\infty,0]$ and on $[8/3,\infty)$; decreasing on $[0,8/3]$; local minimum value of $f(8/3) \approx -9.48$; local maximum value of $f(0) = 0$.

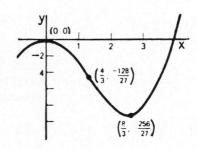

$f''(x) = 2(3x-4)$; $(4/3,-4.74)$ is an inflection point.

13. $f(x) = x(x^3 - 32)$; domain is $\mathbb{R}$; intercepts are at $(0,0)$ and $(3.17,0)$.

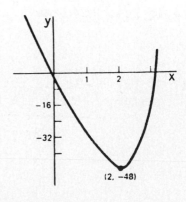

$f'(x) = 4(x-2)(x^2 + 2x + 4)$; decreasing on $(-\infty,2]$; increasing on $[2,\infty)$; global minimum of $(2,-48)$

$f''(x) = 12x^2$; concave up on $\mathbb{R}$.

15. $f(x) = x\sqrt{x-3}$.

Domain: $[3,\infty)$, since the radicand must be greater than or equal to 0.

Symmetry: None.

Intercepts: x-intercept of 3; no y-intercept.

Monotonicity: $f'(x) = \dfrac{3(x-2)}{2(x-3)^{1/2}}$.

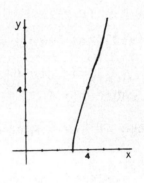

Concavity: $f''(x) = \dfrac{3(x-4)}{4(x-3)^{3/2}}$.

Asymptotes: None.

Summary

(3,4)	Inc and CD
At 4	Inflection
(4,∞)	Inc and CU

x	f(x)	
3	0	(global minimum)
4	4	(inflection point)
5	7.07	

17. $f(x) = x^3(3x-4)$; domain is **R**; intercepts are at $(0,0)$, and $(4/3,0)$.

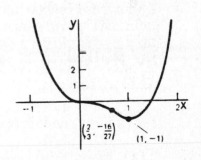

$f'(x) = 12x^2(x-1)$; f is dec. on $(-\infty,1]$; inc. on $[1,\infty)$; global minimum at $(1,-1)$.

$f''(x) = 12x(3x-2)$; f is CU on $(-\infty,0)$ and on $(2/3,\infty)$; f is CD on $(0,2/3)$. Inflection points are $(2/3,-16/27)$ and $(0,0)$.

19.

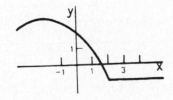

21. Step 1: Let y (inches) be as indicated.

Let A be the cross-sectional area (in^2).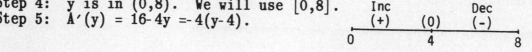

Step 2: Maximize A = (width)(height) = (16-2y)y.

Step 3: $A(y) = (16-2y)y = 16y-2y^2$.

Step 4: y is in (0,8). We will use [0,8].

Step 5: $A'(y) = 16-4y = -4(y-4)$.

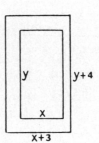

Step 6. A will be maximum where y=4.

Conclusion: 4 inches should be turned up at each side.

23. Minimize A = (x+3)(y+4), if 27 = xy. Then

$A(x) = (x+3)(27x^{-1}+4) = 39+4x+81x^{-1}$ for 0<x.

$A'(x) = 4-81x^{-2} = x^{-2}(2x+9)(2x-9)$. A is minimum where x = 4.5, y = 6.

Use a page of 7.5'' wide by 10'' high.

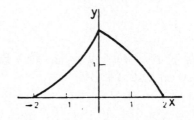

25. $f'(x) = \begin{Bmatrix} (1/2)(x+3), & -2 < x < 0 \\ \text{undefined}, & x = 0 \\ (-1/3)(x+2), & 0 < x < 2 \end{Bmatrix}$.

f is continuous at 0.
Critical points are -2, 0, and 2.
f(-2) = 0, f(0) = 2, f(2) = 0, so
the maximum is 2 and the minimum
is 0.

$f''(x) = \begin{Bmatrix} 1/2, & -2 < x < 0 \\ \text{undefined}, & x = 0 \\ -1/3, & 0 < x < 2 \end{Bmatrix}$.

f is CU on (-2,0) and CD on (0,2).

27.   Let $y = f(x) = x^4 - 6x^3 + 12x^2 - 3x + 1$.

Then $f'(x) = 4x^3 - 18x^2 + 24x - 3$; $f''(x) = 12(x^2 - 3x + 2) = 12(x-1)(x-2)$.

```
    CU              CD              CU
    (+)      (0)    (-)      (0)    (+)
  _____  x
           1               2
```

Thus, inflection points occur when x = 1,2.

The corresponding y-values are f(1) = 5 and f(2) = 11, so the inflection points are (1,5) and (2,11).

The slopes of the respective tangent lines are f'(1) = 7 and f'(2) = 5. Therefore, equations of the tangent lines are:

y-5 = 7(x-1)   and y-11 = 5(x-2).   [point-slope form]

y = 7x-2   and   y = 5x+1.   [slope-intercept form]

29.

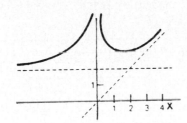

Problem Set 5.1 Antiderivatives (Indefinite Integrals)

1. $4x+C.$

3. $\int (3x^2 + \sqrt{2})\,dx = x^3 + \sqrt{2}x + C.$

5. $(3/5)x^{5/3}+C.$

7. $2x^3 - 3x^2 + C.$

9. $\int (18x^8 - 25x^4 + 3x^2)\,dx = 18\frac{x^9}{9} - 25\frac{x^5}{5} + 3\frac{x^3}{3} + C = 2x^9 - 5x^5 + x^3 + C.$

11. $-x^{-4} + x^{-3} + C.$

13. $2x^2 + 3x + 2x^{-4} + C.$

15. $\int (x^3 + x^{1/2})\,dx = \frac{x^4}{4} + \frac{x^{3/2}}{3/2} + C = \frac{x^4}{4} + \frac{2x^{3/2}}{3} + C.$

17. $(1/5)y^5 + 2y^4 + (16/3)y^3 + C.$

19. $(1/3)x^3 - x^2 - x^{-1} + C.$

21. $\int (3\sin t - 2\cos t)\,dt = 3(-\cos t) - 2(\sin t) + C = -3\cos t - 2\sin t + C.$

23. $\int u^4\,du = (1/5)u^5 + C = (1/5)(3x+1)^5 + C.$ 　　　　　　　[Let $u = 3x+1$]

25. $\int u^7\,du = (1/8)u^8 + C = (1/8)(5x^3 - 18)^8 + C.$ 　　　　[Let $u = 5x^3 - 18$]

27. $\int 3x^4(2x^5+9)^3\,dx = \int (2x^5+9)^3\,\frac{3}{10}\,10x^4\,dx = \int \frac{3}{10}\,u^3\,du$

> Let $u = 2x^5 + 9$
> Then $du = 10x^4\,dx$

$= \frac{3}{10}\,\frac{u^4}{4} + C = \frac{3(2x^5+9)^4}{40} + C.$

29. $\int (1/3)u^6\,du = (1/21)u^7 + C = (1/21)(5x^3 + 3x - 8)^7 + C.$ 　[Let $u = 5x^3 + 3x - 8$]

31. $\int (3/4)u^{1/3}\,du = (9/16)u^{4/3} + C = (9/16)(2t^2 - 11)^{4/3} + C.$ 　　[Let $u = 2t^2 - 11$]

33. $\int \sin^4 x \cos x\,dx = \int u^4\,du = \frac{u^5}{5} + C = \frac{\sin^5 x}{5} + C.$

> Let $u = \sin x$
> Then $du = \cos x\,dx$

35. $\int (1/2)u^5 du = (1/12)u^6 + C = (1/12)\sin^6 x^2 + C.$ [Let $u = \sin x^2$]

37. $\int (x^8 + 3x^6 + 3x^4 + x^2)dx = (1/9)x^9 + (3/7)x^7 + (3/5)x^5 + (1/3)x^3 + C.$

39. $f''(x) = 3x + 1;\ f'(x) = \dfrac{3x^2}{2} + x + C_1;\ f(x) = \dfrac{3}{2}\dfrac{x^3}{3} + \dfrac{x^2}{2} + C_1 x + C_2$

$$= \frac{x^3}{2} + \frac{x^2}{2} + C_1 x + C_2.$$

41. $f'(x) = (2/3)x^{3/2} + C_1;\ f(x) = (4/15)x^{5/2} + C_1 x + C_2.$

43. $f'(x) = (1/2)x^2 - (1/2)x^{-2} + C_1;\ f(x) = (1/6)x^3 + (1/2)x^{-1} + C_1 x + C_2.$

45. To prove this we need to take the derivative with respect to x of the right-hand side of the equation and end up with the integrand of the left-hand side.

$D[f(x)g(x) + C] = f(x)g'(x) + g(x)f'(x) + 0 = f(x)g'(x) + g(x)f'(x).$

47. Let $f(x) = x^2$ and $g(x) = (x-1)^{1/2}$. The integral equals $x^2(x-1)^{1/2} + C.$

49. $(d/dx)[x^{-1}(x^4-1)^{1/2} + C] = x^{-1}(1/2)(x^4-1)^{-1/2}(4x^3) + (x^4-1)^{1/2}(-x^{-2})$

$$= x^{-2}(x^4-1)^{-1/2}[2x^4 - (x^4-1)] = x^{-2}(x^4-1)^{-1/2}(x^4+1).$$

51. $\int f''(x)dx = f'(x) + C = x\left[\dfrac{1}{2\sqrt{x^3+1}}\,3x^2\right] + \sqrt{x^3+1}\ (1) + C$

$$= \frac{3x^3 + 2(x^3+1)}{2\sqrt{x^3+1}} + C = \frac{5x^3+2}{2\sqrt{x^3+1}} + C.$$

53. $(d/dx)[f^m(x)g^n(x) + C] = f^m(x)[ng^{n-1}(x)g'(x)] + g^n(x)[mf^{m-1}(x)f'(x)]$

$$= f^{m-1}(x)g^{n-1}(x)[nf(x)g'(x) + mg(x)f'(x)].$$

55. Note that $|x| = \begin{cases} x, & \text{if } x \geq 0 \\ -x, & \text{if } x < 0 \end{cases}$

$\frac{x^2}{2}$ is an antiderivative of x, and $\frac{-x^2}{2}$ is an antiderivative of -x.

Notice that $\frac{x|x|}{2} = \begin{cases} x^2/2, & \text{if } x \geq 0 \\ -x^2/2, & \text{if } x < 0 \end{cases}$

Thus, $\int |x|\,dx = \frac{x|x|}{2} + C$.

57. (a) $-2\cos(3x-6) + C$.

(b) $-2\sin^2(x/6)\cos(x/6) - 4\cos(x/6) + C$.

(c) $\frac{1}{2}x^2\sin 2x + C$.

Problem Set 5.2   Introduction to Differential Equations

1. $\dfrac{-2x}{2\sqrt{4-x^2}} + \dfrac{x}{\sqrt{4-x^2}} = 0$.

3. $y = C_1\sin x + C_2\cos x$; $\frac{dy}{dx} = C_1\cos x - C_2\sin x$; $\frac{d^2y}{dx^2} = -C_1\sin x - C_2\cos x$.

Therefore, $\frac{d^2y}{dx^2} + y = (-C_1\sin x - C_2\cos x) + (C_1\sin x + C_2\cos x) = 0$.

5. $y = x^3+x+C$. $4 = 1+1+C \Rightarrow C=2$. $y = x^3+x+2$.

7. $2y\,dy = x\,dx$. $y^2 = (1/2)x^2 + C$. $9 = 2+C \Rightarrow C=7$. $y = [(1/2)x^2+7]^{1/2}$.

9. $\frac{dy}{dt} = t^3y^2$; $y=1$ at $t=2$.

$y^{-2}dy = t^3dt$; $\int y^{-2}dy = \int t^3dt$; $-y^{-1} = (1/4)t^4+C$. (general solution)

$y=1$, $t=2 \Rightarrow C=-5$.

$-y^{-1} = \frac{t^4}{4} - 5$ or $y = \frac{-4}{t^4-20}$. (particular solution)

11. $s = t^3 + 2t^2 - t + C.$ $5 = 8 + 8 - 2 + C \Rightarrow C = -9.$ $s = t^3 + 2t^2 - t - 9.$

13. $y = (1/10)(2x+1)^5 + C.$ $6 = 1/10 + C \Rightarrow C = 59/10.$ $y = \dfrac{(2x+1)^5 + 59}{10}.$

15. Slope at (x,y) is $\dfrac{dy}{dx}$, and it is given that this equals $4x$.

$\dfrac{dy}{dx} = 4x \Rightarrow y = 2x^2 + C.$ (general solution)

   $x=1,\ y=2 \Rightarrow C=0.$

Therefore, $y = 2x^2$ is the particular solution.

17. $v(t) = (1/2)t^2 + C.$ $C=2.$ $v(2) = 4.$
$s(t) = (1/6)t^3 + 2t + K.$ $K=0.$ $s(2) = 16/3.$

19. $v(t) = (3/8)(2t+1)^{4/3} + C.$ $C = -3/8.$ $v(2) \approx 2.8312$ cm/sec.

$s(t) = (9/112)(2t+1)^{7/3} - (3/8)t + K.$ $K = 1111/1112.$ $s(2) \approx 12.6049$ cm..

21. $v_0 = 96$ and $s_0 = 0$ [That is, $v=96$ and $s=0$ when $t=0$].
Making use of the results of Example 3, the equations of motion are
$a(t) = -32,\ v(t) = -32t + 96,\ s(t) = -16t^2 + 96t + 0.$

The maximum height is reached when $v=0$.

$0 = -32t + 96 \Rightarrow t=3.$
$s(3) = -16(3)^2 + 96(3) = 144.$ Maximum height reached is 144 feet.

23. $a(t) = -5.28.$ $v(t) = -5.28t + 56.$ $s(t) = -2.64t^2 + 56t + 1000.$

$v(4.5) = 32.24$ ft/sec; $s(4.5) = 1198.54$ ft.

25. $dV/dt = -kS \Rightarrow (d/dt)[(4/3)\pi r^3] = -k(4\pi r^2)$, using formulas for V and S.
$\Rightarrow 4\pi r^2 (dr/dt) = -k(4\pi r^2) \Rightarrow dr/dt = -k \Rightarrow r = -kt + C.$

$r=2$ when $t=0$, so $C=2.$ Therefore, $r = -kt + 2.$

$r=0.5$, when $t=10$, so $k=3/20.$ Therefore, $r = -(3/20)t + 2.$

27. The analysis in Example 5 is valid for escape velocity from any large body. Escape velocity is $\sqrt{2GR}$ where -G is the gravitational acceleration and R is the radius of the large body.

$g \approx 32$ ft/sec^2 = 32/5280 mi/sec^2.

Moon: $\sqrt{2GR} = \sqrt{2[(0.165)(32/5280)](1080)} \approx 1.47$ mi/sec.

Venus: $\sqrt{2GR} = \sqrt{2[(0.85)(32/5280)](3800)} \approx 6.26$ mi/sec.

Jupiter: $\sqrt{2GR} = \sqrt{2[(2.6)(32/5280)](43000)} \approx 36.8$ mi/sec.

Sun: $\sqrt{2GR} = \sqrt{2[(28)(32/5280)](432000)} \approx 383$ mi/sec.

29. $a(t) = k$; $v(t) = kt+45$.

$t = 10/3600$ hr and $v(10/3600) = 60$ mph $\Rightarrow k = 5400$ m/hr^2 = 2.2 ft/sec^2.

31. For t in [0,10]: $a(t) = 6t$; $v(t) = 3t^2$, so $v(10) = 300$.

$$s(t) = t^3, \text{ so } s(10) = 1000.$$

For t > 10: $a(t) = -10$; $v(t) = -10t + C$.   t=10 and v=300 $\Rightarrow$ C=400.

$$s(t) = -5t^2+400t+K. \quad t=10 \text{ and } s=1000 \Rightarrow K=-2500.$$

$$v(t) = 0 \Rightarrow t=40; \quad s(40) = 5500 \text{ m.}$$

33. (a)

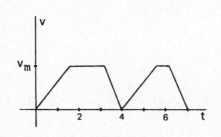

(b) The acceleration and deceleration times are the same between stops C and D as between stops D and E. Thus, since the bus travels at v_m miles per minute while going 0.6 miles farther in 1 more minute between stops C and D than between stops D and E, v_m = 0.6 mi/1min = 0.6 mi/min.

(c) v_m = 0.6. It is also true that v_m = aT where T is the amount of time it takes to accelerate from v=0 to v=v_m. Thus, aT = 0.6.

Now consider the trip between C and D. The total distance traveled is 2 miles.

Thus, 2 = (distance during acceleration) + (distance at 0.6 mi/min)
                                         + (distance during deceleration)
     = 2(distance during acceleration) + (distance at 0.6 mi/min)
     $= 2[(1/2)aT^2] + 0.6[4-2T] = aT^2 + 2.4 - 1.2T$
     = 0.6T + 2.4 - 1.2T [since aT = 0.6]
     = 2.4 - 0.6T

Therefore, T = 2/3.

a(2/3) = 0.6, so a = 0.9 mi/min^2

35. Let V denote volume of water in the tank.

$V = \pi(10/\pi)^2 h = 100h$, so h = V/100.

(a) $\frac{dV}{dt} = k\sqrt{h} = k\sqrt{V/100} = (k/10)V^{1/2}$, V=1600 when t=0   V=0 when t=40.

(b) $10V^{-1/2}dV = kdt$; $20V^{1/2} = kt = C$.

V=1600 when t=0 $\Rightarrow$ C = 800,   so $20V^{1/2} = kt + 800$.

V=0 when t=40 $\Rightarrow$ k = -20, so $20V^{1/2} = -20t + 800$, or $V = (40-t)^2$.

That is $V(t) = (40t)^2$.

(c) $V(10) = (30)^2 + 900$ cc.

37. (a) $v(t) = \begin{cases} -32t & \text{if } 0 < t < 1 \\ -32t + 56 & \text{if } 1 < t < 2.5 \end{cases}$.

(b) $\sqrt{7}/4$ and 7/4.

Problem Set 5.3   Sums and Sigma Notation

1. 2+5+8+11+14 = 40.

3. $\frac{2}{3+1} + \frac{2}{4+1} + \frac{2}{5+1} = \frac{37}{30} \approx 1.2333$.

5. -1+2-4+8-16 = -11.

7. 1+0-3+0+5+0 = 3.

9. $\sum_{i=1}^{98} i$.

11. $\sum_{i=1}^{69} 1/i$.

13. $\displaystyle\sum_{j=1}^{n} a_j.$

15. $\displaystyle\sum_{k=1}^{n} f(c_k).$

17. $2(40) + (50) = 130.$

19. $4(40) - (50) + 2(10) = 130.$

21. $\left[\dfrac{1}{1} - \dfrac{1}{2}\right] + \left[\dfrac{1}{2} - \dfrac{1}{3}\right] + \left[\dfrac{1}{3} - \dfrac{1}{4}\right] + \cdots + \left[\dfrac{1}{40} - \dfrac{1}{41}\right] = \dfrac{1}{1} - \dfrac{1}{41} = \dfrac{40}{41} \approx 0.9756.$

23. $\left[\dfrac{1}{4^2} - \dfrac{1}{3^2}\right] + \left[\dfrac{1}{5^2} - \dfrac{1}{4^2}\right] + \cdots + \left[\dfrac{1}{21^2} - \dfrac{1}{20^2}\right] = \dfrac{1}{21^2} - \dfrac{1}{3^2} = \dfrac{-16}{147} \approx -0.1088.$

25. $3[(1/2)(100)(101)] - 2(100) = 14950.$

27. $\displaystyle\sum_{k=1}^{10} (k^3 - k^2) = \sum_{k=1}^{10} k^3 - \sum_{k=1}^{10} k^2 = \left[\dfrac{10(10+1)}{2}\right]^2 - \dfrac{10(10+1)(20+1)}{6} = 2640.$

29. $2[(1/6)n(n+1)(2n+1)] - 3[(1/2)n(n+1)] + 1(n) = (1/6)n(4n+1)(n-1).$

31. $\displaystyle\sum_{k=1}^{17} (k+2)k.$

33. $\displaystyle\sum_{k=0}^{10} \dfrac{k}{k+1} = \sum_{i=1}^{11} \dfrac{i-1}{i}.$

Let $i = k+1$
Then $k = k-1$
$k = 10 \Rightarrow i = 11$
$k = 0 \Rightarrow i = 1$

35. $\displaystyle\sum_{i=1}^{10} (3i/5)(1/5) = (3/25)[(10)(11)/2] = 33/5 = 6.6.$

37. (a) $\displaystyle\sum_{i=0}^{9} (1/2)(1/2)^i = \dfrac{(1/2) - (1/2)(1/2)^{10}}{1 - (1/2)} = \dfrac{1023}{1024}.$

(b) $\displaystyle\sum_{i=0}^{9} (2)2^i = \dfrac{2 - (2)2^{10}}{1 - 2} = 2046.$

39.
$$S = a \qquad + (a+d) \qquad + (a+2d) \qquad + \cdots + (a+nd)$$
$$S = (a+nd) + [a+(n-1)d] + [a+(n-2)d] + \cdots + a$$
$$\overline{2S = (2a+nd) + (2a+nd) \qquad + (2a+nd) \qquad + \cdots + (2a+nd)} = (n+1)(2a+nd)$$

Then $S = \dfrac{n+1}{2}(2a+nd)$.

41. $\bar{x} = 55/7 \approx 7.8571$; $S^2 \approx 12.4082$.

43. $At^2 + 2Bt + C = t^2 \sum a_i^2 + 2t \sum a_i b_i + \sum b_i^2 = \sum (a_i^2 t^2 + 2a_i b_i t + b_i^2)$

$$= \sum (a_i t + b_i)^2 \geq 0.$$

45.
$$\sum_{i=1}^{n}(a_i - a_{i-1})b_{i-1} = \sum_{i=1}^{n}(a_i b_{i-1} - a_{i-1}b_{i-1}) = \sum_{i=i}^{n} a_i b_{i-1} - \sum_{i=1}^{n} a_{i-1}b_{i-1}$$

$$= \sum_{i=1}^{n} a_i b_{i-1} - \left[\sum_{i=1}^{n} a_i b_i + a_0 b_0 - a_n b_n\right]$$

$$= a_n b_n - a_0 b_0 + \sum_{i=1}^{n} a_i b_{i-1} \sum_{i=1}^{n} a_i b_i = a_n b_n - a_0 b_0 + \sum_{i=1}^{n}(a_i b_{i-1} - a_i b_i)$$

$$= a_n b_n - a_0 b_0 + \sum_{i=1}^{n} a_i(b_{i-1} - b_i) = a_n b_n - a_0 b_0 - \sum_{i=1}^{n} a_i(b_i - b_{i-1}).$$

47. Left: The total number of small squares in a rectangle with n squares in each row and (n+1) squares in each column is n(n+1). If half of them are shaded as in the pattern of the diagram, then the number of shaded squares in n(n+1)/2. And the number of shaded squares is $1 + 2 + \cdots + n$.

Right: Along the bottom row, the number of variously shaded small squares is $1 + 2 + \cdots + n = n(n+1)/2]^2$. Thus, the total number of shaded squares is $[n(n+1)/2]^2$.

On the other hand, the number of small squares in the ith set of shaded squares is 1 for i = 1, and for i > 1 it is

$$[1 + 2 + \cdots + i]^2 - [1 + 2 + \cdots + (i-1)]^2$$
$$= [i(i+1)/2]^2 - [(i-1)i/2]^2 = i^3.$$

Thus, the total number of shaded squares is $1^3 + 2^3 + \cdots + n^3$, so $\sum_{i=1}^{n} i^3 = [n(n+1)/2]^2$.

49. Each successive layer (bottom to top) has one less orange per each dimension of the rectangle. Therefore, the total number of oranges is

$$(10)(16) + (9)(15) + (8)(14) + \cdots + (1)(7) = \sum_{1=1}^{10} i(i+6) = \sum_{1=1}^{10} (i^2+6i)$$

$$= \sum_{1=1}^{10} i^2 + 6 \sum_{1=1}^{10} i = \frac{(10)(11)(21)}{6} + 6 \frac{(10)(11)}{2} = 715; \ 55,625;$$

$$m(m+1)(3n-m+1)/6.$$

Problem Set 5.4 Introduction to Area

1. $A(R_4) = (1)(0.5) + (1.5)(0.5) + (2)(0.5) + (2.5)(0.5) = 3.5.$

3. $A(S_4) = (1.5)(0.5) + (2)(0.5) + (2.5)(0.5) + (3)(0.5) = 4.5.$

5. $A(R_4) = (0.5)(1 + 1.125 + 1.5 + 2.125) = 2.875.$

7. $A(S_3) = (3)(1)+(5)(1)+(7)(1)$
        $= 15.$

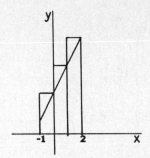

9. $f(x) = x^2+2$ on $[0,2]$. $n=6$. $\Delta x = \dfrac{2-0}{6} = \dfrac{1}{3}$.
    $A(S_6) = [f(1/3)+f(2/3)+f(1)+f(4/3)+f(5/3)+f(2)](1/3)$

        $= [(19/9)+(22/9)+(3)+(34/9)+(43/9)+(6)](1/3)$

        $= (199/9)(1/3)) \approx 7.3704.$

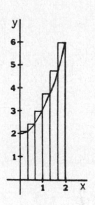

11. $A(S_n) = \sum f(2i/n)(2/n) = \sum(2i/n + 1)\,(2/n) = (4/n^2)\sum i + (2/n)$
        $= (4/n^2)[n(n+1)/2] + 2 = 4 + 2/n.$

    Area $= \lim\limits_{n\to\infty} A(S_n) = 4.$

13. $A(S_n) = \sum[f(1+3i/n)](3/n) = \sum[3(1 + 3i/n) + 1](3/n)$

        $= \sum(12/n + 27i/n^2) = (12/n)n = (27/n^2)\lfloor(1/2)n(n+1)\rfloor$

        $= 12 + (27/2)(1 + 1/n).$

    Area $= \lim\limits_{n\to\infty} A(S_n) = 12 + 27/2 = 25.5.$

15. Let $f(x) = x^3$ on $[0,1]$. Since f is increasing on $[0,1]$, the circumscribed polygon is determined by the union of rectangles whose widths are $\Delta x = (1-0)/n = 1/n$, and whose heights are $f(x_i)$ where x_i is the right end point of the ith subinterval.

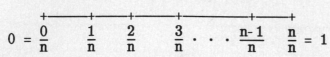

$$0 = \frac{0}{n} \quad \frac{1}{n} \quad \frac{2}{n} \quad \frac{3}{n} \cdots \frac{n-1}{n} \quad \frac{n}{n} = 1$$

[In the following, all summations are from i=1 to i=n.]

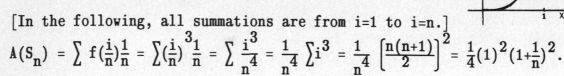

$$A(S_n) = \sum f(\tfrac{i}{n})\tfrac{1}{n} = \sum (\tfrac{i}{n})^3 \tfrac{1}{n} = \sum \frac{i^3}{n^4} = \frac{1}{n^4} \sum i^3 = \frac{1}{n^4}\left[\frac{n(n+1)}{2}\right]^2 = \frac{1}{4}(1)^2(1+\tfrac{1}{n})^2.$$

$$\text{Area} = \lim_{n\to\infty} (S_n) = \tfrac{1}{4}(1)^2(1)^2 = \tfrac{1}{4}.$$

17. The object travels 4 feet, the area under $v = t+1$ over $[0,2]$.

19. (a) $A(S_n) = \sum [f(bi/n)](b/n) = \sum (bi/n)^2(b/n) = (b^3/n^3) \sum i^2$

$$= (b^3/n^3)[(1/6)n(n+1)(2n+1)] = (b^3/6)(1 + (1/n)(2 + 1/n)).$$

Therefore, $A_0^b = \lim_{n\to\infty} A(S_n) = (1/3)b^3$.

(b) $A_a^b = A_0^b = b^3/3 - a^3/3$.

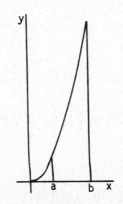

21. (a) $A_0^5 = 5^3/3 = 125/3$.   (b) $A_1^4 = 4^3/3 - 1^3/3 = 21$.

(c) $A_2^5 = 5^3/3 - 2^3/3 = 117/3 = 39$.

---

23. (a) $A_0^2(x^3) = 2^4/4 - 0^4/4 = 4.$ (b) $A_1^2(x^3) = 2^4/4 - 1^4/4 = 3.75.$

(c) $A_1^2(x^5) = 2^6/6 - 1^6/6 = 10.5.$ (d) $A_0^2(x^9) = 2^{10}/10 = 102.4.$

Problem Set 5.5   The Definite Integral

1. $(4)(1.5) + (3)(1) + (-2.25)(1.5) = 5.625.$

3. $\displaystyle\sum_{i=1}^{5} f(\bar{x}_i)\Delta x_i = f(3)(3.75\text{-}3) + f(4)(4.25\text{-}3.75) + f(4.75)(5.5\text{-}4.25) +$

$$+ f(6)(6\text{-}5.5) + f(6.5)(7\text{-}6)$$

$$= (2)(0.75) + (3)(0.5) + (3.75)(1.25) + (5)(0.5) + (5.5)(1)$$

$$= 15.6875.$$

5. $(0.5)(1.28125+1.03125+1.03125+1.28125+1.78125+2.53125) = 4.46875.$

7. $\displaystyle\int_1^3 x^2 dx.$ **9.** $\displaystyle\int_0^3 \frac{x}{1+x} dx.$

11. Partition $[0,4]$ into n subintervals each of length $4/n$, and let the right endpoint be the sample point for each subinterval.

$$\lim_{|P|\to 0}\sum f(\bar{x}_i)\Delta x_i = \lim_{n\to\infty}\sum\left[\frac{32i}{n^2} + \frac{12}{n}\right] = \lim_{n\to\infty}\left[\frac{32}{n}\sum i + 12\right]$$

$$= \lim_{n\to\infty}\left[\frac{32}{n^2}\frac{n(n+1)}{2} + 12\right] = \lim_{n\to\infty}\left[16(1 + 1/n) + 12\right] = 28.$$

13. $\displaystyle\lim_{n\to\infty}\sum\left[(-1 + 3i/n)^2 - 1\right](3/n)$

$$= \lim_{n\to\infty}\left[(-18/n^2)(1/2)(n)(n+1) + (27/n^3)(1/6)(n)(n+1)(2n+1)\right] = 0.$$

15. The integral exists since $f(x) = x^2 - 2x$ is continuous on $[0,4]$. Let P be a regular partition of $[0,4]$ of n subintervals. In each subinterval let the right end point be the sample point for that interval.

$$0 = \frac{0}{n} \quad \frac{4}{n} \quad \frac{8}{n} \quad \frac{12}{n} \cdots \frac{4n-4}{n} \quad \frac{4n}{n} = 4 \qquad \Delta x_i = \frac{4-0}{n} = \frac{4}{n}.$$

(In the following, each summation is from i=1 to i=n; $\bar{x}_i = \frac{4i}{n}$).

Then $\displaystyle \int_0^4 (x^2 - 2x)dx = \lim_{|P| \to 0} \sum f(\bar{x}_i)\Delta x_i = \lim_{n \to \infty} \left[\sum \frac{16i^2}{n^2} - \frac{8i}{n}\right]\frac{4}{n}$

$$= \lim_{n \to \infty} \left[\frac{64}{n^3}\sum i^2 - \frac{32}{n^2}\sum i\right] = \lim_{n \to \infty}\left[\frac{64}{n^3}\frac{n(n+1)(2n+1)}{6} - \frac{32}{n^2}\frac{n(n+1)}{2}\right]$$

$$= \lim_{n \to \infty}\left[\frac{32}{3}(1)(1+\frac{1}{n})(2+\frac{1}{n}) - 16(1)(1+\frac{1}{n})\right] = \frac{32}{3}(1)(1)(2) - 16(1)(1) = \frac{16}{3}.$$

17. Area(I) + Area(II)
    - Area(III) + Area(IV)
    = (1/2)(1)(1) + (1)(2)
    - (1/2)(1)(1) + (1/2)(1)(1)
    = 2.5.

19. Area(I) + Area(II)
    $= (1/4)\pi(2)^2 + (2)(3)$
    $= \pi + 6 \approx 9.1416.$

21. (a) Is, since f is continuous on [-2,2].

    (b) Is, since f is continuous on [-2,2].

    (c) Is not, since f is not bounded on [-2,2].

    (d) Is not, since f is not even defined everywhere on [-2,2]; namely, it is not defined at ±π/2. f is also not bounded on [-2,2].

    (e) Is, since f is bounded on [-2,2] and f has only one discontinuity.

    (f) Is, since f is bounded on [-2,2] and f has only one discontinuity.

23. (a) -3 -2 -1 + 0 + 1 + 2 = -3   (b) 9 + 4 + 1 + 0 + 1 + 4 = 19

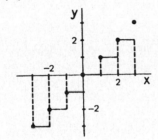   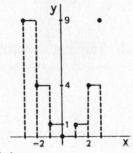

    (c) 6(1/2) = 3.   (d) 6(1^3/3) = 2.

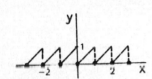   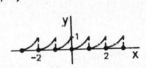

    (e) 2(9/2) = 9.   (f) 0 [By symmetry].

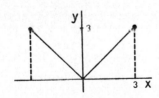   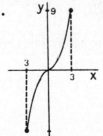

    (g) -1/2 + 0 + 3/2 = 1.   (h) -1^3/3 + 0 + [2^3/3 - 1^3/3] = 2.

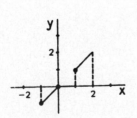   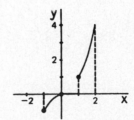

25. $\bar{x}_i$ is the point midway between x_{i-1} and x_i.

$$R_P = (1/2) \sum_{i=1}^{n} (x_i^2 - x_{i-1}^2) = (1/2)(x_n^2 - x_0^2) = (1/2)(b^2 - a^2).$$

Since R_P is a constant, the limit of R_P as $|P| \to 0$ is that constant.

27. 5.24, 6.84, 5.98

29. 0.86375, 0.81778, 0.84182

31. 4

33. 0.6

35. Not integrable

Problem Set 5.6 The Fundamental Theorem of Calculus

1. $\left[(1/4)x^4 \right]_0^2 = 4 - 0 = 4.$

3. $\int_{-1}^{2} (3x^2 - 2x + 3)\,dx = \left[x^3 - x^2 + 3x \right]_{-1}^{2} = (8 - 4 + 6) - (-1 - 1 - 3) = 15.$

5. $\left[-1/w \right]_1^4 = (-1/4) - (-1) = 3/4.$

7. $\left[(2/3)t^{3/2} \right]_0^4 = 16/3 - 0 = 16/3.$

9. $\int_{-4}^{-2} (y^2 + y^{-3})\,dy = \left[\dfrac{y^3}{3} + \dfrac{y^{-2}}{-2} \right]_{-4}^{-2} = \left[\dfrac{-8}{3} - \dfrac{1}{8} \right] - \left[\dfrac{-64}{3} - \dfrac{1}{32} \right] = \dfrac{1783}{96} \approx 18.5729.$

11. $\left[\sin x \right]_0^{\pi/2} = 1 - 0 = 1.$

13. $\left[(2/5)x^5 - x^3 + 5x \right]_0^1 = (2/5 - 1 + 5) - (0) = 22/5 = 4.4.$

15. $\int (x^2+1)^{10}(2x)\,dx = \int u^{10}\,du$

$$= \frac{u^{11}}{11} + C = \frac{(x^2+1)^{11}}{11} + C.$$

Then $\int_0^1 (x^2+1)\,dx = \left[\frac{(x^2+1)^{11}}{11}\right]_0^1 = \frac{2048}{11} - \frac{1}{11} = \frac{2047}{11} \approx 186.0909.$

17. $\int u^{-2}\,du = -1/u + C = -1/(t+2) + C$ \qquad\qquad\qquad [Let u = t+2]

$$\left[-1/(t+2)\right]_{-1}^3 = -1/5 - (-1) = 4/5.$$

19. $\int (1/3)u^{1/2}\,du = (2/9)u^{3/2}+C = (2/9)(3x+1)^{3/2}+C.$ \qquad [Let u = 3x+1]

$$\left[(2/9)(3x+1)^{3/2}\right]_5^8 = 250/9 - 128/9 = 122/9 \approx 13.5556.$$

21. $\int \sqrt{7+2t^2}\,(8t)\,dt = \int 2(7+2t^2)^{1/2}4t\,dt$

$$= \int 2u^{1/2}\,du = \frac{4u^{3/2}}{3} + C = \frac{4(7+2t^2)^{3/2}}{3} + C.$$

Then $\int_{-3}^3 \sqrt{7+2t^2}\,(8t)\,dt = \left[\frac{4(7+2t^2)^{3/2}}{3}\right]_{-3}^3 = \frac{500}{3} - \frac{500}{3} = 0.$

23. $\int -u^2\,du = (-1/3)u^3+C = (-1/3)\cos^3 x+C.$ \qquad\qquad [Let u = cosx]

$$\left[(-1/3)\cos^3 x\right]_0^{\pi/2} = 0 - (-1/3) = 1/3.$$

25. $\left[x^2 - \cos x\right]_0^{\pi/2} = (\pi^2/4) - (-1) = \pi^2/4 + 1 \approx 3.4674.$

27. $\int (2x+1)^{1/2} dx = \int \frac{1}{2}(2x+1)^{1/2} \, 2dx$

$$= \int \frac{1}{2} u^{1/2} du = \frac{1}{2} \frac{2u^{3/2}}{3} + C = \frac{(2x+1)^{3/2}}{3} + C.$$

(We will use this for the second term.)

Then $\int_0^4 [x^{1/2} + (2x+1)^{1/2}] \, dx = \int_0^4 x^{1/2} dx + \int_0^4 (2x+1)^{1/2} dx$

$$= \left[\frac{2x^{3/2}}{3}\right]_0^4 + \left[\frac{(2x+1)^{3/2}}{3}\right]_0^4 = (\frac{16}{3} - 0) + (9 - \frac{1}{3}) = 14.$$

29. $\left[(1/5)x^5 + x^4 + (4/3)x^3\right]_0^1 = (1/5 + 1 + 4/3) - (0) = 38/15 \approx 2.5333.$

31. $\int_0^3 x^2 dx = \left[(1/3)x^3\right]_0^3 = 9\text{-}0 = 9.$ **32.** $\int_0^2 x^3 dx = \left[(1/4)x^4\right]_0^2 = 4\text{-}0 = 4.$

33. $\int_0^\pi \sin x \, dx = \left[-\cos x\right]_0^\pi = 1 - (-1) = 2.$

35. Using n subintervals of equal length and right endpoints as sample points, the sum is a Riemann sum for the integral.

For n = 10, the summation equals $(1/1000)[(1/6)(10)(11)(21)] = 0.385.$

$$\int_0^1 x^2 dx = \left[(1/3)x^3\right]_0^1 = 1/3 - 0 = 1/3 \approx 0.333.$$

To three decimal places, the difference is 0.052.

37. $4 - (-2) = 6.$

39. $\int_0^1 [2g(x) - 3f(x)] dx = 2 \int_0^1 g(x) dx - 3 \int_0^1 f(x) dx = 2(-2) - 3(4) = -16.$

41. $\int_{-2}^{4} (2[x] - 3|x|) = 2\int_{-2}^{4} [x]dx - 3\int_{-2}^{4} |x|dx = 2[-2-1+0+1+2+3] - 3[2+8] = -24.$

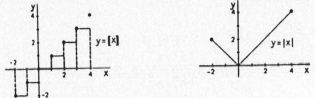

43. $\int_{0}^{b} [x]dx = \int_{0}^{[b]} [x]dx + \int_{[b]}^{b} [x]dx$

$= [0+1+2+\cdots+([b]-1)] + [b](b-[b]) = \frac{1}{2}([b]-1)([b]) + [b](b-[b])$

$= \frac{[b]}{2}(2b-1 - [b])$

Problem Set 5.7 More Properties of the Definite Integral

1. $2+3 = 5.$

3. $\int_{1}^{2} g(x)\, dx$ [Think of going from x=1 to x=2 by going first from x=1 to
x=0, and then going from x=0 to x=2.]

$= \int_{1}^{0} g(x)dx + \int_{1}^{2} g(x)dx = -\int_{0}^{1} g(x)dx + (4) = -(-1) + (4) = 5.$

5. $-[3(3)] = -9.$ **7.** $2(5)-3(4) - -2.$ [See Problem 1.]

9. $\int_{1}^{2} f(x)dx + \int_{2}^{0} f(x)dx = \int_{1}^{0} f(x)dx = -\int_{0}^{1} f(x)dx = -2.$

11. $2x+1.$ **13.** $\sqrt{1+x^4}.$

15. $G(x) = -\int_{\pi/4}^{x} u\, \tan u\, du,$ so $G'(x) = -x\, \tan x.$

17. $[\sin^2 x + \cos(\sin x)]\cos x.$

19. $-(1+x^4)^{1/2}(1) + [1+(x^3)^4]^{1/2}(3x^2) = 3x^2(1+x^{12})^{1/2} - (1+x^4)^{1/2}.$

21. $f'(x) = \dfrac{x}{\sqrt{a^2+x^2}}$. [Theorem D]

$$f''(x) = \frac{\sqrt{a^2+x^2}\,(1) - (x)\,\dfrac{1}{2\sqrt{a^2+x^2}}\,(2x)}{a^2+x^2} = \frac{(a^2+x^2) - x^2}{(a^2+x^2)^{3/2}}$$

$$= \frac{a^2}{(a^2+x^2)^{3/2}} > 0.$$

Therefore, f is concave up on **R**.

23. $\left[(1/3)x^3\right]_0^2 + \left[(1/2)x^2\right]_2^4 = (8/3 - 0) + (8-2) = 26/3.$

25. $\displaystyle\int_0^2 -(x-2)\,dx + \int_2^4 (x-2)\,dx = \left[(-1/2)x^2+2x\right]_0^2 + \left[(1/2)x^2-2x\right]_2^4 = 4.$

27. $\displaystyle\int_0^1 1\,dx = \left[x\right]_0^1 = 1-0 = 1$ and $\displaystyle\int_0^1 (1+x^4)\,dx = \left[x + \frac{x^5}{5}\right]_0^1 = (1+\frac{1}{5}) - 0 = \frac{6}{5}.$

Therefore, $1 \le \displaystyle\int_0^1 \sqrt{1+x^4}\,dx \le \frac{6}{5}.$

29. $\dfrac{\left[x^4\right]_1^3}{3-1} = \dfrac{81-1}{2} = 40.$

31. $\displaystyle\int_{-1}^0 (2-x)\,dx + \int_0^3 (2+x)\,dx = 13.$ Therefore, the average is 13/4.

33. ≈ 4 **35.** ≈ 3

37. True. By the Comparison Property, since $\displaystyle\int_a^b (0)\,dx = 0.$

That is, $f(x) \ge 0 \Rightarrow \displaystyle\int_a^b f(x)\,dx \ge \int_a^b (0)\,dx = 0.$

39. False. For $f(x) = x$, the integral from -1 to 1 is 0, but $f(1) \neq 0$.

41. True.   Subtract the integral on the right from each side and use linearity of the definite integral.

43. $\dfrac{\int_a^b v(t)\,dt}{b-a} = \dfrac{\left[s(t)\right]_a^b}{b-a} = \dfrac{s(b)-s(a)}{b-a}.$

45. Since f is continuous on $[a,c]$, it is continuous on $[a,b]$ and on $[b,c]$, so f is integrable on all three intervals.

To prove that $\displaystyle\int_a^c = \int_a^b + \int_b^c$, we will show that $\displaystyle\int_a^c - \int_a^b - \int_b^c = 0.$

$$\left| \int_a^c - \int_a^b - \int_b^c \right| = \left| \int_a^c - \int_a^b - \int_b^c - \sum_1^n + \sum_1^m + \sum_{m+1}^n \right|$$

$$= \left| \left(\int_a^c - \sum_1^n \right) - \left(\int_a^b - \sum_1^m \right) - \left(\int_b^c - \sum_{m+1}^n \right) \right|$$

$$\leq \left| \left(\int_a^c - \sum_1^n \right) \right| + \left| \left(\int_a^b - \sum_1^m \right) \right| + \left| \left(\int_b^c - \sum_{m+1}^n \right) \right| \leq \frac{\epsilon}{3} + \frac{\epsilon}{3} + \frac{\epsilon}{3} = \epsilon$$

if $|P|$ is sufficiently small. Thus, the constant $\left| \int_a^c - \int_a^b - \int_b^c \right|$

is less than every positive number ϵ, so it must be 0.

47. Since f is continuous on $[a,b]$, f attains extreme values on $[a,b]$. Let x_1 and x_2 be elements of $[a,b]$ such that $f(x_1)$ and $f(x_2)$ are the respective minimum and maximum values of f on $[a,b]$.

Thus, $f(x_1) \leq f(x) \leq f(x_2)$ on $[a,b]$.

Therefore, $f(x_1)(b-a) \le \int_a^b f(x)dx \le f(x_2)(b-a)$.   [Theorem C]

$$f(x_1) \le \frac{\int_a^b f(x)dx}{b-a}. \le f(x_2).$$

Therefore, there is number c between x_1 and x_2 such that

$$f(c) \frac{\int_a^b f(x)dx}{b-a} .   \text{[Intermediate Value Theorem]}$$

Therefore, $\int_a^b f(x)dx = f(c)(b-a)$ for some c in (a,b).

49. -0.046593

Problem Set 5.8 Aids in Evaluating Definite Intgrals

1. $\int (1/3)u^{1/2}du = (2/9)u^{3/2}+C = (2/9)(3x+2)^{3/2}+C.$ [Let u = 3x+2]

3. $\int \cos(3x+2) \ dx = \int \frac{1}{3} \cos(3x+2) \ 3dx$

> Let   u = 3x+2
> Then du = 3dx

$$= \int \frac{1}{3} \cos u \ du = \frac{1}{3} \sin u + C = \frac{1}{3} \sin(3x+2) + C.$$

5. $\int (1/2)u^{1/2}du = (1/3)u^{3/2}+C = (1/3)(x^2+4)^{3/2}+C.$ [Let u = x^2+4]

7. $\int (1/2)\sin u \ du = (-1/2)\cos u + C = (-1/2)\cos(x^2+4) + C.$ [Let u = x^2+4]

9. $\int \frac{x \sin \sqrt{x^2+4}}{\sqrt{x^2+4}} \ dx$

> Let   u = $\sqrt{x^2+4}$
> Then du = $\frac{1}{2\sqrt{x^2+4}}$ 2xdx
> $= \frac{x}{\sqrt{x^2+4}}$

$$= \int \sin \sqrt{x^2+4} \ \frac{x}{\sqrt{x^2+4}}dx$$

$$= \int \sin u \ du = -\cos u + C = -\cos \sqrt{x^2+4} + C.$$

11. $\int (1/2)u^{1/2}du = (1/3)u^{3/2}+C = (1/3)\sin^{3/2}(x^2+4)+C.$ [Let u = $\sin(x^2+4)$]

13. $\int 2u^3 du = (1/2)u^4 + C = (1/2)(\sqrt{t}+4)^4 + C$ [Let $u = \sqrt{t}+4$]

15. $\int_0^1 (3x+1)^3 dx$

$$\boxed{\begin{array}{l} \text{Let} \quad u = 3x+1 \\ \text{Then } du = 3dx \\ \quad x=1 \Rightarrow \quad u=4 \\ \quad x=0 \Rightarrow \quad u=1 \end{array}}$$

$= \int_0^1 \frac{1}{3}(3x+1)^3\; 3dx = \int_1^4 \frac{1}{3} u^3 du = \left[\frac{1}{3}\frac{u^4}{4}\right]_1^4 = \frac{64}{3} - \frac{1}{12} = 21.25.$

17. $\int_9^{13} (1/2)u^{-2} du = \left[-1/2u\right]_9^{13} = -1/26 + 1/18 \approx 0.0171.$ [Let $u = t^2+9$]

19. $\int_1^6 (1/2)u^{-2} du = \left[-1/2u\right]_1^6 = -1/12 + 1/2 = 5/12.$ [Let $u = x^2+4x+1$]

21. $\int_0^{\pi/6} \sin^3\theta \cos\theta \; d\theta$

$$\boxed{\begin{array}{l} \text{Let} \quad u = \sin\theta \\ \text{Then} \quad du = \cos\theta \; d\theta \\ \quad \theta=\pi/6 \Rightarrow \quad u = \sin(\pi/6) = 1/2 \\ \quad \theta=0 \quad\Rightarrow \quad u = \sin(0) = 0 \end{array}}$$

$= \int_0^{1/2} u^3 du = \left[\frac{u^4}{4}\right]_0^{1/2} = \frac{1}{64} = 0.015625.$

23. $\int_{-3}^0 (1/3)\cos u \; du = \left[(\sin u)/3\right]_{-3}^0 = (\sin{-3})/3 \approx -0.0470.$ [Let $u = 3x-3$]

25. $\int_0^\pi (1/2\pi)\sin u \; du = 1/\pi \approx 0.3183.$ (See Problem 24) [Let $u = \pi x^2$]

27. $\int_0^{\pi/2} \sin x \, \sin(\cos x) \; dx$

$$\boxed{\begin{array}{l} \text{Let} \quad u = \cos x \\ \text{Then } du = -\sin x \; dx \\ \quad x=\pi/2 \Rightarrow \quad u = \cos(\pi/2) = 0 \\ \quad x=0 \quad\Rightarrow \quad u = \cos(0) = 1 \end{array}}$$

$= \int_0^{\pi/2} -\sin(\cos x)\;(-\sin x)dx$

$= \int_1^0 -\sin u \; du = \left[\cos u\right]_1^0 = \cos(0) - \cos(1) = 1 - \cos(1) \approx 0.4597.$

29. $\int_2^3 2u^{-3}du = \left[-1/u^2\right]_2^3 = -1/9 + 1/4 = 5/36 \approx 0.1389.$ $\qquad$ [Let $u = \sqrt{t}+1$]

31. $0 + 2\int_0^\pi \cos x\, dx = 2\left[\sin x\right]_0^\pi = 0.$

33. $\int_{-\pi/2}^{\pi/2} \frac{\sin x}{1+\cos x}\, dx = 0$, since $f(x) = \frac{\sin x}{1+\cos x}$ is an odd function.

35. $\int_{-\pi}^\pi (\sin^2 x + 2\sin x\cos x + \cos^2 x)dx = \int_{-\pi}^\pi (1 + \sin 2x)dx = 2\int_0^\pi 1dx + 0 = 2\pi.$

37. $2\int_0^1 x^3 dx + 0 = 2\left[(1/4)x^4\right]_0^1 = 2(1/4 - 0) = 1/2.$

39. f is even: $\displaystyle\int_{-b}^{-a} f(x)dx = \int_{-b}^{-a} f(-x)dx$

$$\begin{array}{ll} \text{Let} & u = -x \\ \text{Then} & du = -dx \\ & x = -a \;\Rightarrow\; u = a \\ & x = -b \;\Rightarrow\; u = b \end{array}$$

$\qquad\qquad = \displaystyle\int_b^a - f(u)du$

$\qquad\qquad = \displaystyle\int_a^b f(u)du = \int_a^b f(x)dx$ [changing the dummy variable].

$\qquad$ f is odd: $\displaystyle\int_{-b}^{-a} f(x)dx = \int_{-b}^{-a} -f(-x)dx$

$\qquad\qquad = \displaystyle\int_b^a f(u)du$

Same substitution as above for f even

$\qquad\qquad = -\displaystyle\int_a^b f(u)du = -\int_a^b f(x)dx$ [changing the dummy variable].

41. $4\int_0^\pi |\cos x|dx = 4\int_0^{\pi/2}\cos x\,dx + \int_0^{\pi/2}-\cos x\,dx = 4[(1-0) + (0+1)] = 8.$

43. $\int_1^{1+\pi}|\sin x|dx$

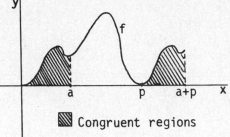

 Congruent regions

$= \int_0^\pi |\sin x|dx = \int_0^\pi \sin x\,dx = 2.$

(See Problem 42.)

45. Average temperature for the 6:00 a.m. to 6:00 p.m. period is

$$\frac{\int_6^{18}T(t)dt}{18-6} = \frac{\int_6^{18}(70 + 8\sin[\frac{\pi}{12}(t-9)])dt}{12}$$

$$= \frac{1}{12}\left[70t - \frac{96}{\pi}\cos\left[\frac{\pi}{12}(t-9)\right]\right]_6^{18}$$

$$= \frac{1}{12}[70(18) - \frac{96}{\pi}\cos(\frac{3\pi}{4})] - [70(6) - \frac{96}{\pi}\cos(\frac{-\pi}{4})]$$

$$= \frac{1}{12}(1260 + \frac{96}{\pi}\frac{\sqrt{2}}{2} - 420 + \frac{96}{\pi}\frac{\sqrt{2}}{2}) = 70 + \frac{8\sqrt{2}}{\pi} \approx 73.6$$

47. $y = k\cos kx.$ $y' = -k^2\sin kx.$ $y'(\pi/2k) = -k^2$ is the slope of the curve at $x = \pi/2k$, and an equation of the tangent there is

$y - 0 = -k^2(x - \pi/2k)$ or $y - -k^2x + k\pi/2.$

Area of shaded region is $2\left[\frac{1}{2}(\frac{\pi}{2k})(\frac{\pi k}{2}) - \int_0^{\pi/2k}k\cos kx\,dx\right]$

$$= 2\left[\frac{\pi^2}{8} - \left[\sin kx\right]_0^{\pi/2k}\right] = \frac{\pi^2-8}{4} \approx 0.4674$$

49. (a) even   (b) 2π
     (c) 0.4597, 0.9194, -0.4597, -0.9194, 0, 0, -0, -0.44484, -0.44484.

Problem Set 5.9   Chapter Review

True-False Quiz

1. True.   See Theorem B, Section 5.1.

3. True. $\dfrac{dy}{dx} = -\sin x$, and $(-\sin x)^2 = 1 - (\cos x)^2$ or $\sin^2 x = 1 - \cos^2 x$.

5. True. $s=0$ when $t=0$ or $t = v_0/16$. $v(t) = -32 + v_0$; $v(v_0/16) = -v_0$.

7. True. $\sum(2i-1) = 2\sum i - \sum 1 = 2\,\dfrac{(100)(101)}{2} - 100 = 10{,}000.$

9. False.   See Problem 22 of Section 5.5.

11. False.   See write-up of Problem 39, Section 5.7.

13. True.   If all the limits in question are real numbers.

15. True.   By the Interval Additive Property (Theorem A of Section 5.7) since $f(x) = \sin^2 x$ is integrable on $[1,7]$.

17. False.   It is that times $2x$.

19. True.   Each equals 4. Can use periodicity and/or symmetry.

21. True.   By the Fundamental Theorem of Calculus. F and G are continuous on $[a,b]$, and both F and G are antiderivatives of $F' = G'$.

23. False.   For $F(x) = x+1$ on $[0,1]$, $f(x) = 1$. However, $F(1) = 2$ but the integral of $f(x)$ from 0 to 1 is 1.

25. False. $-x < x^2$ on $[0,1]$, but 1 and 1/3 are the values of the respective integrals.

27. True.   This is a generalization of the Triangle Inequality. See property 3 of absolute values on Page 15. Could prove it using the Principle of Mathematical Induction.

29. True.   The left-hand side involves Riemann sums for regular partitions of $[0,2]$ with right endpoints as sample points.

Sample Test Problems

1. $\left[(1/4)x^4 - x^3 + 2x^{3/2}\right]_0^1 = 5/4.$

3. $\int y\sqrt{y^2-4} = \int y(y^2-4)^{1/2}dy = \int \frac{1}{2}(y^2-4)^{1/2}2ydy$

$$\boxed{\begin{array}{l} \text{Let} \quad u = y^2-4 \\ \text{Then } du = 2ydy \end{array}}$$

$$= \int \frac{1}{2}u^{1/2}du = \frac{1}{2}\frac{2u^{3/2}}{3} + C = \frac{(y^2-4)^{3/2}}{3} + C.$$

5. $\int_9^{25} (1/4)u^{-1/2}du = \left[(1/2)u^{1/2}\right]_9^{25} = 5/2 - 3/2 = 1.$ \qquad [Let $u = t^4+9$]

7. $y = 2(x+1)^{1/2} + C.$ \quad $18 = 4+C \Rightarrow C=14.$ \quad $y = 2(x+1)^{1/2}+14.$

9. $y^{-4}dy = t^2dt;$ \quad $\int y^{-4}dy = \int t^2dt;$ \quad $\frac{y^{-3}}{-3} = \frac{t^3}{3} + C.$ \quad (general solution)

$$y = 1, \ t=1 \Rightarrow C = -2/3.$$

$$\frac{y^{-3}}{-3} = \frac{t^3}{3} - \frac{2}{3} \text{ or } y = (2-t^3)^{-1/3}. \quad \text{(particular solution)}$$

11. Slope of curve $xy = 2$ at (x,y) is $-2x^{-2}$. Therefore, the family of curves satisfying the condition, $dy/dx = -(-2x^{-2})^{-1} = (1/2)x^2$, are of the form $y = (1/6)x^3 + C.$ \quad $-1/3 = -4/3 + C \Rightarrow C=1.$ \quad $y = (1/6)x^3 + 1.$

13. $a(t) - -32.$ \quad $v(t) - -32t+48.$ \quad $s(t) = -16t^2 + 48t + 448 = 16(t-7)(t+4).$

$s(t) = 0 \Rightarrow t = 7.$ \quad $v(7) = -176.$ The ball will strike the ground in 7 seconds with a velocity of -176 ft/sec.

15. $\sum_{i=1}^{4} f(x_i)\Delta x_i = [f(0.5) + f(1) + f(1.5) + f(2)](0.5)$

$$= [(-0.75) + (0) + (1.25) + (3)](0.5)$$

$$= 1.75.$$

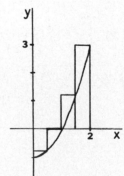

17. $\int_1^4 (2-u^{1/2})^2 du = \int_1^4 (4-4u^{1/2}+u)du = \left[4u-(8/3)u^{3/2}+(1/2)u^2\right]_1^4$

$$= (16 - 64/3 + 8) - (4 - 8/3 + 1/2) = 5/6.$$

19. $\left[5x+x^{-1}\right]_2^4 = (20 + 1/4) - (10 + 1/2) = 39/4 = 9.75.$

21. $\displaystyle\sum_{i=1}^{10} (6i^2-8i) = 6\sum_{i=1}^{10} i^2 - 8\sum_{i=1}^{10} i = \frac{6(10)(10+1)(20+1)}{6} - \frac{8(10)(10+1)}{2} = 1870.$

23. (a) $\displaystyle\sum_{k=2}^{78} (1/k).$ 　　　　 (b) $\displaystyle\sum_{i=1}^{50} ix^{2i}.$

25. (a) $\int_1^0 f(x)dx + \int_0^2 f(x)dx = -(4)+ 2 = -2.$

(b) $-(4) = -4.$

(c) $3(2) = 6.$

(d) $2(-3) - 3(2) = -12.$

(e) $\int_0^2 -f(u)du = -2.$ 　[Let $u = -x$]

27. (a) $\int_{-2}^2 f(x)dx = 2\int_0^2 f(x)dx$ [since f is an even function]

$$= 2(-4) = -8.$$

(b) $\int_{-2}^2 |f(x)|dx = \int_{-2}^2 -f(x)dx$ [since $f(x) \le 0$]

$$= -(-8) = 8. \quad [\text{from (a)}]$$

(c) $\int_{-2}^{2} g(x)dx = 0.$ [since g is an odd function]

(d) $\int_{-2}^{2} [f(x) + f(-x)]dx = \int_{-2}^{2} [f(x) + f(x)]dx = \int_{-2}^{2} 2f(x)dx$

$$= 2\int_{-2}^{2} f(x)dx = 2(-8) \text{ [from (a)]}$$

$$= -16.$$

(e) $\int_{0}^{2} [2g(x) + 3f(x)]dx = 2\int_{0}^{2} g(x)dx + 3\int_{0}^{2} f(x)dx = 2(5) + 3(-4) = -2.$

(f) $\int_{-2}^{0} g(x)dx = \int_{-2}^{0} -g(x)(-1)dx$

$$\boxed{\begin{array}{l} \text{Let} \quad u = -x \\ \text{Then } du = -dx \\ \qquad x=0 \;\Rightarrow\; u=0 \\ \qquad x=-2 \;\Rightarrow\; u=2 \end{array}}$$

$$= \int_{2}^{0} -g(-u)du = \int_{0}^{2} g(-u)du = \int_{0}^{2} -g(u)du = -5.$$

29. $\int_{-1}^{-1} 3x^2 dx = 3c^2 [-1 - (-4)].$

$\left[x^3\right]_{-4}^{-1} = 9c^2.$

$-1 - (-64) = 9c^2$, so $c^2 = 7$.
$c = -\sqrt{7}$ is the value in $[-4,-1].$

31. (a) $\int_{0}^{4} x^{1/2}dx = \left[(2/3)x^{3/2}\right]_{0}^{4} = 16/3 - 0 = 16/3.$

(b) $\int_{1}^{3} x^2 dx = \left[(1/3)x^3\right]_{1}^{3} = 9 - 1/3 = 26/3.$

Problem Set 6.1   The Area of a Plane Region

1. $\int_{-1}^{2} (x^2+1)\,dx.$

3. Slice:

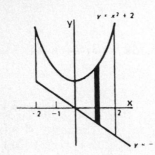

Approximate: $\Delta A \approx [(x^2+2)-(-x)]\Delta x.$

Integrate:

$$A = \int_{-2}^{2} [(x^2+2)-(-x)]\,dx = \int_{-2}^{2} (x^2+x+2)\,dx.$$

5. $\int_{-2}^{1} [(2-x^2)-(x)]\,dx.$

7. $\int_{-2}^{0} (x^3-x^2-6x)\,dx + \int_{0}^{3} -(x^3-x^2-6x)\,dx.$

9. Slice: (Below right)

Approximate: $\Delta A \approx [(x\text{-value at right}) - (x\text{-value at left})](\text{width})$

$$= [(3-y^2)-(y+1)\Delta y.$$

Integrate:

$$A = \int_{-2}^{1} [(3-y^2)-(y+1)]\,dy = \int_{-2}^{1} (-y^2-y+2)\,dy.$$

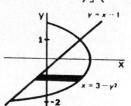

11. $\Delta A \approx = [4-(1/3)x^2]\Delta x.$

$$A = \int_{0}^{3} [4-(1/3)x^2]\,dx = \left[4x-(1/9)x^3\right]_{0}^{3} = 9.$$

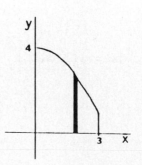

13. $\Delta A \approx [0 - (x^2 - 2x - 3)]\Delta x.$

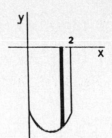

$$A = \int_0^2 (-x^2 + 2x + 3)\Delta x = \left[(-1/3)x^3 + x^2 + 3x\right]_0^2 = 22/3.$$

15. $\Delta A \approx (0 - x^3)\Delta x, \ x \text{ in } [-1, 0].$

$\Delta A \approx x^3 \Delta x, \ x \text{ in } [0, 2].$

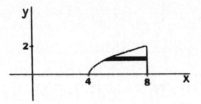

$$A = \int_{-1}^0 -x^3 dx + \int_0^2 x^3 dx$$

$$= \left[\frac{-x^4}{4}\right]_{-1}^0 + \left[\frac{x^4}{4}\right]_0^2 = (0 - \frac{-1}{4}) + (4 - 0) = 4.25.$$

17. $\Delta A \approx [8 - (y^2 + 4)]\Delta y, \ y \text{ in } [0, 2].$

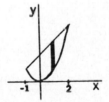

$$A = \int_0^2 (-y^2 + 4)]dy = 16/3.$$

19. $\Delta A \approx [(x+2) - x^2]\Delta x.$

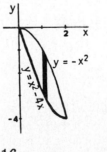

$$A = \int_{-1}^2 (-x^2 + x + 2)dx = 9/2.$$

21. Solve $y = x^2 - 4x$ and $y = -x^2$ simultaneously to find that the curves intersect where $x = 0$ and $x = 2$.

$\Delta A \approx [(-x^2) - (x^2 - 4x)]\Delta x.$

$$A = \int_0^2 [(-x^2) - (x^2 - 4x)]dx$$

$$= \int_0^2 (-2x^2 + 4x)dx = \left[\frac{-2x^3}{3} + 2x^2\right]_0^2 = (\frac{-16}{3} + 8) - 0 = \frac{8}{3}.$$

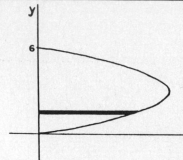

23. $\Delta A \approx (6y - y^2)\Delta y$.

$A \int_0^6 (6y - y^2)dy = 36$.

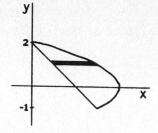

25. $\Delta A \approx [(4 - y^2) - (-y + 2)]\Delta y$.

$A = \int_{-1}^2 (-y^2 + y + 2)dy = 9/2$.

(See Problem 24.)

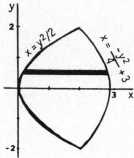

27. Solve $y^2 - 2x = 0$ and $y^2 + 4x - 12 = 0$ simultaneously to find that the curves intersect where $y = -2$ and $y = 2$.

$\Delta A \approx \left[\left[\dfrac{-y^2}{4} + 3\right] - \left[\dfrac{y^2}{2}\right]\right]\Delta y$.

$A = \int_{-2}^2 \left[\dfrac{-3y^2}{4} + 3\right]dy$

$= 2\int_0^2 \left[\dfrac{-3y^2}{4} + 3\right]dy$ (since the integrand is an even function)

$2\left[\dfrac{-y^3}{4} + 3y\right]_0^2 = 2[(-2 + 6) - 0] = 8$.

29. $\Delta A \approx [(x + 6) - (-1/2)x]\Delta x$, x in $[-4, 0]$.

$\Delta A \approx [(x + 6) - (x^3)]\Delta x$, x in $[0, 2]$.

$A = \int_{-4}^0 [(3/2)x + 6]dx + \int_0^2 (-x^3 + x + 6)dx$

$= 12 + 10 = 22$.

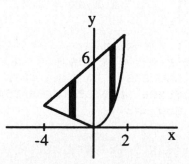

31. $v(t) = 3(t-6)(t-2)$

Auxiliary axis for $v(t)$:

```
        (+)    (0)    (-)    (0)    (+)
   ┣━━━━━━━━━━━━━━━━━━━━━━━━━━━━━━━━┫ x
   -1      2             6       9
```

$\text{Displacement} = \int_{-1}^{9} v(t)\, dt = 3\int_{-1}^{9}(t^2-8t+12)dt = 3\left[t^3/3 - 4t^2 + 12t\right]_{-1}^{9}$

$= 3[(27) - (-49/3)] = 130.$

$\text{Distance traveled} = \int_{-1}^{9}|v(t)|t$

$= \int_{-1}^{2}(t^2-8t+12)dt + \int_{2}^{6} -(t^2-8t+12)dt + \int_{6}^{9}(t^2-8t+12)dt$

$= 3[(32/3)-(-49/3)] + 3[(0)-(-32/3)] + 3[(27)-(0)] = 194.$

33. 1st Question: Let z denote the time it will take to travel to s=12.

Solve $12 = \int_{0}^{z} v(t)dt$ for z.

$12 = \int_{0}^{z}(2t-4)dt = \left[t^2-4t\right]_{0}^{z} = z^2-4z.$

$z^2-4z-12 = 0; \quad (z-6)(z+2) = 0; \quad z = 6.$

That is, it will take 6 seconds to reach s=12.

2nd Question: Now let z denote the time it takes to travel 12 cm.

Auxiliary axis for $v(t)$:

```
        (-)    (0)    (+)
   ┣━━━━━━━━━━━━━━━━━━━━┫ x
   0      2
```

During the first two seconds the object will travel to the left a total distance of

$\int_{0}^{2}|2t-4|dt - \int_{0}^{2} -(2t-4)dt - \left[-t^2+4t\right]_{0}^{2} = 4 \text{ cm}.$

Therefore, it takes more than 2 seconds to travel 12 cm.

Solve $12 = \int_0^2 |v(t)|\,dt$ for z.

$$12 = \int_0^z |2t-4|\,dt = \int_0^2 -(2t-4)\,dt + \int_2^z (2t-4)\,dt$$

$$= 4 + \left[t^2-4t\right]_2^z = 4 + [z^2-4z+4]$$

$z^2-4z-4 = 0$. Now use the quadratic formula to obtain $z = 2 + 2\sqrt{2} \approx 4.8284$

That is, it will take about 4.8284 seconds to travel a total of 12 cm.

35. The equations of the lines are $y = x+6$ and $y = -x+6$.

$A = (1/2)(6)(3) = 9.$

$$B = C = \int_0^2 [(x+6) - (-x+6)]\,dx + \int_2^3 [(x+6) - (x^2)]\,dx = 37/6.$$

$$D = 2\int_0^2 [(-x+6) - (x^2)]\,dx = 44/3.$$

Check: $A+B+C+D = 2\left[(3)(9) - \int_0^3 x^2\,dx\right] = 2[27-9] = 36.$

Also, $A+B+C+D = 9 + 37/6 + 37/6 + 44/3 = 36.$

37. Using symmetry:

$$A = 2 \int_{\frac{\pi}{6}}^{\frac{5\pi}{6}} \left(\sin x - \frac{1}{2}\right)\,dx + 2 \int_{\frac{5\pi}{6}}^{\frac{3\pi}{2}} \left(\frac{1}{2} - \sin x\right)\,dx$$

$$A = 2 \left[= 2\cos x - \frac{x}{2}\right]_{\frac{\pi}{6}}^{\frac{5\pi}{6}} + 2 \left[\frac{x}{2} + \cos x\right]_{\frac{5\pi}{6}}^{\frac{3\pi}{2}}$$

$$= 2\left[\left[\frac{\sqrt{3}}{2} - \frac{5\pi}{12}\right] - \left[-\frac{\sqrt{3}}{2} - \frac{\pi}{12}\right]\right] + 2\left[\left[\frac{3\pi}{4} + 0\right] - \left[\frac{5\pi}{12} - \frac{\sqrt{3}}{2}\right]\right]$$

$$= 2\left[\sqrt{3} - \frac{4\pi}{12}\right] + 2\left[\frac{4\pi}{12} + \frac{\sqrt{3}}{2}\right]$$

$$= 2\,\frac{3\sqrt{3}}{2}$$

$$= 3\sqrt{3}$$
$$= 5.1962$$

Problem Set 6.2   Volumes of Solids: Slabs, Disks, Washers

1. $\Delta V \approx \pi(x^2+1)^2\Delta x$. $V = \int_0^2 \pi(x^4+2x^2+1)dx = 206\pi/15 \approx 43.1445$.

3. (a) $\Delta V \approx \pi(4-x^2)^2\Delta x$.

$$V = \int_0^2 \pi(4-x^2)^2dx = \pi\int_0^2 (16-8x^2+x^4)dx$$

$$= \pi\left[16x - \frac{8x^3}{3} + \frac{x^5}{5}\right]_0^2 = \frac{256\pi}{15} \approx 53.6165.$$

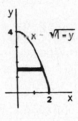

(b) $\Delta V \approx \pi(\sqrt{4-y})^2\Delta y = \pi(4-y)\Delta y$.

$$V = \int_0^4 \pi(4-y)dy = \pi\left[4y - \frac{y^2}{2}\right]_0^4$$

$$= 8\pi \approx 25.1327.$$

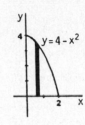

5. $\Delta V \approx \pi\left[(1/4)x^2\right]^2\Delta x$.

$$V = \int_0^4 (\pi/16)x^4dx = 64\pi/5 \approx 40.2124.$$

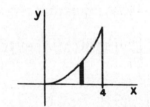

7. $\Delta V \approx \pi(1/x)^2\Delta x$.

$$V = \int_1^4 \pi x^{-2}dx = 3\pi/4 \approx 2.3562.$$

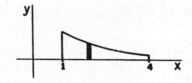

9. $\Delta V \approx \pi(\sqrt{4-x^2})^2\Delta x$.

$$V = \pi\int_{-1}^2 (4-x^2)dx = \pi\left[4x - \frac{x^3}{3}\right]_{-1}^2$$

$$= 9\pi \approx 28.2743.$$

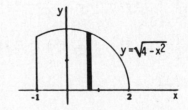

11. $\Delta V \approx \pi(y^2)^2 \Delta y.$

$$v = \int_0^2 \pi y^4 dy = 32\pi/5 \approx 20.1062.$$

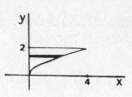

13. $\Delta V \approx \pi(y^{1/2})^2 \Delta y.$

$$V = \int_0^4 \pi y dy = 8\pi \approx 25.1327.$$

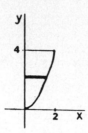

15. $\Delta V \approx \pi(y^{3/2})^2 \Delta y = \pi y^3 \Delta y.$

$$V = \pi \int_0^4 y^3 dy = \pi \left[\frac{y^4}{4}\right]_0^4$$

$$= 64\pi \approx 201.0619.$$

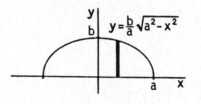

17. $\Delta V \approx \pi[(b/a)(a^2-x^2)^{1/2}]^2 \Delta x.$

$$V = 2 \int_0^a \pi(b^2/a^2)(a^2-x^2) dx = (4/3)\pi ab^2.$$

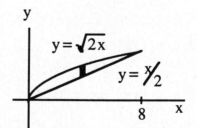

19. $\Delta V \approx \pi[(\sqrt{2x})^2 - (x/2)^2] \Delta x.$

$$V = \int_0^8 \pi[2x - (1/4)x^2] dx = 64\pi/3 \approx 67.0206.$$

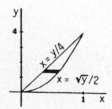

21. $\Delta V \approx \pi \left[\left[\frac{\sqrt{y}}{2}\right]^2 - \left[\frac{y}{4}\right]^2\right] \Delta y.$

$$V = \frac{\pi}{16} \int_0^4 (4y - y^2) dy = \frac{\pi}{16} \left[2y^2 - \frac{y^3}{3}\right]_0^4$$

$$= \frac{\pi}{16}(32/3) = 2\pi/3 \approx 2.0944.$$

23. [Same figure as in Problem 24, except the slices are square slabs, so $\Delta V \approx$ (Area of square) (thickness of square slab).]

$$\Delta V \approx [2(4-x^2)^{1/2}]^2 \Delta x. \quad V = 2\int_0^2 4(4-x^2)\,dx = 128/3.$$

25. $\Delta V \approx [(\cos x)^{1/2}]^2 \Delta x.$

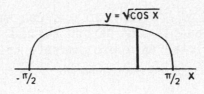

$$V = 2\int_0^{\pi/2} \cos x \, dx = 2.$$

27. See figure to the right for location of axes and cylinders.

$\Delta V \approx$ (area of square) (thickness of square piece)

$$= \left(\sqrt{1-y^2}\right)^2 \Delta y = (1-y^2)\Delta y.$$

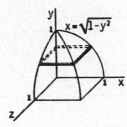

$$V = \int_0^1 (1-y^2)\,dy = \left[y - \frac{y^3}{3}\right]_0^1 = 2/3.$$

29. (a) $\Delta V \approx \pi(4-y^{2/3})^2 \Delta y.$

$$V = \int_0^8 \pi(16 - 8y^{2/3} + y^{4/3})\,dy$$

$$= 1024\pi/35 \approx 91.9140.$$

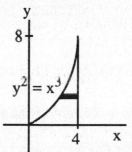

(b) $\Delta V \approx \pi\lfloor 8^2 - (8 - x^{3/2})^2 \rfloor \Delta x.$

$$V = \int_0^4 \pi(16x^{3/2} - x^3)\,dx = 704\pi/5 \approx 442.3362.$$

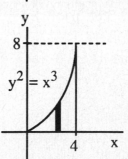

31. The first integral is the area of the quarter circle of radius 2 with center at the origin. Therefore, the integral in Example 4 equals

$$2\left[(1/4)\pi(2)^2\right] + \left[4y- (1/3)y^3\right]_0^2 = 2\pi + 16/3 \approx 11.6165.$$

33. Lay the cylinder on its side so you can see its base. The first figure is a view of the base. The second figure is a view from the side of the reclined cylinder with an approximately triangular slice featured.

$$\frac{y}{r} = \frac{z}{h}$$

$$\text{Volume} = 2\int_0^r \frac{1}{2} zy\,dx = 2\int_0^r \frac{1}{2}\left(\frac{hy}{r}\right) y\,dx$$

$$= \frac{h}{r}\int_0^r y^2\,dx = (\tan\theta)\int_0^r (r^2-x^2)\,dx \qquad (1)$$

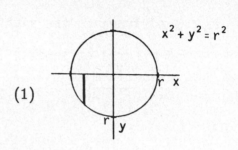

$$= (\tan\theta)\left[r^2 x - \frac{x^3}{3}\right]_0^r$$

$$= (\tan\theta)\left[r^3 - \frac{r^3}{3}\right] = \frac{2r^3\tan\theta}{3} \qquad (2)$$

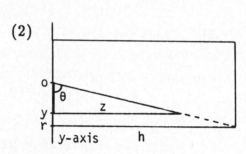

35. For a circular base:

$$\frac{r}{R} = \frac{d(0,P)}{d(0,Q)} = \frac{y}{h}, \text{ so } \frac{y^2}{h^2} = \frac{\pi r^2}{\pi R^2} = \frac{B}{A}.$$

Note that B is a function of y.

$$\text{Volume} = \int_0^h B\,dy = \int_0^h \frac{A}{h^2} y^2\,dy = \left[\frac{Ay^3}{3h^2}\right]_0^h$$

$$= \frac{Ah^3}{3h^2} = \frac{Ah}{3}.$$

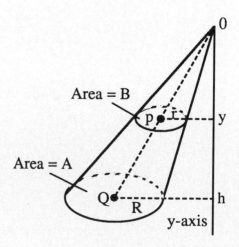

A "hand-waving" argument for an arbitrary base will be given at this point. Partition the base into a union of disjoint circles with areas A_α. Volume will then be h/3 times A, the "sum" of the A_α.

(a) Volume = $(h/3)(\pi r^2)$ = $\pi r^2 h/3$.

(b)

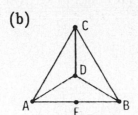 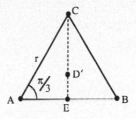 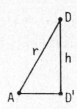

| View from Above | Base | Side View, Half Section |

$$d(C,E) = \frac{\sqrt{3}r}{2} \qquad d(A,D') = \frac{2}{3}\left(\frac{\sqrt{3}r}{2}\right) = \frac{\sqrt{3}r}{3}$$

$$\text{Area} = \frac{1}{2}\, r\, \frac{\sqrt{3}r}{2} = \frac{\sqrt{3}r^2}{4} \qquad h = \sqrt{r^2 - \left(\frac{\sqrt{3}r}{3}\right)^2} = \frac{\sqrt{6}r}{3}$$

Thus, Volume = $(1/3)(\sqrt{3}r^2/4)(\sqrt{6}r/3) = \sqrt{2}r^3/12$.

Problem Set 6.3   Volumes of Solids of Revolution: Shells

1. $\Delta V \approx 2\pi x(4/x)\Delta x$.

$$V = \int_1^4 8\pi dx = 24\pi \approx 75.3982.$$

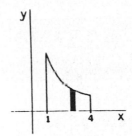

3. $\Delta V \approx 2\pi x \sqrt{x}\Delta x = 2\pi x^{3/2}\Delta x$.

$$V = 2\pi\int_0^4 x^{3/2}\, dx = 2\pi\left[\frac{2x^{5/2}}{5}\right]_0^4$$

$$= 2\pi\left[\frac{64}{5} - 0\right] = \frac{128\pi}{5} \approx 80.4248.$$

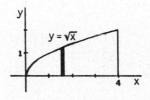

5. $\Delta V \approx 2\pi(4-x)x^{1/2}\Delta x.$

$$V = \int_0^4 2\pi(4x^{1/2}-x^{3/2})dx = 256\pi/15 \approx 53.6165.$$

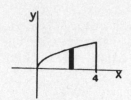

7. $\Delta V \approx 2\pi x([(1/4)x^3+2]-[2-x])\Delta x.$

$$V = \int_0^2 2\pi[(1/4)x^4+x^2]dx = 128\pi/15 \approx 26.8083.$$

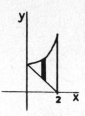

9. $\Delta V \approx 2\pi(y)(y^2)\Delta y = 2\pi y^3\Delta y.$

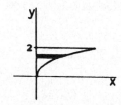

$$V = 2\pi\int_0^2 y^3 dy = 2\pi\left[\frac{y^4}{4}\right]_0^2$$

$$= 8\pi \approx 25.1327.$$

x = y²

11. $\Delta V \approx 2\pi(2-y)y^2\Delta y.$

$$V = \int_0^2 2\pi(2y^2-y^3)dy = 8\pi/3 \approx 8.3776.$$

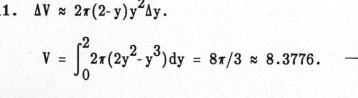
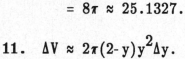

13. (a) $\int_a^b \pi[f^2(x)-g^2(x)]dx.$ (b) $\int_a^b 2\pi x[f(x)-g(x)]dx.$

(c) $\int_a^b 2\pi(x-a)[f(x)-g(x)]dx.$ (d) $\int_a^b 2\pi(b-x)[f(x)-g(x)]dx.$

15. (a) $A = \int_1^3 x^{-3}dx.$

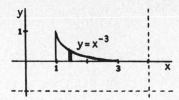

(b) $V = \int_1^3 2\pi x(x^{-3})dx.$ (Shells)

(c) $V = \int_1^3 \pi[x^{-3}-(-1)]^2 - [0-(-1)]^2 dx.$ (Washers)

(d) $V = \int_1^3 2\pi(4-x)(x^{-3})dx.$ (Shells)

17. $\Delta V \approx 2\pi y[y^{1/2}-(1/32)y^3]\Delta y.$ (Same figure as for Problem 18.)

$$V = \int_0^4 2\pi[y^{3/2}-(1/32)y^4]dy = 64\pi/5 \approx 40.2124.$$

19. Half of the solid in question can be obtained by revolving the region shown about x-axis.

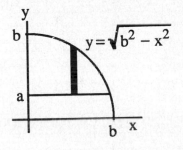

$\Delta V \approx \pi \left([(b^2-x^2)^{1/2}]^2-a^2\right)\Delta x.$

$$V = 2\int_0^{(b^2-a^2)^{1/2}} \pi(b^2-a^2-x^2)dx = (4/3)\pi(b^2-a^2)^{3/2}.$$

21. $\sin(x^2) = \cos(x^2) \Rightarrow \tan(x^2) = 1$

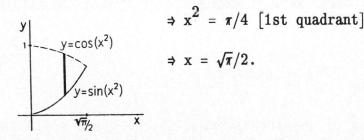

$\Rightarrow x^2 = \pi/4$ [1st quadrant]

$\Rightarrow x = \sqrt{\pi}/2.$

Let $u = x^2$
Then $du = 2xdx$
      $x = \sqrt{\pi}/2 \Rightarrow u=\pi/4$
      $x = 0 \Rightarrow u=0$

177

$$\text{Volume} = \int_0^{\sqrt{\pi}/2} 2\pi x [\cos(x^2) - \sin(x^2)] dx$$

$$= \pi \int_0^{\pi/4} [\cos u - \sin u] du = \pi \left[\sin u + \cos u \right]_0^{\pi/4}$$

$$= \pi \left[\left[\frac{\sqrt{2}}{2} + \frac{\sqrt{2}}{2} \right] - (0 + 1) \right] = \pi [\sqrt{2} - 1] \approx 1.3013.$$

23. (a) $\text{Volume} = \int_0^1 \pi [(x)^2 - (x^2)^2] dx$ [Washers]

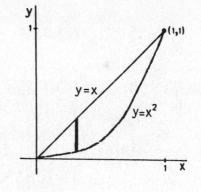

$$= \pi \left[\frac{x^3}{3} - \frac{x^5}{5} \right]_0^1 = \frac{2\pi}{15} \approx 0.4189.$$

(b) $\text{Volume} = \int_0^1 2\pi x (x - x^2) dx$ [Shells]

$$= 2\pi \left[\frac{x^3}{3} - \frac{x^4}{4} \right]_0^1 = \frac{\pi}{6} \approx 0.5236.$$

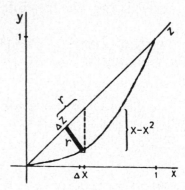

(c) $r^2 + r^2 = (x - x^2)^2;\ r^2 = (1/2)(x^2 - 2x^3 + x^4).$

$(\Delta x)^2 (\Delta x)^2 = (\Delta z)^2;\ \Delta z = \sqrt{2}\Delta x.$

Thus, $\text{Volume} = \int_0^1 \pi (1/2)(x^2 - 2x^3 + x^4)(\sqrt{2} dx)$

$$= \frac{\pi\sqrt{2}}{2} \left[\frac{x^3}{3} - \frac{x^4}{2} + \frac{x^5}{5} \right]_0^1 = \frac{\pi\sqrt{2}}{60} \approx .0740.$$

25. Let $A(x)$ denote the area of the intersection of the given solid and a concentric sphere of radius x, for x in $[0, r]$. Then, since $A(x)$ and S are similar, $\frac{A(x)}{S} = \frac{x^2}{r^2}$, so $A(x) = \frac{Sx^2}{r^2}$, and

$$\text{Volume} = \int_0^r A(x)\,dx = \frac{S}{r^2}\int_0^r x^2\,dx = \frac{S}{r^2}\frac{r^3}{3} = (1/3)rS.$$

Problem Set 6.4   Length of a Plane Curve

1. $\displaystyle\int_1^4 (1 + 3^2)^{1/2}\,dx = 3\sqrt{10}.$ $\qquad\qquad [(4\text{-}1)^2+(17\text{-}8)^2]^{1/2} = 3\sqrt{10}.$

3. $\dfrac{dy}{dx} = 3x^{1/2}.$

$$L = \int_{1/3}^7 \sqrt{1 + (3x^{1/2})^2}\,dx$$

> Let $\quad u = 1+9x$
> Then $du = 9dx$
> $\qquad\quad x = 7 \;\Rightarrow\; u=64$
> $\qquad\quad x = 1/3 \;\Rightarrow\; u=4$

$$= \int_{1/3}^7 \frac{1}{9}(1+9x)^{1/2}9dx = \int_4^{64}\frac{1}{9}u^{1/2}du$$

$$= \left[\frac{2u^{3/2}}{27}\right]_4^{64} = \frac{1008}{27} \approx 37.3333.$$

5. $dy/dx = -x^{-1/3}(4\text{-}x^{2/3})^{1/2};\;\; L = \displaystyle\int_1^8 \sqrt{1+x^{-2/3}(4\text{-}x^{2/3})}\;dx = \int_1^8 2x^{-1/3}dx = 9.$

7. $dx/dy = (1/4)y^3\text{-}y^{-3};\;\; L = \displaystyle\int_{-2}^{-1}(1+[(1/4)y^3\text{-}y^{-3}]^2)^{1/2}dy$

$$= \int_{-2}^{-1}[(-1/4)y^3\text{-}y^{-3}]dy = 21/16 = 1.3125.$$

9. $x = t^3,\; y = t^2,\; t$ in $[0,4].$

t	x	y
0	0	0
1	1	1
2	8	4
3	27	9
4	64	16

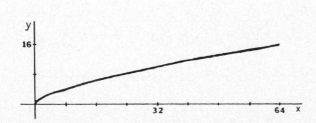

179

$$\frac{dx}{dt} = 3t^2, \; \frac{dy}{dt} = 2t.$$

$$L = \int_0^4 \sqrt{(3t^2)^2 + (2t)^2} \; dt = \int_0^4 \sqrt{9t^4 + 4t^2} \; dt$$

$$= \int_0^4 \sqrt{9t^2 + 4} \; t \, dt = \int_0^4 \frac{1}{18} (9t^2 + 4)^{1/2} \, 18t \, dt$$

> Let $\quad u = 9t^2 + 4$
> Then $du = 18t \; dt$
> $\qquad t = 4 \;\Rightarrow\; u = 148$
> $\qquad x = 1/3 \Rightarrow u = 4$

$$= \int_4^{148} \frac{1}{18} u^{1/2} \; du = \left[\frac{u^{3/2}}{27}\right]_4^{148} = \frac{296\sqrt{37} - 8}{27} \approx 66.3888.$$

11. $x^2 + (y+3)^2 = 9\sin^2 t + 9\cos^2 t = 9$; the graph is a circle with radius 3, so $L = 2\pi(3) = 6\pi \approx 18.8496$. The arc length integration formula will give the same value. (See the write-up for Problem 12.)

13. $dx/dt = 3a\sin^2 t \cos t$; $dy/dt = -3a\cos^2 t \sin t$.

$$L = 4\int_0^{\pi/2} [9a^2\sin^4 t \cos^2 t + 9a^2\cos^4 t \sin^2 t]^{1/2} dt$$

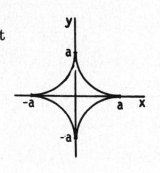

$$= 4\int_0^{\pi/2} 3a\sin t \cos t \; dt \quad \text{[Let } u = \sin t\text{]}$$

$$= 4\int_0^1 3au \; du = 6a.$$

15. $$\left[\frac{dx}{d\theta}\right]^2 + \left[\frac{dy}{d\theta}\right]^2 = [a(1-\cos\theta)]^2 + [a\sin\theta]^2 = a^2(1 - 2\cos\theta + \cos^2\theta) + a^2\sin^2\theta$$

$$= a^2(1 - 2\cos\theta + \cos^2\theta + \sin^2\theta) = a^2(1 - 2\cos\theta + 1)$$
$$= 2a^2(1 - \cos\theta) = 2a^2\left[2\sin^2\left(\tfrac{\theta}{2}\right)\right] \quad \text{[Half-angle identity]}$$
$$= 4a^2\sin^2\left(\tfrac{\theta}{2}\right).$$

Then $L = \int_0^{2\pi} \sqrt{4a^2\sin^2(\tfrac{\theta}{2})}\,d\theta = \int_0^{2\pi} 2a\sin(\tfrac{\theta}{2})\,d\theta = \left[-4a\cos(\tfrac{\theta}{2})\right]_0^{2\pi} = 8a.$

17. (a) $\dfrac{dy}{dx} = \sqrt{x^3-1}$, so length $= \int_1^2 \sqrt{1+(x^3-1)}\;dx = \int_1^2 x^{3/2}dx = \left[(2/5)x^{5/2}\right]_1^2$

$$= (2/5)[(2)^{5/2}-1] \approx 1.8627$$

(b) $\dfrac{dx}{dt} = 1-\cos t;\quad \dfrac{dy}{dt} = \sin t.$

$\text{Length} = \int_0^{4\pi} \sqrt{(1-\cos t)^2 + (\sin t)^2}\;dt = \sqrt{2}\int_0^{4\pi} \sqrt{1-\cos t}\;dt$

$= \sqrt{2}\int_0^{4\pi} \sqrt{2\sin^2(t/2)}\;dt = 2\int_0^{4\pi} |\sin(t/2)|\,dt$

$= 2\int_0^{2\pi} \sin(t/2)\,dt + 2\int_{2\pi}^{4\pi} -\sin(t/2)\,dt$

$= 2\left[-2\cos(t/2)\right]_0^{2\pi} + 2\left[2\cos(t/2)\right]_{2\pi}^{4\pi} = -4[-1-1] + 4[1-(-1)] = 16.$

19. $dy/dx = 6;\quad A = \int_0^1 2\pi(6x)[1+6^2]^{1/2}dx = 6\pi\sqrt{37} \approx 114.6574.$

21. $dy/dx = x^2;\quad A = \int_1^{\sqrt{7}} 2\pi(1/3)x^3(1+x^4)^{1/2}dx = \int_2^{50} (\pi/6)u^{1/2}du \quad [u = 1+x^4]$

$$= 248\sqrt{2}\pi/9 \approx 122.4261.$$

23. $\dfrac{dx}{dt} = 1,\ \dfrac{dy}{dt} = 3t^2.$

Then $A = \int_0^1 2\pi(t^3)\sqrt{(1)^2+(3t^2)^2}\;dt = \int_0^1 2\pi t^3(1+9t^4)^{1/2}dt$

[Use substitution of $u = 1+9t^4$ if needed to see the result.]

$$= \left[\frac{\pi(1+9t^4)^{3/2}}{27}\right]_0^1 = \frac{\pi(10\sqrt{10}-1)}{27} \approx 3.5631.$$

181

25. (a) Arc length of sector of radius ℓ is the circumference of the base of the cone with radius r, so $\ell\theta = 2\pi r$; hence, $\theta = 2\pi r/\ell$.

(b) Let S = lateral surface area of the cone = area of sector. Then
$S = (1/2)\ell^2\theta = (1/2)\ell^2(2\pi r/\ell) = \pi r\ell$.

(c) $\dfrac{\ell_1}{r_1} = \dfrac{\ell_1+\ell}{r_2}$, from which $\ell_1 = \ell r_1(r_2-r_1)^{-1}$.

Then, with the result obtained in (b),

$$S_{frustrum} = S_{large\ cone} - S_{small\ cone}$$
$$= \pi r_2(\ell+\ell_1) - \pi r_1\ell_1 = \pi r_2\ell + \pi(r_2-r_1)\ell_1$$

$$= \pi r_2\ell + \pi(r_2-r_1)[\ell r_1(r_2-r_1)^{-1}] = \pi(r_2+r_1)\ell = 2\pi[(r_1+r_2)/2]\ell.$$

27. (a) By direct substitution into $A = \displaystyle\int_a^b 2\pi y\ ds$.

(b)
$$A = 2\sqrt{2}\ \pi\ a^2\int_0^{2\pi} (1 - \cos t)^{3/2}\ dt$$

$$= 2\sqrt{2}\ \pi\ a^2\int_0^{2\pi} (2\ \sin^2 \tfrac{t}{2})^{3/2}\ dt$$

$$= 2\sqrt{2}\ \pi\ a^2\int_0^{2\pi} 2^{3/2}\sin^3 \tfrac{t}{2}\ dt$$

$$= 8\pi\ a^2\int_0^{2\pi} \sin \tfrac{t}{2}\left[1 - \cos^2 \tfrac{t}{2}\right]dt$$

$$= 8\pi a^2\left[-2\cos \tfrac{t}{2} - \tfrac{2}{3}\cos^3 \tfrac{t}{2}\right]\Big|_0^{2\pi}$$

$$= 8\pi a^2\left\{\left[(-2)(-1) + \tfrac{2}{3}(-1)^3\right] - \left[-2(1) + \tfrac{2}{3}(1)\right]\right\}$$

$$= \frac{64\pi a^2}{3}$$

29. (a)   (b)

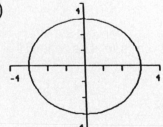

(c)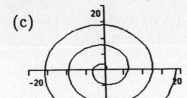

(d)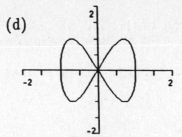

(e)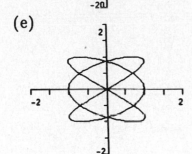

(f)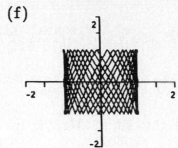

31. $f(x) = x$ $L = 1.41421$
      $f(x) = x^2$ $L = 1.47894$
      $f(x) = x^4$ $L = 1.60023$
      $f(x) = x^{10}$ $L = 1.75441$
      $f(x) = x^{100}$ $L = 1.95167$

      $f(x) = x^{10,000}$ $L = 2$

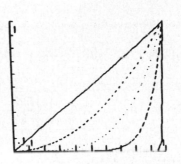

Problem Set 6.5 Work

1. $8 = k(1/2) \Rightarrow k=16$; $F(x) = 16x$; $W = \int_0^{1/2} 16x \, dx = 2$ ft-lb.

3. $F = 200$ dynes when $x = 2$ cm, so $200 = k(2)$ for some k (Hooke's Law). Thus, $k = 100$, so for this spring $F(x) = 100x$.

Therefore, the work done is $W = \int_0^4 100x \, dx = \left[50x^2\right]_0^4 = 800$ ergs.

5. $W = \int_0^d kx \, dx = (1/2) \, kd^2$ (beginning from its natural length).

7. $W = \int_0^2 11s \, ds = 22$ ft-lb.

9. Approx. volume of a horizontal slab is (length)(width)(thickness), or

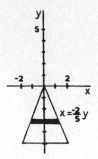

$$\Delta V \approx (10) \left(-\frac{4}{5} y\right) \Delta y \approx -8y\Delta y,$$

so approx. weight is $62.4(-8y\Delta y)$, or $-499.2y\Delta y$. The approximate distance to lift it is $(5-y)$.

$$\Delta W \approx (-499.2y\Delta y)(5-y) = -499.2(5y-y^2)\Delta y.$$

$$W = -499.2 \int_{-5}^0 (5y-y^2)dy = -499.2 \left[\frac{5y^2}{2} - \frac{y^3}{3}\right]_{-5}^0 = 52,000 \text{ ft-lbs.}$$

11. $\Delta W \approx (62.4)\left[(10)(3 + 2\,\frac{3y}{8})\Delta y\right](9-y)$

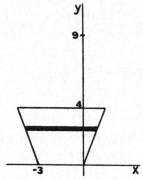

$$= 468(36+5y-y^2)\Delta y.$$

$$W = 468 \int_0^4 (36+5y-y^2)dy = 76,128 \text{ ft-lb.}$$

13. $\Delta W \approx (50)\left[\pi(5)^2\Delta y\right](0-y)$

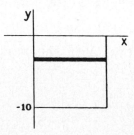

$$= 1250\pi y\Delta y.$$

$$W \approx -1250\pi \int_{10}^0 y\,dy = 62500\pi \approx 196,349.5 \text{ ft-lb.}$$

15. $F(x) = Af(x)$, so $\Delta W \approx Af(x)\Delta x$, and then $W = \int_{x_2}^{x_1} Af(x)dx$.

17. $V = 2x$, so x=8 when v=16, and x=1 when v=2.

$p = 40(16)^{1.4}v^{-1.4} = 40(16)^{1.4}(2x)^{-1.4} = 40(8)^{1.4}x^{-1.4}$.

$W = (2)\int_{1}^{8} 40(8)^{1.4}x^{-1.4}dx = -200(8)^{1.4}[8^{-0.4}-1] \approx 2,075.83$ in-lb.

19. W_{load} = (500 ft)(200 lb) = 100,000 ft-lb.

$\Delta W_{cable} \approx$ (weight of short piece)(distance lifted)
$$= (2\Delta y)(-y) = -2y\Delta y.$$
$\Delta W_{cable} = \int_{-500}^{0} -2ydy = 250,000$ ft-lb.

Work for load and cable is 350,000 ft-lb.

21. Lifting force is $f(x) = kx^{-2}$. $5000 = k(4000)^{-2}$ implies $k = 8(10)^{10}$. Therefore, $f(x) = 8(10)^{10}x^{-2}$.

$W = 8(10)^{10}\int_{4000}^{4200} x^{-2}dx = 8(10)^{10}\left[\frac{-1}{x}\right]_{4000}^{4200} \approx 952,400$ miles-lbs.

[or 5,029,000,000 ft-lb].

23. Total weight is 600-3t = 600-3(y/2) [since y = 2t]
$$= 600-1.5y$$

Work $= \int_{0}^{80} (600-1.5y)dy = \left[600y-0.75y^2\right]_{0}^{80} = 43,200$ ft-lb.

25. Work $= \displaystyle\int_{-10}^{10} 62.4\pi(100-y^2)(y+40)\,dy$

$$+ \int_{-40}^{-10} 62.4\pi(0.5)^2(y+40)\,dy$$

$$= 62/4\pi \int_{-10}^{10} (4000+100y-40y^2-y^3)\,dy$$

$$+ 15.6\pi \int_{-40}^{-10} (40y + y^2/2)\,dy$$

$$= 3{,}335{,}020\pi \approx 10{,}477{,}274 \text{ ft-lb.}$$

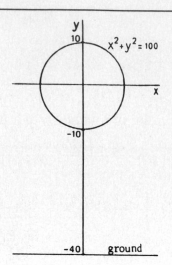

27. Work done in pushing buoy down two feet is

$$\int_0^2 \delta\left[\frac{1}{3}\,\pi\left[\frac{a(8+z)}{8}\right]^2(8+z) - \frac{1}{3}\,\pi(a)^2(8)\right]dz$$

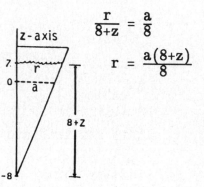

$$\frac{r}{8+z} = \frac{a}{8}$$

$$r = \frac{a(8+z)}{8}$$

$$= \int_0^2 \delta\,\frac{1}{3}\,\pi a^2(8)\left[\frac{(8+z)^3}{(8)^3} - 1\right]dz$$

$$= \frac{300}{512} \int_0^2 \left[(8+z)^3 - 512\right]dx$$

$$= \frac{300}{512}\left[\frac{(8+z)^4}{4} - 512z\right]_0^2 = \frac{8475}{32} \approx 264.84 \text{ ft-lb.}$$

Problem Set 6.6 Moments, Center of Mass.

1. $\sum x_i m_i = (2)(4)+(-2)(6)+(1)(9) = 5$; $\sum m_i = 4+6+9 = 19$; $\bar{x} = 5/19.$

3. $\bar{x} = \dfrac{M \ (\text{moment})}{m \ (\text{mass})}$.

$\Delta M \approx [x][\delta(x)\Delta x] = x\sqrt{x}\Delta x = x^{3/2}\Delta x. \quad \Delta m = \delta(x)\Delta x = \sqrt{x}\Delta x.$

$M = \displaystyle\int_0^9 x^{3/2} \ dx = \left[\dfrac{2x^{5/2}}{5}\right]_0^9 = 97.2. \qquad m = \displaystyle\int_0^9 x^{1/2} dx = \left[\dfrac{2x^{3/2}}{3}\right]_0^9 = 18.$

Thus, $\bar{x} = (97.2)/18 = 5.4$ units from that end to the center of mass.

5. $M_y = \displaystyle\sum x_i m_i = 11; \quad M_x = \sum y_i m_i = -3; \quad m = \sum m_i = 17.$

$\bar{x} = M_y/m = 11/17; \quad \bar{y} = M_x/m = -3/17. \quad (11/17, -3/17)$ is center of mass.

7. Centroid of upper rectangle is $(0, 1/2)$ and its area is 4, so moment about y-axis is $(0)(4) = 0$, and moment about x-axis is $(1/2)(4) = 2$.

Centroid of lower rectangle is $(-1/2, -1/2)$ and its area is 3, so moment about each axis is $(-1/2)(3) = -3/2$.

Thus, $M_y = 0 + (-3/2) = -3/2; \quad M_x = 2 + (-3/2) = 1/2;$ total area is 7.

Therefore, $\bar{x} = (-3/2)/7 = -3/14; \quad \bar{y} = (1/2)/7 = 1/14.$
Centroid of region is $(-3/14, 1/14)$.

9. For a homogeneous rectangular lamina, the centroid is at the intersection of the diagonals of the rectangle.

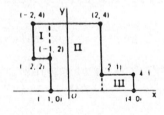

Rectangle I: The centroid is $(-3/2, 3)$ and the area is 2, so the moment about the y-axis is $(-3/2)(2) = -3$, and the moment about the x-axis is $(3)(2) = 6$.

Rectangle II: The centroid is $(1/2, 2)$ and the area is 12, so moment about the y-axis is $(1/2)(12) = 6$, and the moment about the x-axis is $(2)(12) = 24$.

Rectangle III: The centroid is (3,1/2) and the area is 2, so moment about the y-axis is (3)(2) = 6, and the moment about the x-axis is (1/2)(2) = 1.

Entire Region: Moment about y-axis = (-3)+(6)+(6) = 9.
Moment about x-axis = (6)+(24)+(1) = 31.
Area = (2)+(12)+(2) = 16.

$$\overline{x} = \frac{9}{16}, \quad \overline{y} = \frac{31}{16}. \quad \text{Centroid: } (9/16, \ 31/16).$$

11. $\overline{x} = 0$ (by symmetry).

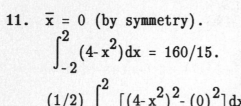

$$\int_{-2}^{2} (4-x^2)dx = 160/15.$$

$$(1/2)\int_{-2}^{2}[(4-x^2)^2 - (0)^2]dx = 256/15.$$

$\overline{y} = 256/160 = 1.6$. Centroid is (0,1.6).

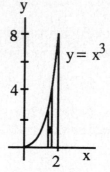

13. $\displaystyle\int_{0}^{2}(x^3-0)dx = 4. \quad \int_{0}^{2}x(x^3-0)dx = 32/5.$

$$(1/2)\int_{0}^{2}[(x^3)^2 - 0^2]dx = 64/7.$$

$\overline{x} = (32/5)/4 = 8/5$; $\overline{y} \ (64/7)/4 = 16/7$. Centroid is (8/5,16/7).

15. $y = 2x-4$ and $y = 2\sqrt{x}$ intersect where $x = 4$.

$$\int_{0}^{4}[2\sqrt{x} - (2x-4)]dx = \int_{0}^{4}(2x^{1/2} - 2x+4)dx$$

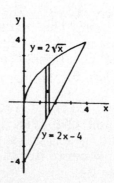

$$= \left[\frac{4x^{3/2}}{3} - x^2 + 4x\right]_{0}^{4} = \frac{32}{3}. \quad \text{(Area)}$$

$$\int_{0}^{4}x[2\sqrt{x} - (2x-4)]dx = \int_{0}^{4}(2x^{3/2} - 2x^2 - 2x^2 + 4x)dx$$

$$= \left[\frac{4x^{5/2}}{5} - \frac{2x^3}{3} + 2x^2\right]_{0}^{4} = \frac{224}{15}.$$

$$\frac{1}{2} \int_0^4 [(2\sqrt{x})^2 - (2x-4)^2] dx = \int_0^4 (-2x^2 + 10x - 8) dx$$

$$= \left[\frac{-2x^3}{3} + 5x^2 - 8x \right]_0^4 = \frac{16}{3}.$$

Then, $\bar{x} = \frac{224/15}{32/3} = 1.4$; $\bar{y} = \frac{16/3}{32/3} = 0.5$, so the centroid is $(1.4, 0.5)$.

17. $\bar{y} = 0$ (by symmetry).

$$\int_{-2}^2 (4-y^2) dy = 32/3.$$

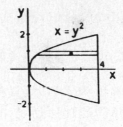

$$(1/2) \int_{-2}^2 [(4)^2 - (y^2)^2] dy = 128/5.$$

$\bar{x} = (128/5)/(32/3) = 2.4.$ Centroid is $(2.4, 0)$.

19. From Problem 13, area of the region is 4 and centroid is $(8/5, 16/7)$.
Centroid travels $2\pi(8/5) = 3.2\pi$.
By Pappus's Theorem, volume is $(4)(3.2\pi) = 12.8\pi \approx 40.2124$.

By shells, volume is $\int_0^2 2\pi(x)(x^2) dx = 12.8\pi$.

21. Centroid is at $(0, \bar{y})$

In revolving the shown region about
the x-axis, the centroid travels $2\pi\bar{y}$.
The area of the region is $(1/2)\pi a^2$.
The volume of the sphere is $(4/3)\pi a^3$.

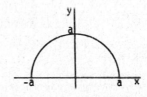

By Pappus: $(4/3)\pi a^3 = [(1/2)\pi a^2][2\pi\bar{y}]$, so $\bar{y} = \frac{4a}{3\pi} \approx 0.4244a$. The
centroid is $(0, 4a/3\pi)$.

23. $2\pi \int_c^d (e-y)w(y) dy$

25. (a) Area(P) $= 4n[(1/2)r\cos(\pi/2n)r\sin(\pi/2n)]$
          $= 2nr^2\sin(\pi/2n)\cos(\pi/2n)$

Volume $= [\text{Area}(P)][\text{Distance traveled by centroid of } P]$
$= [2nr^2\sin(\pi/2n)\cos(\pi/2n)][2\pi r\cos(\pi/2n)]$
$= 4\pi nr^3\sin(\pi/2n)\cos^2(\pi/2n)$

(b) Part (a): $\lim\limits_{n\to\infty}[4\pi nr^3\sin(\pi/2n)\cos^2(\pi/2n)]$

$= \lim\limits_{n\to\infty}\left[2\pi^2 r^3\dfrac{\sin(\pi/2n)}{(\pi/2n)}\right] = 2\pi^2 r^3(1)(1) = (\pi r^2)(2\pi r)$

$=$ (area in circle)(distance traveled by its centroid).

Part (b): $\lim\limits_{n\to\infty}[nr^3\delta\sin(\pi/n)] = \lim\limits_{n\to\infty}\left[\pi r^3\delta\dfrac{\sin(\pi/n)}{(\pi/n)}\right] = \pi r^3\delta(1)$

$= (\delta r)(\pi r^2) =$ (force on centroid)(area in circle).

27. R: Let $g(x) = x$ for x in $[0,1]$.

S: Let $f(x) = \begin{cases} 0, & \text{if } 0 \le x < 0.9 \\ 0.9, & \text{if } 0.9 \le x \le 1 \end{cases}$

Then $\bar{y}_S = \dfrac{1}{3}(1) = \dfrac{1}{3}$ [by Problem 24(a)], but $\bar{y}_R = 0.45$.

Now use this example to approximate result for any regions R and S.

Problem Set 6.7   Chapter Review

True-False Quiz

1. False.   That integral equals 0. The correct integral is $\displaystyle\int_0^\pi |\cos|dx$.

3. False.   Ex: For $f(x) = \sin x$, $g(x) = \cos x$, $a=0$, $b=\pi/2$, the integral equals zero, but the area of the region is not zero. The correct integral is $\displaystyle\int_a^b |f(x)-g(x)|dx$.

5. True.   Without loss of generality, consider the bases to be in the xy-plane, and $0 \le z \le h$ for the solid. Then the volumes equal

$\displaystyle\int_0^h A(z)dz$, where $A(z)$ is the area of the cross sections at z.

7. **False.** To use the method of washers, one would need to solve for x.

9. **False.** (Intuitive) Place the unit circle with center at the origin. Then define a curve that starts at (1,0), smoothly moves up and down touching the circle at the top and bottom in each cycle as x approaches zero. Have the curve complete one cycle as x goes from 2^{-n} to 2^{-n-1} (for $n = 0,1,2,3,\cdots$). Thus as x approaches zero the lengths of the cycles approaches infinity.

   A particular counterexample (which doesn't satisfy all the conditions in the intuitive approach) is $f(x) = (1-x)\sin x(1/x)$, for x in (0,1).

11. **False.** (Intuitive) There is a greater proportion of water toward the top of the cone, so for the cone a greater proportion of water has to be lifted less. (In fact, the ratio is 1/2.)

13. **True.** $\sum(x_i-\bar{x})m_i = \sum x_i m_i - \bar{x}\sum m_i = \bar{x}M - \bar{x}M = 0.$

15. **True.** Centroid is $(\pi/2,\bar{y})$. Volume is $(2)[2\pi(\pi/2)] = 2\pi^2$.

17. **True.** (Intuitive) The density is the same going each way from the midpoint, so the wire is balanced at the midpoint.

Sample Test Problems

1. $\int_0^1 (x-x^2)\,dx = 1/6.$

3. $\int_0^1 2\pi x(x-x^2)\,dx = 2\pi \int_0^1 (x^2-x^3)\,dx = 2\pi \left[\frac{x^3}{3} - \frac{x^4}{4}\right]_0^1 = \frac{\pi}{6} \approx 0.5236.$

5. $\int_0^1 2\pi(2-x)(x-x^2)\,dx = \pi/2 \approx 1.5708.$

7. $V(S_1) = (1/6)[2\pi(1/10)] = \pi/30.$ $V(S_2) = (1/6)[2\pi(1/2)] = \pi/6.$
   $V(S_3) = (1/6)[2\pi(11/10)] = 11\pi/30.$ $V(S_4) = (1/6)[2\pi(3/2)] = \pi/2.$

9. $W = \int_0^8 (62.4)\pi(5)^2(10-y)\,dy$

$= 1560\pi \left[10y - \dfrac{y^2}{2}\right]_0^8 = 74880\pi \approx 235,242$ ft-lb.

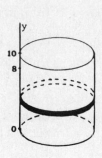

11. See graph with Problem 12.

(a) $\int_0^3 (3x-x^2)\,dx = 4.5$. (b) $\int_0^9 [y^{1/2} - (1/3)y]\,dy = 4.5$.

13. $V = \int_0^3 \pi(9x^2-x^4)\,dx = 32.4\pi$. By Pappus, $V = (4.5)[2\pi(3.6)] = 32.4\pi$.

15. $dy/dx = x^2 - (1/4)x^{-2}$

$L = \int_1^3 [1 + (x^2 - x^{-2}/4)^2]^{1/2}\,dx = \int_1^3 [x^2 + (1/4)x^{-2}]\,dx = 53/6 \approx 8.8333$.

17. $\Delta V \approx (\sqrt{4-x^2})^2 \Delta x = (4-x^2)\Delta x$.

$V = \int_{-2}^2 (4-x^2)\,dx = 2\int_0^2 (4-x^2)\,dx$

$= 2\left[4x - \dfrac{x^3}{3}\right]_0^2 = \dfrac{32}{3}$.

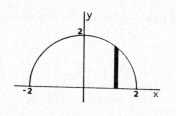

19. $\int_a^b \pi[f^2(x) - g^2(x)]\,dx$.

21. $M_y = \int_a^b \delta x [f(x) - g(x)] dx.$

    $M_x = \int_a^b (\delta/2) [f^2(x) - g^2(x)] dx.$

23. Surface area = (left surface area) + (right surface area) +

                 + (outside surface area) + (inside surface area)

        $= \pi [f^2(a) - g^2(a)] + \pi [f^2(b) - g^2(b)] +$

        $+ \int_a^b 2\pi f(x) \sqrt{1 + [f'(x)]^2} dx + \int_a^b 2\pi g(x) \sqrt{1 + [g'(x)]^2} dx.$

Problem Set 7.1   The Natural Logarithm Function

1. (a) $\ln 2 + \ln 3 \approx 1.792.$
   (b) $\ln 3 - \ln 2 \approx = 0.406.$
   (c) $4 \ln 3 \approx 4.396.$
   (d) $(1/2) \ln 2 \approx 0.346.$
   (e) $-2 \ln 6 \approx -3.584.$
   (f) $\ln 3 + 4 \ln 2 \approx 3.871.$

3. $D_x \ln(x^2 - 5x+6) = \dfrac{1}{x^2-5x+6}\,(2x-5) = \dfrac{2x-5}{x^2-5x+6}.$

5. $D_x[4 \ln(x-5)] = 4/(x-5).$

7. $(x)(1/x) + (\ln x)(1) = 1 + \ln x$

9. $y = 3 \ln x + (\ln x)^3.$

$\dfrac{dy}{dx} = 3\,\dfrac{1}{x} + 3\,(\ln x)^2\,\dfrac{1}{x} = \dfrac{3(1 + \ln^2 x)}{x}.$

11. $\dfrac{1}{x+(x^2-1)^{1/2}}\left[1 + \dfrac{1}{2(x^2-1)^{1/2}}\,2x\right] = (x^2-1)^{-1/2}.$

13. $f(x) = (1/3)\ln x;\ f'(x) = 1/3x;\ f'(100) = 1/300.$

15. $\displaystyle\int \dfrac{4}{2x+1}\,dx = \int 2\,\dfrac{1}{2x+1}\,dx$

$\boxed{\begin{array}{l}\text{Let } u = 2x+1 \\ \text{Then } du = 2dx\end{array}}$

$= \displaystyle\int 2\,\dfrac{1}{u}\,du = 2\,\ln|u| + C = 2\,\ln|2x+1| + C$

17. $2\displaystyle\int (1/u)\,du = 2\,\ln|u| + C = 2\ln(x^2+x+5) + C.$ [Let $u = x2+x+5$]

19. $\displaystyle\int u\,du = (1/2)\,u^2 + C = (1/2)\ln^2 x + C.$ [Let $u = \ln x$]

21. $\displaystyle\int_0^3 \frac{x^3}{x^4+1}\,dx = \frac{1}{4}\int_0^3 \frac{1}{x^4+1}\,4x^3\,dx$

$$\boxed{\begin{array}{l} \text{Let}\quad u = x^4 + 1 \\ \text{Then}\quad du = 4x^3 dx \\ \qquad x=3 \Rightarrow u=82 \\ \qquad x=0 \Rightarrow u=1 \end{array}}$$

$$= \frac{1}{4}\int_1^{82}\frac{1}{u}\,du = \frac{1}{4}\Big[\ln|u|\Big]_1^{82} = \frac{\ln 82 - \ln 1}{4} = \frac{\ln 82}{4}$$

$$\approx 1.1017.$$

23. $\ln\big[(x+1)^2/x\big].$

25. $\ln\big[x^2(x-2)/(x+2)\big],\ x > 2.$

27. $\ln y = \ln(x+11) - (1/2)\ln(x^3-4).$
Then, taking the derivative of each side with respect to x, obtain:

$$\frac{1}{y}\frac{dy}{dx} = \frac{1}{x+11} - \frac{1}{2}\,\frac{1}{x^3-4}\,3x^2 = \frac{1}{x+11} - \frac{3x^2}{2(x^3-4)} = \frac{-x^3-33x^2-8}{2(x+11)(x^3-4)}.$$

Therefore, $\displaystyle\frac{dy}{dx} = y\,\frac{(-x^3-33x^2-8)}{2(x+11)(x^3-4)} = \frac{(x+11)(-x^3-33x^2-8)}{2(x+11)(x^3-4)^{3/2}} = \frac{x^3+33x^2+8}{-2(x^3-4)^{3/2}}.$

29. $\ln y = (1/2)\ln(x+13) - \ln(x-4) - (1/3)\ln(2x+1).$

$$\frac{1}{y}\,y' = \frac{1}{2(x+13)}\quad \frac{1}{x-4}\quad \frac{2}{3(2x+1)}.$$

$$y' = y\left[\frac{1}{2(x+13)} - \frac{1}{x-4} - \frac{2}{3(2x+1)}\right] = \frac{-10x^2-219x+118}{6(x+13)^{1/2}(x-4)^2(2x+1)^{4/3}}.$$

31. y-axis symmetry

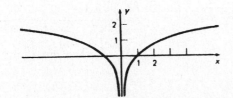

33. y = -lnx (reflection of
y = ln x in the x-axis).

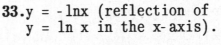
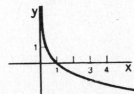

35. $y = \ln[(\cos x)(\sec x)]$

   $= \ln(1) = 0.$

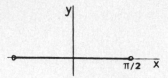

37. $D_f = (0,\infty)$. $f'(x) = (2x^2)(1/x) + (\ln x)(4x) - 2x = 4x \ln x = 0$ if $x = 1$. Use an auxiliary axis for f' and conclude that f has a local minimum value of $f(1) = -1$.

39. $\ln 4 > 1 \Leftrightarrow m \ln 4 > m \Leftrightarrow \ln 4^m > m.$ $[m > 0]$

   Thus, given any $m > 0$, $\ln x > m$ for all $x > 4^m$, since $\ln$ is an increasing function.

   Therefore, $\displaystyle\lim_{x \to \infty} \ln x = \infty.$

41. $\ln x - \ln(1/3) = 2 \ln x;\ \ln x = \ln 3;\ x = 3.$

43. It is a Riemann sum for $f(x) = 1/(1+x)$ on $[0,1]$. Thus, the limit is

   $$\int_0^1 1/(1+x)\ dx = \Big[\ln(x+1)\Big]_0^1 = \ln 2 = \ln 1 = \ln 2 \approx .6931.$$

45. (a) $f(x) = c[\ln(ax-b) - \ln(ax+b)]$

   $f'(x) = c\left[\dfrac{a}{ax-b} - \dfrac{a}{ax+b}\right],$ so $f'(1) = c\left[\dfrac{a}{a-b} - \dfrac{a}{a+b}\right] = c\,\dfrac{2ab}{(a^2-b^2)} = 1.$

   [The above depends on $a \neq \pm b$; if $a = \pm b$; $f'(1)$ doesn't exist.]

   (b) $f'(x) = \cos^2 u\,\dfrac{du}{dx} = \cos^2[\ln(x^2+x-1)]\,\dfrac{2x+1}{x^2+x-1},$ so $f'(1) = 3.$

47. Volume = $\int_1^4 2\pi x(x^2+4)^{-1}dx$ [Shells]

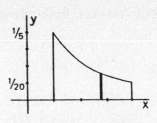

$$= \int_5^{20} \pi u^{-1}du = \pi\Big[\ln(u)\Big]_5^{20}$$

Let $u = x^2+4$
Then $du = 2xdx$
      $x=4 \Rightarrow u=20$
      $x=1 \Rightarrow u=5$

$$= \pi(\ln 20 - \ln 5) = \pi\ln(20/5)$$

$$= \pi\ln 4 \approx 4.3552$$

49. Let $\{1,2,\cdots,n\}$ be a partition of $[1,n]$.
Area (inscribed polygon)
  < Area (under curve)
  < Area (circumscribed polygon).

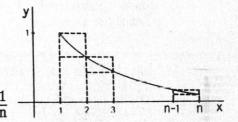

$$\frac{1}{2} + \frac{1}{3} + \cdots + \frac{1}{n} < \int_1^n \frac{1}{x}dx < 1 + \frac{1}{2} + \frac{1}{3} + \cdots + \frac{1}{n-1}$$

$$\frac{1}{2} + \frac{1}{3} + \cdots + \frac{1}{n} < \ln(n) < 1 + \frac{1}{2} + \frac{1}{3} + \cdots + \frac{1}{n-1} + \frac{1}{n}$$

51. (a) Absolute max point: $\left(\pi/2, \ln(5/2)\right) = (1.57079, 0.916)$
$\left(5\pi/2, \ln(5/2)\right) = 7.854, 0.916)$
Absolute min point: $(3\pi/2, \ln(1/2)) = (4.712, -0.693)$.

(b) $(3.87, -0.18)$, $(5.55, -0.18)$.

(c) 4.04218.

53. (a) 0.13889

(b) 0.26035

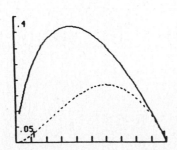

Problem Set 7.2   Inverse Functions and Their Derivatives

1. $f^{-1}(2) \approx 4$.

3. $f(x) = 2$ for three values of x,
    so f does not have an inverse.

5. $f^{-1}(2) \approx -2$.

7. $f'(x) = -15x^4 - 1 < 0$ for all x,
    so f is decreasing on R.

9. [See Example 2, Page 109] $f'(x) = \sec^2 x$ which exists and is greater than zero on $(-\pi/2, \pi/2)$, so f is increasing on that interval.

11. $f'(x) = 2(x+1) < 0$ for $x < -1$ and f is continuous on $(-\infty, -1]$, so f is decreasing on $(-\infty, -1]$.

13. $f'(x) = (x^2+2)^{1/2} > 0$ for all x, so f is increasing on R.

15. $f(x) = 3x-1$.

    Step 1: Let $y = 3x-1$. Then $x = \frac{y+1}{3}$.

    Step 2: $f^{-1}(y) = \frac{y+1}{3}$.

    Step 3: $f^{-1}(x) = \frac{x+1}{3}$.

    $f^{-1}(f(x)) = f^{-1}(3x-1) = \frac{(3x-1)+1}{3} = x$.

    $f(f^{-1}(x)) = f(\frac{x+1}{3}) = 3(\frac{x+1}{3}) - 1 = x$.

17. $y = (2x+5)^{1/2}$; $y^2 = 2x+5$, $y \geq 0$; $x = (1/2)(y^2-5)$; $f^{-1}(y) = (1/2)(y^2-5)$;

    Therefore, $f^{-1}(x) = (1/2)(x^2-5)$, $x \geq 0$.

    $f^{-1}(f(x)) = f^{-1}((2x+5)^{1/2}) = (1/2)([(2x+5)^{1/2}]^2 - 5) = x$.

    $f(f^{-1}(x)) = f((1/2)(x^2-5)) = (2[(1/2)(x^2-5)] + 5)^{1/2} = |x| = x$, since $x \geq 0$.

19. $y = (x-5)^{-1}$; $x = 5+y^{-1}$; $f^{-1}(y) = 5+y^{-1}$; $f^{-1}(x) = 5+x^{-1}$.

$$f^{-1}(f(x)) = f^{-1}([x-5]^{-1}) = 5 + [x-5] = x.$$

$$f(f^{-1}(x)) = f(5+x^{-1}) = ([5+x^{-1}] - 5)^{-1} = x.$$

21. $f(x) = x^2$, $x \le 0$.
Step 1: Let $y = x^2$. Then $x = -\sqrt{y}$, (since $x \le 0$).
Step 2: $f^{-1}(y) = -\sqrt{y}$.
Step 3: $f^{-1}(x) = -\sqrt{x}$.

$f^{-1}(f(x)) = f^{-1}(x^2) = -\sqrt{x^2} = -|x| = -(-x) = x$, (since the domain of f contains only nonpositive numbers).
$$f(f^{-1}(x)) = f(-\sqrt{x}) = (-\sqrt{x})^2 = x.$$

23. $y = (x-4)^3$; $x = y^{1/3}+4$; $f^{-1}(y) = y^{1/3}+4$; $f^{-1}(x) = x^{1/3}+4$.

$$f^{-1}(f(x)) = f^{-1}([x-4]^3) = ([x-4]^3)^{1/3}+4 = x.$$

$$f(f^{-1}(x)) = f(x^{1/3}+4) = ([x^{1/3}+4]-4)^3 = x.$$

25. $y = (2x-2)/(x+3)$; $x = (-3y-2)/(y-2)$; $f^{-1}(x) = (-3x-2)/(x-2)$.

$$f^{-1}(f(x)) = f^{-1}(\frac{2x-2}{x+3}) = \frac{-3(2x-2)-2(x+3)}{(2x-2)-2(x+3)} = x.$$

$$f(f^{-1}(x)) = f(\frac{-3x-2}{x-2}) = \frac{2(-3x-2)-2(x-2)}{(-3x-2)+3(x-2)} = x.$$

27. $f(x) = \dfrac{x^3+1}{x^3+2}$.

Step 1: Let $y = \dfrac{x^3+1}{x^3+2}$. Then $y(x^3+2) = x^3+1$.

$$yx^3+2y = x^3+1.$$
$$yx^3-x^3 = -2y+1.$$
$$x^3(y-1) = -2y+1.$$
$$x^3 = \frac{-2y+1}{y-1}, \text{ so } x = \sqrt[3]{\frac{-2y+1}{y-1}}.$$

Step 2: $f^{-1}(y) = \sqrt[3]{\dfrac{-2y+1}{y-1}}$.

Step 3: $f^{-1}(x) = \sqrt[3]{\dfrac{-2x+1}{x-1}}$.

$$f^{-1}(f(x)) = f^{-1}\left[\dfrac{x^3+1}{x^3+2}\right] = \sqrt[3]{\dfrac{-2\,\dfrac{x^3+1}{x^3+2}+1}{\dfrac{x^3+1}{x^3+2}-1}} = \sqrt[3]{\dfrac{-2(x^3+1)+(x^3+2)}{(x^3+1)-(x^3+2)}} = \sqrt[3]{x^3} = x.$$

Similar work with a complex fraction yields $f(f^{-1}(x)) = -x/(-1) = x$.

29. $f'(x) = 4x+1$. Use an auxiliary axis for f' and conclude f is decreasing on $(-\infty,-1/4]$ and f is increasing on $[-1/4, \infty)$.

Restrict x to $[-1/4, \infty)$: $y = 2x^2+x-4$; $x = -1/4 + (1/4)(33+8y)^{1/2}$;
$$f^{-1}(x) = -1/4 + (1/4)(33+8y)^{1/2}.$$

Restrict x to $(-\infty, -1/4]$: $y = 2x^2+x-4$; $x = -1/4 + (1/4)(33+8y)^{1/2}$;
$$f^{-1}(x) = -1/4 + (1/4)(33+8y)^{1/2}.$$

31. $(f^{-1})'(3) \approx 1/4$.   33. $(f^{-1})'(3) \approx 1/4$.

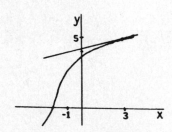   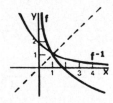

35. $y=2$ when $x=1$, and $f'(x) = 15x^4+1$, so $(f^{-1})'(2) = 1/f'(1) = 1/16$.

37. $y=2$ when $x = \pi/4$, and $f'(x) = 2\sec^2 x$, so $(f^{-1})'(2) = 1/f'(\pi/4) = 1/4$.

39. Let $y = g(x)$ and $z = f(y)$. Then $z = f\circ g(x)$.
For each z, there is a unique y since f^{-1} exists.
For each y, there is a unique x since g^{-1} exists.
Therefore, for each z, there is a unique x, so $(f\circ g)^{-1} = h^{-1}$ exists.
We need to show that $g^{-1} \circ f^{-1}$ is $(f\circ g)^{-1}$.
$[(g^{-1}\circ f^{-1})\circ(f\circ g)](x)$ $(g^{-1}\circ f^{-1}\circ f\circ g)(x) = (g^{-1}\circ g)(x) = x$
$[(f\circ g)\circ(g^{-1}\circ f^{-1})](x) = (f\circ g\circ g^{-1}\circ f^{-1})(x) = (f\circ f^{-1})(x) = x$
Therefore, $h^{-1} = g^{-1}\circ f^{-1}$.

COOL ASI?

41. $f'(x) = \sqrt{1+\cos^2 x} > 0$, so f is strictly increasing.

   (a) $(f^{-1})'(A) = 1/f'(\pi/2) = 1$

   (b) $(f^{-1})'(B) = 1/f'(5\pi/6) = \frac{2\sqrt{7}}{7} \approx 0.7559$

   (c) $(f^{-1})'(0) = 1/f'[f^{-1}(0)] = 1/f'(0) = 1/\sqrt{2} \approx 0.7071$

43. $A = \int_0^1 f(x)dx = 2/5.$

    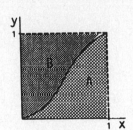

    $B = \int_0^1 f^{-1}(y)dy = 1 - 2/5 = 3/5.$

45. Let $y = f(x)$. Then $y = x^{p-1}$, $x = y^{\frac{1}{p-1}}$, so $f^{-1}(y) = y^{\frac{1}{p-1}}$.

   Now consider that $\dfrac{1}{p-1} = \dfrac{\frac{1}{p}}{1 - \frac{1}{p}} = \dfrac{1 - \frac{1}{q}}{\frac{1}{q}} = q-1$, so $f^{-1}(y) = y^{q-1}$.

   By Problem 44, $ab < \int_0^a x^{p-1}dx + \int_0^b y^{q-1}dy = \left[\dfrac{x^p}{p}\right]_0^a + \left[\dfrac{y^q}{q}\right]_0^b = \dfrac{a^p}{p} + \dfrac{a^q}{q}.$

Problem Set 7.3   The Natural Exponential Function

1. (a) 20.0855.   (b) 0.0183.   (c) 8.1662.   (d) 544.5719.

   (e) 4.1133.   (f) 0.4692.   (g) 1.2014.

3. $e^{2\ln x} = e^{\ln x^2} = x^2$, $x > 0$.   5. $\sin x$.

7. $\ln(x^2) + \ln(e^{-2x}) = 2\ln x - 2x.$

9. $e^{\ln 2 + \ln x} = e^{\ln(2x)} = 2x$, $x>0$.

11. $2e^{2x+1}.$ 13. $\dfrac{e^{\sqrt{x+1}}}{2\sqrt{x+1}}.$

15. $y = e^{\ln x} = x$, so $D_x y = 1.$

17. $x^2 e^x + e^x 2x = x e^x (x+2)$.

19. $e^{\sqrt{x}}(1/2)x^{-1/2} + (1/2)(e^x)^{-1/2}e^x = \dfrac{e^{\sqrt{x}} + \sqrt{xe^x}}{2\sqrt{x}}$.

21. $\dfrac{d}{dx}(e^{xy} + y) = \dfrac{d}{dx}(2)$.

$e^{xy}(x\dfrac{dy}{dx} + y) = 0$; $e^{xy} \neq 0$; so $x\dfrac{dy}{dx} + y = 0$; $\dfrac{dy}{dx} = \dfrac{-y}{x}$.

23. (a) Reflection of $y = x^e$ (b) Reflection of $y = e^x$
        in the x-axis.   in the y-axis.

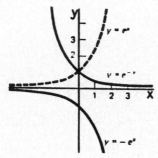

25. $f'(x) = -xe^{-x}(x-2)$; $f''(x) = e^{-x}(x^2 - 4x + 2)$.
    f is decreasing on $(-\infty, 0]$ and on $[2, \infty)$ and
    increasing on $[0,2]$. f is concave up on
    $(-\infty, 0.59)$ and on $(3.41, \infty)$ and concave down
    on $(0.59, 3.41)$, approximately. [The exact
    values are $2 \pm \sqrt{2}$.] Some points are $(0,0)$,
    $(-1, 2.72)$, $(0.59, 0.19)$, $(3.41, 0.38)$, and $(2,054)$.

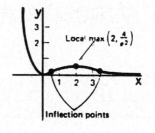

27. $f(x) = e^{-x^2}$.
    Domain: R.
    Symmetry: With resspect to the y-axis. [f is an even function.]
    Intercepts: No x-intercepts; y-intercept is $f(0) = 1$.
    Monotonicity: $f'(x) = e^{-x^2}(-2x) = -2xe^{-x^2}$.

	Inc (+)	(0)	Dec (-)
		0	x

Concavity: $f''(x) = e^{-x^2}(-2) + (-2x)e^{-x^2}(-2x) = e^{-x^2}(4x^2 - 2)$

$\qquad = e^{-x^2}(2x + \sqrt{2})(2x - \sqrt{2})$.

	CU (+)	(0)	CD (-)	(0)	CU (+)
		-0.71		0.71	x

Asymptotes: $y=0$ (horizontal) since $\lim\limits_{x \to \pm\infty} e^{-x^2} = \lim\limits_{x \to \pm\infty} \dfrac{1}{e^{x^2}} = 0$.

Summary	
$(-\infty, -0.71)$	Inc and CU
At -0.71	Inflection
$(-0.71, 0)$	Inc and CD
At 0	Levels off
$(0, 0.71)$	Dec and CD
At 0.71	Inflection
$(0.71, \infty)$	Dec and CU

x	$f(x)$
0	1 (Maximum)
0.71	0.61 (Inflection Point)
1	0.37
2	0.02

29. $\displaystyle\int (1/3)\, c^u du = (1/3)e^u + C = (1/3)e^{3x+1}.$ [Let $u = 3x+1$]

31. $\displaystyle\int (1/2)e^u du = (1/2)e^u + C = (1/2)e^{x^2+6x} + C.$ [Let $u = x^2+6x$]

33. $\displaystyle\int \frac{e^{-1/x}}{x^2}\, dx = \int e^{-x^{-1}}(x^{-2})\, dx$

$$\boxed{\begin{array}{l} \text{Let} \quad u = -x^{-1} \\ \text{Then } du = x^{-2} dx \end{array}}$$

$$= \int e^u du = e^u + C = e^{-1/x} + C.$$

35. $(1/2)\displaystyle\int_3^5 e^u du = (1/2)\left[e^u\right]_3^5 = (1/2)(e^5 - e^3) \approx 64.1638.$ [Let $u = 2x+3$]

37. (By disks) $\Delta V \approx \pi(e^x)^2 \Delta x.$

$V = \displaystyle\int_0^{\ln 3} \pi e^{2x}\, dx = \left[(\pi/2)e^{2x}\right]_0^{\ln x}$

$= 4\pi \approx 12.5664.$

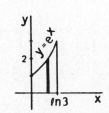

39. The line has slope $\dfrac{1/e - 1}{1-0} = \dfrac{1-e}{e}$, and y-intercept 1. An equation of the line is $y = (\dfrac{1-e}{e})x+1$.

$\Delta A \approx \left[[(\dfrac{1-e}{e})x+1] - e^{-x} \right] \Delta x$.

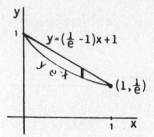

$A = \displaystyle\int_0^1 [(\dfrac{1-e}{e})x+1 - e^{-x}]\,dx = \left[(\dfrac{1-e}{e})\dfrac{x^2}{2} + x + e^{-x} \right]_0^1 = \dfrac{3-e}{2e} \approx 0.0518.$

41. (a) $10! = 3,628,800$, $[2\pi(10)]^{1/2}(10/e)^{10} \approx 3,598,696$.
    (b) $60! \approx [2\pi(60)]^{1/2}(60/e)^{60} \approx 8.3094 \times 10^{81}$.

43. $dx/dt = e^t(\cos t + \sin t)$, so $(dx/dt)^2 = e^{2t}(1+\sin 2t)$.
    $dy/dt = e^t(-\sin t + \cos t)$, so $(dy/dt)^2 = e^{2t}(1-\sin 2t)$.

$L = \displaystyle\int_0^{\pi} [(dx/dt)^2 + (dy/dt)^2]^{1/2}\,dt = \int_0^{\pi} \sqrt{2}\,e^t\,dt = \left[\sqrt{2}\,e^t \right]_0^{\pi} \approx 31.3117.$

45. (a) $\displaystyle\lim_{x\to 0^+} \dfrac{\ln x}{1+(\ln x)^2} = \lim_{x\to 0^+} \dfrac{\frac{1}{\ln x}}{\frac{1}{(\ln x)^2}+1} = \dfrac{0}{0+1} = 0.$

$\displaystyle\lim_{x\to\infty} \dfrac{\ln x}{1+(\ln x)^2} = \lim_{x\to\infty} \dfrac{\frac{1}{\ln x}}{\frac{1}{(\ln x)^2}+1} = \dfrac{0}{0+1} = 0.$

(b) $f'(x) = \dfrac{[1+(\ln x)^2](\frac{1}{x}) - (\ln x)[2(\ln x)(\frac{1}{x})]}{[1+(\ln x)^2]^2}$

$= \dfrac{1 + (\ln x)^2 - 2(\ln x)^2}{x[1 + (\ln x)^2]^2} = \dfrac{1 - (\ln x)^2}{x[1 + (\ln x)^2]^2}.$

$f'(x) = 0$ if $\ln x = \pm 1$; i.e., if $x = e$ or $x = e^{-1}$.

Auxiliary axis for $f'(x)$:

$$\underset{\substack{0 \qquad\qquad 1/e \qquad\qquad e \qquad\qquad x}}{(-) \quad (0) \quad (+) \quad (0) \quad (-)}$$

$f(1/e) = -0.5$ is the minimum value of f.
$f(e) = 0.5$ is the maximum value of f.

(c) $F'(x) = f(x^2)(2x) = \dfrac{2x \ln(x^2)}{1+[\ln(x^2)]^2}$; $F'(\sqrt{e}) = \dfrac{2\sqrt{e}(1)}{1+(1)^2} = \sqrt{e} \approx 1.6487$.

47. Consider the area of the region R bounded by $f(x) = e^x$ and the x-axis for x in $[0,1]$.

Let $\{0, \frac{1}{n}, \frac{2}{n}, \cdots, \frac{n}{n}\}$ be a partition of $[0,1]$. f is increasing on $[0,1]$. Therefore, the area of the corresponding circumscribed polygon for R is $\dfrac{e^{1/n}}{n} + \dfrac{e^{2/n}}{n} + \cdots + \dfrac{e^{n/n}}{n}$.

f is integrable on $[0,1]$,

so $\lim\limits_{n\to\infty} \left[\dfrac{e^{1/n}}{n} + \dfrac{e^{2/n}}{n} + \cdots + \dfrac{e^{n/n}}{n}\right] = \int_0^1 e^x dx = \left[e^x\right]_0^1 = e-1 \approx 1.7183$

49. (a) 3.10968.   (b) 0.9099.

51. 4.26141

53. It is like $-x$; it is like $2\ln x$.

Problem Set 7.4   General Exponential and Logarithmic Functions

1. $3^x = 9$, so $x = 2$.   **3.** $x = 9^{3/2} = 27$.

5. $x = 3(10)^{1/2} \approx 9.4868$.

7. $(x+1)/x = 2^2 = 4$, $x > 0$; $x = 1/3$.

9. $\log_5 13 = \dfrac{\ln 13}{\ln 5} \approx 1.5937$.

11. $(1/5)[\ln(8.16)]/(\ln 11) \approx 0.1751$.

13. $x = (\ln 19)/(\ln 2) \approx 4.2479$.

15. $\ln 4^{3x-1} = \ln 5$; $(3x-1)\ln 4 = \ln 5$; $3x \ln 4 - \ln 4 = \ln 5$;

$x = \dfrac{\ln 5 + \ln 4}{3\ln 4} = \dfrac{\ln 20}{\ln 64} \approx 0.7203$.

17. $5^{x^2}(\ln 5)(2x) = 5^{x^2}2x\ln 5.$

19. $1/(\ln 2)$, since $\log_2 e^x = (\ln e^x)/(\ln 2) = x/(\ln 2).$

21. $D_x[2^x\ln(x+5)] = (2^x)(\frac{1}{x+5})+[\ln(x+5)](2^x\ln 2) = \dfrac{2^x[1 + (x+5)\ln 2(x+5)]}{x+5}.$

23. $\displaystyle\int(1/2)2^u du = (2^u)/2\ln 2 + C = (2^{x^2})/(\ln 4) + C.$ [Let $u = x^2$]

25. $2\displaystyle\int_1^2 5^u du = 2\left[5^u/\ln 5\right]_1^2 = 40/\ln 5 \approx 24.8534.$ [Let $u = \sqrt{x}$]

27. $y = 10^{x^2} + x^{20}.$
$\dfrac{dy}{dx} = 10(x^2)(\ln 10)(2x) + 20x^{19} = 2x(\ln 10)10^{(x^2)} + 20x^{19}.$

29. $(\pi+1)x^\pi + (\pi+1)^x\ln(\pi+1).$

31. $y = e^{(\ln x)[\ln(x^2+1)]}.$
$dy/dx = e^{(\ln x)[\ln(x^2+1)]}\left[\dfrac{2x\ln x}{x^2+1} + \dfrac{\ln(x^2+1)}{x}\right] = (x^2+1)^{\ln x}\left[\dfrac{2x\ln x}{x^2+1} + \dfrac{\ln(x^2+1)}{x}\right].$

33. $f(x) = x^{\sin x} = e^{(\sin x)\ln x}.$
$f'(x) = e^{(\sin x)\ln x}[(\sin x)(1/x) + (\ln x)(\cos x)] = x^{\sin x}(\dfrac{\sin x}{x} + \ln x \cos x).$
Then $f'(1) = 1^{\sin 1}(\dfrac{\sin 1}{1} + \ln 1 \cos 1) = \sin 1 \approx 0.8415.$

35. $\log_{1/2}x = -\log_2 x.$ [See work in Problem 36.]

37. $E = 10^{[(5.54/0.67) - \log_{10}(0.37)]} \approx 5.0171 \times 10^8$ kilowatt-hours.

39. Let a denote the frequency of C. $\bar{C}$ is the 12th note after C, so the frequency of $\bar{C}$ is ar^{12}. We are also given that the frequency of $\bar{C}$ is $2a$. Therefore, $ar^{12} = 2a$; $r^{12} = 2$; $r = \sqrt[12]{2}$. $\bar{C}$ is the 3rd note after A, so the frequency of $\bar{C}$ is $440r^3 = 440(\sqrt[4]{2}) \approx 523.$

41. If $y = Ab^x$, $\ln y = \ln A + x\ln b = (\ln b)x + (\ln A)$. Therefore, plotting $\ln y$ against x on semilog graph paper, $y = Ab^x$ shows up as a line with slope $\ln b$ and y-intercept $\ln A$.

If $y = Ax^b$, $\ln y = \ln A + b\ln x = b(\ln x) + (\ln A)$. Therefore, plotting $\ln y$ against $\ln x$ on loglog graph paper, $y = Ax^b$ shows up as a line with slope b and y-intercept $\ln A$.

43. $D_f = D_g = (0,\infty)$. $f(x) = x^{x^2}$, so $f(x) = g(x)$ only if x is 1 or 2. Using the result of Problem 42,

$$f'(x) = x(x^x)^{x-1}(x^x + x^x\ln x) + (x^x)^x[\ln(x^x)](1)$$

$$= x(x^x)^x(1+\ln x) + (x^x)^x\ln(x^x) = (x^x)^x(x + x\ln x + x\ln x)$$

$$= (x^x)^x x(1+2\ln x) = x^{x^2+1}(1+2\ln x).$$

$$g'(x) = (x^x)(x^{x^x-1})(1) + x^{(x^x)}(\ln x)(x^x+x^x\ln x)$$

$$= x^{x^x+x-1} + x^{x^x+x}(\ln x)(1+\ln x) = x^{x^x+x-1}[1+x(\ln x)(1+\ln x)]$$

$$= x^{x^x+x-1}(1 + x\ln x + x\ln^2 x).$$

45. (a) $\ln[f(x)] = a\ln x - x\ln a = \left[a\dfrac{\ln x}{x} - \ln a\right]x \to -\infty$ as $x \to \infty$.

[See Problem 42(b) of Section 7.1 for $\dfrac{\ln x}{x} \to 0$ as $x \to \infty$.]

That is, $\lim\limits_{x \to \infty} \ln[f(x)] = -\infty$, so $\lim\limits_{x \to \infty} f(x) = 0$.

(b) $f'(x) = \dfrac{(a^x)(ax^{a-1}) - (x^a)(a^x\ln a)}{(a^x)^2} = \dfrac{ax^{a-1} - x^a\ln a}{a^x} = \dfrac{x^{a-1}(a-x\ln a)}{a^x}$

$f'(x) = 0$ if $x = 0$ (one-sided), or if $x = \dfrac{a}{\ln a}$.

Auxiliary axis for $f'(x)$:

	(+)	(0)	(−)	
0		a/ℓna		x

Thus, $f(x)$ is maximum at $x = \dfrac{a}{\ln a}$.

(c) Case a=e: From (b), $f(x) = \dfrac{x^e}{e^x}$ has a maximum at $\dfrac{e}{\ln e} = e$. $f(e) = 1$.

Thus, $\dfrac{x^e}{e^x} = 1$ only if x=e.

Case a≠e: $f(a)=1$, and f has a maximum at $x = \dfrac{a}{\ln a} \ne a$, so $f\left(\dfrac{a}{\ln a}\right) > 1$.
Hence, $f(x) = 1$ has two positive solutions.
That is, $\dfrac{x^a}{a^x} = 1$, and $x^a = a^x$, has two positive solutions.

(d) From the a=e case of (c), $\dfrac{x^e}{e^x} < 1$ for $x \ne e$.

In particular, $\dfrac{\pi^e}{e^\pi} < 1$, so $\pi^e < e^\pi$.

47. 1, (0.36788,0.6922)

49. 20.2259

Problem Set 7.5 Exponential Growth and Decay

1. $y = 4e^{-5t}$.

3. The solution is $y = y_o e^{0.006t}$ [y_p is the value of y when t=0 -- see the
first derivation in this section.] Although we don't know the value of y
when t=0, we do know that y=2 if t=10. Substituting those values we can
solve for y_o.

$2 = y_o^{0.006(10)}$, so $y_o = 2e^{-0.06} \approx 1.8835$.

Therefore, $y = 2e^{-0.06}e^{0.006t}$, or $y \approx 1.8835e^{0.006t}$.

5. $y = 10000e^{kt}$. At (10,24000), $24000 = 10000e^{10k}$, so $k = 0.1\ln(2.4)$.
Therefore $y(t) = 10000(2.4)^{t/10}$; $y(25) = 10000(2.4)^{2.5} \approx 89,234$.

7. $30000 = 10000e^{[0.1\ln 2.4]t}$; $t = (10\ln 3)/(\ln 2.4)$ 12.5488 days.

9. End of 1 year: $(1.032)(4.5) \approx 4.644$ million.

    End of 2 years: (1.032)(pop at end of 1 year) $= (1.032)^2(4.5) \approx 4.793$ million.

    End of 10 years: $(1.032)^{10}(4.5) \approx 6.166$ million.

    End of 100 years: $(1.032)^{100}(4.5) \approx 105.0$ million.

11. $y = 10e^{kt}$; $5 = 10e^{810k}$, so $k = (-\ln 2)/810$.

    $y(t) = 10e^{[(-\ln 2)/810]t}$; $y(300) = 10e^{(-300\ln 2)/810} \approx 7.7358$ gm.

13. Let $y = 1e^{kt}$. $1/2 = e^{5730k}$, so $k = (-\ln 2)/5730$.
    When $y = 0.70$, $t = [5730\ln(0.70)]/(-\ln 2) \approx 2949$, so the fort burned about 2,950 years ago.

15. $\dfrac{dT}{dt} = k(T-75)$; $\dfrac{1}{T-75}\, dT = k\,dt$; $\ln(T-75) = kt + C$.

    $T=300$ when $t=0$: $\ln(300-75) = k(0) + C \Rightarrow C = \ln 225$.
    $\qquad \ln(T-75) = kt + \ln 225$.

    $\qquad T-75 = e^{kt+\ln 225} = e^{kt}e^{\ln 225} = e^{kt}225$; thus, $T = 225e^{kt}+75$.

    $T=200$ when $t=0.5$: $200 = 225e^{k(0.5)}+75 \Rightarrow k = 2\ln\left[\dfrac{125}{225}\right] = 2\ln(5/9)$.

    Therefore, $T = 225e^{[2\ln(5/9)]t}+75 = 225(25/81)^t+75$.

    After 3 hours: $T = 225(25/81)^3+75 \approx 81.62°F$.

17. (a) $A(2) = 375(1 + 0.095/1)^{1(2)} \approx \$449.63$.
    (b) $A(2) = 375(1 + 0.095/12)^{12(2)} \approx \$453.13$.
    (c) $A(2) = 375(1 + 0.095/365)^{365(2)} \approx \$453.46$.
    (d) $A(2) = 375e^{0.095(2)} \approx \$453.47$.

19. (a) $2A_0 = A_0(1 + 0.12/12)^{12t}$, so $t = (\ln 2)/[12\ln(1.01)] \approx 5.8051$ years, or 5 years, 10 months.

    (b) $2A_0 = A_0e^{0.12t}$, so $t = (\ln 2)/(0.12) \approx 5.7762$ years, or 5 years, 9 months, 9 days, 10 hours, 35 minutes, 2.4 seconds.

21. $1984 - 1626 = 358$ years.
    $A(358) = 24e^{0.06(358)} \approx 51,513,043,900$. [51.5 billion dollars]

23. $2A_0 = A_0(1 + 0.01p)^t$.
    Therefore, $t = (\ln 2)/[\ln(1+0.01p)] \approx (\ln 2)/(0.01p) \approx 69.31/p$ years.

25. $y = 16/(1+3e^{-0.0298t})$.

t	y
0	4
25	6.60
50	9.55
75	12.11
100	13.88
125	14.92

27. (a) $\lim\limits_{x\to 0}(1-x)^{1/x} = \lim\limits_{z\to 0}((1+z)^{1/z})^{-1} = e^{-1} = 1/e.$ [Let $z = -x$]

(b) $\lim\limits_{x\to 0}(1+3x)^{1/x} = \lim\limits_{z\to 0}(1+z)^{3/z} = \lim\limits_{z\to 0}\left[(1+z)^{1/z}\right]^3 = e^3.$ [Let $z=3x$]

(c) $\lim\limits_{n\to\infty}\left[\dfrac{n+2}{n}\right]^n = \lim\limits_{z\to 0}(1+z)^{2/z} = \lim\limits_{z\to 0}\left[(1+z)^{1/z}\right]^2 = e^2.$ [Let $z = \dfrac{2}{n}$]

(d) $\lim\limits_{n\to\infty}\left[\dfrac{n-1}{n}\right]^{2n} = \lim\limits_{z\to 0^-}(1+z)^{-2/z} = \lim\limits_{z\to 0}\left[(1+z)^{1/z}\right]^{-2} = e^{-2}.$ [Let $z = \dfrac{-1}{n}$]

29. Let $t=0$ in 1985.
Then it is given that $a = 1.2\%$, $b = 0.06$ million, $y_0 = 10$ million.
The solution is $y = (10 + \dfrac{0.06}{0.012})e^{0.012t} - \dfrac{0.06}{0.012} = 15e^{0.012t} - 5.$

When $t=25$ (value of t in 2010), $y = 15e^{0.012(25)} - 5 \approx 15.25$, so in the year 2010 the population will be about 15,250,000.

31. There is enough land to supply food for $(13,500,000 \text{ mi}^2)(640 \text{ acre/mi}^2)(2 \text{ persons/acre}) = 17,280,000,000$ persons. Let $t=0$ in 1975. Then $y = 5,000,000,000e^{0.019t}.$

$17,280,000,000 = 5,000,000,000e^{0.109t}$; $3.46 = e^{0.019t}$;
$t = \dfrac{\ln(3.46)}{0.019} \approx 62.69.$

Under the assumptions given, it will be about 63 years before the world reaches the maximum population.

33. 2000:   6.56   vs   6.60
         2040:   14.49   vs 11.17
         2090:   38.99   vs 14.58

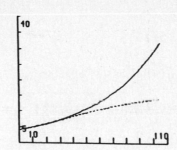

Problem Set 7.6 The Inverse Trigonometric Functions

1. $\pi/3$.

3. $-\pi/4$, since $\sin(-\pi/4) = -\sqrt{2}/2$ and $-\pi/4$ is on the interval $[-\pi/2, \pi/2]$.

5. $-\pi/3$.   **7.** $2\pi/3$.

9. $\sec^{-1}(-2) = \cos^{-1}(-1/2) = 2\pi/3$, since $\cos(2\pi/3) = -1/2$ and $2\pi/3 \in [0, \pi]$.

11. 0.541.   **13.** 0.1262.   **15.** -1.3780. **17.** 0.3048.

19. 2.0177.   **21.** 0.6075.   **23.** -1.1598.

25. $\sin^{-1}(x/8) = \cos^{-1}(64-x^2)^{1/2}/8 = \tan^{-1}x(64-x^2)^{-1/2} = \sec^{-1}8(64-x^2)^{-1/2}$.

27. $\theta = \sin^{-1}(5/x) = \cos^{-1}(\sqrt{x^2-25}/x)$

$\qquad = \tan^{-1}(5/\sqrt{x^2-25}) = \sec^{-1}(x/\sqrt{x^2-25})$

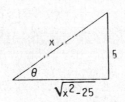

29. $\sin^{-1}3(x^2+9)^{-1/2} - \sin^{-1}(x^2+1)^{-1/2} = \sin^{-1}2x(x^2+9)^{-1/2}(x^2+1)^{-1/2}$.

$\cos^{-1}x(x^2+9)^{-1/2} - \cos^{-1}x(x^2+1)^{-1/2} = \cos^{-1}(x^2+3)(x^2+9)^{-1/2}(x^2+1)^{-1/2}$.

$\tan^{-1}(3/x) - \tan^{-1}(1/x) = \tan^{-1}2x(x^2+3)^{-1}$.

$\sec^{-1}(x^2+3)^{-1}(x^2+9)^{1/2}(x^2+1)^{1/2}$.

31. $1 - 2\sin^2[\sin^{-1}(-2/3)] = 1 - 2[-2/3]^2 = 1/9.$

33. $\sin[\cos^{-1}(3/5) + \cos^{-1}(5/13)]$

$= \sin[\cos^{-1}(3/5)]\cos[\cos^{-1}(5/13)] + \cos[\cos^{-1}(3/5)]\sin[\cos^{-1}(5/13)]$

$= \sqrt{1-(3/5)^2}\,(5/13) + (3/5)\sqrt{1-(5/13)^2} = \frac{4}{5}\frac{5}{13} + \frac{3}{5}\frac{12}{13} = \frac{56}{65} \approx 0.8615.$

35. $\tan(\sin^{-1}x) = \dfrac{\sin(\sin^{-1}x)}{\cos(\sin^{-1}x)} = \dfrac{x}{(1-x^2)^{1/2}}.$

37. $\cos(2\sin^{-1}x) = 1-2\sin^2(\sin^{-1}x) = 1-2x^2.$

39. $\cos(\tan^{-1}x) = \dfrac{1}{\sec(\tan^{-1}x)} = \dfrac{1}{\sqrt{1+x^2}}.$

41. Let $y = \tan^{-1}x.$

(a) $\displaystyle\lim_{x\to\infty} \tan^{-1}x = \lim_{y\to\pi/2^-} y = \pi/2.$ (b) $\displaystyle\lim_{x\to-\infty} \tan^{-1}x = \lim_{y\to-\pi/2^+} y = -\pi/2.$

43.

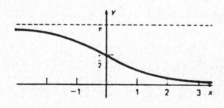

45. In each case, use the three-step approach (introduced in Section 7.2) for finding the inverse of a function.

(a) Restrict $2x$ so that $0 \le 2x \le \pi$; i.e., so that x is in $[0,\ \pi/2]$. Let $y = 3\cos2x.$

$y/3 = \cos2x; \ 2x = \cos^{-1}(y/3); \ x = (1/2)\cos^{-1}(y/3).$ [Step 1]
$f^{-1}(y) = (1/2)\cos^{-1}(y/3).$ [Step 2]
$f^{-1}(x) = (1/2)\cos^{-1}(x/3).$ [Step 3]

(b) Restrict $3x$ so that $-\pi/2 \le 3x \le \pi/2$; i.e., x is in $[-\pi/6,\ \pi/6]$. Let $y = 2\sin3x.$

$y/2 = \sin3x; \ 3x = \sin^{-1}(y/2); \ x = (1/3)\sin^{-1}(y/2).$ [Step 1]
$f^{-1}(y) = (1/3)\sin^{-1}(y/2).$ [Step 2]
$f^{-1}(x) = (1/3)\sin^{-1}(x/2).$ [Step 3]

(c) Restrict x to $(-\pi/2,\ \pi/2)$.
Let $y = (1/2)\tan x$.

$2y = \tan x;\quad x = \tan^{-1}2y.$ [Step 1]
$f^{-1}(y) = \tan^{-1}2y.$ [Step 2]
$f^{-1}(x) = \tan^{-1}2x.$ [Step 3]

(d) Restrict x so that $-\pi/2\leq(1/x)\leq\pi/2$; i.e., $x \leq -2/\pi$ or $x \geq 2/\pi$; i.e.
x is in $(-\infty,\ -2/\pi]\cup[2/\pi,\infty)$.
Let $y = \sin(1/x)$.

$\sin^{-1}y = 1/x;\quad x = 1/\sin^{-1}y.$ [Step 1]
$f^{-1}(y) = 1/\sin^{-1}y.$ [Step 2]
$f^{-1}(x) = 1/\sin^{-1}x.$ [Step 3]

47. $\tan(4x) = \tan(2x+2x) = \dfrac{2\tan 2x}{1 - \tan^2 2x} = \dfrac{4(1-\tan^2 x)\tan x}{(1-\tan^2 x)^2 - 4\tan^2 x}$

$$= \frac{4(1-\tan^2 x)\tan x}{1 - 6\tan^2 x + \tan^4 x}.$$

Therefore, $\tan[4\tan^{-1}(1/5)] = \dfrac{4[1-(1/5)^2](1/5)}{1-6(1/5)^2+(1/5)^4} = \dfrac{120}{119}.$

[The results above will now be used.]

$\tan[4\tan^{-1/2}(1/5)-\tan^{-1}(1/239)] = \dfrac{\tan[4\tan^{-1}(1/5)] - \tan[\tan^{-1}(1/239)]}{1 + \tan[4\tan^{-1}(1/5)]\tan[\tan^{-1}(1/239)]}$

$$= \frac{(120/119) - (1/239)}{1 + (120/119)(1/239)} = 1.$$

Thus, $4\tan^{-1/2}(1/5)-\tan^{-1}(1/239) = \tan^{-1/2}(1) = \pi/4.$

49.

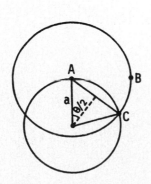
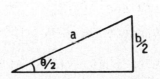

$$a^2 - (b/2)^2 = \frac{\sqrt{4a^2-b^2}}{2}$$

Shaded area = Area(upper larger semicircle) + 2[Area(Sector ABC) - Area(Segment AC)]

$$= (1/2)\pi b^2 + 2[(1/2)b^2(\theta/2) - (1/2)a^2(\theta - \sin\theta)]$$

$$= (\pi/2)b^2 + b^2\sin^{-1}(\tfrac{b/2}{a}) - a^2[2\sin^{-1}(\tfrac{b/2}{a}) - 2\sin(\theta/2)\cos(\theta/2)]$$

$$= (\pi/2)b^2 + b^2\sin^{-1}(\tfrac{b}{2a})$$
$$\qquad - 2a^2\sin^{-1}(\tfrac{b}{2a}) - 2a^2\sin[\sin^{-1}(\tfrac{b}{2a})]\cos[\sin^{-1}(\tfrac{b}{2a})]$$

$$= (\pi/2)b^2 + (b^2 - 2a^2)\sin^{-1}(\tfrac{b}{2a}) - 2(\tfrac{b}{2a})\frac{\sqrt{4a^2 - b^2}}{2a}$$

$$= \frac{\pi b^2}{2} + (b^2 - 2a^2)\sin^{-1}(\tfrac{b}{2a}) - \frac{b\sqrt{4a^2 - b^2}}{2a^2}.$$

51. $\arccos x = \dfrac{\pi}{2} - \arcsin x$

Problem Set 7.7 Derivatives of Trigonometric Functions

1. $2\cos(x-2)[-\sin(x-2)] = -2\sin(x-2)\cos(x-2) = -\sin 2(x-2).$

3. $\dfrac{dy}{dx} = (\cot x)(-\csc x \cot x) + (\csc x)(-\csc^2 x) = -\csc x(\cot^2 x + \csc^2 x).$

5. $e^{\cot x}(-\csc^2 x) = -e^{\cot x}\csc^2 x.$

7. $\dfrac{dy}{dx} = (\tan x)e^{\tan x}(\sec^2 x) + (e^{\tan x})(\sec^2 x) = e^{\tan x}(1+\tan x)\sec^2 x.$

9. $\dfrac{dy}{dx} = \dfrac{1}{\sqrt{1-(x^2)^2}}(2x) = \dfrac{2x}{\sqrt{1-x^4}}.$

11. $\dfrac{dy}{dx} = 7\dfrac{-1}{(1-2x)^{1/2}}\dfrac{1}{2(2x)^{1/2}}2 = \dfrac{-7}{[2x(1-2x)]^{1/2}}.$

13. $\dfrac{e^x}{1+e^{2x}}.$

15. $\dfrac{dy}{dx} = (x)\,\dfrac{1}{1+x^2} + (\tan^{-1}x)(1) = \dfrac{x}{1+x^2} + \tan^{-1}x.$

17. $\dfrac{dy}{dx} = \cos(\tan^{-1}x)(1+x^2)^{-1} = (1+x^2)^{-1/2}(1+x^2)^{-1} = (1+x^2)^{-3/2}.$ [See Problem 39, Section 7.6.]

19. $\dfrac{dy}{dx} = \dfrac{2}{x(x^4-1)^{1/2}}.$

21. $\dfrac{dy}{dx} = \dfrac{1}{1 + \left[\dfrac{1-x}{1+x}\right]^2}\ \dfrac{(1+x)(-1)-(1-x)(1)}{(1+x)^2} = \dfrac{-2}{(1+x)^2+(1-x)^2} = \dfrac{-2}{2+2x^2} = \dfrac{-1}{1+x^2}.$

23. $\displaystyle\int (1/2)\sin u\ du = (-1/2)\cos u + C = (-1/2)\cos(x^2) + C.$ [Let $u = x^2$]

25. $\displaystyle\int -1/u\ du = -\ln|u| + C = -\ln|\cos x| + C.$ [Let $u = \cos x$]

27. $\displaystyle\int \dfrac{\sec^2 x}{\tan x}\ dx = \int \dfrac{1}{u}\ du = \ln|u| + C = \ln|\tan x| + C.$

| Let $u = \tan x$ |
| Then $du = \sec^2 x\ dx$ |

29. $(1/2)\displaystyle\int_1^{e^2} \cos u\ du = (1/2)\Big[\sin u\Big]_1^{e^2} \approx .02619$ [Let $u = e^{2x}$]

31. $\Big[\sin^{-1}x\Big]_0^{\sqrt{2}/2} = \sin^{-1}(\sqrt{2}/2)\ \sin^{-1}(0) = \pi/4 - 0 = \pi/4 \approx 0.7854.$

33. $\displaystyle\int_{-1}^1 \dfrac{1}{1+x^2}\ du = \Big[\tan^{-1}x\Big]_{-1}^1 = \tan^{-1}(1) - \tan^{-1}(-1) = \pi/4 - (-\pi/4) = \pi/2.$

35. $(1/2)\displaystyle\int (1+u^2)^{-1}\ du = (1/2)\tan^{-1}u + C = (1/2)\tan^{-1}2x + C.$ [Let $u = 2x$]

37. $\displaystyle\int \dfrac{1}{[1-(x/a)^2]^{1/2}}\ \dfrac{1}{a}\ dx = \int \dfrac{1}{(1-u^2)^{1/2}}\ du = \sin^{-1}u + C = \sin^{-1}(x/a) + C.$

39. $\dfrac{d}{dx}\left[\dfrac{1}{a}\tan^{-1}(x/a) + C\right] = \dfrac{1}{a}\ \dfrac{1}{1+(x/a)^2}\ \dfrac{1}{a} = \dfrac{1}{a^2}\ \dfrac{1}{1+(x^2/a^2)} = \dfrac{1}{a^2+x^2}.$

215

41. $\dfrac{x}{2} \dfrac{1}{2(a^2-x^2)^{1/2}} (-2x) + (a^2-x^2)^{1/2} \dfrac{1}{2} + \dfrac{a^2}{2} \dfrac{1}{(1-x^2/a^2)^{1/2}} \dfrac{1}{a} = (a^2-x^2)^{1/2}.$

43. $y = 4(x^2+1)^{-1}.$

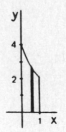

$A = \displaystyle\int_0^1 4(x^2+1)^{-1}\,dx = 4\left[\tan^{-1}x\right]_0^1$

$= 4\,(\tan^{-1}1 - \tan^{-1}0) = 4(\pi/4 - 0) = \pi.$

45. $\beta(b) = \tan^{-1}(12/b) - \tan^{-1}(2/b),\ b > 0$

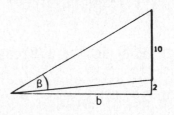

$\beta'(b) = \dfrac{1}{1+(144/b^2)} \dfrac{-12}{b^2} - \dfrac{1}{1+(4/b^2)} \dfrac{-2}{b^2}$

$= \dfrac{-12}{b^2+144} + \dfrac{2}{b^2+4} = \dfrac{-12b^2-48+2b^2+288}{(b^2+144)(b^2+4)}$

$= \dfrac{-10(b^2-24)}{(b^2+144)(b^2+4)}$. Auxiliary axis for β':

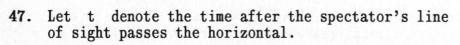

β is maximum where $b = \sqrt{24} \approx 4.8990$ ft.   [about $4'10.8''$]

47. Let t denote the time after the spectator's line of sight passes the horizontal.

$y = 60\tan\theta;\ dy/dt = 60\sec^2\theta\ (d\theta/dt).$

At $t=6$, $y=15(6) = 90$, $\theta = \tan^{-1}(90/60)$, and

since $dy/dt = 15$, $d\theta/dt = (1/4)\cos^2[\tan^{-1}(3/2)]$
$= 1/3 \approx 0.0769$ rad/sec. [$4.41°$ per second]

49. $\theta = \tan^{-1}(x/2)$, so $d\theta/dt = 2(4+x^2)^{-1}(dx/dt).$

$dx/dt = 5\pi$, so when $x=1$,

$d\theta/dt = 2\pi$ rad/min. [1 rev/min]

51. Given: $\dfrac{dx}{dt} = -2$ km/sec.

Find: $\dfrac{d\theta}{dt}\Big|_{x=9376}$

$\sin\left(\dfrac{\theta}{2}\right) = \dfrac{6376}{x}$, so $\theta = 2\sin^{-1}\left(\dfrac{6376}{x}\right)$.

$$\dfrac{d\theta}{dt}\Big| = \dfrac{2}{\sqrt{1-(6376/x)^2}}\, \dfrac{-6376}{x^2}\, \dfrac{dx}{dt} = \dfrac{25504}{x\,\sqrt{x^2-(6376)^2}}.$$

$$\dfrac{d\theta}{dt}\Big|_{x=9376} = \dfrac{25504}{(9376)\sqrt{(9376)^2-(6376)^2}} \approx 0.0003957.$$

θ is increasing at about 0.0003957 rad/sec.   [0.07123 deg/sec]

53. (0.251, 0.258)

Problem 7.8   The Hyperbolic Functions and Their Inverses

1. $\cosh x + \sinh x = \dfrac{e^x + e^{-x}}{2} + \dfrac{e^x - e^x}{2} = e^x.$

3. $\sinh x \cosh y + \cosh x \sinh y = \dfrac{e^x - e^{-x}}{2}\, \dfrac{e^y + e^{-y}}{2} + \dfrac{e^x + e^{-x}}{2}\, \dfrac{e^y - e^{-y}}{2}$

$$= \dfrac{e^{x+y} + e^{x-y} - e^{-x+y} - e^{-x-y} + e^{x+y} - e^{x-y} + e^{-x+y} - e^{-x-y}}{4}$$

$$= \dfrac{2e^{x+y} - 2e^{-(x+y)}}{4} = \dfrac{e^{x+y} - e^{-(x+y)}}{2} = \sinh(x+y).$$

5. $\tanh(x+y) = \dfrac{\sinh(x+y)}{\cosh(x+y)} = \dfrac{\sinh x \cosh y + \cosh x \sinh y}{\cosh x \cosh y + \sinh y \sinh y} = \dfrac{\tanh x + \tanh y}{1 + \tanh x \tanh y}.$

7. $D_x y = 2\sinh x \cosh x = \sinh 2x.$

9. $D_xy = [\sinh(x^2-1)](2x) = 2x\sinh(x^2-1)$.

11. $D_xy = (x^2)(\cosh x) + (\sinh x)(2x) = x^2\cosh x + 2x\sinh x$.

13. $D_xy = (\sinh 4x)(\sinh 2x)2 + (\cosh 2x)(\cosh 4x)4 = 2\sinh 4x \sinh 2x + 4\cosh 2x \cosh 4x$.

15. $D_xy = \dfrac{1}{\sqrt{(x^3)^2-1}}(3x^2) = \dfrac{3x^2}{\sqrt{x^6-1}}, \quad x > 1$.

17. $D_xy = \dfrac{2x}{(4x^2+1)^{1/2}} + \sinh^{-1}(2x)$.

19. $-\sin x \cosh(\cos x)$.

21. $\int x \cosh(x^2+3)\, dx = \int \frac{1}{2}\cosh(x^2+3)\, 2x\, dx$

$$\boxed{\begin{array}{l}\text{Let}\quad u = x^2+3 \\ \text{Then}\ du = 2x\, dx\end{array}}$$

$$= \int \frac{1}{2}\cosh u\, du = \frac{1}{2}\sinh u + C = \frac{1}{2}\sinh(x^2+3) + C.$$

23. $\int \sinh u\, du = \cosh u + C = \cosh(e^x) + C.$ $\qquad$ [Let $u = e^x$]

25. $A = \int_0^{\ln 3}\cosh(2x)\, dx = \left[(1/2)\sinh(2x)\right]_0^{\ln 3}$

$= (1/2)\sinh(2\ln 3) - (1/2)\sinh(0)$

$= (1/2)\sinh(\ln 9) = (1/2)(1/2)(e^{\ln 9} - e^{-\ln 9})$

$= (1/4)(9 - 1/9) = 20/9 \approx 2.2222.$

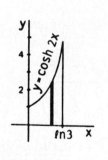

27. Using the Method of Disks:

$\text{Volume} = \int_0^1 \pi(\cosh x)^2\, dx$

$= \frac{\pi}{2}\int_0^1 (1+\cosh 2x)\, dx$ [from hint]

$= \frac{\pi}{2}\left[x + \frac{1}{2}\sinh 2x\right]_0^1 = \frac{\pi}{2}\left(1 + \frac{1}{2}\sinh 2\right) \approx 4.4193.$

29. $y = a\cosh(x/a);\; dy/dx = \sinh(x/a);\; d^2y/dx^2 = (1/a)\cosh(x/a).$

$$(\delta/H)\,[1+(dy/dx)^2]^{1/2} = (1/a)\,[1+\sinh^2(x/a)]^{1/2}$$

$$= (1/a)\,[\cosh^2(x/a)]^{1/2} = (1/a)\cosh(x/a) = d^2y/dx^2.$$

31. (a)

(b)   Volume = (Cross-sectional area)(100)

$$= \left[\,2\int_0^{24}[37-24\cosh(x/24)]\,dx\right](100)$$

$$= 200\left[37x-576\sinh(x/24)\right]_0^{24} = 200[888-\sinh(1)] \approx 42{,}217 \text{ ft}^3.$$

(c)   Surface Area = (Length of arch) (100)

$$= \left[2\int_0^{24}[1+\sinh^2(x/24)]^{1/2}dx\right](100)$$

$$= 200\left[24\sinh(x/24)\right]_0^{24} = 200(24\sinh 1) \approx 5{,}641 \text{ ft}^2.$$

33. (a)   [See Problem 1.]

$$\sinh(rx) + \cosh(rx) = e^{rx} = [e^x]^r = [\sinh(x) + \cosh(x)]^r.$$

(b)   [See Problem 2.]

$$\cosh(rx) - \sinh(rx) = e^{-rx}\,]\,[e^{-x}]^r = [\cosh(x) - \sinh(x)]^r.$$

37. They are inverses.

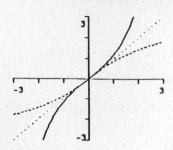

Problem Set 7.9   Chapter Review

True-False Quiz

1. False.   It is not defined for $x=0$.

3. True.   The integral equals $\ln(e^3) = 3\ln(e) = 3(1) = 3$.

5. True.   It is the same as the range of $\ln$, $\mathbf{R}$.

7. False.   False. C^{ex}: $(\ln e)^4 = 1$, but $4\ln e = 4$.

9. True. $(f \circ g)(x) = f(\ln[x-4]) = 4 + e^{\ln(x-4)} = 4 + (x-4) = x$.

11. True. $\lim\limits_{x\to 0^+} [\ln(\sin x) - \ln x] = \lim\limits_{x\to 0^+} \ln\,[(\sin x)/x] = \ln\left[\lim\limits_{x\to 0^+}\,\frac{\sin x}{x}\right] = \ln 1 = 0$.

13. False. $\ln\pi$ is a constant, so $(d/dx)\,(\ln\pi) = 0$.

15. True.   Direct application of the Power Rule (base is a variable and the exponent is a constant).

17. False.   Misapplication of Exponential Function Rule.

19. False. C^{ex}: $\text{Arcsin}(\sin\pi) = \text{Arcsin}(0) = 0 \neq \pi$. [It is true iff x is in $[-\pi/2,\pi/2.]$

21. False. $\cosh(\ln 3) = \dfrac{e^{\ln 3} + e^{-\ln 3}}{2} = \dfrac{3 + 3^{-1}}{2} = \dfrac{5}{3}$.

23. False. $\sin^{-1}(\cosh x)$ is defined only for x=1 since the range of cosh is $[1, \infty)$ but the domain of $\sin^{-1}$ is $[-1,1]$.

25. False.   Neither does.   [They both satisfy $y'' + y = 2y$ or $y'' - y = 0$.]

27. False. C^{ex}: $\ln(x-3)$ is not defined for x = 0.
    [Note: It is true, however, for all x such that $|x| > 3$.]

29. False. $A_0(1 + 0.12/12)^{12} \approx 1.1268 A_0(e^{0.11}) \approx 1.1163 A_0$.

Sample Test Problems

1. $D_x[4\ln x - \ln 2] = 4/x$.

3. $D_x\left[e^{x^2-4x}\right] = e^{x^2-4x}(2x-4) = (2x-4)e^{x^2-4x}$.

5. $D_x(\tan x) = \sec^2 x$.   7. $\dfrac{dx}{dy} = \dfrac{\text{sech}^2\sqrt{x}}{\sqrt{x}}$

9. $D_x[\sinh^{-1}(\tan x)] = \dfrac{1}{\sqrt{\tan^2 x + 1}}(\sec^2 x) = \dfrac{\sec^2 x}{|\sec x|} = |\sec x|$.

11. $D_x[\cos^{-1}(1/e^x)] = D_x[\cos^{-1}(e^{-x})] = \dfrac{-1(-e^{-x})}{(1-e^{-2x})^{1/2}} = \dfrac{1}{(e^{2x}-1)^{1/2}}$.

13. $\dfrac{dy}{dx} = \dfrac{15 e^{5x}}{e^{5x}+1}$.

15. $D_x[\cos(e^{\sqrt{x}})] = [-\sin(e^{\sqrt{x}})](e^{\sqrt{x}})\dfrac{1}{2\sqrt{x}} = \dfrac{-e^{\sqrt{x}}\sin(e^{\sqrt{x}})}{2\sqrt{x}}$.

17. $\dfrac{-2}{(1-x)^{1/2}} \dfrac{1}{2x^{1/2}} = \dfrac{-1}{[x(1-x)]^{1/2}}.$

19. $D_x[2\csc\sqrt{x}] = -2\csc\sqrt{x}\,\cot\sqrt{x}\,[1/(2\sqrt{x})] = (-\csc\sqrt{x}\,\cot\sqrt{x})/\sqrt{x}$.

21. $D_x(4\tan5x\,\sec5x) = (4\tan5x)\,[(\sec5x\,\tan5x)(5)] + (\sec5x)\,[(4\sec^2 5x)(5)]$

$$= 20\sec5x\,\tan^2 5x + 20\sec^3 5x.$$

23. $D_x[e^{(1+x)\ln x}] = e^{(1+x)\ln x}[(1+x)(1/x)+\ln x] = x^{1+x}[(1+x)/x + \ln x].$

25. $D_x[(1/3)e^{3x-1} + C] = (1/3)e^{3x-1}(3) = e^{3x-1}.$

27. $\displaystyle\int e^x\sin(e^x)dx = \int\sin u\,du = -\cos u + C = -\cos(e^x) + C.$ $\boxed{\begin{array}{l}\text{Let}\quad u = e^x\\ \text{Then } du = e^x dx\end{array}}$

$$D_x[-\cos(e^x) + C] = [\sin(e^x)](e^x) = e^x\sin(e^x)\ .$$

29. $\displaystyle\int e^{-1}(1/u)du = e^{-1}\ln|u| + C = e^{-1}\ln(e^{x+3}+1) + C.$ $\qquad$ [Let $u = e^{x+3}+1$]

$$D_x[e^{-1}\ln(e^{x+3}+1) + C] = e^{-1}[1/(e^{x+3}+1)]e^{x+3} = (e^{x+2})/(e^{x+3}+1).$$

31. $\displaystyle\int 2(1-u^2)^{-1/2}du = 2\sin^{-1}u + C = 2\sin^{-1}2x + C.$ $\qquad$ [Let $u = 2x$]

$$D_x[2\sin^{-1}2x + C] = 2[1-(2x)^2]^{-1/2}(2) = 4(1-4x^2)^{-1/2}.$$

33. $\displaystyle\int\dfrac{-1}{x+x(\ln x)^2}\,dx = \int\dfrac{-1}{1+(\ln x)^2}\,\dfrac{1}{x}\,dx$ $\qquad$ $\boxed{\begin{array}{l}\text{Let}\quad u = \ln x\\ \text{Then } du = (1/x)dx\end{array}}$

$$= \int\dfrac{-1}{1+u^2}\,du = -\tan^{-1}u + C = -\tan^{-1}(\ln x) + C.$$

$$D_x[-\tan^{-1}(\ln x) + C] = \dfrac{-1}{1+(\ln x)^2}\,\dfrac{1}{x} = \dfrac{-1}{x+x(\ln x)^2}.$$

35. $f'(x) = \cos x - \sin x.$

$f''(x) = -\sin x - \cos x.$

f is increasing on $[-\pi/2, \pi/4]$ and decreasing on $[\pi/4, \pi/2]$; $f(\pi/4) =$ $\sqrt{2}$ is a global maximum and $f(-\pi/2) = -1$ is a global minimum.

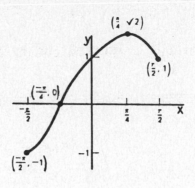

f is concave up on $(-\pi/2, -\pi/4)$ and concave down on $(-\pi/4, \pi/2)$; $(-\pi/4, 0)$ is an inflection point.

$(0,1)$ and $(\pi/2, 1)$ are other points.

37. (a) $f'(x) = 5x^4 + 6x^2 + 4 > 0$ on $\mathbf{R}$ so f is increasing and has an inverse on $\mathbf{R}$.

(b) $f^{-1}(7) = 1$ since $f(1) = 7$.   (c) $g'(7) = 1/f'(1) = 1/15$.

39. $A(1) = 100\left(1 + \dfrac{0.12}{n}\right)^n$ for n compounding periods per year for 1 year.

$A(1) = 100e^{0.12}$ for continuous compounding for 1 year.

(a) n=1: $A(1) = 100\left(1 + \dfrac{0.12}{1}\right)^1 = 112$ dollars.

(b) n=12: $A(1) = 100\left(1 + \dfrac{0.12}{12}\right)^{12} \approx 112.68$ dollars.

(c) n=365: $A(1) = 100\left(1 + \dfrac{0.12}{365}\right)^{365} \approx 112.75$ dollars.

(d) continuous: $A(1) = 100e^{0.12} \approx 112.75$ dollars.

41. $y = (\cos x)^{\sin x} = e^{(\sin x)(\ln \cos x)}$.

$\dfrac{dy}{dx} = e^{(\sin x)(\ln \cos x)}\left[(\sin x)\dfrac{1}{\cos x}(-\sin x) + (\ln \cos x)(\cos x)\right]$

$= \dfrac{(\cos x)^{\sin x}[\cos^2 x\,\ln(\cos x) - \sin^2 x]}{\cos x}.$

At $(0,1)$ $dy/dx = 0$, so the slope is 0.

Therefore, the equation of the tangent line at $(0,1)$ is $y=1$.

Problem Set 8.1   Integration by Substitution

1. $(1/5)(x-1)^5 + C.$

3. $\int x(x^2+1)^4 dx = \frac{1}{2}(x^2+1)^4 \ 2x dx$

$$
\boxed{\begin{array}{l} \text{Let} \quad u = x^2+1 \\ \text{Then } du = 2xdx \end{array}}
$$

$$
= \int \frac{1}{2} u^4 du = \frac{u^5}{10} + C = \frac{(x^2+1)^5}{10} + C.
$$

5. $\tan^{-1}x + C.$

7. $\int (1/2u)du = (1/2)\ln|u| + C = (1/2)\ln(x^2+1) + C.$ $[\text{Let } u = x^2+1]$

9. $\int 3t\sqrt{2+t^2} \ dt = \int \frac{3}{2}(2+t^2)^{1/2} \ 2t dt$

$$
\boxed{\begin{array}{l} \text{Let} \quad u = 2+t^2 \\ \text{Then } du = 2tdt \end{array}}
$$

$$
= \int \frac{3}{2} u^{1/2} \ du = u^{3/2} + C = (2+t^2)^{3/2} + C.
$$

11. $\int \sec z \ \tan z \ dz = \sec z + C.$

13. $\int 2\cos u \ du = 2 \sin u + C = 2\sin(x^{1/2}) + C.$ $[\text{Let } u = x^{1/2}]$

15. $\int_0^{\pi/2} \frac{\cos x}{1+\sin^2 x} \ dx = \int_0^1 \frac{1}{1+u^2} \ du$

$$
\boxed{\begin{array}{l} \text{Let} \quad u = \sin x \\ \text{Then } du = \cos x \ dx \\ \quad x = \pi/2 \ \Rightarrow \ u=1 \\ \quad x = 0 \ \ \Rightarrow \ u=0 \end{array}}
$$

$$
= \left[\tan^{-1}u\right]_0^1 = \tan^{-1}(1) - \tan^{-1}(0) = \frac{\pi}{4} - 0 = \frac{\pi}{4} \approx 0.7854.
$$

17. $\int [2x-1+(x+1)^{-1}]dx = x^2-x+\ln|x+1| + C.$

19. $\int (\tan x)^{1/2} \sec^2 x \, dx = \int u^{1/2} du = (2/3)u^{3/2} + C$ [Let $u = \tan x$]

$$= (2/3)(\tan x)^{3/2} + C.$$

21. $\int \frac{\cos(\ln 4x^2)}{x} \, dx = \int \frac{1}{2} \cos(\ln 4x^2) \frac{2}{x} \, dx$

$$\boxed{\begin{array}{l} \text{Let} \quad u = \ln(4x^2) \\ \text{Then } du = \dfrac{1}{4x^2} \, 8x \, dx \\ \qquad\quad = (2/x) \, dx \end{array}}$$

$$= \int \frac{1}{2} \cos u \, du = \frac{\sin u}{2} + C = \frac{\sin(\ln 4x^2)}{2} + C.$$

23. $\int 5(1 - u^2)^{-1/2} du = 5\sin^{-1} u + C = 5\sin^{-1}(e^x) + C.$ [Let $u = e^x$]

25. $\int (-5/2)u^{-1/2} du = -5u^{1/2} + C = -5(1 - e^{2x})^{1/2} + C.$ [Let $u = 1 - e^{2x}$]

27. $\int_0^1 x \, 10^{x^2} dx = \frac{1}{2} \int_0^1 10^{x^2} \, 2x dx$

$$\boxed{\begin{array}{l} \text{Let} \quad u = x^2 \\ \text{Then } du = 2x dx \\ \qquad x=1 \;\Rightarrow\; u=1 \\ \qquad x=0 \;\Rightarrow\; u=0 \end{array}}$$

$$= \frac{1}{2} \int_0^1 10^u du = \frac{1}{2} \left[\frac{10^u}{\ln 10}\right]_0^1 \frac{10-1}{2\ln 10} \approx 1.9543.$$

29. $\int (1 - \cot x) dx = x - \ln|\sin x| + C.$

31. $\int (1/2)\tan u \, du = (-1/2)\ln|\cos u| + C$ [Let $u = z^2 + 4z - 3$]

$$= (1/2)\ln|\cos(z^2 + 4z - 3)| + C.$$

33. $\int e^x \sec(e^x) \, dx = \int \sec u \, du = \ln|\sec u + \tan u| + C$ $\boxed{\begin{array}{l} \text{Let} \quad u = e^x \\ \text{Then } du = e^x dx \end{array}}$

$$= \ln|\sec(e^x) + \tan(e^x)| + C. \quad \text{[See Example 9 for 2nd step.]}$$

35. $\int (\sec^2 x + e^{\sin x} \cos x) dx = \tan x + e^{\sin x} + C.$ [See Problem 12.]

37. $\int (1/3)u^{-2} du = (-1/3)u^{-1} + C = (-1/3)\csc(t^3 - 2) + C.$ [Let $u = \sin(t^3 - 2)$]

39. $\int \dfrac{t^2 \cos^2(t^3-2)}{\sin^2(t^3-2)}\, dt = \int \frac{1}{3}\cot^2(t^3-2)\, 3t^2 dt$

$$\boxed{\begin{array}{l} \text{Let} \quad u = t^3-2 \\ \text{Then } du = 3t^2 dt \end{array}}$$

$= \int\frac{1}{3}\cot^2 u\ du = \int\frac{1}{3}(\csc^2 u - 1)du = \dfrac{-\cot u - u}{3} + C = \dfrac{\cot(t^3-2) + t^3-2}{-3} + C$

$= \dfrac{t^3+\cot(t^3-2)}{-3} + C_1.\quad$ [where $C_1 = C + \frac{2}{3}$]

41. $\int (1/2)e^u du = (1/2)e^u + C = (1/2)e^{\tan^{-1}2t} + C.$ \quad [Let $u = \tan^{-1}2t$]

43. $\int (1/6)(16-u^2)^{-1/2}du = (1/6)\sin^{-1}(u/4) + C$ \quad [Let $u = 3y^2$]

$= (1/6)\sin^{-1}(3y^2/4) + C.$

45. $\int \dfrac{\sec x\ \tan x}{1+\sec^2 x}\, dx + \int\dfrac{1}{1+u^2}du + \tan^{-1}u+C = \tan^{-1}(\sec x)+C.$ $\boxed{\begin{array}{l}\text{Let} \quad u = \sec x \\ \text{Then } du = \sec x\tan x dx\end{array}}$

47. $\int (1/3)(4-u^2)^{-1/2}du = (1/3)\sin^{-1}(u/2) + C$ \quad [Let $u = e^{3t}$]

$= (1/3)\sin^{-1}(e^{3t}/2) + C.$

49. $\int_{0}^{1}\dfrac{1}{016+u^2}\, du = \frac{1}{4}\left[\tan^{-1}\left(\frac{u}{4}\right)\right]_0^1 = \frac{1}{4}\tan^{-1}\left(\frac{1}{4}\right) \approx 0.06124.$ \quad [Let $u = \cos x$]

51. $\int\dfrac{1}{x^2+2x+5}\, dx = \int\dfrac{1}{(x+1)^2+4}\, dx = \frac{1}{2}\tan^{-1}\left(\frac{x+1}{2}\right) + C.$ \quad [Could let $u = x+1$.]

53. $\int\dfrac{1}{(3x+3)^2+1}\, dx = \int\frac{1}{3}\dfrac{1}{u^2+1}\, du = \frac{1}{3}\tan^{-1}u + C$ \quad [Let $u = 3x+3$]

$= (1/3)\tan^{-1}(3x+3) + C.$

55. $\int (1/18u)du = (1/18)\ln|u| + C$ \quad [Let $u = 9x^2+18x+10$]

$= (1/18)\ln(9x^2+18x+10) + C.$

57. $\displaystyle\int \frac{1}{t\sqrt{2t^2-9}}\, dt = \int \frac{1}{(\sqrt{2}t)\sqrt{(\sqrt{2}t)^2-9}}\, \sqrt{2}dt$

$$\boxed{\begin{array}{l}\text{Let}\quad u = \sqrt{2}t \\ \text{Then } du = \sqrt{2}dt\end{array}}$$

$$= \int \frac{1}{u\sqrt{u^2-9}}\, du = \frac{1}{3}\sec^{-1}\left(\frac{|u|}{3}\right) + C = \frac{1}{3}\sec^{-1}\left(\frac{|\sqrt{2}t|}{3}\right) + C.$$

59. $(2/135)(9x-4)(3x+2)^{3/2}+C.$ [Formula 96: a=3, b=2, u=x]

61. $(1/24)\ln|(4x+3)/(4x-3)| + C.$ [Formula 18: a=3, u=4x]

63. $\displaystyle\int x^2\sqrt{9-2x^2}\, dx = \frac{1}{2\sqrt{2}}\int(\sqrt{2}x)^2\sqrt{9-(\sqrt{2}x)^2}\, \sqrt{2}dx$ [Formula 57: a=3, u=$\sqrt{2}$x]

$$= \frac{1}{2\sqrt{2}}\left[\frac{\sqrt{2}x}{8}(4x^2-9)\sqrt{9-2x^2} + \frac{81}{8}\sin^{-1}\left(\frac{\sqrt{2}x}{3}\right) + C\right]$$

$$= \frac{x(4x^2-9)\sqrt{9-2x^2}}{16} + \frac{81}{16\sqrt{2}}\sin^{-1}\left(\frac{\sqrt{2}x}{3}\right) + K.$$

65. $(1/\sqrt{3})\ln[\sqrt{3}x + (5+3x^2)^{1/2}] + C.$ [Formula 45: a=$\sqrt{5}$, u=$\sqrt{3}$x]

67. $\displaystyle\int \frac{1}{[(t+1)^2-4]^{1/2}}\, dt = \ln|(t+1) + (t^2+2t-3)^{1/2}| + C.$ [45: a=2, u=t+1]

69. $\displaystyle\int \frac{\sin t\, \cos t}{\sqrt{3}\sin t + 5}\, dt = \int\frac{u}{3u+5}\, du$

$$\boxed{u = \sin t;\ du = \cos t\, dt}$$
[Formula 98: a=3, b=5]

$$= \frac{2}{27}(3u-10)\sqrt{3u+5} + C = \frac{2}{27}(3\sin t - 10)\sqrt{3\sin t + 5} + C.$$

71. $dy/dx = -\tan x.$

$$\text{Length} = \int_0^{\pi/4}(1+\tan^2 x)^{1/2}dx = \int_0^{\pi/4}\sec x\, dx = \left[\ln|\sec x + \tan x|\right]_0^{1/4}$$

$$= \ln(2 + 1) \approx 0.8814.$$

73. $\displaystyle\int_{-\pi}^{\pi} \frac{(u+\pi)|\sin(u+\pi)|}{1+\cos^2(u+\pi)}\,du = \int_{-\pi}^{\pi} \frac{u|\sin u|}{1+\cos^2 u}\,du + \pi\int_{-\pi}^{\pi} \frac{|\sin u|}{1+\cos^2 u}\,du$

$\displaystyle = 0 + 2\pi\int_0^{\pi} \frac{\sin u}{1+\cos^2 u}\,du = 2\pi\int_1^{-1} \frac{-1}{1+z^2}\,dz.$ $\qquad$ [Let $z = \cos u$]

$\displaystyle = -2\pi\Big[\arctan z\Big]_1^{-1} = \pi^2 \approx 9.8698$

Problem Set 8.2   Some Trigonometric Integrals

1. $\displaystyle\int (1/2)(1+\cos 2x)\,dx = (1/2)x + (1/4)(\sin 2x) + C.$

3. $\displaystyle\int \cos^3 dx = \int (1-\sin^2 x)\cos x\,dx = \int (1-u^2)\,du$ $\qquad$ $\boxed{\begin{array}{l}\text{Let } u = \sin x \\ \text{Then } du = \cos x\,dx\end{array}}$

$\displaystyle = u - \frac{u^3}{3} + C = \sin x = \frac{\sin^3 x}{3} + C.$

5. $\displaystyle\int_0^{\pi/2} (1-\cos^2 t)^2 \sin t\,dt = \int_1^0 -(1-u^2)^2\,du$ $\qquad$ [Let $u = \cos t$]

$\displaystyle = \Big[u - (2/3)u^3 + (1/5)u^5\Big]_0^1 = 8/15 \approx 0.5333.$

7. $\displaystyle\int \tan y\,(\sec^2 y - 1)\,dy = \int u\,du + \ln|\cos y|$ $\qquad$ [1st term: Let $u = \tan y$]

$\displaystyle = (1/2)u^2 + \ln|\cos y| + C = (1/2)\tan^2 y + \ln|\cos y| + C.$

9. $\displaystyle\int \sin^7 3x\,\cos^2 3x\,dx = \int (1-\cos^2 3x)^3\cos^2 3x\,\sin 3x\,dx$ $\qquad$ $\boxed{\begin{array}{l}\text{Let } u = \cos 3x \\ \text{Then } du = -3\sin 3x\,dx\end{array}}$

$\displaystyle = \int \frac{-1}{3}(1-u^2)^3 u^2\,du = \int\Big[\frac{-u^2}{3} + u^4 - u^6 + \frac{u^8}{3}\Big]du = \frac{-u^3}{9} + \frac{u^5}{5} - \frac{u^7}{7} + \frac{u^9}{27} + C$

$\displaystyle = \frac{-\cos^3 3x}{9} + \frac{\cos^5 3x}{5} - \frac{\cos^7 3x}{7} + \frac{\cos^9 3x}{27} + C.$

11. $\int(1-\sin^2\theta)\sin^{-2}\theta\,\cos\theta\,d\theta = \int(1-u^2)u^{-2}du$ [Let $u = \sin\theta$]

$\qquad = -u^{-1}-u+C = -\csc\theta - \sin\theta + C.$

13. $\int[(1/2)\sin 4t]^4 dt = (1/16)\int[(1/2)(1-\cos 8t)]^2 dt$

$\qquad = (1/64)\int[1 - 2\cos 8t + (1/2)(1+\cos 16t)]dt$

$\qquad = (1/64)[(3/2)t - (1/4)\sin 8t + (1/32)\sin 16t] + C.$

15. $\int\tan^3 3y\,\sec^3 3y\,dy$
$\boxed{\begin{array}{l}\text{Let}\quad u = \sec 3y\\ \text{Then } du = 3\sec 3y\tan 3y\,dy\end{array}}$

$\qquad = \int(\sec^2 3y - 1)\sec^2 3y\,\sec 3y\,\tan 3y\,dy$

$\qquad = \int\frac{1}{3}(u^2-1)u^2 du = \int\left[\frac{u^4}{3} - \frac{u^2}{3}\right]du = \frac{u^5}{15} - \frac{u^3}{9} + C = \frac{\sec^5 3y}{15} - \frac{\sec^3 3y}{9} + C.$

17. $\int -u^2 du = (-1/3)u^3 + C = (-1/3)\csc^3 x + C.$ [Let $u = \csc x$]

19. $\int u^{-3}du = (-1/2)u^{-2}+C = (-1/2)\cot^2 t + C.$ [Let $u = \tan t$]

21. $\int\sin 4y\,\cos 5y\,dy = \int\frac{1}{2}[\sin 9y + \sin(-1)y]dy = \int\frac{1}{2}(\sin 9y - \sin y)dy$

$\qquad = \frac{1}{2}\frac{\cos 9y}{9} + \cos y + C = \frac{\cos 9y}{-18} + \frac{\cos y}{2} + C.$

23. $\int\cot^2 2x(\csc^2 2x - 1)dx = \int[\cot^2 2x\,\csc^2 2x - (\csc^2 2x - 1)]dx$

$\qquad = \int(-1/2)u^2 du + (1/2)\cot 2x + x$ [1st term: let $u = \cot 2x$]

$\qquad = (-1/6)u^3 + (1/2)\cot 2x + x + C = (-1/6)\cot^3 2x + (1/2)\cot 2x + x+C.$

25. $\int(\tan^2 7x + 1)\sec^2 7x\,dx = \int(1/7)(u^2+1)du$ [Let $u = \tan 7x$]

$\qquad = (1/21)u^3 + (1/7)u + C = (1/21)\tan^3 7x + (1/7)\tan 7x + C.$

27. $\int \cot^3 x\, dx = \int \cot x\,(\csc^2 x - 1)dx$

$\qquad = \int \cot x\, \csc^2 x\, dx - \int \cot x\, dx$

$\boxed{\text{Let} \quad y = \cot x \\ \text{Then } du = -\csc^2 x\, dx}$

$\qquad = \int -u\, du - \ln|\sin x| = \dfrac{-u^2}{2} - \ln|\sin x| + C = \dfrac{-\cot^2 x}{2} - \ln|\sin x| + C.$

29. $\int (\tan^2 x + 2 + \cot^2 x)dx = \int (\sec^2 x + \csc^2 x)dx = \tan x - \cot x + C.$

31. (Method of disks)

$\qquad \text{Volume} = \displaystyle\int_0^\pi \pi(x+\sin x)^2 dx$

$\qquad = \pi\left[(1/3)x^3 + 2\sin x - 2x\cos x + x/2 - (1/4)\sin 2x\right]_0^\pi$

$\qquad = \pi\left[\dfrac{\pi^3}{3} + 0 + 2\pi + \dfrac{\pi}{2} - 0 - (0 + 0 - 0 + 0 - 0)\right] = \dfrac{\pi^4}{3} + \dfrac{3\pi}{2} \approx 37.1821.$

33. (a) $\dfrac{1}{\pi}\displaystyle\int_{-\pi}^\pi f(x)\,\sin(mx)\,dx = \dfrac{1}{\pi}\displaystyle\int_{-\pi}^\pi \left[\sum_{n=1}^N a_n \sin(nx)\right]\sin(mx)\,dx$

$\qquad = \displaystyle\sum_{n=1}^N \left[\dfrac{1}{\pi}a_n\displaystyle\int_{-\pi}^\pi \sin(nx)\,\sin(mx)\,dx\right]$

$\qquad = a_m\displaystyle\int_{-\pi}^\pi \sin(mx)\,\sin(mx)\,dx \quad \text{if } m \le N \quad \text{[Other terms equal zero.]}$

$\qquad = a_m \pi.$ The result follows.

[Of course, if $m > N$, all terms equal zero.]

(b) $\frac{1}{\pi}\int_{-\pi}^{\pi} f^2(x)dx = \frac{1}{\pi}\int_{-\pi}^{\pi}\left[\sum_{n=1}^{N} a_n \sin(nx)\right]^2 dx$

$= \frac{1}{\pi}\int_{-\pi}^{\pi}\left[\sum_{n=1}^{N} a_n^2 \sin^2(nx)\right]dx$ [Other terms in expansion equal zero.]

$= \sum_{n=1}^{N}\left[a_n^2 \int_{-\pi}^{\pi} \sin^2(nx)dx\right] = \sum_{n=1}^{N} a_n^2 (\pi) = \pi\sum_{n=1}^{N} a_n^2.$ Result follows.

35. Note: $\cos(\pi/4) = \sqrt{2}/2,$

$\cos(\pi/8) = \sqrt{(1/2)[1+\cos(\pi/4)]} = \sqrt{(1/2)(1 + \sqrt{2}/2)} = (1/2)\sqrt{2+\sqrt{2}}$

$\cos(\pi/16) = \sqrt{(1/2)[1+\cos(\pi/8)]} = \sqrt{(1/2)[1 + \sqrt{2+\sqrt{2}}/2]}$

$= (1/2)\sqrt{2+\sqrt{2+\sqrt{2}}},$ etc.

Therefore, the right-hand side, in the statement of the problem, equals

$\lim_{n\to\infty}\left[\cos(\pi/4)\ \cos(\pi/8)\ \cos(\pi/16)\ \cdots\ \cos[(\pi/2)/2^n]\right]$

$= \frac{\sin(\pi/2)}{\pi/2} = \frac{2}{\pi}.$ [Problem 34 with $x = \pi/2$ for next to last step]

Problem Set 8.3 Rationalizing Substitutions

1. $\int (u^2-3)u\ 2udu = (2/5)u^5 - 2u^3 + C$ [Let $u^2 = x+3$]

$= (2/5)(x+3)^{5/2} - 2(x+3)^{3/2} + C.$

3. $\int \frac{t}{\sqrt{2t+7}}\, dt = \int \frac{u^2-7}{2u}\, u\,du$

> Let $u = \sqrt{2t+7}$; $u^2 = 2t+7$; $t = \frac{u^2-7}{2}$
> Then $2u\,du = 2dt$
> $u\,du = dt$

$= \int \left[\frac{u^2}{2} - \frac{7}{2}\right]du = \frac{u^3}{6} - \frac{7u}{2} + C = \frac{(2t+7)^{3/2}}{6} - \frac{7(2t+7)^{1/2}}{2} + C.$

5. $\int_1^2 \frac{2u}{u+2}\, du = 2\int_1^2 \left[1 - 2/(u+2)\right]du = 2\left[u - 2\ln|u+2|\right]_1^2$ [Let $u^2 = x$]

$= 2 + 4\ln(3/4) \approx 0.8493.$

(handwritten): 2
$2[2-2\ln 4] - 2[1-2\ln 3]$
$4 - 4\ln 4 - 2 + 4\ln 3$

7. $\int 7\left[\frac{u^2-1}{2}\right]u^3\, du = \frac{u^7}{2} - \frac{7u^5}{10} + C = \frac{(2t+1)^{7/2}}{2} - \frac{7(2t+1)^{5/2}}{10} + C.$ [Let u^2 $= 2t+1$]

9. $\int \frac{\sqrt{1-x^2}}{x}\, dx = \int \frac{\cos u}{\sin u}\cos u\, du$

> Let $x = \sin u$, $u \neq 0$, u in $[-\pi/2, \pi/2]$
> Then $dx = \cos u\, du$

$= \int \frac{1-\sin^2 u}{\sin u}\, du = \int (\csc u - \sin u)du$

$= -\ln|\csc u + \cot u| + \cos u + C$

$= -\ln\left|\frac{1}{x} + \frac{\sqrt{1-x^2}}{x}\right| + \sqrt{1-x^2} + C = \sqrt{1-x^2} - \ln\frac{1+\sqrt{1-x^2}}{|x|} + C.$

11. $\int \frac{1}{3\tan u\, 3\sec u}\, 3\sec^2 u\, du = \int \frac{1}{3}\csc u\, du$ [Let $x = 3\tan u$, u in $(-\pi 2, \pi/2)$]

$= \frac{-1}{3}\ln|\csc u + \cot u| + C = \frac{-1}{3}\ln \frac{[(x^2+9)^{1/2} + 3]}{|x|} + C.$

13. $\int_{\cos^{-1}(0.8)}^{\pi/3} \frac{4\sec u\, \tan u}{16\sec^2 u\, 4\tan u}\, du$ [Let $x = 4\sec u$, u in $[0, \pi/2)$]

$\frac{1}{16}\int_{\cos^{-1}(0.8)}^{\pi/3} \cos u\, du = \frac{1}{16}\left[\sin u\right]_{\cos^{-1}(0.8)}^{\pi/3} = \frac{1}{16}\left[\frac{\sqrt{3}}{2} - 0.6\right] \approx 0.0166.$

15. $\int \frac{t}{\sqrt{4-t^2}}\, dt \int \frac{-1}{2} u^{-1/2}\, du = -u^{1/2} + C = -\sqrt{4-t^2} + C.$

> Let $u = 4-t^2$
> Then $du = -2t\,dt$

17. $\int (4\sin u - 3)\,du = -4\cos u - 3u + C$ (Let $x = 2\sin u$, u in $[-\pi/2, \pi/2]$)

        $= -2(4-x^2)^{1/2} - 3\sin^{-1}(x/2) + C.$

19. $\int (1/2)(u-4)u^{-3/2}\,du = u^{1/2} + 4u^{-1/2} + C = (y^2+8)(y^2+4)^{-1/2} + C.$ $[u = y^2+4]$

21. $\displaystyle\int \frac{1}{\sqrt{x^2+2x+5}}\,dx = \int \frac{1}{\sqrt{(x+1)^2+4}}\,dx$

                                     | Let $x+1 = 2\tan u$, u in $(-\pi/2, \pi/2)$
                                       | Then $dx = 2\sec^2 u\,du$

     $\displaystyle = \int \frac{1}{\sqrt{4\tan^2 u + 4}}\, 2\sec^2 u\,du = \int \frac{2\sec^2 u}{2\sec u}\,du$

     $\displaystyle = \int \sec u\,du = \ln|\sec u + \tan u| + C$

     $\displaystyle = \ln\left|\frac{\sqrt{x^2+2x+5}}{2} + \frac{x+1}{2}\right| + C$

     $\displaystyle = \ln\left(\sqrt{x^2+2x+5} + x+1\right) - \ln 2 + C = \ln\left(\sqrt{x^2+2x+5} + x+1\right) + K.$

23. $\displaystyle\int \frac{6(2\tan u - 1)\sec^2 u}{2\sec u}$ $[$Let $x+1 = 2\tan u$, u in $(-\pi/2, \pi/2)]$

     $\displaystyle = \int (6\tan u\, \sec u - 3\sec u)\,du = 6\sec u - 3\ln|\sec u + \tan u| + C$

     $= 3(x^2+2x+5)^{1/2} - 3\ln[(x^2+2x+5)^{1/2} + x+1] - 3\ln 2 + C$

     $= 3(x^2+2x+5)^{1/2} - 3\ln[(x^2+2x+5)^{1/2} + x+1] + K.$

25. $\displaystyle\int 3\cos u\, 3\cos u\,du = \int (9/2)(1+\cos 2u)\,du$ (Let $x+2 = 3\sin u$, u in $[-\pi/2, \pi/2]$)

     $\displaystyle = \frac{9}{2}\left(u + \frac{\sin 2u}{2}\right) + C = \frac{9}{2}(u + \sin u\, \cos u) + C$

     $\displaystyle = \frac{9}{2}\sin^{-1}\left(\frac{x+2}{3}\right) + \frac{(x+2)(5-4x-x^2)^{1/2}}{2} + C.$

27. $\int \dfrac{1}{\sqrt{4x-x^2}}\,dx = \int \dfrac{1}{\sqrt{4-(x-2)^2}}\,dx = \int \dfrac{1}{\sqrt{4-u^2}}\,du$ $\boxed{\begin{array}{l}\text{Let}\quad u = x-2\\ \text{Then } du = dx\end{array}}$

$$= \sin^{-1}\left(\frac{u}{2}\right) + C = \sin^{-1}\left(\frac{x-2}{2}\right) + C.$$

29. $\int \dfrac{2x+2}{x^2+2x+2}\,dx - \int \dfrac{1}{x^2+2x+2}\,dx$ $\begin{array}{l}[\text{1st term, let } u = x^2+2x+2;\\ \quad\text{2nd term, let } v = x+1]\end{array}$

$$= \int 1/u\,du - \int (v^2+1)^{-1}dv = \ln|u| - \tan^{-1}v + C$$

$$= \ln(x^2+2x+2) - \tan^{-1}(x+1) + C.$$

31. $V = \displaystyle\int_0^1 \pi(x^2+2x+5)^{-2}dx$ $[\text{Let } x+1 = 2\tan u]$

$$= \pi \int_{\tan^{-1}(0.5)}^{\pi/4} \frac{2\sec^2 u}{(4\sec^2 u)^2}\,du = \frac{\pi}{8}\int_{\tan^{-1}(0.5)}^{\pi/4} \cos^2 u\,du$$

$$= \frac{\pi}{16}\left[u + \frac{\sin 2u}{2}\right]_{\tan^{-1}(0.5)}^{\pi/4} = (\pi/16)[\pi/4 - \tan^{-1}(0.5) + 0.1] \approx 0.0828.$$

33. (a) $\int \dfrac{x}{x^2+9}\,dx = \int \dfrac{1}{2}\dfrac{1}{u}\,du = \frac{1}{2}\ln|u| + C = \ln(\sqrt{x^2+9}) + C.$ $\boxed{\begin{array}{l}\text{Let}\quad u = x^2+9\\ \text{Then } du = 2xdx\end{array}}$

(b) $\int \dfrac{x}{x^2+9}\,dx = \int \dfrac{3\tan u}{9\tan^2 u + 9}\,3\sec^2 u\,du$ $\boxed{\begin{array}{l}\text{Let}\quad x = 3\tan u\\ \text{Then } dx = 3\sec^2 u\,du\end{array}}$

$$= \int \frac{9\tan u\,\sec^2 u}{9\sec^2 u}\,du = \int \tan u\,du$$

$$= -\ln|\cos u| + K = -\ln\frac{3}{\sqrt{x^2+9}} + K$$

$$= -\ln 3 + \ln(\sqrt{x^2+9}) + K = \ln(\sqrt{x^2+9}) + C.$$

35. (a) $\int \dfrac{u^2}{u^2-4}\, du = \int \left[1 + \dfrac{4}{u^2-4}\right] du = u - \ln\left|\dfrac{u+2}{u-2}\right| + C$ [2nd term, Formula 18]

$$= (4-x^2)^{1/2} - \ln\left|\dfrac{2 + (4-x^2)^{1/2}}{2 - (4-x^2)^{1/2}}\right| + C = (4-x^2)^{1/2} - 2\ln\dfrac{2+(4-x^2)^{1/2}}{|x|} + C.$$

(b) $\int \dfrac{4\cos^2 u}{2\sin u}\, du = 2(\csc u - \sin u)du = 2(-\ln|\csc u + \cot u| + \cos u) + C$

$$= -2\ln\left|\dfrac{2+(4-x^2)^{1/2}}{x}\right| + (4-x^2)^{1/2} + C.$$

37. The area of the square is a^2.

(a) Area of lune $= 2\displaystyle\int_0^a \left[\sqrt{a^2-x^2} - (\sqrt{2a^2-x^2} - a)\right] dx$

$$= 2\left[\left[\dfrac{a^2}{2}\sin^{-1}\left[\dfrac{x}{a}\right] + \dfrac{x}{2}\sqrt{a^2-x^2}\right] - \left[\dfrac{2a^2}{2}\sin^{-1}\left[\dfrac{x}{\sqrt{2}a}\right] + \dfrac{x}{2}\sqrt{2a^2-x^2} - ax\right]\right]_0^a$$

$$= a^2.$$

(b) Area of lune = (Area of semicircle of radius a) -
     - (Area of segment of circle of radius $\sqrt{2}a$ and central angle $\pi/2$)

$$= (1/2)\pi a^2 - [(1/2)(\sqrt{2}a)^2(\pi/2) - (1/2)(2a)(a)] = a^2.$$

[Note: Area of segment = Area of sector - Area of triangle.]

39. $\dfrac{dy}{dx} = \dfrac{\sqrt{a^2-x^2}}{-x}$; y=0 when x=a. $dy = \dfrac{\sqrt{a^2-x^2}}{-x}\, dx$.

$y = \displaystyle\int \dfrac{\sqrt{a^2-x^2}}{-x}\, dx = \int \dfrac{a\cos t}{-a\sin t}\, a\cos t\, dt = a\int \dfrac{1-\sin^2 t}{\sin t}\, dt$ $\boxed{\begin{array}{l}\text{Let } \ x = a\sin t \\ \text{Then } dx = a\cos t\, dt\end{array}}$

$= -a\displaystyle\int (\csc t - \sin t)dt = -a[-\ln|\csc(t) + \cot(t)| + \cos(t)] + C$

$= a\ln|\csc(t) + \cot(t)| - a\cos(t) + C = a\ln\left[\dfrac{a}{x} + \dfrac{\sqrt{a^2-x^2}}{-x}\right] - \sqrt{a^2-x^2}$

y=0 when x=a $\Rightarrow$ C=0. $y = a\ln\left[\dfrac{a}{x} + \dfrac{\sqrt{a^2-x^2}}{-x}\right] - \sqrt{a^2-x^2}.$

Problem Set 8.4   Integration by Parts

1. $xe^x - \int e^x dx = xe^x - e^x + C.$ 　　　　　　　[Let $u = x$; $dv = e^x dx$]

3. $\int x\,\sin3x\,dx = \dfrac{-x\cos3x}{3} - \int \dfrac{-1}{3}\cos3x\,dx$

> Let $u = x$ and $dv = \sin3x\,dx$
> Then $du = dx$ and $v = \dfrac{-1}{3}\cos3x$

$\qquad = \dfrac{-x\cos3x}{3} + \dfrac{\sin3x}{9} + C.$

5. $x\tan^{-1}x - \int x(1+x^2)^{-1}dx = x\tan^{-1}x - (1/2)\ln(1+x^2) + C.$ 　　[$u = \tan^{-1}x$; $dv = dx$]

7. $(1/5)t\,\tan5t - \int(1/5)\tan5t\,dt$ 　　　　　[Let $u = t$; $dv = \sec^2 5t\,dt$]

$\qquad = (1/5)t\,\tan5t + (1/25)\ln|\cos5t| + C.$

9. $\int \sqrt{x}\,\ln x\,dx$

> Let $u = \ln x$ and $dv = x^{1/2}dx$
> Then $du = \dfrac{1}{x}\,dx$ and $v = \dfrac{2x^{3/2}}{3}$

$\qquad = \dfrac{2x^{3/2}\ln x}{3} - \int \dfrac{2}{3}x^{1/2}dx$

$\qquad = \dfrac{2x^{3/2}\ln x}{3} - \dfrac{4x^{3/2}}{9} + C.$

11. $\frac{1}{2}x^2\tan^{-1}x - \frac{1}{2}\int\dfrac{x^2}{1+x^2}dx = \frac{1}{2}x^2\tan^{-1}x - \dfrac{x}{2} + \frac{1}{2}\tan^{-1}x + C.$ 　[Let $u = \tan^{-1}x$; $dv = x\,dx$]

13. $(1/2)w^2\ln w - \int w/2\,dw = (1/2)w^2\ln w - (1/4)w^2 + C.$ 　　[$u = \ln w$; $dv = w\,dw$]

15. $\int_{\pi/6}^{\pi/2} x\,\csc^2 x\,dx$

> Let $u = x$ and $dv = \csc^2 x\,dx$
> Then $du = dx$ and $v = -\cot x$

$\qquad = \Big[-x\cot x\Big]_{\pi/6}^{\pi/2} - \int_{\pi/6}^{\pi/2} -\cot x\,dx = \Big[-x\cot x + \ln|\sin x|\Big]_{\pi/6}^{\pi/2}$

$\qquad = (0+0) - \left[\dfrac{-\pi\sqrt{3}}{6} + \ln[1/2]\right] = \dfrac{\pi\sqrt{3}}{6} + \ln2 \approx 1.6000.$

17. $\dfrac{xa^x}{\ln a} - \displaystyle\int \dfrac{a^x}{\ln a}\, dx = \dfrac{xa^x}{\ln a} - \dfrac{a^x}{(\ln a)^2} + C.$ [Let $u = x$; $dv = a^x dx$]

19. $x^2 e^x - \displaystyle\int 2xe^x dx = x^2 e^x - 2xe^x + \displaystyle\int 2e^x dx$

1st: $u = x^2$; $dv = e^x dx$
2nd: $u = 2x$; $dv = e^x dx$

$\quad = x^2 e^x - 2xe^x + 2e^x + C = e^x(x^2 - 2x + 2) + C.$

21. $\displaystyle\int e^t \cos t\, dt = e^t \sin t - \displaystyle\int e^t \sin t\, dt$

Let $\quad u = e^t$ and $dv = \cos t\, dt$
Then $du = e^2 dt$ and $v = \sin t$
Let $\quad u = e^t$ and $dv = \sin t\, dt$
Then $du = e^t dt$ and $v = -\cos t$

$\quad = e^t \sin t - (-e^t \cos t + \displaystyle\int e^t \cos t\, dt)$

$\quad = e^t \sin t + e^t \cos t - \displaystyle\int e^t \cos t\, dt.$

Therefore, $2\displaystyle\int e^t \cos t\, dt = e^t \sin t + e^t \cos t + C.$

Hence, $\displaystyle\int e^t \cos t\, dt = \dfrac{e^t(\sin t + \cos t)}{2} + K.$

23. $x\sin(\ln x) - \displaystyle\int \cos(\ln x)\, dx$

1st: $u = \sin(\ln x)$; $dv = dx$
2nd: $u = \cos(\ln x)$; $dv = dx$

$\quad = x\sin(\ln x) - x\cos(\ln x) - \displaystyle\int \sin(\ln x)\, dx.$

Therefore, $2\displaystyle\int \sin(\ln x)\, dx = x\sin(\ln x) - x\cos(\ln x) + C.$

Therefore, $\displaystyle\int \sin(\ln x)\, dx = (1/2)[\sin(\ln x) - x\cos(\ln x)] + K.$

25. $A = \displaystyle\int_1^e \ln x\, dx = 1$ square unit. [See Example 2.]

27. $\text{Area} = \int_0^9 3xe^{-x/3}dx$

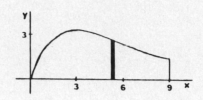

$$\boxed{\begin{array}{l} \text{Let} \quad u = 3x \quad \text{and} \quad dv = e^{-x/3}dx \\ \text{Then } du = 3dx \quad \text{and} \quad v = -3e^{-x/3} \end{array}}$$

$$= \left[-9xe^{-x/3}\right]_0^9 - \int_0^9 -9e^{-x/3}dx$$

$$= \left[-9xe^{-x/3} - 27e^{-x/3}\right]_0^9$$

$$= \left[-9xe^{-x/3}(x+3)\right]_0^9 = (-9e^{-3}[12]) - (-9e^0[3]) = 27 - 108e^{-3} \approx 21.6230$$
$$\text{square units.}$$

29. $\cos^{n-1}x \sin x + (n-1)\int \cos^{n-2}x \sin^2 x \, dx \quad [u = \cos^{n-1}x; \; dv = \cos x \, dx]$

$$= \cos^{n-1}x \sin x + (n-1)\int [\cos^{n-2}x - \cos^n x]dx$$

$$= \cos^{n-1}x \sin x + (n-1)\int \cos^{n-2}x \, dx - (n-1)\int \cos^n x \, dx.$$

Therefore, $n\int \cos^n x \, dx = \cos^{n-1}x \sin x + (n-1)\int \cos^{n-2}x \, dx.$

Therefore, $\int \cos^n x \, dx = \dfrac{\cos^{n-1}x \sin x}{n} + \dfrac{n-1}{n}\int \cos^{n-2}x \, dx, \; n \geq 2.$

31. $\dfrac{6}{7} \dfrac{4}{5} \dfrac{2}{3}\int_0^{\pi/2} \sin x \, dx = \dfrac{6}{7} \dfrac{4}{5} \dfrac{2}{3}\left[-\cos x\right]_0^{\pi/2} = \dfrac{6}{7} \dfrac{4}{5} \dfrac{2}{3} 1 = \dfrac{16}{35} \approx 0.4571.$

33. $\int x^n e^x dx = x^n e^x - n\int x^{n-1}e^x dx.$

$$\boxed{\begin{array}{l} \text{Let} \quad u = x^n \qquad \text{and} \quad dv = e^x dx \\ \text{Then } du = nx^{n-1}dx \quad \text{and} \quad v = e^x \end{array}}$$

$$\int x^3 e^x dx = x^3 e^x - 3\int x^2 e^x dx = x^3 e^x - 3(x^2 e^x - 2\int xe^x dx)$$

$$= x^3 e^x - 3x^2 e^x + 6(xe^x - 1\int e^x dx) = x^3 e^x - 3x^2 e^x + 6xe^x - 6e^x + C.$$

35. $\sec x\ dx + d(\sec x\ \tan x) = \sec x\ dx + (\sec x\ \sec^2 x + \tan x\ \sec x\ \tan x)dx$

$$= [\sec x\ (1+\tan^2 x) + \sec^3 x]dx = 2\sec^3 x\ dx.$$

$$\int \sec^3 x\ dx = (1/2)\left[\int \sec x\ dx + \int d(\sec x\ \tan x)\right]$$

$$= (1/2)\left[\ln|\sec x + \tan x| + \sec x\ \tan x\right] + C.$$

37. **(a)** $\text{Area(nth)} = \displaystyle\int_{(2n-2)\pi}^{(2n-1)\pi} x\sin x\ dx = \left[-x\cos x\right]_{(2n-2)\pi}^{(2n-1)\pi} + \int_{(2n-2)\pi}^{(2n-1)\pi} \cos x\ dx$

$$= (4n-3)\pi\left[\sin x\right]_{(2n-2)\pi}^{(2n-1)\pi} = (4n-3)\pi.$$

(b) $\text{Volume(2nd)} = \displaystyle\int_{2\pi}^{3\pi} 2\pi x(x\sin x)dx \qquad \text{[Shells]}$

$$= 2\pi\left[-x^2\cos x - 2x\sin x + 2\cos x\right]_{2\pi}^{3\pi} \qquad \text{[From Example 4]}$$

$$= 2\pi(13\pi^2 - 4) \approx 781.0305$$

39. $\dfrac{G_n}{n} = \dfrac{\sqrt[n]{(n+1)(n+2)\cdots(n+n)}}{\sqrt[n]{n^n}} = \sqrt[n]{\left(1 + \dfrac{1}{n}\right)\left(1 + \dfrac{2}{n}\right)\cdots\left(1 + \dfrac{n}{n}\right)}$

$\ln\left[\dfrac{G_n}{n}\right] = \dfrac{1}{n}\left[\ln\left(1 + \dfrac{1}{n}\right) + \ln\left(1 + \dfrac{2}{n}\right) + \cdots + \ln\left(1 + \dfrac{n}{n}\right)\right]$, which is a Riemann

sum of $f(x) = \ln x$ on $[1,2]$ with n subintervals each of length $\Delta x_k = \dfrac{1}{n}$,
and right endpoints of subintervals for sample points. Therefore,

$$\ln\left[\lim_{n\to\infty}\left[\dfrac{G_n}{n}\right]\right] = \lim_{n\to\infty}\left[\ln\left[\dfrac{G_n}{n}\right]\right]$$

$$= \lim_{n\to\infty}\left[\dfrac{1}{n}\left[\ln\left(1 = \dfrac{1}{n}\right) + \ln\left(1 + \dfrac{2}{n}\right) + \cdots + \ln\left(1 + \dfrac{n}{n}\right)\right]\right]$$

$$= \int_1^2 \ln x\ dx = 2\ln 2 - 1 = \ln 4 - 1. \qquad \text{[See Example 4.]}$$

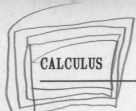

Therefore, $\lim\limits_{n\to\infty}\left[\dfrac{G_n}{n}\right] = e^{\ln 4} - 1 = \dfrac{e^{\ln 4}}{e} = \dfrac{4}{e}$.

Problem Set 8.5   Integration of Rational Functions

1. $\displaystyle\int\left[\dfrac{1}{x} - \dfrac{1}{x+2}\right]dx = \ln|x| - \ln|x+2| + C.$

3. $\dfrac{5x+3}{x^2-9} = \dfrac{5x+3}{(x+3)(x-3)} = \dfrac{A}{x+3} + \dfrac{B}{x-3} = \dfrac{A(x-3) + B(x+3)}{(x+3)(x-3)}.$

Then $5x+3 = A(x-3) + B(x+3)$; $x=-3 \Rightarrow A=2$; $x=3 \Rightarrow B=3$.

Therefore, $\displaystyle\int\dfrac{5x+3}{x^2-9}\,dx = \int\left[\dfrac{2}{x+3} + \dfrac{3}{x-3}\right]dx = 2\ln|x+3| + 3\ln|x-3| + C.$

5. $\displaystyle\int\left[\dfrac{3}{x+4} - \dfrac{2}{x-1}\right]dx = 3\ln|x+4| - 2\ln|x-1| + C.$

7. $\displaystyle\int\left[\dfrac{4}{2x-1} - \dfrac{1}{x+5}\right]dx = 2\ln|2x-1| - \ln|x+5| + C.$

9. $\dfrac{2x^2+x-4}{x^3-x^2-2x} = \dfrac{2x^2+x-4}{x(x-2)(x+1)} = \dfrac{A}{x} + \dfrac{B}{x-2} + \dfrac{C}{x+1} = \dfrac{A(x-2)(x+1)+Bx(x+1)+Cx(x-2)}{x(x-2)(x+1)}.$

Then $2x^2+x-4 = A(x-2)(x+1) + Bx(x+1) + Cx(x-2)$;

$x=0 \Rightarrow A=2$; $x=2 \Rightarrow B=1$; $x=-1 \Rightarrow C=-1.$

$\displaystyle\int\dfrac{2x^2+x-4}{x^3-x^2-2x}\,dx = \int\left[\dfrac{2}{x} + \dfrac{1}{x-2} - \dfrac{1}{x+1}\right]dx = 2\ln|x| + \ln|x-2| - \ln|x+1| + C.$

11. $\displaystyle\int\left[3x - 3 + \dfrac{8}{x+2} + \dfrac{1}{x-1}\right]dx = (3/2)x^2 - 3x + 8\ln|x+2| + \ln|x-1| + C.$

13. $\displaystyle\int\left[\dfrac{1}{x-3} + \dfrac{4}{(x-3)^2}\right]dx = \ln|x-3| - 4(x-3)^{-1} + C.$

15. $\dfrac{3x^2-21x+32}{x^3-8x^2+16x} = \dfrac{3x^2-21x+32}{x(x-4)^2} = \dfrac{A}{x} + \dfrac{B}{x-4} + \dfrac{C}{(x-4)^2} = \dfrac{A(x-4)^2 + Bx(x-4) + Cx}{x(x-4)^2}$

Then $3x^2-21x+32 = A(x-4)^2 + Bx(x-4) + Cx;$

$x=0 \Rightarrow A=2;\ x=4 \Rightarrow C=1;\ x=1 \Rightarrow B=1$ (using $A=2, C=-1$).

$\displaystyle\int \dfrac{3x^2-21x+32}{x^3-8x^2+16x}\ dx = \int\left[\dfrac{2}{x} + \dfrac{1}{x-4} - \dfrac{1}{(x-4)^2}\right] dx = 2\ln|x| + \ln|x-4| + \dfrac{1}{x-4} + C.$

17. $\displaystyle\int\left[\dfrac{-2}{x} + \dfrac{4x}{x^2+4} + \dfrac{1}{x^2+4}\right]dx = -2\ln|x| + 2\ln(x^2+4) + (1/2)\tan^{-1}(x/2) + C.$

19. $\displaystyle\int\left[\dfrac{2}{x+3} - \dfrac{1}{x-2} - \dfrac{1}{x^2+1}\right]dx = 2\ln|x+3| - \ln|x-2| - \tan^{-1}x + C.$

21. $\dfrac{x^3-4x}{(x^2+1)^2} = \dfrac{Ax+B}{x^2+1} + \dfrac{Cx+D}{(x^2+1)^2} = \dfrac{(Ax+B)(x^2+1) + (Cx+D)}{(x^2+1)^2}.$

Then $x^3-4x = (Ax+B)(x^2+1) + (Cx+D) = Ax^3+Bx^2+(A+C)x+(B+D).$

Then $A=1,\ B=0,\ A+C=-4$ so $C=-5$ [since $A=1$], $B+D=0$ so $D=0$ [since $B=0$].

$\displaystyle\int \dfrac{x^3-4x}{(x^2+1)^2}\ dx = \int\left[\dfrac{1x+0}{x^2+1} + \dfrac{-5x+0}{(x^2+1)^2}\right]dx = \int\left[\dfrac{1}{2}\dfrac{2x}{x^2+1} - \dfrac{5}{2}\dfrac{2x}{(x^2+1)^2}\right]dx$

$\qquad = \dfrac{1}{2}\ln(x^2+1) + \dfrac{5}{2}\dfrac{1}{x^2+1} + C$ $\qquad$ [May wish to use $u = x^2+1$.]

23. $\displaystyle\int_4^6\left[\dfrac{3}{x+4} - \dfrac{2}{x-3}\right]dx = \left[3\ln|x+4| - 2\ln|x-3|\right]_4^6 = \ln(125/576) \approx -1.5278.$

25. (a) $\displaystyle\int \dfrac{1}{(x-a)(x-b)}\ dx = \int k\,dt.\quad \dfrac{1}{b-a}\ln\left|\dfrac{x-b}{x-a}\right| = kt+C.$

$x=0,\ t=0 \Rightarrow C = \dfrac{\ln(b/a)}{b-a},$ so $\dfrac{1}{b-a}\ln\left|\dfrac{x-b}{x-a}\right| = kt + \dfrac{\ln(b/a)}{b-a},$

or $x = \dfrac{ab\left[1 - e^{(b-a)kt}\right]}{a - be^{(b-a)kt}}.$

(b) $t \to \infty \Rightarrow \ln \dfrac{x-b}{x-a} \to \infty \Rightarrow \dfrac{x-b}{x-a} \to \infty \Rightarrow x \to a$.

(c) $a=2$, $b=4$, $x=1$, $t=20 \Rightarrow k = (\ln 1.5)/40$.

Then when $t = 60$, $x = 38/23 \approx 1.6622$ gm.

(d) $\displaystyle \int (x-a)^{-2} dx = \int k\, dt$, so $(x-a)^{-1} = kt + C$.

$x=0$, $t=0 \Rightarrow C = 1/a$, so $(x-1)^{-1} = kt + 1/a$ or $x = \dfrac{a^2 kt}{akt+1}$.

27. (a) $\dfrac{dy}{dt} = ky(10-y)$; $\dfrac{1}{y(10-y)}\, dy = k\, dt$; $\dfrac{1}{y(10-y)}\, dy = k\, dt$.

$\dfrac{1}{y(10-y)} = \dfrac{1/10}{y} + \dfrac{1/10}{10-y}$.   [partial fraction decomposition]

Then $(1/10)[\ln(y) - \ln(10-y)] = kt + C$ or $\dfrac{y}{10-y} = e^{10kt}\overline{C}$.

When $t=0$ (1925), $y=2$, so $\overline{C} = 1/4$. Tthen $\dfrac{4y}{10-y} = e^{10kt}$, and solving

for y obtain $y = \dfrac{10}{1 + 4e^{-10kt}}$.

When $t=50$ (1975), $y=4$, so $k = \dfrac{\ln(8/3)}{500}$ and then $y = \dfrac{10}{1 + 4(3/8)^{0.02t}}$.

Then when $y=9$, $t = \dfrac{-50\ln 36}{\ell n(3/8)} \approx 182.68$; i.e. in about the year 2108.

29. Use the result of Problem 28 with $m=-B$, $M=A$, and $y_0=B$. Then simplify

the right-hand side, obtaining $y = \dfrac{B^2 - AB + 2ABe^{k(A+B)t}}{A - B + 2Be^{k(A+B)t}}$.

Problem Set 8.6   Chapter Review

True-False Quiz

1. True.   Since $du = 2x\, dx$ and x is a factor of the integrand.

3. False.   It will be more efficient to use the substitution $u = 9 + x^4$.

5. True. $x^2 - 3x + 5$ does not factor over the reals since the discriminant is negative.

7. True.

9. True.   See Section 8.2, Type 2 with m and n both even.

11. False.   The substitution $u = -x^2 - 4x$ will be simpler.

13. True.   Then let u = sinx.

15. True.   Let u = lnx and $dv = x^2 dx$.

17. False.   Degree of numerator is not less than degree of denominator.

19. True.   A=2, B=-1, C=0.

21. True.   See the first few sentences of Section 8.4.

23. False. $x^2 + 1$ cannot.

25. True.   If the degree of the polynomial is n, n $\geq$ 0, there are at most n solutions.

Sample Test Problems

1. $(1/2)\displaystyle\int_9^{25} u^{-1/2} du = (1/2)\left[2u^{1/2}\right]_9^{25} = 2.$ [Let $u = 9 + t^2$]

3. $\displaystyle\int_0^{\pi/2} e^{\cos x}\sin x \, dx = \int_1^0 -e^u du = \left[-e^u\right]_1^0 = (-1+e) \approx 1.7183.$

$$[u = \cos x; \ du = -\sin x \, dx; \ x=\pi/2 \Rightarrow u=0; \ x=0 \Rightarrow u=1]$$

5. $\displaystyle\int\left[y^2 - y + 2 - \frac{2}{y+1}\right] dy = (1/3)y^3 - (1/2)y^2 + 2y - 2\ln|y+1| + C.$

7. $\displaystyle\int 1/2u \, du = (1/2)\ln|u| + C = (1/2)\ln|y^2 - 4y + 2| + C.$ [Let $u = y^2 - 4y + 2$]

9. $\displaystyle\int \frac{e^{2t}}{e^t - 2} dt = \int\left[e^t + \frac{2e^t}{e^t - 2}\right] dt = e^t + \int \frac{2e^t}{e^t - 2} dt = e^t + \int \frac{2}{u} du = e^t + 2\ln|u| + C.$

$\qquad = e^t + 2\ln|e^t - 2| + C.$ [$u = e^t - 2; \ du = e^t dt$]

11. $\int \dfrac{1}{\sqrt{2}\,[9-(x-1)^2]^{1/2}}\,dx = \dfrac{1}{\sqrt{2}}\sin^{-1}(\dfrac{x-1}{3}) + C.$

13. $(1/\sqrt{3})\ln(\sqrt{3}y + \sqrt{2+3y^2}) + C.$ [$y = (2/3)^{1/2}\tan u$; or use Formula 45]

15. $\int \dfrac{\tan x}{\ln|\cos x|}\,dx = \int \dfrac{-1}{u}\,du = -\ln|u| + C = -\ln|\,\ln|\cos x|\,| + C.$

$\qquad\qquad\qquad\qquad\qquad\qquad\qquad\qquad [u = \ln|\cos x|;\; du = -\tan x\,dx]$

17. $\cosh x + C.$

19. $x(-\cot x - x) - \int(-\cot x - x)\,dx \qquad [u = x;\; dv = \cot^2 x\,dx = (\csc^2 x - 1)\,dx]$

$\qquad = -x\cot x - x^2 + \ln|\sin x| + x^2/2 + C = x\cot x + \ln|\sin x| - x^2/2 + C.$

21. $\int \dfrac{\ln t^2}{t}\,dt = \int \dfrac{2\ln|t|}{t}\,dt = \int 2u\,du = u^2 + C = \ln^2|t| + C.$ $u = \ln|t|;\; du = \dfrac{1}{t}\,dt$

23. $\int 3e^y \sin 9y\,dy \qquad\qquad\qquad\qquad\qquad\qquad\qquad\qquad [y = t/3]$

$\qquad = 3e^y \sin 9y - \int 27e^y \cos 9y\,dy \qquad\qquad\qquad [u = \sin 9y;\; dv = 3e^y dy]$

$\qquad = 3e^y \sin 9y - 27e^y \cos 9y - \int 243 e^y \sin 9y\,dy \qquad [u = \cos 9y;\; dv = 27e^y dy]$

Therefore, $246\int e^y \sin 9y\,dy = 3e^y \sin 9y - 27e^y \cos 9y + C.$

Therefore, $\int 3e^y \sin 9y\,dy = (9/246)e^y[\sin 9y - 9\cos 9y] + K.$

$\qquad\qquad\qquad = (3/82)e^{t/3}[\sin 3t - 9\cos 3t] + K.$

25. $\int (1/2)(\sin 2x + \sin x)\,dx = (-1/4)\cos 2x - (1/2)\cos x + C.$

27. $\int \tan^3 2x \sec 2x\,dx = \int(\sec^2 2x - 1)\sec 2x \tan 2x\,dx = \int \dfrac{1}{2}(u^2 - 1)\,du$

$\qquad = \dfrac{1}{2}\left[\dfrac{u^3}{3} - u\right] + C = \dfrac{\sec^3 2x}{6} - \dfrac{\sec 2x}{2} + C.$

$\boxed{\begin{array}{l} u = \sec 2x \\ du = 2\sec 2x \tan 2x\,dx \end{array}}$

244

29. $\int \tan^{3/2}x \ (1+\tan^2 x) \ \sec^2 x \ dx = \int u^{3/2}(1+u^2)du$ [Let u = tanx]

$\qquad = (2/5)u^{5/2}+(2/9)u^{9/2}+C = (2/5)\tan^{5/2}x+(2/9)\tan^{9/2}x+C.$

31. $\int(-1/2)u^{1/2}du = -u^{1/2}+C = -(9-e^{2y})^{1/2}+C.$ [Let u = 9- e^{2y}]

33. $\int e^{\ln(3\cos x)}dx = \int 3\cos x \ dx = 3\sin x + C.$

35. $\int \dfrac{1}{4(1+u^2)} \ du = (1/4)\tan^{-1}u + C = (1/4)\tan^{-1}(e^{4x}) + C.$ [Let u = e^{4x}]

37. $\int 2(u^2-5)du = (2/3)u^3-10u+C$ [Let u^2 = w+5, u > 0]

$\qquad = (2/3)(w+5)^{1/2}(w-10)+C.$

39. $\int \dfrac{\sin y \ \cos y}{9+\cos^4 y} \ dy = \int \dfrac{1}{9+(\cos^2 y)^2} \ (\sin y \ \cos y)dy$ $\boxed{\begin{array}{l} u = \cos^2 y \\ du = -2\sin y \ \cos y \ dy \end{array}}$

$\qquad = \int \dfrac{-1}{2} \ \dfrac{1}{9+u^2} \ du = \dfrac{-1}{2} \ \dfrac{1}{3} \ \tan^{-1}\left[\dfrac{u}{3}\right] + C = \dfrac{-1}{6}\tan^{-1}\left[\dfrac{\cos^2 y}{3}\right] + C.$

41. $\int\left[\dfrac{1}{x} + \dfrac{2}{x^2} - \dfrac{2x}{2(x^2+3)} + \dfrac{2}{x^2+3}\right]dx = \ln|x| - \dfrac{2}{x} - \dfrac{1}{2} \ln|x^2+3| + \dfrac{2}{\sqrt{3}} \ \tan^{-1}\left[\dfrac{x}{\sqrt{3}}\right]+ C.$

43. $A = \displaystyle\int_{\sqrt{3}}^{3\sqrt{3}} \dfrac{18}{x^2(x^2+9)^{1/2}} \ dx = 2\int_{\pi/6}^{\pi/3} \csc u \ \cot u \ du$ [Let x = 3tanu]

$\qquad = 2\Big[-\csc u\Big]_{\pi/6}^{\pi/3} = (4/3)(3 - \sqrt{3}) \approx 1.6906 \text{ square units.}$

45.

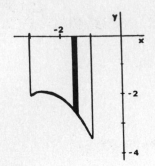

$$\text{Volume} = \int_{-3}^{-1} \pi \left[\frac{6}{x\sqrt{x+4}}\right]^2 dx = 36\pi \int_{-3}^{-1} \frac{1}{x^2(x+4)}\, dx$$

$$= 36\pi \int_{-3}^{-1} \left[\frac{-1/16}{x} + \frac{1/4}{x^2} + \frac{1/16}{x+4}\right] dx \quad \text{[using partial fractions]}$$

$$= 36\pi \left[\frac{-\ln|x|}{16} - \frac{1}{4x} + \frac{\ln(x+4)}{16}\right]_{-3}^{-1}$$

$$= 36\pi \left[\left[\frac{-\ln(1)}{16} + \frac{1}{4} + \frac{\ln(3)}{16}\right] - \left[\frac{-\ln(3)}{16} + \frac{1}{12} + \frac{\ln(1)}{16}\right]\right]$$

$$= \frac{3\pi(4 + 3\ln3)}{2} \approx 34.3808$$

Problem Set 9.1   Indeterminate Forms of Type 0/0

Each step where l′Ĥopital′s Rule is used will be indicated by (L)

1.   ⓛ
$$= \lim_{x \to 0} \frac{\cos x - 2}{1} = -1.$$

3. $\lim\limits_{x \to 0} \dfrac{x - 2\sin x}{\tan x}$ ⓛ $= \lim\limits_{x \to 0} \dfrac{1 - 2\cos x}{\sec^2 x} = \dfrac{1-2}{1} = -1.$

5.   ⓛ
$$= \lim_{x \to -1} \frac{2x+5}{2x-4} = \frac{-1}{2}.$$

7. ∞ since $\left[\dfrac{\to 1}{\to 0^+}\right].$

9. $\lim\limits_{x \to \pi/2} \dfrac{\ln(\sin x)}{(\pi/2) - x}$ ⓛ $= \lim\limits_{x \to \pi/2} \dfrac{\cot x}{-1} = \dfrac{0}{-1} = 0.$

11.   ⓛ
$$= \lim_{t \to 1} \frac{(1/2)t^{-1/2} - 1}{1/t} = \frac{-1}{2}.$$

13.   ⓛ
$$= \lim_{x \to 0} \frac{-3\tan 3x}{4x}$$ ⓛ $= \lim\limits_{x \to 0} \dfrac{-9\sec^2 3x}{4} = \dfrac{-9}{4}.$

15. $\lim\limits_{x \to 0} \dfrac{\tan x - x}{\sin x - x}$ ⓛ $= \lim\limits_{x \to 0} \dfrac{\sec^2 x - 1}{\cos x - 1}$ ⓛ $= \lim\limits_{x \to 0} \dfrac{2\sec^2 x \,\tan x}{-\sin x} = \lim\limits_{x \to 0}(-2\sec^3 x) = -2.$

17. $\;\overset{\text{(L)}}{=}\; \lim_{x\to 0^+} \dfrac{2x}{\cos x - 1} \;\overset{\text{(L)}}{=}\; \lim_{x\to 0^+} \dfrac{2}{-\sin x} = -\infty.$

19. $\;\overset{\text{(L)}}{=}\; \lim_{x\to 0} \dfrac{(1+x^2)^{-1}-1}{24x^2} = \lim_{x\to 0} \dfrac{-x^2}{24x^2+24x^4} = \lim_{x\to 0} \dfrac{-1}{24+24x^2} = \dfrac{-1}{24}.$

21. $\lim_{x\to 0^+} \dfrac{1-\cos x - x\sin x}{2-2\cos x - \sin^2 x} \;\overset{\text{(L)}}{=}\; \lim_{x\to 0^+} \dfrac{\sin x - x\cos x - \sin x}{2\sin x - 2\sin x\cos x} = \lim_{x\to 0^+} \dfrac{-x\cos x}{2\sin x - \sin 2x}$

$\;\overset{\text{(L)}}{=}\; \lim_{x\to 0^+} \dfrac{x\sin x - \cos x}{2\cos x - 2\cos 2x} = -\infty \quad \text{since} \quad \left[\dfrac{\to 1}{\to 0^+}\right].$

Note: $2\cos x - 2\cos 2x > 0$ as $x\to 0^+$ since $x < 2x$ and $\cos$ is a decreasing function on $[0,\pi]$.

23. $\;\overset{\text{(L)}}{=}\; \lim_{x\to 0} \dfrac{(1+\sin x)^{1/2}}{1} = 1.$

25. At that stage we did not know that $D(\sin x) = \cos x$. That is what we were trying to prove.

27. (a) $\lim_{t\to 0^+} \dfrac{(1/2)(1-\cos t)(\sin t)}{(1/2)(1)^2(t) - (1/2)(\cos t)(\sin t)} = \lim_{t\to 0^+} \dfrac{\sin t - \sin t\cos t}{t - \sin t\cos t}$

$= \lim_{t\to 0^+} \dfrac{\sin t - (1/2)\sin 2t}{t - (1/2)\sin 2t} = \lim_{t\to 0^+} \dfrac{2\sin t - \sin 2t}{2t - \sin 2t}$

$\;\overset{\text{(L)}}{=}\; \lim_{t\to 0^+} \dfrac{2\cos t - 2\cos 2t}{2 - 2\cos 2t} \;\overset{\text{(L)}}{=}\; \lim_{t\to 0^+} \dfrac{-2\sin t + 4\sin 2t}{4\sin 2t}$

$\;\overset{\text{(L)}}{=}\; \lim_{t\to 0^+} \dfrac{-2\cos t + 8\cos 2t}{8\cos 2t} = \dfrac{-2+8}{8} = \dfrac{3}{4}.$

(b) $\displaystyle\lim_{t\to 0^+} \frac{(1/2)(\cos t)(\sin t) - (1/2)(\cos t)^2(t)}{(1/2)(1)^2(t) - (1/2)(\cos t)(\sin t)}$

$\displaystyle = \lim_{t\to 0^+} \frac{(1/4)\sin 2t - (1/4)(1+\cos 2t)(t)}{(1/2)t - (1/4)\sin 2t}$

$\displaystyle = \lim_{t\to 0^+} \frac{\sin 2t - t(1+\cos 2t)}{2t - \sin 2t}$

$\overset{\text{\textcircled{L}}}{=} \displaystyle\lim_{t\to 0^+} \frac{2\cos 2t - [(1+\cos 2t) + t(-2\sin 2t)]}{2 - 2\cos 2t}$

$\displaystyle = \lim_{t\to 0^+} \frac{\cos 2t + 2t\sin 2t - 1}{2 - 2\cos 2t} \overset{\text{\textcircled{L}}}{=} \lim_{t\to 0^+} \frac{-2\sin 2t + (2\sin 2t + 4t\cos 2t)}{4\sin 2t}$

$\displaystyle = \lim_{t\to 0^+} \frac{t\cos 2t}{\sin 2t} \overset{\text{\textcircled{L}}}{=} \lim_{t\to 0^+} \frac{\cos 2t - 2t\sin 2t}{2\cos 2t} = \frac{1-0}{2} = \frac{1}{2}.$

29. $\displaystyle\lim_{a\to b^+}(A)$ should be $4\pi b^2$. [surface area of sphere of radius b]

$\displaystyle\lim_{a\to b^+} \frac{a\,\arcsin(\sqrt{a^2-b^2}/a)}{\sqrt{a^2-b^2}} \overset{\text{\textcircled{L}}}{=} \lim_{a\to b^+} \frac{\sqrt{a^2-b^2}\,\arcsin(\sqrt{a^2-b^2}/a) + b}{a}$ [after simplifying]

$\displaystyle = \frac{0+b}{b} = 1.$

Therefore, $\displaystyle\lim_{a\to b^+}(A) = 2\pi b^2 + 2\pi(b)b(1) = 4\pi b^2.$

Problem Set 9.2   Other Indeterminate Forms

1. $\displaystyle\lim_{x\to\infty} \frac{100\ln x}{x} \overset{\text{(L)}}{=} \lim_{x\to\infty} \frac{100/x}{1} = 0.$

3. $\displaystyle\lim_{x\to\infty} \frac{x^{10}}{e^x} \overset{\text{(L)}}{=} \lim_{x\to\infty} \frac{10x^9}{e^x} \overset{\text{(L)}}{=} \lim_{x\to\infty} \frac{90x^8}{e^x} \overset{\text{(L)}}{=} \lim_{x\to\infty} \frac{720x^7}{e^x} \overset{\text{(L)}}{=} \lim_{x\to\infty} \frac{5040x^6}{e^x}$

$\displaystyle \overset{\text{(L)}}{=} \lim_{x\to\infty} \frac{30240x^5}{e^x} \overset{\text{(L)}}{=} \lim_{x\to\infty} \frac{151200x^4}{e^x} \overset{\text{(L)}}{=} \lim_{x\to\infty} \frac{604800x^3}{e^x} \overset{\text{(L)}}{=} \frac{1814400x^2}{e^x}$

$\displaystyle \overset{\text{(L)}}{=} \lim_{x\to\infty} \frac{3628800x}{e^x} \overset{\text{(L)}}{=} \lim_{x\to\infty} \frac{10!}{e^x} = 0.$

5. $\displaystyle = \overset{\text{(L)}}{\lim_{x\to\pi/2}} \frac{\sec x\,\tan x}{\sec^2 x} = \lim_{x\to\pi/2} (\sin x) = 1.$

7. $\displaystyle = \overset{\text{(L)}}{\lim_{x\to\infty}} \frac{1/(x\ln x)}{1/x} = \lim_{x\to\infty} \frac{1}{\ln x} = 0.$

9. $\displaystyle\lim_{x\to0^+} \frac{\cot x}{\ln x} \overset{\text{(L)}}{=} \lim_{x\to0^+} \frac{-\csc^2 x}{1/x} = \lim_{x\to0^+} \frac{-x}{\sin^2 x} \overset{\text{(L)}}{=} \lim_{x\to0^+} \frac{-1}{2\sin x\cos x} = -\infty. \quad \left[\begin{array}{c} \to -1 \\ \to 0^+ \end{array}\right]$

11. $\displaystyle\lim_{x\to0} 4x\ln x = \lim_{x\to0} \frac{4\ln x}{x^{-1}} \overset{\text{(L)}}{=} \lim_{x\to0} \frac{4/x}{-x^{-2}} = \lim_{x\to0} (-4x) = 0.$

13. $\displaystyle\lim_{x\to0} \frac{1-\cos x}{\sin x} \overset{\text{(L)}}{=} \lim_{x\to0} \frac{\sin x}{\cos x} = 0.$

15. $\displaystyle\lim_{x\to0^+} (2x)^{x^2} = \lim_{x\to0^+} \exp(x^2\ln2x) = \exp\left[\lim_{x\to0^+} \frac{\ln2 + \ln x}{x^{-2}}\right]$ $\begin{array}{l}\text{[since exp is a}\\ \text{cont. function]}\end{array}$

$\overset{\text{Ⓛ}}{=} \exp\left[\lim_{x\to0^+} \frac{1/x}{-2x^{-3}}\right] = \exp\left[\lim_{x\to0} \frac{x^2}{2}\right] = \exp(0) = e^0 = 1.$

17. 0. [not an indeterminate form]

19. $\displaystyle\lim_{x\to0} e^{(2/x)\ln(x+e^{x/2})} = \exp\lim_{x\to0} \frac{2\ln(x+e^{x/2})}{x} \overset{\text{Ⓛ}}{=} \exp\lim_{x\to0} \frac{2[1+(1/2)e^{x/2}]}{x + e^{x/2}}$

$= \exp[3/1] \approx 20.0855.$

21. 1 [A determinate form].

23. $\displaystyle\lim_{x\to\infty} \exp\frac{\ln x}{x} = \exp\left[\lim_{x\to\infty} \frac{\ln x}{x}\right] \overset{\text{Ⓛ}}{=} \exp\lim_{x\to\infty} \frac{1/x}{1} = \exp(0) = 1.$

25. 1. [not an indeterminate form]

27. $\displaystyle\lim_{x\to0^+} (\sin x)^x = \lim_{x\to0^+} \exp[x\ln(\sin x)] = \exp\lim_{x\to0^+} \frac{\ln\sin x}{x^{-1}} \overset{\text{Ⓛ}}{=} \exp\lim_{x\to0^+} \frac{\cot x}{-x^{-2}}$

$= \exp\lim_{x\to0^+} \frac{x^2}{-\tan x} \overset{\text{Ⓛ}}{=} \exp\lim_{x\to0^+} \frac{2x}{-\sec^2 x} = \exp\frac{0}{-1} = e^0 = 1.$

29. $\displaystyle= \lim_{x\to0} \frac{x-\sin x}{x\sin x} \overset{\text{Ⓛ}}{=} \lim_{x\to0} \frac{1-\cos x}{x\cos x + \sin x} \overset{\text{Ⓛ}}{=} \lim_{x\to0} \frac{\sin x}{-x\sin x + \cos x + \cos x} = 0.$

31. ∞. [not an indeterminate form]

33. $\displaystyle\lim_{x\to0} (\cos x)^{1/x} = \lim_{x\to0} \exp\frac{\ln\cos x}{x} = \exp\left[\lim_{x\to0} \frac{\ln\cos x}{x}\right] \overset{\text{Ⓛ}}{=} \exp\left[\lim_{x\to0} \frac{-\tan x}{1}\right]$

$= \exp(0) = e^0 = 1.$

35. Doesn't exist. $e^{\cos x}$ oscillates between e^{-1} and e.

37. 0. [not an indeterminate form] **38.** $-\infty$. [not an indeterminate form]

39. $\displaystyle \lim_{x \to \infty} \frac{\int_1^x \sqrt{1+e^{-t}}\, dt}{x} \;\text{\textcircled{L}}\; = \lim_{x \to \infty} \frac{\sqrt{1+e^{-x}}}{1} = 1.$

41. (a) $\displaystyle \lim_{n \to \infty} a^{1/n} = \lim_{x \to \infty} a^{1/x} = \lim_{x \to \infty}\left[\exp(\tfrac{1}{x}\ln a)\right] = \exp\left[\lim_{x \to \infty}\frac{\ln a}{x}\right] = e^0 = 1.$

  (b) $\displaystyle \lim_{n \to \infty} n^{1/n} = \lim_{x \to \infty} x^{1/x} = \lim_{x \to \infty}\left[\exp(\tfrac{1}{x}\ln x)\right] = \exp\left[\lim_{x \to \infty}\frac{\ln x}{x}\right] = e^0 = 1.$
  [Example 3]

  (c) $\displaystyle \lim_{n \to \infty} \frac{a^{1/n}-1}{1/n} = \lim_{x \to 0^+} \frac{a^x-1}{x} \;\text{\textcircled{L}}\; = \lim_{x \to 0^+} \frac{a^x \ln a}{1} = \ln a.$ Let $x = 1/n$

  (d) $\displaystyle \lim_{n \to \infty} \frac{n^{1/n}-1}{1/n} = \lim_{x \to \infty} \frac{x^{1/x}-1}{x^{-1}} \;\text{\textcircled{L}}\; = \lim_{x \to \infty} \frac{x^{1/x}(\ln x)(-x^{-2}) + (1/x)x^{[1/x - 1]}}{-x^{-2}}$

  $= \displaystyle \lim_{x \to \infty}\left[x^{1/x}\ln x - x^{1/x}\right] = \lim_{x \to \infty} x^{1/x}(\ln x - 1) = \infty.$

  $\left[\text{Since } \displaystyle \lim_{x \to \infty} x^{1/x} = 1. \quad (\text{Problem } 23)\right]$

43. Let $f(x) = x^{1/x} = \exp\left[\dfrac{\ln x}{x}\right]$.

  $f'(x) = \dfrac{x^{1/x}(1-\ln x)}{x^2} = 0$ if $x = e$.

  $f'(x) > 0$ on $(0,e)$; $f'(x) < 0$ on (e,∞).

$$f''(x) = \frac{x^{1/x}\left[(\ln x - 1)(\ln x - 1 + 2x) - x\right]}{x^4} = 0 \text{ if } x \approx 0.58 \text{ or } x \approx 4.37.$$

$\lim\limits_{x \to \infty} x^{1/x} = 0$ [Problem 23], so y=1 is a horizontal asymptote.

$\lim\limits_{x \to 0^+} x^{1/x} = 0.$

x	$x^{1/x}$
0.58	0.39 (inflection)
e	1.44 (global max.)
4.37	1.40 (inflection)
10	1.25
2	1.41
1	1
0.50	0.25
0.25	0.004

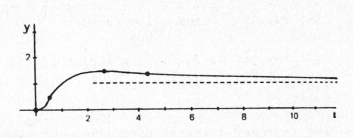

45. $\dfrac{1^k + 2^k + \cdots + n^k}{n^{k+1}} = \left[\left(\frac{1}{n}\right) + \left(\frac{2}{n}\right)^k + \cdots + \left(\frac{n}{n}\right)^k\right]\dfrac{1}{n}$ is a Riemann sum of

$f(x) = x^k$ on $[0,1]$ with n subintervals of equal length and right endpoints as sample points.

Therefore, $\lim\limits_{n \to \infty} \dfrac{1^k + 2^k + \cdots + n^k}{n^{k+1}} = \displaystyle\int_0^1 x^k dx = \left[\dfrac{x^{k+1}}{k+1}\right]_0^1 = \dfrac{1}{k+1}.$

47. (a) 3.16228   (b) 4.16277   (c) 4.56222

49. Absolute max point: $(25, 1.2335 \times 10^{24})$, no absolute min point.

Problem Set 9.3 Improper Integrals: Infinite Limits

1. $\displaystyle\int_1^b e^x dx = \left[e^x\right]_1^b = e^b - e.$ Thus, $\displaystyle\int_1^\infty e^x dx$ diverges since $\lim\limits_{b \to \infty}(e^b - e) = \infty.$

3. $\int xe^{-x^2}dx = \int \frac{e^u}{-2}\,du = \frac{e^u}{-2} + C.$ $[u = -x^2;\ du = -2xdx]$

Then $\int_4^\infty xe^{-x^2}dx = \lim_{b\to\infty}\int_4^b xe^{-x^2}dx = \lim_{b\to\infty}\left[\frac{e^{-x^2}}{-2}\right]_4^b = \frac{-1}{2}\lim_{b\to\infty}\left[e^{-b^2} - e^{-16}\right]$

$$= \frac{-1}{2}(0 - e^{-16}) = \frac{1}{2e^{16}} \approx 5.6268(10^{-8}).$$

5. $\lim_{b\to\infty}\int_3^b (9+x^2)^{-1/2}x\,dx = \lim_{b\to\infty}[(b^2+9)^{1/2} - 18^{1/2}] = \infty.$ [diverges]

7. $\lim_{b\to\infty}\int_1^b x^{-1.01}dx = \lim_{b\to\infty}[-100b^{-0.01} + 100] = 100.$

9. The improper integral diverges. See Example 6 ($p = 0.99 < 1$).

11. $\lim_{b\to\infty}\int_2^b \frac{1}{x\ln x} = \lim_{b\to\infty}[\ln(\ln b) - \ln(\ln 2)] = \infty.$ [diverges]

13. $\lim_{b\to\infty}\int_2^b \frac{1}{x(\ln x)^2}\,dx = \lim_{b\to\infty}[(\ln 2)^{-1} - (\ln b)^{-1}] = (\ln 2)^{-1} \approx 1.4427.$

15. $\int_a^0 (2x-1)^{-3}dx = \left[\frac{(2x-1)^{-2}}{-4}\right]_a^0 = \frac{1 - (2a-1)^{-2}}{-4} = \frac{1}{4(2a-1)^2} - \frac{1}{4}.$

[Use a substitution of $u = 2x-1$ if necessary to see the above.]

Then, $\int_{-\infty}^0 (2x-1)^{-3}dx = \lim_{a\to\infty}\frac{1}{4(2a-1)^2} - \frac{1}{4} = \frac{-1}{4}.$

17. $\lim_{b\to\infty}\int_0^b x(x^2+4)^{-1/2}dx = \lim_{b\to\infty}[(b^2+4)^{1/2} - 2] = \infty.$ [diverges]

19. $\lim_{a\to\infty}\int_a^0 (x^2+2x+5)^{-1}dx + \lim_{b\to\infty}\int_0^b (x^2+2x+5)^{-1}dx$

$$= \lim_{a\to\infty}\left[\frac{1}{2}\tan^{-1}\left(\frac{1}{2}\right) - \frac{1}{2}\tan^{-1}\left(\frac{a+1}{2}\right)\right] + \lim_{b\to\infty}\left[\frac{1}{2}\tan^{-1}\left(\frac{b+1}{2}\right) - \frac{1}{2}\tan^{-1}\left(\frac{1}{2}\right)\right]$$

$$= -(-\pi/4) + \pi/4 = \pi/2 \approx 1.5708.$$

21. $\int_0^b \text{sech}x\ dx = \left[\tan^{-1}(\sinh x)\right]_0^b = \tan^{-1}(\sinh b) - 0.$

Then, $\int_0^\infty \text{sech}x\ dx = \lim_{b\to\infty} \tan^{-1}(\sinh b) = \pi/2.$

Thus, $\int_{-\infty}^\infty \text{sech}x\ dx = \pi$ since the integrand defines an even function.

23. $\lim_{b\to\infty}\int_0^b e^{-x}\cos x\ dx = \lim_{b\to\infty}\left[\dfrac{\sin b - \cos b}{2e^b} + \dfrac{1}{2}\right] = \dfrac{1}{2}.$

25. $\lim_{b\to\infty}\int_1^b \dfrac{2}{4x^2 - 1}\ dx = \lim_{b\to\infty}\left[\ln\sqrt{\dfrac{2b-1}{2b+1}} + \ln(\sqrt{3})\right] = \ln(\sqrt{3}) \approx 0.5493.$

27. F = k/x would be the force required to counteract the force due to gravity. At x = 3960 miles, F = m lbs (weight of object at surface of earth is m lbs), so k = 3960m, and the force formula for the object is F = 3960m/x. Therefore, the work required to send the object out of earth's gravitational field would be

$$\int_{3960}^\infty (3960m/x)\,dx = \lim_{b\to\infty}\int_{3960}^b (3960m/x)\,dx = \lim_{b\to\infty}\left[3960m\ \ln x\right]_{3960}^b$$

$$= \lim_{b\to\infty}(3960m\ \ln b\ - \ 3960m\ \ln 3960) = \infty.$$

29. $FP = \lim_{b\to\infty}\int_0^b 100{,}000 e^{-0.08t}\,dt = 100{,}000 \lim_{b\to\infty}\left[(-25/2)e^{-0.08t} + (25/2)\right]$

$= \$1{,}250{,}000.$

31. (a) $\lim_{b\to\infty}\int_0^b a e^{-ax}\,dx = \lim_{b\to\infty}[-e^{-ab} + 1] = 1.$

(b) $\lim_{b\to\infty}\int_0^b x\ a e^{-ax}\,dx = \lim_{b\to\infty}[-be^{-ab} - (1/a)e^{-ab} + 1/a] = 1/a.$

33. (a) $\int_0^b \sin x \, dx = \left[-\cos x \right]_0^b = -\cos b + 1$; $\int_0^\infty \sin x \, dx$ and, hence, $\int_{-\infty}^\infty \sin x \, dx$

diverges since $\lim_{b \to \infty}(-\cos b + 1)$ doesn't exist.

(b) $\lim_{a \to \infty} \int_{-a}^a \sin x \, dx = \lim_{a \to \infty}(0) = 0$. [since $\sin x$ defines an odd function]

35. The region used in **THE PARADOX OF GABRIEL'S HORN** section was shown to yield a finite volume when revolved about the x-axis. However, when revolved about the y-axis, the volume is $\int_1^\infty 2\pi x(1/x)dx$, which diverges.

37. 0.99, 3.69043. 4.50074, 4.60517, 4.71286

39. 0.68269, 0.9545, 0.9973, 0.99994

Problem Set 9.4 Improper Integrals: Infinite Integrands

1. $\lim_{s \to 1^+}\int_s^2 (x-1)^{-1/3}dx = \lim_{s \to 1^+} [3/2 - (3/2)(s-1)^{2/3}] = 3/2.$

3. The asymptotic discontinuity is at x=1.

$$\int_0^1 \frac{1}{\sqrt{1-x^2}} \, dx = \lim_{t \to 1^-} \int_0^t \frac{1}{\sqrt{1-x^2}} \, dx = \lim_{t \to 1^-} \left[\sin^{-1}x \right]_0^t = \lim_{t \to 1^-} (\sin^{-1}t - 0) = \pi/2.$$

5. Diverges since $\lim_{s \to 0^+}\int_s^2 x^{-4}dx = \lim_{s \to 0^+} [-1/24 + (1/3)s^{-3}] = \infty.$

7. $\lim_{t \to 3^-} \int_2^t (3-x)^{-2/3}dx + \lim_{s \to 3^+}\int_s^4 (3-x)^{-2/3}dx$

$= \lim_{t \to 3^-} [-3(3-t)^{1/3} + 3] + \lim_{s \to 3^+} [3 + 3(3-s)^{1/3}] = 3+3 = 6.$

9. The asymptotic discontinuity is at x=-2. Use a u = (x^2-4) substitution to find that an antiderivative of $x(x^2-4)^{-2/3}$ is $(3/2)(x^2-4)^{1/3}$.

$$\int_{-3}^{0} x(x^2-4)^{-2/3}dx = \int_{-3}^{-2} x(x^2-4)^{-2/3}dx + \int_{-2}^{0} x(x^2-4)^{-2/3}dx$$

$$= \lim_{t\to -2^-}\left[(3/2)(x^2-4)^{1/3}\right]_{-3}^{t} + \lim_{s\to -2^+}\left[(3/2)(x^2-4)^{1/3}\right]_{s}^{0}$$

$$= \lim_{t\to -2^-}\frac{3[(t^2-4)^{1/3}-(5)^{1/3}]}{2} + \lim_{s\to -2^+}\frac{3[(-4)^{1/3}-(s^2-4)^{1/3}]}{2}$$

$$= \frac{-3[(5)^{1/3}+(4)^{1/3}]}{2} \approx -4.9461.$$

11. Diverges since $\lim_{t\to -1^-}\int_{-2}^{t}(x+1)^{-4/3}dx = \lim_{t\to -1^-}[-3(t+1)^{-1/3}-3] = \infty.$

13. $\lim_{t\to(\pi/4)^-}\int_{0}^{t}\tan 2x\, dx = \lim_{t\to(\pi/4)^-}[(1/2)\ln|\sec 2t|] = \infty.$ [diverges]

15. The asymptotic discontinuity is at x=0.

$$\int_{0}^{1}\frac{\ln x}{x}dx = \lim_{s\to 0^+}\int_{s}^{1}\frac{\ln x}{x}dx = \lim_{s\to 0^+}\left[\frac{\ln^2 x}{2}\right]_{s}^{1} = \lim_{s\to 0^+}\left[0-\frac{\ln^2 s}{2}\right] = -\infty,$$

so the improper integral diverges. [Use u = lnx substitution at the second step if necessary to see the result.]

17. The integral equals 0 since the integrand defines an odd function, and the corresponding integral on [0,3) converges (see Problem 10].

19. In example 4 of this section it was shown that the first integral diverges if p > 1. In Example 6 of Section 3 it was shown that the second integral diverges if p ≤ 1.

21. Area $= \int_0^8 (x-8)^{-2/3} dx = \lim_{t \to 8^-} \int_0^t (x-8)^{-2/3} dx$

$= \lim_{t \to 8^-} \left[3(x-8)^{1/3} \right]_0^t = \lim_{t \to 8^-} \left[3(t-8)^{1/3} + 6 \right]$

$= 6$ square units.

23. (a) Area $= \lim_{s \to 0^+} \int_s^1 x^{-2/3} dx = \lim_{s \to 0^+} \int_s^1 \left[3 - 3s^{1/3} \right] = 3.$

(b) Volume $= \lim_{s \to 0^+} \int_s^1 \pi (x^{-2/3})^2 dx = \lim_{s \to 0^+} \left[-3\pi + 3\pi s^{-1/3} \right] = \infty.$

25. It is not an improper integral since (1) the limits of integration are finite, and (2) the integrand is bounded on $(0,1]$. However, it is not a proper integral either since the integrand is not defined at 0. This can be remedied, however, since the function defined by the integrand has a removable discontinuity at 0.

27. $\int_1^\infty e^{-x} dx = \lim_{b \to \infty} \int_1^b e^{-x} dx = \lim_{b \to \infty} \left[-e^{-x} \right]_1^b = \lim_{b \to \infty} \left[\frac{-1}{e^b} + \frac{1}{e} \right] = 1/e.$

The fact that this improper integral converges, along with the hint and the convergence test, implies that the integral in question converges.

29. (a) $\lim_{x \to \infty} \frac{x^{n+1}}{e^x} = 0 \Rightarrow$ for some $M > 0$, $0 < \frac{x^{n+1}}{e^x} \leq 1$ for $x \geq M$

$\Rightarrow 0 < \frac{x^{n-1}}{e^x} \leq x^{-2}$ for $x \geq M.$

(b) $\int_M^\infty x^{-2} dx$ converges (see Example 6 of Section 3).

Therefore, the integral in question converges since on $[1,M]$ it is a proper integral and on $[M,\infty)$ it converges by (a) and the Comparison Test.

31. (a) $\Gamma(1) = \int_0^\infty e^{-x}dx = \lim_{b\to\infty}\int_0^b -e^{-x}dx = \lim_{b\to\infty}[-e^{-b} + 1] = 1.$

(b) $\Gamma(n+1) = \int_0^\infty x^n e^{-x}dx = \lim_{b\to\infty}\int_0^b x^n e^{-x}dx$ [Let $u = x^n$; $dv = e^{-x}dx$]

$= \lim_{b\to\infty}\left[-b^n e^{-b} + n\int_0^b x^{n-1} e^{-x}dx\right] = 0 + n\int_0^\infty x^{n-1} e^{-x}dx = n\Gamma(x).$

(c) (By induction):   (i) $\Gamma(1+1) = 1\Gamma(1) = 1(1) = 1!$
                  (ii)   Assume $\Gamma(k+1) = k!$
                  (iii) $\Gamma([k+1]+1) = (k+1)\Gamma(k+1) = (k+1)k! = (k+1)!$

33. (a) Let $y = \sqrt{\dfrac{1-x}{x}}$ and solve for x: $x = (1+y^2)^{-1}.$

$\text{Area} = \int_0^\infty (1+y^2)^{-1}dy = \lim_{b\to\infty}\int_0^b (1+y^2)^{-1}dy$

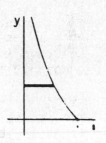

$= \lim_{b\to\infty}\left[\arctan(y)\right]_0^b = \lim_{b\to\infty}[\arctan(b) - 0] = \dfrac{\pi}{2}.$

(b) Let $y = \sqrt{\dfrac{1+x}{1-x}}$ and solve for x: $x = 1 - 2(1+y^2)^{-1}.$

$\text{Area} = \int_1^\infty \left[1 - [1 - 2(1+y^2)^{-1}]\right]dy = 2\int_1^\infty (1+y^2)^{-1}dy$

$= 2(\pi/2) = \pi.$

Problem Set 9.5   Chapter Review

True-False Quiz

1. True.   Use l'Hôpital's Rule 100 times.

3. False.   It equals 1000000. (Use l'Hôpital's Rule four times, or divide numerator and denominator by x^4.)

5. False. $f(x) = x^2$, $g(x) = x$ provides a counterexample.

7. True.   First use continuity of $f(z) = z^n$ at $z=1$.

9. True.   Let $-M < 0$ be given. Then $g(x) > 2M$, and $-3/2 < f(x) < -1/2$ for x in some deleted neighborhood of a, so $f(x)g(x) < -M$ for x in the deleted neighborhood.

11. False. $f(x) = 3\sin x$, $g(x) = \sin x$ provides a counterexample.

13. True.   Take exp of each side and use continuity of exp at $x=2$.

15. True.   Use l'Hôpital's Rule n times, where n is the degree of $p(x)$.

17. False. $f(x) = x^2+3x+1$, $g(x) = x^2+x+2$ provides a counterexample.

19. True.   See Problem 19 of Section 9.4.

21. True. $\displaystyle\int_{-a}^{0} f(x)dx = \int_{0}^{a} f(x)dx$ for all $a \geq 0$, so $\displaystyle\int_{-\infty}^{0} f(x)dx$ converges.

23. True.   It equals $\displaystyle\lim_{b\to\infty} [f(b)-f(0)] = -f(0)$.

25. False. The function defined by the integrand has a removable discontinuity at $x=0$. and is continuous on $(0,\pi/4]$.

Sample Test Problems

1. $\displaystyle\overset{(L)}{=} \lim_{x\to 0} \frac{4}{\sec^2 x} = 4.$

3. $\displaystyle\lim_{x\to 0} \frac{\sin x - \tan x}{x^2/3} \overset{(L)}{=} \lim_{x\to 0} \frac{\cos x - \sec^2 x}{2x/3} \overset{(L)}{=} \lim_{x\to 0} \frac{-\sin x - 2\sec^2 x \tan x}{2/3} = 0.$

5. $\displaystyle\lim_{x\to 0} \frac{2x}{\tan x} \overset{(L)}{=} \lim_{x\to 0} \frac{2}{\sec^2 x} = 2.$ **7.** $\displaystyle\overset{(L)}{=} \lim_{t\to\infty} \frac{1/t}{2t} = \lim_{t\to\infty} \frac{1}{2t^2} = 0.$

9. $\lim\limits_{x\to 0^+} (\sin x)^{1/x} = \lim\limits_{x\to 0^+} e^{(1/x)\ln(\sin x)} = \lim\limits_{x\to 0^+} \exp \dfrac{\ln(\sin x)}{x} = 0$

      since $\lim\limits_{x\to 0^+} \dfrac{\ln(\sin x)}{x} = -\infty$.

11. $\lim\limits_{x\to 0^+} \exp(x\ln x) = \exp \lim\limits_{x\to 0^+} x\ln x = \exp(0) = e^0 = 1.$ [See Problem 10.]

13. $\lim\limits_{x\to 0^+} \dfrac{\ln x}{x^{-1/2}} \overset{\textcircled{L}}{=} \lim\limits_{x\to 0^+} \dfrac{x^{-1}}{(-1/2)x^{-3/2}} = \lim\limits_{x\to 0^+} -2x^{1/2} = 0.$

15. $\lim\limits_{x\to 0^+} \left[\dfrac{1}{\sin x} - \dfrac{1}{x}\right] = \lim\limits_{x\to 0^+} \dfrac{x-\sin x}{x\sin x} \overset{\textcircled{L}}{=} \lim\limits_{x\to 0^+} \dfrac{1-\cos x}{x\cos x + \sin x}$

      $= \lim\limits_{x\to 0^+} \dfrac{\sin x}{-x\sin x + \cos x + \cos x} = \dfrac{0}{2} = 0.$

17. $\lim\limits_{x\to \pi/2} \exp\dfrac{(\ln \sin x)}{\cot x} \overset{\textcircled{L}}{=} \exp \lim\limits_{x\to \pi/2} \dfrac{\cot x}{-\csc^2 x} = \exp \lim\limits_{x\to \pi/2} (-\sin x \cos x) = 1.$

19. $\lim\limits_{b\to\infty} \left[\dfrac{-1}{b+1} + 1\right] = 1.$

21. $\displaystyle\int_{-\infty}^1 e^{2x}dx = \lim_{a\to -\infty} \int_a^1 e^{2x}dx = \lim_{a\to\infty} \left[\dfrac{e^{2x}}{2}\right]_a^1 = \lim_{a\to -\infty} \dfrac{e^2 - e^{2a}}{2} = \dfrac{e^2-0}{2} = \dfrac{e^2}{2} \approx 3.6945.$

23. $\lim\limits_{b\to\infty} [\ln(b+1) - 0] = \infty.$ [The improper integral diverges.]

25. $\displaystyle\lim_{b\to\infty}\int_1^b \left[\dfrac{1}{x^2} - \dfrac{1}{1+x^2}\right]dx = \lim_{b\to\infty} \left[\left(\dfrac{-1}{b} - \tan^{-1}b\right) - \left(-1 - \dfrac{\pi}{4}\right)\right] = 1 - \dfrac{\pi}{4} \approx 0.2146.$

27. There is an asymptotic discontinuity at x=-3/2.

$$\int_{-3/2}^{0} \frac{1}{2x+3} \, dx = \lim_{s \to -3/2^+} \int_{s}^{0} \frac{1}{2x+3} \, dx$$

> Let $u = 2x+3$
> Then $du = 2dx$
> $x=0 \Rightarrow u=3$
> $x=s \Rightarrow u=2s+3$

$$= \lim_{s \to -3/2^+} \int_{2s+3}^{3} \frac{1}{u} \, du = \lim_{v \to 0^+} \int_{v}^{3} \frac{1}{u} \, du \qquad [\text{letting } v = 2s+3]$$

which diverges [See Example 3 of Section 9.4]. Therefore, the improper integral in question diverges.

29. $\displaystyle \lim_{b \to \infty} \left[\frac{-1}{\ln b} + \frac{1}{\ln 2} \right] = \frac{1}{\ln 2} \approx 1.4427.$

31. $\displaystyle \lim_{t \to 4^-} [-3(4-t)^{1/3} + 3] = \lim_{s \to 4^+} [3 + 3(4-s)^{1/3}] = 3+3 = 6.$

33. $\displaystyle \int \frac{x}{x^2+1} \, dx = \int \frac{1}{2} \frac{1}{u} \, du = \frac{\ln|u|}{2} + C = \frac{\ln(x^2+1)}{2} + C.$

> Let $u = x^2+1$
> Then $du = 2xdx$

Then, $\displaystyle \int_{0}^{\infty} \frac{x}{x^2+1} \, dx = \lim_{b \to \infty} \int_{0}^{b} \frac{x}{x^2+1} \, dx = \lim_{b \to \infty} \left[\frac{\ln(x^2+1)}{2} \right]_{0}^{b}$

$$= \lim_{b \to \infty} \left[\frac{\ln(b^2+1)}{2} - 0 \right] = \infty.$$

Therefore, $\displaystyle \int_{0}^{\infty} \frac{x}{x^2+1} \, dx$ diverges, so $\displaystyle \int_{-\infty}^{\infty} \frac{x}{x^2+1} \, dx$ diverges.

35. If $x > 0$, $0 < (x^6+x)^{1/2} < x^{-3}$. $\displaystyle \int_{1}^{\infty} x^{-3} dx$ converges.

Therefore, $\displaystyle \int_{1}^{\infty} (x^6+x)^{-1/2} dx$ converges.

37. If $x > 3$, $0 < \frac{1}{x} < \frac{\ln x}{x}$. $\displaystyle \int_{3}^{\infty} \frac{1}{x} \, dx$ diverges.

Therefore, $\displaystyle \int_{3}^{\infty} \frac{\ln x}{x} \, dx$ diverges.

Problem Set 10.1   Taylor's Approximation to Functions

1. 1.723034.

3. 0.1193318.

5. 118.11261.

7. $P_4(x) = 1 + 2x + \dfrac{4x^2}{2!} + \dfrac{8x^3}{3!} + \dfrac{16x^4}{4!}$. $f(0.23) \approx 1.58389$.

9. $\begin{aligned} f(x) &= \sin 2x. & f(0) &= 0. \\ f'(x) &= 2\cos 2x. & f'(0) &= 2. \\ f''(x) &= -4\sin 2x & f''(0) &= 0. \\ f^{(3)}(x) &= -8\cos 2x. & f^{(3)}(0) &= -8. \\ f^{(4)}(x) &= 16\sin 2x & f^{(4)}(0) &= 0. \end{aligned}$

$$P_4(x) = 0 + 2x + \frac{0x^2}{2!} + \frac{-8x^3}{3!} + \frac{0x^4}{4!} = 2x - \frac{4x^3}{3}.$$

$f(0.23) \approx P_4(0.23) \approx 0.44378$.

11. $P_4(x) = x - \dfrac{x^2}{2} + \dfrac{x^3}{3} - \dfrac{x^4}{4}$. $f(0.23) \approx 0.20691$.

13. $P_4(x) = x - (1/3)x^3$. $f(0.23) \approx 0.22594$.

15. Let $f(x) = e^x$. $f^{(n)}(x) = e^x$, $f^{(n)}(2) = e^2$ for each $n = 0, 1, 2, \cdots$.

$$P_3(x) = e^2 + e^2(x-2) + \frac{e^2(x-2)^2}{2!} + \frac{e^2(x-2)^3}{3!}.$$

17. $P_3(x) = 1 + 2(x - \pi/4) + \dfrac{4(x - \pi/4)^2}{2!} + \dfrac{16(x - \pi/4)^3}{3!}$.

19. $P_3(x) = \dfrac{\pi}{4} + \dfrac{1}{2}(x-1) - \dfrac{1}{2}\dfrac{(x-1)^2}{2!} + \dfrac{1}{2}\dfrac{(x-1)^3}{3!}$.

21. Let $f(x) = x^3 - 2x^2 + 3x + 5.$ $f(1) = 7.$

$\quad\quad\quad f'(x) = 3x^2 - 4x + 3.$ $f'(1) = 2.$

$\quad\quad\quad f''(x) = 6x - 4.$ $f''(1) = 2.$

$\quad\quad\quad f^{(3)}(x) = 6.$ $f^{(3)}(1) = 6.$

$$P_3(x) = 7 + 2(x-1) + \frac{2(x-1)^2}{2!} + \frac{6(x-1)^3}{3!}$$

$$= 7 + (2x-2) + (x^2 - 2x + 1) + (x^3 - 3x^2 + 3x - 1) = x^3 - 2x^2 + 3x + 5 = f(x).$$

23. $P_n(x) = 1 + x + x^2 + x^3 + \cdots + x^n.$ $P_4(x) = 1 + x + x^2 + x^3 + x^4.$

(a) $P_4(0.1) \approx 1.1111$ and $f(0.1) \approx 1.1111.$

(b) $P_4(0.5) \approx 1.9375$ and $f(0.5) = 2.$

(c) $P_4(0.9) = 4.0951$ and $f(0.9) = 10.$

(d) $P_4(2) = 31$ and $f(2) = -1.$

25. Area $= (1/2)r^2 t - (1/2)[2r\sin(t/2)r\cos(t/2)]$

$\quad\quad\quad\quad = (1/2)r^2 t = (1/2)r^2 \sin t = (1/2)r^2(t - \sin t)$

$\quad\quad\quad\quad = (1/2)r^2[t - (t - t^3/6)]$ [Maclaurin of order 3 for sin t]

$\quad\quad\quad\quad = (1/2)r^2(t^3/6) = r^2 t^3/12.$

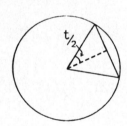

27. (a) $(1 + \frac{r}{12})^{12n} = 2 \Rightarrow 12n \ln(1 + \frac{r}{12}) = \ln 2$

$$\Rightarrow n = (\ln 2)\left[\frac{1}{12\ln(1 + \frac{r}{12})}\right]$$

(b) $P_2(x) = x - \dfrac{x^2}{2}$ for $\ln(1+x)$. [From Problem 11]

Therefore, $\ln\left(1 + \dfrac{r}{12}\right) \approx P_2\left(\dfrac{r}{12}\right) = \dfrac{r}{12} - \dfrac{r^2}{288}$.

Then, $n = (\ln 2)\left[\dfrac{1}{12\ln\left(1 + \dfrac{r}{12}\right)}\right] \approx 0.693\left[\dfrac{1}{r - r^2/24}\right]$

$= 0.693\left[\dfrac{24}{24r - r^2}\right] = 0.693\left[\dfrac{1}{r} + \dfrac{1}{24-r}\right]$

$\approx 0.693\left[\dfrac{1}{r} + \dfrac{1}{24}\right] \approx \dfrac{0.693}{r} + 0.029$

(c)

r	n [Exact to 3rd decimal place]	n [Approximation from part (b)]	n [Rule of 70]
0.05	13.892	13.889	14.000
0.10	6.960	6.959	7.000
0.15	4.650	4.649	4.667
0.20	3.495	3.494	3.500

29. (a) $f(x) - \sin(\exp(x)$

Terms of series:
.841471
$.540302 * x^1$
$-.150584 * x^2$
$-.420735 * x^3$
$-.322931 * x^4$

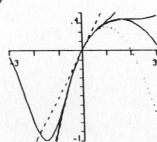

(b) $f(x) = \sin(x)/(2+\sin(x))$

Terms of series:
$1/2 * x^1$
$-1/4 * x^2$
$1/24 * x^3$
$1/48 * x^4$

31. (a) $2x^3 - 3x^2 + x$ (b) $x - 3x^2 + 2x^3 - x^4$

    (c) $x - 3x^2 + 2x^3 - x^4$ (d) $x - 3x^2 + 2x^3 - x^4$

    (e) $1 + x + x^2 + x^3 + x^4$ (f) $x - \frac{1}{3}x^3$

    (g) $1 + x^2 + x^4$ (h) $1 + 2x + x^2 + \frac{2}{3}x^3 + x^4$

Problem Set 10.2 Estimating the Errors

1. $|e^c + e^{-c}| \leq e^2 + 1 < 8.39.$

3. $\left|\dfrac{2}{\cos c}\right| = \dfrac{2}{|\cos c|} \leq \dfrac{2}{.5} = 4.$

    (For c in $[2\pi/3, \pi]$, the smallest $|\cos c|$ can be is 0.5.)

5. $|e^{-c}/(c+4)| = e^{-c}/|c+4| < e^8/2 < 1491.$

7. $|(c^2 + \cos c)/(16 \ln c)| \leq (c^2 + |\cos c|)/(16 \ln 2) < 1.$

9. Let $f(x) = \ln(1+x)$. $f'(x) = (1+x)^{-1}$, $f''(x) = -(1+x)^{-2}$,

    $f^{(3)}(x) = 2(1+x)^{-3}$, $f^{(4)}(x) = -3!(1+x)^{-4}$, $f^{(5)}(x) = 4!(1+x)^{-5}$,

    $f^{(6)}(x) = -5!(1+x)^{-6}$, $f^{(7)}(x) = 6!(1+x)^{-7}.$

    $R_6(x) = \dfrac{f^{(7)}(c)\, x^7}{7!} = \dfrac{6!x^7}{7!(1+c)^7} = \dfrac{x^7}{7(1+c)^7}$ for some c between 0 and x.

    $|R_6(0.5)| = \dfrac{(0.5)^7}{7(1+c)^7} < \dfrac{(0.5)^7}{7} < 0.001117$ for the c between 0 and 0.5.

11. $|R_6(0.5)| = |(0.5)^7/(7! \cos c)| < (0.5)^7/7! < 0.000001551.$

13. $R_n(x) = \dfrac{f^{(n+1)}(c)}{(n+1)!} x^{n+1} = \dfrac{e^c x^{n+1}}{(n+1)!}.$ $R_n(1) = \dfrac{e^c}{(n+1)!}$ for some c in $(0,1)$.

    $|R_n(1)| < 3/(n+1)! \leq 0.000005$, if $(n+1)! \geq 600,000$, or $n \geq 9.$

15. Let $f(x) = (1+x)^{-1/2}.$ $f(0) = 1.$
$f'(x) = (-1/2)(1+x)^{-3/2}.$ $f'(0) = -1/2.$
$f''(x) = (3/4)(1+x)^{-5/2}.$ $f''(0) = 3/4.$
$f^{(3)}(x) = (-15/8)(1+x)^{-7/2}.$ $f^{(3)}(0) = -15/8.$
$f^{(4)}(x) = (105/16)(1+x)^{-9/2}.$

$$P_3(x) = 1 + (-1/2)x + \frac{(3/4)x^2}{2!} + \frac{(-15/8)x^3}{3!} = 1 - \frac{x}{2} + \frac{3x^2}{8} - \frac{5x^3}{16}.$$

$$|R_3(x)| = \frac{f^{(4)}(c)x^4}{4!} \text{ for some c between 0 and x, hence in } (-0.05, 0.05),$$

$$= \frac{105x^4}{16(1+c)^{9/2}(4!)} \le \frac{35(0.05)^4}{128(0.95)^{9/2}} < 0.0000022.$$

17. $|R_4(x)| = |\cos c|(x^5/5!), \quad 0 < c < x \le 0.5.$

$$\le (1)[(0.5)^5/5!] = (0.5)^5/120.$$

Therefore, sinx is between $x - (1/6)x^3 \pm (0.5)^5/120.$

$$\int_0^{0.5} \sin x \, dx = \int_0^{0.5} [x - (1/6)x^3]dx \approx 0.1223958$$

$$\text{with } |Error| < \int_0^{0.5} (0.5)^5/120 \, dx = (0.5)^6/120 < 0.0001303.$$

19. $P_4(x) = -1 + 0(x-1) - (2/2!)(x-1)^2 + (6/3!)x^3 + (24/4!)x^4$

$$= -1 - (x-1)^2 + (x-1)^3 + (x-1)^4. \quad R_4(x) = 0 \text{ since } f^{(5)}(x) = 0.$$

21. Let
$$f(x) \quad = \sin x. \qquad f(\pi/4) \quad = \sqrt{2}/2.$$
$$f'(x) \quad = \cos x. \qquad f'(\pi/4) \quad = \sqrt{2}/2.$$
$$f''(x) \quad = -\sin x. \qquad f''(\pi/4) \quad = -\sqrt{2}/2.$$
$$f^{(3)}(x) \quad = -\cos x \qquad f^{(3)}(\pi/4) \quad = -\sqrt{2}/2.$$
$$f^{(4)}(x) \quad = \sin x.$$

$\sin 43° = \sin(43\pi/180)$ and $(43\pi/180 - \pi/4) = -\pi/90$.

$$P_3(x) = (\sqrt{2}/2) + (\sqrt{2}/2)(x-\pi/4) + \frac{(-\sqrt{2}/2)(x-\pi/4)^2}{2} + \frac{(-\sqrt{2}/2)(x-\pi/4)^3}{3!}.$$

$$\sin(43\pi/180) \approx \frac{\sqrt{2}}{2} + \frac{\sqrt{2}(-\pi/90)}{2} - \frac{\sqrt{2}(-\pi/90)^2}{4} - \frac{\sqrt{2}(-\pi/90)^3}{12} \approx 0.731354.$$

$$R_3(x) = \frac{f^{(4)}(c)(x-\pi/4)^4}{4!} = \frac{(\sin c)(x-\pi/4)^4}{4!} \text{ for some c between x and } \pi/4.$$

$$|R_3(43\pi/180)| \le \frac{(\sqrt{2}/2)(-\pi/90)^4}{24} < 0.0000000438.$$

23. Next term is 0, so consider

$$|R_{10}(x)| = \frac{|f^{(11)}(c)|}{11!} x^{11} \le \frac{x^{11}}{11!} \le \frac{(\pi/2)^{11}}{11!} < \frac{(1.6)^{11}}{11!} < 5(10^{-6}).$$

25. Each of the first n derivatives of x^{n+1} has a factor of x, so each of the first n derivatives of $x^{n+1}f(x)$ has a factor of x.
Thus, $g^{(k)}(0) = p^{(k)}(0)$ for $k = 0,1,2,\cdots,n$, so the Maclaurin polynomial of g(x) of order n is the Maclaurin polynomial of p(x) of order n, viz., p(x) itself since $\deg[p(x)] \le n$.

Problem Set 10.3 Numerical Integration

1. (T): $\frac{1/8}{2}(1 + \frac{128}{81} + \frac{32}{25} + \frac{128}{121} + \frac{8}{9} + \frac{128}{169} + \frac{32}{49} + \frac{128}{225} + \frac{1}{4}) \approx 0.5023.$

(S): $\frac{1/8}{3}(1 + \frac{256}{81} + \frac{32}{25} + \frac{256}{121} + \frac{8}{9} + \frac{256}{169} + \frac{32}{49} + \frac{256}{225} + \frac{1}{4}) \approx 0.5000.$
Exact: 0.5.

3. $f(x) = x$.

$h = \frac{4-0}{8} = 0.5$.

i	x_i	$f(x_i)$	$2f(x_i)$	$4f(x_i)$
0	0	0		
1	0.5	$\sqrt{0.5}$	$2\sqrt{0.5}$	$4\sqrt{0.5}$
2	1	1	2	
3	1.5	$\sqrt{1.5}$	$2\sqrt{1.5}$	$4\sqrt{1.5}$
4	2	$\sqrt{2}$	$2\sqrt{2}$	
5	2.5	$\sqrt{2.5}$	$2\sqrt{2.5}$	$4\sqrt{2.5}$
6	3	$\sqrt{3}$	$2\sqrt{3}$	
7	3.5	$\sqrt{3.5}$	$2\sqrt{3.5}$	$4\sqrt{3.5}$
8	4	$\sqrt{2}$		

Trapezoidal: $\frac{0.5}{2} [0 + 2(\sqrt{0.5} + 1 + \sqrt{1.5} + \sqrt{2} + \sqrt{2.5} + \sqrt{3}$

$+ \sqrt{3.5}) + 2] \approx 5.2650$.

Parabolic: $\frac{0.5}{3} [0 + 4\sqrt{0.5} + 2 + 4\sqrt{1.5} + 2\sqrt{2} + 4\sqrt{2.5} + 2\sqrt{3}$

$+ 4\sqrt{3.5} + 2] \approx 5.3046$.

Exact: $\int_0^4 x^{1/2} \, dx = \left[\frac{2x^{3/2}}{3}\right]_0^4 = \frac{16}{3} = 5.333 \cdots$.

5. $n=2$: $\frac{\pi/2}{2} (0 + 2 + 0) = \frac{\pi}{2} \approx 1.5708$.

$n=6$: $\frac{\pi/6}{2} (0 + 1 + \sqrt{3} + 2 + \sqrt{3} + 1 + 0) = (2 + \sqrt{3})(\pi/6) \approx 1.9541$.

$n=12$: $\frac{\pi/12}{2} [4\sin(\pi/12) + 4 + 2\sqrt{2} + 2\sqrt{3} + 4\sin(5\pi/12)] \approx 1.9886$.

7. $\frac{1}{30}\left[\frac{4}{1} + \frac{16}{1.01} + \frac{8}{1.04} + \frac{16}{1.09} + \frac{8}{1.16} + \frac{16}{1.25} + \frac{8}{1.36} + \frac{16}{1.49} + \frac{8}{1.64} + \frac{16}{1.81} + 2\right]$

$= 3.1415926$, the same as the exact value of π to 7 decimal places.

9. Let $f(x) = e^{-x^2}$. $f'(x) = -2xe^{-x^2}$, $f''(x) = 2e^{-x^2}(2x^2-1)$.

$$|E_n| = \frac{(1.2-0)^3|f''(c)|}{12n^2} = \frac{(1.728)2e^{-c^2}|2c^2-1|}{12n^2} \text{ for some c in } (0,1.2)$$

$$\leq \frac{(1.728)(2)(1)(1.88)}{12n^2} = \frac{0.54144}{n^2} \leq 0.01 \text{ if } n^2 \geq 54.144, \text{ or } n \geq 8.$$

$$h = \frac{1.2-0}{8} = 0.15.$$

$$\begin{array}{c|ccccccccc} & 0 & 0.15 & 0.3 & 0.45 & 0.6 & 0.75 & 0.9 & 1.05 & 1.2 \end{array}$$

$$\frac{0.15}{2}[e^0 + 2(e^{-0.0225} + e^{-0.09} + e^{-0.2025} + e^{-0.36} + e^{-0.5625} + e^{-0.81} +$$

$$+ e^{-1.1025}) + e^{-1.44}] \approx 0.8057.$$

11. $|E_n| = \dfrac{1}{12n^2}\dfrac{1 + \sin^2 c}{4(\sin c)^{3/2}} < \dfrac{2}{48n^2(\sin 1)^{3/2}} < \dfrac{0.06}{n^2} < 0.01 \text{ if } n \geq 3.$

$$\frac{1/3}{2}[\sqrt{\sin 1} + 2\sqrt{\sin(4/3)} + 2\sqrt{\sin(5/3)} + \sqrt{\sin 2}] \approx 0.973.$$

13. $|E_n| = \dfrac{2^5}{180n^4}\dfrac{48}{|(1-c)^5|} < \dfrac{(32)(48)}{180n^4} < 0.005 \text{ if } n \geq 7.$ We will use $n = 8.$

$$\frac{2/8}{3}[(\frac{3}{-1}) + 4(\frac{3.25}{-1.25}) + \cdots + 4(\frac{4.75}{-2.75}) + (\frac{5}{-3})] \approx -4.197.$$

15. $\displaystyle\int_{m-h}^{m+h}(ax^2+bx+cx)dx = \left[\frac{ax^3}{3} + \frac{bx^2}{2} + cx\right]_{m-h}^{m+h}$

$$= \frac{a(m+h)^3}{3} + \frac{b(m+h)^2}{2} + c(m+h) - \frac{a(m-h)^3}{3} - \frac{b(m-h)^2}{2} - c(m-h)$$

$$= \frac{6am^2h + 2ah^3 + 6bmh + 6ch}{3} = \frac{h}{3}[a(6m^2+2h^2) + b(6m) + 6c].$$

$\dfrac{h}{3}[f(m-h) + 4f(m) + f(m+h)]$

$$= \frac{h}{3}[a(m-h)^2+b(m-h)+c] + [4(am^2+bm+c)] + [a(m+h)^2+b(m+h)+c]$$

$$= \frac{h}{3}(6am^2 + 2ah^2 + 6bm + 6c) = \frac{h}{3}[a(6m^2+2h^2) + b(6m) + 6c].$$

17. $|E_n| = \dfrac{(2-1)^3}{12n^2}\dfrac{2}{c^3} < \dfrac{(1)(2)}{(12n^2)(1)} = \dfrac{1}{6n^2} < 10^{-10}$ if $n \geq 40,825$.

19. If k is odd, the integral equals 0. If [-a,a] is partitioned into n (an even number) of intervals of length h, the value of the expression in Simpson's Rule is 0 since the terms consist of 2f(0) or of 4f(0) (each equals 0), along with some terms that appear in pairs of 2f(rh) and 2f(-rh) and the remaining terms that appear in pairs of 4f(rh) and 4f(-rh). Since f is an odd function, these pairs of terms cancel out.

21. About 4507 ft^2

23. h = 20.

Area $\approx \dfrac{20}{3}[0 + 4(7) + 2(12) + 4(18) + 2(20) + 4(20) + 2(17) + 4(10) + 0]$
= 2120 square feet.

About $(2120 \text{ ft}^2) \left(\dfrac{4 \text{ mi}}{1 \text{ hr}}\right) \left(\dfrac{24 \text{ hr}}{1 \text{ day}}\right) \left(\dfrac{5280 \text{ ft}}{1 \text{ mi}}\right)$ = About 1,074,585,600 ft^3.

25.

n=10	value	error
Trapezoid	48.7466	.34656
Midpoint	48.2268	-.17324
Simpson	48.4	.00003

n=20	value	error
Trapczoid	48.4867	.08666
Midpoint	48.3567	-.04333
Simpson	48.4	0.

27.

n=10	value	error
Trapezoid	3.13993	-.00167
Midpoint	3.14243	.00083
Simpson	3.14159	0.

n=20	value	error
Trapezoid	3.14118	-.00042
Midpoint	3.1418	.00021
Simpson	3.14159	0.

29. (a) 25   (b) 46   (c) 300   (d) 8   (e) 1.60027, 0.00017

Problem Set 10.4 Solving Equations Numerically

1. Let $f(x) = x^3 + 3x - 6$. $f(1) = -2$; $f(2) = 8$.

    The successive values of m_n are 1.5, 1.25, 1.375, 1.3125, 1.28125, 1.296875, 1.2890625, 1.28515625. The sign of f changes between the last two, so the root is between them. Therefore, the root is 1.29 to two-decimal-place accuracy.

3. Let $f(x) = \cos x - e^{-x}$ on $[1,2]$. $f'(x) = \sin x + e^{-x}$. Note that f is decreasing on $[1,2]$ since $-\sin x$ is between -0.84 and -1 while e^{-x} is between 0.37 and 0.13, so $f'(x) < 0$ on $[1,2]$.
    $f(1) = \cos 1 - 1/e > 0$, and $f(2) = \cos 2 - 1/e^2 < 0$.

n	h_n	m_n	$f(m_n)$
1	0.5	1.5	-0.2
2	0.25	1.25	0.03
3	0.125	1.375	-0.06
4	0.0625	1.3125	-0.01
5	0.03125	1.28125	0.008
6	0.015625	1.296875	-0.003
7	0.0078125	1.2890625	0.002
8	0.00390625	1.29296875	-0.0002
9	0.991953125	1.291015625	0.001

A real root is 1.29 accurate to two decimal places, since $m_9 \pm h_9$ is in the interval $[1.285, 1.295]$.

5. The root is between -1 and 0.

    Successive truncated values of x_n are
    0.5, -0.2, -0.264, -0.26793, -0.2679419, -0.26794919. The last two are the same through eight decimal places so we feel confident that -0.26795 has the desired accuracy.

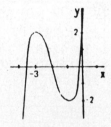

7. Successive truncated values of x_n are 1.6, 1.556, 1.55714559, 1.55714559. The last two are the same through 8 decimal places so we feel confident that 1.55715 has the desired accuracy.

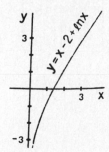

9. Let $f(x) = \cos x - x$. $f'(x) = -\sin x - 1$, $f''(x) = -\cos x$.

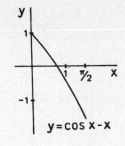

$y = \cos x - x$

x	$f(x)$
0	1
$\pi/2$	$-\pi/2$

There is a root between 0 and $\pi/2$; let $x_1 = 0.8$.

$$x_{n+1} = x_n - \frac{f(x_n)}{f'(x_n)} = x_n - \frac{x_n - \cos(x_n)}{1 + \sin(x_n)} .$$

n	x_n
1	0.8
2	0.739853306
3	0.739085263
4	0.739085133

The root is approximately 0.73909.

11. There are roots between 0 and 1, and between 3 and 4.
Successive truncated values of x_n are

0.3, 0.35, 0.3470, 0.347108347, and 0.347108349 for the root between 0 and 1; and 3.7, 3.655, 3.6529, 3.65289165, 3.65289165 for the root between 3 and 4. In each case the last two are the same through 8 decimal places so we feel confident that 0.34711 and 3.65289 have the desired accuracy.

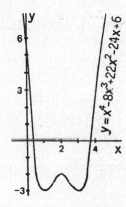

$y = x^4 - 8x^3 + 22x^2 - 24x + 6$

13. The root is between 0 and 1. Successive values of x_n are

approximately 0.5, 0.527, 0.52658, 0.52657957, 0.52657957. The last two are the same through 8 decimal places so we feel confident that 0.52658 has the desired accuracy.

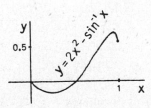

$y = 2x^2 - \sin^{-1} x$

273

15. Let $f(x) = x^3 - 6$. $f'(x) = 3x^2$.
$f(x) = 0$ for some x between 1 and 2 (closer to 2). Let $x_1 = 1.75$.

$$x_{n+1} = x_n - \frac{f(x_n)}{f'(x_n)} = x_n - \frac{x_n^3 - 6}{3x_n^2} = x_n - x_n/3 + 2/x_n^2.$$

n	x_n
1	1.75
2	1.819727891
3	1.817124327
4	1.817120593
5	1.817120593

There root is approximately 1.81712

Therefore, $\sqrt[3]{6} \approx 1.81712$.

17. Let $f(x) = (\sin x)/x$. Then $f'(x) = (x\cos x - \sin x)/x^2$ which equals 0 if $x\cos x - \sin x = 0$. Let $g(x) = x\cos x - \sin x$. $g(\pi) = -\pi$, $g(2\pi) = 2\pi$, so $f'(x)$ is 0 somewhere in $(\pi, 2\pi)$. Successive truncated values of x_n are 4.5, 4.49341, 4.493409458, 4.493409458. Therefore, f is minimum at (approximately) 4.493409458. The minimum value is approximately $f(4.493409458) \approx -0.21723362$.

19. Let $f(x) = x^2 - 2$. Then $f'(x) = 2x$ (so m=2) and $f''(x) = 2$ (so M=2).
$|x_6 - \sqrt{2}| \le (4/2)[(2/4)|1.5 - \sqrt{2}|]^{32} \le 2[(1/2)(1/2)]^{32} = 2^{-63} < 10^{-19}$.

21. Let $f(x) = \frac{1 + \ln x}{x}$. $f'(x) = \frac{-\ln x}{x^2}$.

$$x_{n+1} = x_n - \frac{f(x_n)}{f'(x_n)} = x_n - \frac{[1 + \ln(x_n)]/x_n}{[-\ln(x_n)]/x_n^2} = x_n - \frac{x_n + x_n\ln(x_n)}{-\ln(x_n)}.$$

$$= x_n + x_n/\ln(x_n) + x_n = 2x_n + x_n/\ln(x_n)$$

For $x_1 = 1.2$:

n	x_n
1	1.2
2	8.98
3	22.05
4	51.24
5	115.50

For $x_1 = 0.5$:

n	x_n
1	0.5
2	0.278652479
3	0.339231168
4	0.364671303
5	0.367837678
6	0.367879434
7	0.367879411

The root is approximately 0.36788.

f′(1.2) is small and negative, and f(1.2) is positive, so the tangent at x=1.2 intersects the x-axis "far" to the right of x=1.2. And the slope remains positive but is decreasing so the tangents at x_n intersect the x-axis farther and farther to the right.

23. (a) $2000 = \dfrac{100}{i}\left[1 - \dfrac{1}{(1+i)^{24}}\right]$; $20i = 1 - \dfrac{1}{(1+i)^{24}}$; $20i(1+i)^{24}$

$$= (1 + i)^{24} - 1;$$

$$20i(1+i)^{24} - (1+i)^{24} + 1 = 0.$$

(b) $i_{n+1} = i_n - \dfrac{f(i_n)}{f'(i_n)}$

$$= i_n \frac{20i_n(1+i_n)^{24} - (1+i_n)^{24}+1}{20(1+i_n)^{24} + 20i_n(24)(1+i_n)^{23} - 24(1+i_n)^{23}}$$

which simplifies to the expression given in the statement of the problem.

(c)

n	i_n	
1	0.012	
2	0.016529706	
3	0.0152650720	
4	0.015132285	Thus, i ≈ 0.01513 per month, so
5	0.015132287	Annual rate = 1200i% ≈ 18.16%.

25. Experiment with calculator and different values of a

27. -1.87939, 0.3473, 1.53209

29. -2.08204, 0.09251, 0.91314, 1.620915, 1.85411

Problem Set 10.5   Fixed-Point Methods

1. The successive truncated values of x_n are 1, 0.01, 0.097, 0.082, 0.0848, 0.0843, 0.08446, 0.084455, 0.084458, 0.0844579, 0.0844579. 0.08446 is the desired result.

3. $x_{n+1} = 2.5 + x_n$.

n	x_n	n	x_n
1	1	8	2.158267035
2	1.870828693	9	2.158301887
3	2.090652696	10	2.158309961
4	2.142580849	11	2.158311831
5	2.154664904	12	2.158312265
6	2.157467243	13	2.158312365
7	2.158116596		Root: 2.15831.

5. (a)

(b) 0.6, 0.48, 0.4992, 0.49999872, 5.0000000.

(c) $x = 2x - 2x^2$; $2x^2 - x = 0$; $x(2x - 1) = 0$;

   $x = 0$ or $x = 1/2$.

(d) $g'(x) = 2 - 4x$; $g'(1/2) = 0$.

7. (a)

(b) $x_{n+1} = 2 \sin(\pi x_n)$
   0.9, 0.6, 1.8, -0.8 (for $x_1 = 0.9$).
   0.1, 0.6, 1.8, -0.8 (for $x_1 = 0.1$).

(c) $g'(x) = 2\pi\cos(\pi x)$. The fixed points occur at 0 and near ±0.9. Near those points $|g(x)|$ is greater than 1.

9. (a) $x = 2 \sin\pi x$

   $x + 5x = 5x + 2\sin\pi x$

   $x = \dfrac{5x + 2\sin\pi x}{6}$.

(b) Let $g(x) = \dfrac{5x + 2\sin\pi x}{6}$

   $x_{n+1} = \dfrac{5x_n + 2\sin(\pi x_n)}{6}$

n	x_n
1	0.9
2	0.853005664
3	0.859357050
4	0.858666231
5	0.858744163
6	0.858735405
7	0.885736389
8	0.858736279
Root:	0.85874.

(c) $g'(x) = \dfrac{5 + 2\pi\cos\pi x}{6}$.

$g'(x)$ is near 0 in an interval which includes 0.9 and the fixed point. In particular, at the fixed point it is about -0.11242172.

11. The positive root is between 1 and 2, but $|D_x(x^3-x^2-1)| = |3x^2-2x|$ is between 1 and 8 for x in (1,2), so the algorithm doesn't work well for x $= x^3-x^2-1$.

$-x = -x^3+x^2+1$; $5x = -x^3+x^2+6x+1$; $x = (1/5)(-x^3+x^2+6x+1)$.

$|D_x[(1/5)(-x^3+x^2+1)]| = (1/5)|-3x^2+2x+6| < 1$ for x in (1,2). Successive truncated values of x_n are 1.7, 1.835, 1.8396, 1.83925, 1.839289, 1.8392864, 1.8392867. The root is approximately 1.83929.

13. $x_n = 0.5(x_n + \pi/x_n)$.

Successive truncated values of x_n: 1.8, 1.7726, 1.77245386, 1.77245385.

$\sqrt{\pi} \approx 1.77245$.

$g'(x) = (1/2)(1 - ax^{-2})$, so $g'(\sqrt{a}) = 0$.

15. (a) $10000 = \dfrac{R}{0.015}[1-(1+0.015)^{-48}]$.

$R = \dfrac{150}{1-(1.015)^{-48}} \approx 293.75$.

(b) $10000 = \dfrac{300}{i}[1-(1+i)^{-48}]$.

$i = 0.03[1-(1+i)]^{-48}$. Let $g(i) = 0.03[1-(1+i)^{-48}]$.

$g'(i) = 1.44(1+i)^{-49} < 1$ if $i > .008$.

$i_n = 0.03[1-(1+i_n)^{-48}]$, $i_1 = 0.015$.

The convergence is slow. On one calculator it finally stabilized at 0.015990923 beginning at the 36th step. (By starting at 0.016 it was also slow but decreased, reaching 0.015990923 at the 26th step.) It became apparent a bit earlier that 0.01599 would be the correct value accurate to five decimal places. The first few values are:

n	i_n
1	0.015
2	0.015319149
3	0.015539025
4	0.015688550
5	0.015789330
6	0.015856848
7	0.015901897
8	0.015931875
9	0.015951786

17. (a) r is a root $\Leftrightarrow$ $r = r - f(r)/f'(r)$ $\Leftrightarrow$ $f(r) = 0$.

(b) Let r be a solution of $x = g(x)$ and $g'(r) = 0$. Apply Newton's Method to $f(x) = x - g(x)$.

$$x_{n+1} = x_n - f(x_n)/f'(x_n) = x_n - [x_n - g(x_n)]/(1-0) = x_n - x_n + g(x_n).$$

19. If $x = 2.5(x-x^2) = g(x)$, $|g'(x)| = 2.5|1-2x| < 1$ for x in $[0.5, 0.65]$, so the Fixed Point Method works.

$$x_{n+1} = 2.5(x_n - x_n^2)$$

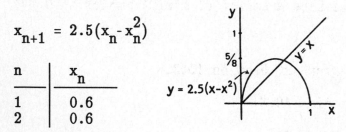

n	x_n
1	0.6
2	0.6

The solution is $x = 0.6$. [$x_1 = 0.6$ was a lucky start. Starting with $x_1 = 0.59$ yields $x_{24} = 0.600000001$.]

21. $x = 3.1(x-x^2)$ can be easily solved algebraically to obtain roots of 0 and 21/31. $|f'(21/31)| = 1.1 > 1$, so the Fixed Point Method fails to work here.

Trying the Fixed Point Method with x_1 = 0.6, one calculator obtained five-decimal-place stabilization for attractors $r_1 \approx 0.55801$ and $r_2 \approx 0.76457$, near n=46.

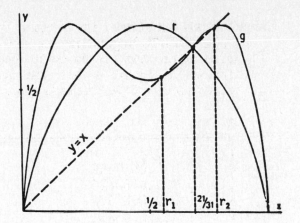

23. The attractors are approximately 0.50088, 0.87500, 0.38282, 0.82694. They are solutions of $x = h(x) = g \circ g(x)$; i.e., of $x = f \circ f \circ f \circ o(x)$ where $|h'(x)| < 1$.

Problem Set 10.6 Chapter Review

1. True.   Evaluate each.

3. True. $P_2(x) = f(0) + f'(0) + \dfrac{f''(0)x^2}{2} = 0 + 0x + 0x^2 = 0$ is the second-order Maclaurin polynomial of f(x). Note that the second-order polynomial is not a polynomial of degree 2.

5. True.   All odd power terms have a factor of sin(0) which is 0.

7. True.   Let n=0, a < b = x.

9. False.   See the first paragraph of Section 10.3.

11. True.   See Problem 16, Section 10.3.

13. True. $|e^{-x^2} + x^2 + \sin(x+1)| \le e^{-x^2} + x^2 + |\sin(x+1)| \le 1 + 4 + 1 = 6$.

15. True.   By Intermediate Value Theorem since 0 is between f(a) and f(b).

17. False.   See Problem 22, Section 10.4.

19. False.   At least it won't work using $x_{n+1} = 5(x_n - x_n^2) + 0.01$, but it will work using $x_{n+1} = (1/6)(10x_n - 5x_n^2 + 0.01)$.

Sample Test Problems

1. $p(x) = ((3x-2)x - 5) \ x + 7)x - 3; \ p(2.31) \approx 47.2586.$

3. $g(x) \quad\quad = x^3 - 2x^2 + 5x - 7. \quad\quad g(2) \quad\quad = 3.$
   $g'(x) \quad\quad = 3x^2 - 4x + 5. \quad\quad g'(2) \quad\quad = 9.$
   $g''(x) \quad\quad = 6x - 4. \quad\quad\quad g''(2) \quad\quad = 8.$
   $g^{(3)}(x) \quad = 6. \quad\quad\quad\quad\quad g^{(3)}(2) = 6.$

   $P_4(x) \quad\quad = 3 + 9(x-2) + \dfrac{8(x-2)^2}{2} + \dfrac{6(x-2)}{3!}$

   $\quad\quad\quad\quad = 3 + (9x-18) + (4x^2 - 16x + 16) + (x^3 - 6x^2 + 12x - 8) = x^3 - 2x^2 + 5x - 7.$

5. $P_4(x) = 1/2 - (1/4)(x-1) + (1/8)(x-1)^2 - (1/16)(x-1)^3 + (1/32)(x-1)^4.$

7. $P_4(x) = x^2 - (1/3)x^4.$

   $|R_5(x)| = |(1/6!)(32 \cos 2c)x^6|$, for some c between 0 and x.

   If $|x| < 0.2$, $|R_5(x)| < (1/6!)(32)(0.2)^6 < 0.000003.$

9. $\displaystyle\int_{0.8}^{1.2} P_5(x) \ dx = \int_{0.8}^{1.2} \left[(x-1) - \frac{(x-1)^2}{2} + \frac{(x-1)^3}{3} - \frac{(x-1)^4}{4} + \frac{(x-1)^5}{5} \right] dx$

   $\displaystyle = \left[\frac{(x-1)^2}{2} - \frac{(x-1)^3}{6} + \frac{(x-1)^4}{12} - \frac{(x-1)^5}{20} + \frac{(x-1)^6}{30} \right]_{0.8}^{1.2}$

   $\displaystyle = \left[\frac{(0.2)^2}{2} - \frac{(0.2)^3}{6} + \frac{(0.2)^4}{12} - \frac{(0.2)^5}{20} + \frac{(0.2)^6}{30} \right]$

   $\displaystyle \quad\quad - \left[\frac{(-0.2)^2}{2} - \frac{(-0.2)^3}{6} + \frac{(-0.2)^4}{12} - \frac{(-0.2)^5}{20} + \frac{(-0.2)^6}{30} \right]$

   $= -2(0.2)^3 - 2(0.2)^5 \approx -0.00269867.$

   Therefore, $\displaystyle\int_{0.8}^{1.2} \ln x \ dx \approx -0.00269867.$

   $|R_5(x)| < \dfrac{(0.2)^6}{6(0.8)^6}.$ [from Problem 8]

   Therefore, the error of the approximation for the integral is less than

   $\displaystyle\int_{0.8}^{1.2} \frac{(0.2)^6}{6(0.8)^6} \ dx = \left[\frac{(0.2)^6 x}{6(0.8)^6} \right]_{0.8}^{1.2} = \frac{(0.2)^6 (1.2 - 0.8)}{6(0.8)^6} < 0.00001628.$

11. $\dfrac{(0.4)/8}{3}[\ln(0.8) + 4\ln(0.85) + 2\ln(0.9) + \cdots + 4\ln(1.15) + \ln(1.2)] \approx -0.00269939.$

$|E_8| = \dfrac{(1.2 - 0.8)^5}{180(8)^2} \dfrac{6}{c^4} < \dfrac{(0.4)^5}{737280} \dfrac{6}{(0.8)^4} < 0.00000021.$

13. The successive truncated values of x_n are 0.5, 0.29, 0.2818, 0.281784604, 0.281784602. The approximate solution is 0.281785.

15. Observe from the graph that the point of intersection $y = x$ and $y = \tan x$ in $(\pi, 2\pi)$ is a little to the left of $3\pi/2$, so we will let $x_1 = 4.6$.

Let $f(x) = x - \tan x$.

$\qquad f'(x) = 1 - \sec^2 x.$

$\qquad x_{n+1} = x_n - \dfrac{f(x_n)}{f'(x_n)}$

$\qquad\qquad = x_n - \dfrac{x_n - \tan(x_n)}{1 - \sec^2(x_n)}.$

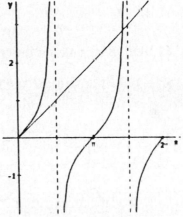

n	x_n
1	4.6
2	4.545732122
3	4.506145588
4	4.494171630
5	4.493412197
6	4.493409458
7	4.493409458

The solution is 4.49341 to five-decimal place accuracy.

17. Let $x_1 = -3.2$. Successive truncated values of x_n are -3.2, -3.8305, -3.183063012, -3.183063012. The desired solution is -3.183063 accurate to six decimal places.

Problem Set 11.1 Infinite Sequences

1. 1, 2/3, 3/5, 4/7, 5/9. Converges. $a_n = \dfrac{1}{2 - 1/n} \to \dfrac{1}{2}$.

3. 5/2, 17/3, 37/6, 65/11, 101/18. Converges. $a_n = \dfrac{4 + (1/n^2)}{1 - (2/n) + (3/n^2)} \to 4$.
    [2.5, 5.67, 6.17, 5.91, 5.61]

5. 5/3, 6/9, 7/19, 8/33, 9/51. Converges. $a_n = \dfrac{1/n + 4/n^2}{2 + 1/n^2} \to 0$.

7. -1/2, 2/3, -3/4, 4/5, -5/6. Diverges since the even-numbered terms converge to +1 and the odd-numbered terms converge to -1.

9. 1, 0, -1/3, 0, 1/5. Converges to 0 by the Squeeze Theorem since -1/n and 1/n converge to 0 and $-1/n \le a_n \le 1/n$.

11. $e/1$, $e^2/4$, $e^3/9$, $e^4/16$, $e^5/25$.
     [2.72, 1.85, 2.23, 3.41, 5.94]

     Diverges since $\lim\limits_{n\to\infty} \dfrac{e^n}{n^2} = \lim\limits_{x\to\infty} \dfrac{e^x}{x^2} \overset{\text{Ⓛ}}{=} \lim\limits_{x\to\infty} \dfrac{e^x}{2x} \overset{\text{Ⓛ}}{=} \lim\limits_{x\to\infty} \dfrac{e^x}{2} = \infty$.

13. $-\pi/4$, $\pi^2/16$, $-\pi^3/64$, $\pi^4/256$, $-\pi^5/1024$.
     [0.79, 0.20, -0.05, 0.01, -0.003]

     Converges. $a_n = (-\pi/4)^n$ and $-1 < -\pi/4 < 1$.

15. 1.9, 1.81, 1.729, 1.6561, 1.59049. Converges. $1 + (0.9)^n \to 1+0 = 1$.

17. 0/1, (ln2)/2, (ln3)/3, (ln4)/4, (ln5)/5. Converges.
     [0, 0.35, 0.37, 0.35, 0.32]

     $\lim\limits_{n\to\infty} \dfrac{\ln n}{n} = \lim\limits_{n\to\infty} \dfrac{\ln x}{x} \overset{\text{Ⓛ}}{=} \lim\limits_{x\to\infty} \dfrac{1/x}{1} = 0$.

19. $2, (3/2)^2, (4/3)^3, (5/4)^4, (6/5)^5$. Converges to e.   [See Problem 30
   $[2, 2.25, 2.37, 2.44, 2.49]$ Section 9.2]

21. $a_n = \dfrac{n}{n+1} = \dfrac{1}{1 + (1/n)} \to \dfrac{1}{1 + 0} = 1.$

23. $a_n = \dfrac{(-1)^n n}{2n-1}$. Diverges since even-numbered terms converge to 1/2 and
   odd-numbered terms converge to -1/2.

25. $a_n = \dfrac{n}{n^2 - (n-1)^2} = \dfrac{n}{2n-1} = \dfrac{1}{2 - 1/n} \to \dfrac{1}{2}.$

27. $a_n = n \sin(\frac{1}{n})$. $\lim\limits_{n\to\infty} n \sin(1/n) = \lim\limits_{n\to\infty} \dfrac{\sin(1/n)}{1/n} = \lim\limits_{x\to\infty} \dfrac{\sin(1/x)}{1/x}$

   $= \lim\limits_{z\to 0^+} \dfrac{\sin z}{z} = 1.$ [z=1/x.]

29. $a_n - \dfrac{2^n}{n^2} \cdot \lim\limits_{n\to\infty} \dfrac{2^n}{n^2} - \lim\limits_{x\to\infty} \dfrac{2^x}{x^2} \overset{L}{-} \lim\limits_{x\to\infty} \dfrac{2^x \ln 2}{2x} \overset{L}{-} \lim\limits_{x\to\infty} \dfrac{2^x (\ln 2)^2}{2} = \infty.$
   Sequence diverges.

31. $1/2, 5/4, 9/8, 13/16$. The sequence is bounded below by 0.

   $\dfrac{a_{n+1}}{a_n} = \dfrac{4n+1}{8n-6} < 1$ if $n \geq 2$, so the sequence is decreasing for $n \geq 2$.

33. $3/4, 2/3, 5/8, 3/5$ [or 0.75, 0.67, 0.62, 0.60]. The sequence is bounded
   below by 0 and is decreasing since each factor is less than 1.
   (Multiplying a positive number by a positive number that is less than 1
   yields a smaller positive number.)

35. $1, 3/2, 7/4, 15/8$. $a_n = \dfrac{2^n - 1}{2^{n-1}} = 2 - \dfrac{1}{2^{n-1}}$, so bounded above by 2.

   $\dfrac{a_{n+1}}{a_n} = \dfrac{2 + a_n}{2a_n} > 1$, if $2a_n < 2+a_n$; i.e., if $a_n < 2$. (Shown above).

37. Successive truncated values of u_n are 1.7, 2.1, 2.27, 2.29, 2.301,
   2.3024, 2.30271, 2.30276, 2.302772, 2.3027750, 2.30277560, 2.302775631,
   2.302775636, 2.302775637, 2.3027775638. It seems that the desired result
   is 2.3028.

39. Continue where the hint leaves off, and *note that u is nonnegative.*

$$u^2 = 3+u, \quad u^2-u-3 = 0, \quad u = \frac{1 + \sqrt{13}}{2} \approx 2.3028.$$

41. Successive truncated values of u_n are 0,1, 1.1, 1.111, 1.1116, 1.11176, 1.111780, 1.1117818, 1.1117819, 1.11178200, 1.111782011, 1.111782011. The desired value seems to be 1.1118.

43. The sum is the Riemann sum for $f(x) = \sin x$ on $[0,1]$ for a regular partition with n subintervals of length $1/n$ and right endpoints as sample points. Therefore, the value is $-\cos 1 + 1 \approx 0.4597$.

45. Choose N to be any integer greater than $(1/\epsilon)$.

Then $n > N \Rightarrow n+1 > (1/\epsilon) \Rightarrow \frac{1}{n+1} < \epsilon \Rightarrow \left| \frac{n}{n+1} - 1 \right| < \epsilon.$

47. x in $S \Rightarrow x^2 < 2 \Rightarrow x < \sqrt{2}$, so $\sqrt{2}$ is an upper bound.

$B < \sqrt{2} \Rightarrow B < q < \sqrt{2}$ for some rational number q
$\Rightarrow q^2 < 2 \Rightarrow q$ is in S, so B is not an upper bound of S.

Therefore, $\sqrt{2} = \text{lub}(S)$ in the real number system.

Since $\sqrt{2}$ is irrational, it is conceivable that there is a rational number that is lub(S) in the rational number system. The argument above (about B) shows this is impossible for any rational number B that is less than $\sqrt{2}$. It is also impossible for any rational number B that is greater than $\sqrt{2}$ since

$B > \sqrt{2} \Rightarrow B > q > \sqrt{2}$ for some rational number q
$\Rightarrow q > x$ for each x in S $\Rightarrow q$ is an upper bound of S.
so B is not the lub of S.

49. Let $\{b_n\}$ be bounded by B. That is, $-B \leq b_n \leq B$ for all n.

Let $\epsilon > 0$ be given. Then $\epsilon/|B|$ is a positive number. Now choose N such that $n > N \Rightarrow |a_n| < \epsilon \;/\; |B| \Rightarrow |a_n||b_n| < (\epsilon/|B|)|B| = \epsilon.$

51. No. $\{n\}$ and $\{-n\}$ both diverge but $\{n+(-n)\} = \{0\}$ converges to 0.

53. $A_n = (1/2)n(n+1)[\pi(1/2)^2] = (1/8)n(n+1)\pi$.

   Base of triangle $= 2[d(A,B)] + (n-1)(1)$

   $\qquad\qquad\qquad = 2(\sqrt{3}/2) + n-1 = n-1+\sqrt{3}$.

   $B_n = (1/2)(n-1+\sqrt{3})[(\sqrt{3}/2)(n-1+\sqrt{3})]$

   $\qquad = (\sqrt{3}/4)(n-1+\sqrt{3})^2$.

   Thus, $\lim\limits_{n\to\infty} \dfrac{A_n}{B_n} = \dfrac{\pi}{2\sqrt{3}}$.

55. 1.64872   57. 0.13534   59. 1

Problem Set 11.2   Infinite Series

1. Converges. $\dfrac{1/5}{1-1/5} = \dfrac{1}{4}$.

   [Geometric — $a=r=1/5$]

3. Converges. $2\sum\limits_{k=0}^{\infty}(1/3)^k + 3\sum\limits_{k=0}^{\infty}(1/6)^k = 2\dfrac{1}{1-(1/3)} + 3\dfrac{1}{1-(1/6)} = 6.6$

5. Diverges since $(k-3)/k \to 1 \neq 0$.   [nth-Term Test for Divergence]

7. Converges. $s_1 = 1 - 1/2$, $s_2 = 1 - 1/3$, $\cdots$, $s_n = [1 - 1/(n+1)] \to 1$, so sum is 1.

9. Diverges by the nth-term test. $a_{n+1} > a_n$ if $n > 9$ since

$$\dfrac{a_{n+1}}{a_n} = \dfrac{(n+1)!}{10^{n+1}}\dfrac{10^n}{n!} = \dfrac{(n+1)n!}{10^n 10^1}\dfrac{10^n}{n!} = \dfrac{n+1}{10} > 1 \text{ if } n > 9.$$

11. Converges. $\dfrac{(e/\pi)^2}{1-e/\pi} \approx 5.5562$.   [Geometric — $a = (e/\pi)^2$, $r=e/\pi$]

13. Collapsing series: $s_n = \dfrac{3}{(n+1)^2} - 3$ converges to -3.

15. $0.2 + 0.02 + 0.002 + \cdots = \dfrac{0.2}{1-0.1} = \dfrac{2}{9}$.   [Geometric — $a=0.2$, $r=0.1$]

17. $0.013 + 0.000013 + 0.000000013 + \cdots = \dfrac{0.013}{1-0.001} = \dfrac{13}{999}$.
[Geometric — a=0.013, r=0.001]

19. $0.4 + 0.09 + 0.009 + 0.0009 + \cdots = 0.4 + \dfrac{0.09}{1-0.1} = \dfrac{1}{2}$.
[After 0.4, geometric — a=0.09, r =0.01]

21. $r + r(1-r) + r(1-r)^2 + \cdots = \dfrac{r}{1-(1-r)} = 1$.
[Geometric — 1st term = r, common ratio = (1-r), $|1-r| < 1$]

23. $\ln[k/(k+1)] = \ln(k) - \ln(k+1)$, so the partial sums are collapsing and s_n = ln1 - ln(n+1) = -ln(n+1) $\to \infty$, so the series diverges.

25. $100 + 2(100)(2/3) + 2(100)(2/3)^2 + \cdots = 100 + \dfrac{400/3}{1-2/3} = 500$.

27. 1 billion + (1 billion)(3/4) + (1 billion)$(3/4)^2 + \cdots = \dfrac{1\ \text{billion}}{1-(3/4)} = 4$ billion. [Geometric — a = 1 billion, r = 3/4]
The total increase in spending is 4 billion dollars (including the 1 billion put in by the government).

29. At each time step the area of the next inside square is one-half of the area of the present square. Then one-eighth of that square is colored. The amount colored is then

$$1/8 + (1/8)(1/2) + (1/8)(1/2)^2 + \cdots = \dfrac{1/8}{1-1/2} = \dfrac{1}{4}.$$

31. $(3/4) + (3/4)(1/4)^2 + (3/4)(1/4)^4 + \cdots = \dfrac{3/4}{1-(1/16)} = \dfrac{4}{5}$.

The original triangle doesn't need to be equilateral since at each step the four triangles are congruent.

33. Achilles will run $100 + 10 + 1 + 0.1 + \cdots = \dfrac{100}{1-0.1} = \dfrac{1000}{9}$ yds.

[111 and 1/9 yards]

35. Proof of the contrapositive:

$\sum ca_k$ converges for $c \neq 0 \Rightarrow \sum a_k = \sum(1/c)(ca_k) = (1/c)\sum ca_k$ converges.

37. Note that blocks are being added to the pile from the bottom.

(a) The blocks will not topple if the center of mass is above the base. For the six blocks illustrated, with origin at the center of the bottom block, the center of mass would have to have x-coordinate between $-1/2$ and $1/2$. The x-coordinates of the centers of mass of the individual blocks (bottom to top) are:

$$0, \ \frac{1}{10}, \ \frac{1}{10} + \frac{1}{8}, \ \frac{1}{10} + \frac{1}{8} + \frac{1}{6}, \ \frac{1}{10} + \frac{1}{8} + \frac{1}{6} + \frac{1}{4}, \ \frac{1}{10} + \frac{1}{8} + \frac{1}{6} + \frac{1}{4} + \frac{1}{2}.$$

Therefore, the x-coordinate of the center of mass of the stack is the average of those numbers.

$$\frac{\frac{5}{10} + \frac{4}{8} + \frac{3}{6} + \frac{2}{4} + \frac{1}{2}}{6} = \frac{5(1/2)}{6} < \frac{1}{2} \qquad \text{so the six blocks do not tumble.}$$

For n blocks, one obtains $\dfrac{(n-1)(1/2)}{n} < \dfrac{1}{2}$ so the n blocks do not tumble.

(b) Indefinitely since $\frac{1}{2} + \frac{1}{4} + \frac{1}{6} + \frac{1}{8} + \frac{1}{10} + \cdots$ diverges. [Problem 36]

39. Let $\{s_n\}$, $\{t_n\}$ and $\{u_n\}$ be the respective sequences of partial sums. Then $u_n = s_n + t_n$, $\{s_n\}$ diverges, and $\{t_n\}$ converges. Therefore, $\{u_n\}$, and hence, $\Sigma(a_n + b_n)$ diverges. [See Problem 50, Section 11.1.]

41. Vertical: $1 + \frac{1}{2} + \frac{1}{4} + \frac{1}{8} + \cdots$

Horizontal: $\frac{1}{2} + \frac{2}{4} + \frac{3}{8} + \frac{4}{16} + \cdots$

(a) [Horizontals] $= \displaystyle\sum_{k=1}^{\infty} \frac{k}{2^k} = \sum_{k=1}^{\infty} \left(\frac{1}{2}\right)^{k-1}$ [The verticals]

$$= \frac{1}{1-(1/2)} = 2 \quad \text{[Geometric Series]}$$

(b) $\bar{x} = \dfrac{(0)(1) + (1)(1/2) + (2)(1/4) + (3)(1/8) + \cdots}{(1) + (1/2) + (1/4) + (1/8) + \cdots} = 1.$

43. (a) $A = \dfrac{C}{1-e^{-kt}}$ [Geometric series with a=C, $r=e^{-kt}$]

(b) C=2. When t=6, $Ce^{-kt} = (1/2)C$. Therefore, $k = (\ln 2)/6$.
Substituting C=2, t=12, k=(ln2)/6, obtain A = 8/3.

45. $\dfrac{1}{f_k f_{k+1}} - \dfrac{1}{f_{k+1} f_{k+2}} = \dfrac{f_{k+2} - f_k}{f_k f_{k+1} f_{k+2}} = \dfrac{f_{k+1}}{f_k f_{k+1} f_{k+2}} = \dfrac{1}{f_k f_{k+2}}$.

Thus, $\displaystyle\sum_{k=1}^{\infty} \dfrac{1}{f_k f_{k+2}} = \sum_{k=1}^{\infty} \left[\dfrac{1}{f_k f_{k+1}} - \dfrac{1}{f_{k+1} f_{k+2}}\right] \overset{*}{=} \dfrac{1}{f_1 f_2} = \dfrac{1}{(1)(1)} = 1$.

*Collapsing series, so $S_n = \dfrac{1}{f_1 f_2} - \dfrac{1}{f_{n+1} f_{n+2}} \to \dfrac{1}{f_1 f_2}$, since $f_k \to \infty$ as $k \to \infty$.

Problem 11.3 Positive Series: The Integral Test

1. $\displaystyle\int_1^{\infty} (x+2)^{-1} dx = \lim_{b \to \infty} [\ln(b+2) - \ln 3] = \infty$, so the series diverges.

3. $f(x) = \dfrac{x}{1+x^2}$ satisfies the conditions of the Integral Test, since

$$f'(x) = \dfrac{1-x^2}{(1+x^2)^2} < 0 \text{ for } x > 1.$$

$\displaystyle\int_1^{\infty} \dfrac{x}{1+x^2} dx = \lim_{b \to \infty} \left[(1/2)\ln(1+x^2)\right]_1^b = \lim_{b \to \infty} [(1/2)\ln(1+b^2) - (1/2)\ln 2] = \infty.$

The improper integral diverges, so the series diverges.

5. $\displaystyle\int_1^{\infty} (x+2)^{-1/2} dx = \lim_{b \to \infty} [2(b+2)^{1/2} - 2(3)^{1/2}] = \infty$, so the series diverges.

7. $\displaystyle\int_1^{\infty} (10x+3)^{-1} dx = \lim_{b \to \infty} \dfrac{\ln(10b+3) - \ln(13)}{10} = \infty$, so the series diverges.

9. $f(x) = (4+3x)^{-3/2}$ satisfies the conditions of the Integral Test. The series converges since

$$\int_1^\infty (4+3x)^{-3/2} dx = \lim_{b \to \infty} \left[\frac{-2}{3(4+3x)^{1/2}} \right]_1^b = \lim_{b \to \infty} \left[\frac{-2}{3(4+3b)^{1/2}} + \frac{2}{3(7)^{1/2}} \right] = \frac{2}{3\sqrt{7}}.$$

11. $\int_1^\infty xe^{-x^2} dx = \lim_{b \to \infty} (-1/2) \left[e^{-b^2} - e^{-1} \right] = (\frac{1}{2})e^{-1}$, so the series converges.

13. Diverges since $(k^2+1)/(k^2+5) \to 1 \neq 0$.   [nth-Term Test]

15. Diverges since $\left[\left[\frac{1}{2} \right]^k + \frac{k-1}{2k+1} \right] \to 0 + \frac{1}{2} \neq 0$.   [n-th Term Test]

17. Diverges by nth-Term Test. $\sin(k\pi/2)$ has values $1, 0, -1, 0, 1, \cdots$,

19. Converges by Integral Test.

21. $f(x) = \dfrac{\tan^{-1} x}{1+x^2}$ satisfies the conditions of the Integral Test since

$$f'(x) = \frac{1 - 2x\tan^{-1} x}{(1+x^2)^2} < 0 \text{ for } x > 1.$$

$$\int_1^\infty \frac{\tan^{-1} x}{1+x^2} dx = \lim_{b \to \infty} \left[\frac{(\tan^{-1} x)^2}{2} \right]_1^b = \lim_{b \to \infty} \left[\frac{(\tan^{-1} b)^2}{2} - \frac{(\tan^{-1} 1)^2}{2} \right]$$

$$= \frac{(\pi/2)^2}{2} - \frac{(\pi/4)^2}{2}, \text{ so the series converges.}$$

23. Error $< \int_5^\infty xe^{-x} dx = \lim_{b \to \infty} \left[(-b-1)e^{-b} + 6e^{-5} \right] = 6e^{-5} < 0.0405.$

25. Error $= \displaystyle\sum_{k=6}^\infty \frac{1}{1+k^2} < \int_5^\infty (1+x^2)^{-1} dx = \lim_{b \to \infty} \left[\tan^{-1} b - \tan^{-1} 5 \right]$

$$= \pi/2 - \tan^{-1} 5 \approx 0.1974.$$

27. Case $p < 0$: $(\ln n)^{-p} > 1$ if $n > 3$, so $\dfrac{1}{n(\ln n)^p} = \dfrac{(\ln n)^{-p}}{n} > \dfrac{1}{n}$.

Therefore, the partial sums of the given series are larger than the respective partial sums of the harmonic series, which diverges to ∞, so the given series diverges.

Note that $f(x) = \dfrac{(\ln x)^{-p}}{x}$ satisfies the conditions of the Integral.

Test on $[2, \infty)$ since $f'(x) = \dfrac{(\ln x)^{-p-1}(-p - \ln x)}{x^2} < 0$ if $p \geq 0$.

Case $p = 1$: $\displaystyle\int_2^\infty \dfrac{(\ln x)^{-1}}{x}\,dx = \lim_{b \to \infty}\left[\ln|\ln x|\right]_2^b = \infty$. The series diverges.

Other cases: $\displaystyle\int_2^\infty \dfrac{(\ln x)^{-p}}{x}\,dx = \lim_{b \to \infty}\left[\dfrac{(\ln x)^{1-p}}{1-p}\right]_2^b = \lim_{b \to \infty}\dfrac{(\ln b)^{1-p} - (\ln 2)^{1-p}}{1-p}$

which converges if $p > 1$, and diverges if $p < 1$.

Summary: The series converges iff $p > 1$.

29. $\ln(n+1) = \displaystyle\int_1^{n+1}(1/x)\,dx < 1 + \dfrac{1}{2} + \dfrac{1}{3} + \cdots + \dfrac{1}{n}$.

$1 + \dfrac{1}{2} + \dfrac{1}{3} + \cdots + \dfrac{1}{n} < 1 + \displaystyle\int_1^n(1/x)\,dx = 1 + \ln(n)$.

31. The limit exists by the Monotonic Sequence Theorem. [Theorem D, Section 11.1] Whether or not the limit is rational is a famous unsolved problem.

33. $0.5772 + \ln(n+1) > 20 \Rightarrow \ln(n+1) > 19.4228$
$\Rightarrow n+1 > 272{,}404{,}867$
$\Rightarrow n > 272{,}404{,}866$.   [272,400,600 in 11.2(38)]

35. (i) A_n is increasing since each term (area of a region) is positive.

(ii) As in Problem 34, slide (without rotating) the shaded lunes into the heavily outlined triangle so that the right-hand corners are coincident with the upper corner of the lune already in the triangle. That each lune will fit inside the triangle is guaranteed by the fact that f is increasing (so right-hand corner is the high point of each lune). That no two lunes will overlap is guaranteed by the fact that f is concave down (so tangent lines are above the curve which means that the tangent line at the right-hand corner of each region is above the bottom edge of the next region to the right).

Therefore, $\{A_n\}$ is a monotonic sequence which is bounded above. Hence, $\lim_{n\to\infty} A_n$ exists.

Problem Set 11.4   Positive Series: Other Tests

1. Diverges. Use Harmonic Series. $\dfrac{n}{n^2+2n+3}\dfrac{n}{1} = \dfrac{1}{1 + 2/n + 3/n^2} \to 1 > 0.$

3. Converges. Use p = 3/2 series. $\dfrac{1}{n\sqrt{n+1}}\dfrac{n^{3/2}}{1} = \dfrac{1}{1+(1/n)} \to 1 > 0.$

[Note: Multiplying by $\dfrac{n^{3/2}}{1}$ is equivalent to dividing by $\dfrac{1}{n^{3/2}}$.]

5. Converges. $\dfrac{8^{n+1}}{(n+1)!}\cdot\dfrac{n!}{8^n} = \dfrac{8}{n+1} \to 0 < 1.$

7. Diverges. $\dfrac{(n+1)!}{(n+1)^{100}}\dfrac{n^{100}}{n!} = \dfrac{(n+1)n^{100}}{(n+1)^{100}} = \dfrac{n+1}{(1 + 1/n)^{100}} \to \infty.$

9. Converges. $\dfrac{(n+1)^3}{[2(n+1)]!}\dfrac{(2n)!}{n^3} = \dfrac{(n+1)^3(2n)!}{(2n+2)(2n+1)(2n)!n^3} = \dfrac{(n+1)^3}{(2n+2)(2n+1)n^3}$

$$= \dfrac{[1+(1/n)]^3}{(2n+2)(2n+1)(1)} \to 0 < 1.$$

11. Diverges by nth-Term Test. $n/(n+200) \to 1$.

13. $\dfrac{n+3}{n^{5/2}} \dfrac{n^{3/2}}{1} = 1 + \dfrac{3}{n} \to 1 > 0$. [Converges by Limit Comparison Test using the p=3/2 series.]

15. Converges by Ratio Test. $\dfrac{(n+1)^2}{(n+1)!} \dfrac{n!}{n^2} = \dfrac{(n+1)^2}{(n+1)n^2} = \dfrac{n+1}{n^2} = \dfrac{1}{n} + \dfrac{1}{n^2} \to 0 < 1$.

17. $\dfrac{4n^3+3n}{n^5-4n^2+1} \dfrac{n^2}{1} = \dfrac{4 + 3/n^2}{1 - 4/n^3 + 1/n^5} \to 4 > 0$. [Converges by Limit Comparison Test using the p=2 series.]

19. $\dfrac{1}{n(n+1)} < \dfrac{1}{n^2}$. [Converges by Ordinary Comparison Test using the p=2 series.]

21. It is $\sum \dfrac{n+1}{n(n+2)(n+3)}$. [Converges by the Limit Comparison Test using the p=2 series.]

$\dfrac{n+1}{n(n+2)(n+3)} \dfrac{n^2}{1} = \dfrac{n(n+1)}{(n+2)(n+3)} = \dfrac{[1][1+(1/n)]}{[1+(2/n)][1+(3/n)]} \to 1 > 0$.

23. This is the same series as in Problem 8.

25. Converges. [It is the p=3/2 series.]

27. Diverges by nth-Term Test. $\dfrac{1}{2+\sin^2 n}$ has values between 1/3 and 1/2, so does not converge to 0.

29. $\dfrac{4 + \cos n}{n^3} \leq \dfrac{5}{n^3}$. [Converges by Ordinary Comparison Test using $\Sigma(5/n^3)$, which converges by linearity with p=3 series.]

31. $\dfrac{(n+1)^{n+1}}{(2n+2)!} \dfrac{(2n)!}{n^n} = \dfrac{1}{2(2n+1)} \left(1 + \dfrac{1}{n}\right)^n \to (0)(e) = 0 < 1$. [Converges by Ratio Test.]

33. $\sum \dfrac{4^n}{n!}$ converges by Ratio Test. $\dfrac{4^{n+1}}{(n+1)!} \dfrac{n!}{4^n} = \dfrac{4}{n+1} \to 0 < 1$.

$\sum \dfrac{n}{n!}$ converges by the Ratio Test. $\dfrac{n+1}{(n+1)!} \dfrac{n!}{n} = \dfrac{1}{n} \to 0 < 1$.

Therefore, using linearity of Σ, the given series converges.

35. $a_n > 0$ and Σa_n converges $\Rightarrow a_n > 0$ and $a_n \to 0$

$\Rightarrow 0 < a_n < 1$ for n sufficiently large

$\Rightarrow a_n^2 < a_n$ for n sufficiently large

$\Rightarrow \Sigma a_n^2$ converges by Ordinary Comparison Test.

37. $a_n \geq 0$, $b_n > 0$, $a_n/b_n \to 0 \Rightarrow 0 < a_n < b_n$ for n sufficiently large. This, along with Σb_n converges, yields Σa_n converges by the Ordinary Comparison Test.

39. Diverges by the Limit Comparison Test using the Harmonic Series.

$\dfrac{a_n}{1/n} = na_n \to 1 > 0$.

41. $R < 1$: Let r be a number between R and 1 $[0 \leq R < r < 1.]$

Then $(a_n)^{1/n} < r$, so $a_n < r^n$, for n sufficiently large.

Therefore, Σa_n converges by the Ordinary Comparison Test since

Σr^n in a converging geometric series.

$R > 1$: Let r be a number between R and 1 $[1 < r < R]$.

Then $(a_n)^{1/n} > r$, so $a_n > r^n$, for n sufficiently large.

Therefore, Σa_n diverges by the Ordinary Comparison Test since

Σr^n is a diverging geometric series.

43. (a) $\ln\left(1 + \dfrac{1}{n}\right) = \ln\left(\dfrac{n+1}{n}\right) = \ln(n+1) - \ln(n)$. It is a collapsing series.

$s_n = \ln(n+1)$, which diverges, so the series diverges.

(b) $\ln \dfrac{(n+1)^2}{n(n+2)} = 2\ln(n+1) - \ln(n) - \ln(n+2)$. It is a collapsing series.

$s_n = \ln \dfrac{2(n+1)}{n+2}$, which converges to $\ln 2$, so series converges to $\ln 2$.

(c) $\quad [\ln(n)]^{\ln(n)} = e^{[\ln(n)]\ln[\ln(n)]} = \left[e^{\ln(n)}\right]^{\ln[\ln(n)]} = n^{\ln[\ln(n)]}$

$$> n^2, \text{ for } n \text{ sufficiently large.}$$

The series converges by the Comparison Test, using the p=2 series.

(d) $\quad \ln[\ln(n)]^{\ln(n)} = e^{[\ln(n)]\ln[\ln(\ln(n))]} = \left[e^{\ln(n)}\right]^{\ln[\ln(\ln(n))]}$

$$= n^{\ln[\ln(\ln(n))]} > n^2, \text{ for } n \text{ sufficiently large.}$$

The series converges by the Comparison Test, using the p=2 series.

(e) Simplify and use l'Hôpital's Rule (4 times) to show $\dfrac{1/[\ln(n)]^4}{1/n} \to \infty$.

Conclude by Problem 39, using the harmonic series, that the series diverges.

(f) Simplify and use l'Hôpital's Rule (twice) to show $\dfrac{[\ln(n)/n]^2}{n^{-3/2}} \to 0$.

Conclude by Problem 37, using the p=3/2 series, that the series converges.

45. Use the Limit Comparison Test with a p-series.

$$\frac{1}{n^p}\left[1 + \frac{1}{2^p} + \cdots + \frac{1}{n^p}\right]\frac{n^p}{1} = 1 + \frac{1}{2^p} + \cdots + \frac{1}{n^p},$$ which converges iff p > 1.
Therefore, the given series converges iff p > 1.

Problem Set 11.5 Alternating Series, Absolute Convergence

1. Conditions are satisfied. |Error| < 2/[3(10)+1] < 0.0646.

3. $\dfrac{1}{\ln(n+1)} \to 0$ and $\dfrac{1}{\ln(n+1)}$ is a decreasing sequence. Series converges.

$|Error| < \dfrac{1}{\ln(10+1)} < 0.4171.$

5. Conditions are satisfied for n ≥ 3. |Error| < (ln10)/10 < 0.2303.

7. $\Sigma[(3/4)^n]$ is a converging geometric series. [r=3/4]

9. By Absolute Ratio Test. $\dfrac{n+1}{2^{n+1}}\dfrac{2^n}{n} = \dfrac{n+1}{2n} = \dfrac{1+(1/n)}{2} \to \dfrac{1}{2} < 1.$

11. $\Sigma[1/(n^2+n)]$ converges by the Ordinary Comparison Test using the p=2 series since $1/(n^2+n) < 1/n^2.$

13. Conditionally convergent. It converges by the Alternating-Series Test but $\Sigma(1/5n)$ diverges since $\Sigma(1/n)$ diverges.

15. Divergent by nth-Term Test. $\dfrac{n}{10n+1} \to \dfrac{1}{10}$ so the terms do not approach 0.

17. Conditionally convergent. It converges by the Alternating-Series Test, but the series of absolute values diverges (Example 4, page 469).

19. Absolute convergent. $\dfrac{(n+1)^4}{2^{n+1}}\dfrac{2^n}{n^4} = \dfrac{(1 + 1/n)^4}{2} \to \dfrac{1}{2} < 1.$

21. Conditionally convergent. Shown to converge in Problem 4,

    but $\displaystyle\sum \dfrac{n}{n^2+1}$ diverges by the Limit Comparison Test, using the Harmonic

    Series. $\dfrac{n}{n^2+1}\dfrac{n}{1} = \dfrac{n^2}{n^2+1} = \dfrac{1}{1+(1/n^2)} \to 1 > 0.$

23. Conditionally convergent. This is (-1) times the Alternating Harmonic Series.

25. Absolutely convergent by the Ordinary Comparison Test using the p=3/2 series since $|\sin n| \le 1.$

27. Conditionally convergent. Converges by the Alternating-Series Test, but

    $\displaystyle\sum \dfrac{1}{[n(n+1)]^{1/2}}$ diverges by the Limit Comparison Test, using the Harmonic

    Series. $\dfrac{1}{[n^2+n]^{1/2}}\dfrac{n}{1} = \dfrac{1}{[1+(1/n)]^{1/2}} \to 1 > 0.$

29. Diverges by the nth-Term Test since $|a_{n+1}|/|a_n| \to 3.$

31. This is the contrapositive of the Absolute Convergence Test.

33. $1 + \frac{1}{3} + \frac{1}{5} + \cdots = \sum \frac{1}{2n-1}$ which diverges by the Limit Comparison Test, using the Harmonic Series. $\frac{1}{2n-1} \frac{n}{1} = \frac{n}{2n-1} = \frac{1}{2-(1/n)} \rightarrow \frac{1}{2} > 0$.

$-\frac{1}{2} - \frac{1}{4} - \frac{1}{6} - \cdots = \sum (-1/2) \frac{1}{n}$ diverges since $\Sigma(1/n)$ diverges.

35. Let Σa_k be the original series and let s_n be the sequence of partial sums of the rearrangement described in (a) through (c). [Note that the procedure described in (a) through (c) is possible by Problem 33.] Since the terms of the original series converge to 0, the terms of the rearrangement converge to 0.

Let $\epsilon > 0$ be given. For some N, $n > N \Rightarrow -\epsilon < a_k < \epsilon$. Then let M be an integer greater than N such that all a_k for $k \leq N$ have been used and $S_M - 1.3$ and $S_{M+1} - 1.3$ have opposite signs if one of them is 0. Then $n > M \Rightarrow |s_n - 1.3| < \epsilon$. Thus, s_n converges to 1.3, so the rearrangement converges to 1.3.

37. Follow the basic procedure described in Problem 35.

39. The even-numbered partial sums of the series formed are partial sums of the series $\sum \left[\frac{1}{n} - \frac{1}{n^2}\right]$ which is the series $\sum \left[\frac{n-1}{n^2}\right]$, and this series diverges by the Limit Comparison Test, using the Harmonic Series.

$\frac{n-1}{n^2} \frac{n}{1} = 1 - \frac{1}{n} \rightarrow 1 > 0$.

41. $0 \leq (a_k \pm b_k)^2 = a_k^2 \pm 2a_k b_k + b_k^2$; $a_k^2 + b_k^2 \geq 2|a_k b_k|$. Therefore, $\Sigma a_k b_k$ converges absolutely since $\Sigma(a_k^2 + b_k^2) = \Sigma a_k^2 + \Sigma b_k^2$ converges.

43. $\int_{(n-1)\pi}^{n\pi} \frac{|\sin x|}{x} dx > \int_{(n-1)\pi}^{n\pi} \frac{|\sin x|}{n\pi} dx = \frac{|\cos(n\pi) - \cos(n\pi-\pi)|}{n\pi} = \frac{2}{\pi} \frac{1}{n}$.

The series diverges by Ordinary Comparison Test, using harmonic series.

45. $\dfrac{1}{n+1} + \dfrac{1}{n+2} + \cdots + \dfrac{1}{n+n} = \dfrac{1}{n}\left[\dfrac{1}{1+\dfrac{1}{n}} + \dfrac{1}{1+\dfrac{2}{n}} + \cdots + \dfrac{1}{1+\dfrac{n}{n}}\right]$, which is the

Riemann sum of $f(x) = \dfrac{1}{1+x}$ on $[0,1]$ using n subintervals of equal length and right endpoints as sample points. Hence, the alternating harmonic series equals $\displaystyle\int_0^1 \dfrac{1}{1+x}dx = \left[\ln(1+x)\right]_0^1 = \ln2$.

Problem Set 11.6   Power Series

1. $\displaystyle\sum \dfrac{(-1)^{n+1}x^n}{n(n+1)} \cdot \left|\dfrac{x^{n+1}}{(n+1)(n+2)} \dfrac{n(n+1)}{x^n}\right| = \dfrac{n|x|}{n+2} \to |x| < 1$, if x is in $(-1,1)$.

At $x = -1$:   Convergence by the Ordinary Comparison Test using the p=2 series.

At $x = 1$:   Convergence by the Alternating-Series Test.

Convergence Set: $[-1,1]$.

3. $\displaystyle\sum \dfrac{(-1)^n x^{2n+1}}{(2n+1)!} \cdot \left|\dfrac{x^{2n+3}}{(2n+3)!} \dfrac{(2n+1)!}{x^{2n+1}}\right| = \dfrac{x^2}{(2n+3)(2n+2)} \to 0$ for all x.

Convergence Set: **R**.

5. $\displaystyle\sum nx^n \cdot \left|\dfrac{(n+1)x^{n+1}}{nx^n}\right| = \dfrac{(n+1)|x|}{n} \to |x| < 1$, if x is in $(-1,1)$.

At $x = -1$ and at $x=1$: Divergence by nth-Term Test.

Convergence Set: $(-1,1)$.

7. $1 + \sum \frac{(-1)^n x^n}{n}$. $\left|\frac{x^{n+1}}{n+1} \frac{n}{x^n}\right| = \frac{n|x|}{n+1} \to |x| < 1$, if x is in (-1,1).

At x=-1: Harmonic Series. At x=1: Alternating Harmonic Series.

Convergence Set: (-1,1].

9. $1 + \sum_{n=1}^{\infty} \frac{(-1)^n x^n}{n(n+2)}$. $\left|\frac{x^{n+1}}{(n+1)(n+3)} \frac{n(n+2)}{x^n}\right| = \frac{n(n+2)|x|}{(n+1)(n+3)} \to |x| < 1$, for x in (-1,1).

At x=-1: $\sum \frac{1}{n(n+2)}$ converges by the Ordinary Comparison Test, using the p=2 series.

At x=1: $\sum \frac{(-1)^n}{n(n+2)}$ converges by the Absolute Convergence Test, using the the x=-1 series.

Convergence Set: [-1,1].

11. $\sum \frac{(-1)^n x^n}{2^n}$.$\left|\frac{x^{n+1}}{2^{n+1}} \frac{2^n}{x^n}\right| \to \frac{|x|}{2} < 1$, if x is in (-2,2).

At x=-2, x=2: Divergence by nth-Term Test. Convergence Set: (-2,2).

13. $\sum \frac{2^n x^n}{n!}$. $\left|\frac{2^{n+1} x^{n+1}}{(n+1)!} \frac{n!}{2^n x^n}\right| = \frac{2|x|}{n+1} \to 0$ for all x. Convergence Set: **R**.

15. $\sum \frac{(x-1)^{n+1}}{n+1}$. $\left|\frac{(x-1)^{n+2}}{n+2} \frac{n+1}{(x-1)^{n+1}}\right| = \frac{(n+1)|x-1|}{n+2} \to |x-1| < 1$, for x in (0,2).

At x=0: Convergence since it is an alternating harmonic series.

At x=2: Divergence since it is a harmonic series.

Convergence Set: [0,2).

17. $\sum \frac{(x+1)^n}{2^n}$. $\left|\frac{(x+1)^{n+1}}{2^{n+1}} \frac{2^n}{(x+1)^n}\right| = \frac{|x+1|}{2} < 1$, if x is in (-3,1).

19. $\sum \frac{(x+5)^n}{n(n+1)}$. $\left|\frac{(x+5)^{n+1}}{(n+1)(n+2)} \frac{n(n+1)}{(x+5)^n}\right| \to |x+5| < 1$, for x in (-6,-4).

At x=-4: Convergence by Ordinary Comparison Test using the p=2 series.
At x=-6: Convergence by Absolute Convergence Test, with x=-4 case.
Convergence Set: [-6,-4].

21. By the nth-Term Test.

23. Use Absolute Ratio Test. $x^2/2 < 1$, if $x^2 < 2$. The radius of convergence is $\sqrt{2}$.

25. This is a geometric series with a=1, and r = (x-3). Therefore, the sum is $S(x) = \dfrac{1}{1-(x-3)} = \dfrac{1}{4-x}$ and is valid if $|x-3| < 1$; i.e., x in (2,4).

27. (a) $\left| \dfrac{(3x+1)^{n+1}}{(n+1)2^{n+1}} \dfrac{n\,2^n}{(3x+1)^n} \right| = \dfrac{|3x+1|\,n}{2(n+1)} \to \dfrac{|3x+1|}{2} < 1$ if $-1 < x < \dfrac{1}{3}$.

At x = -1: Alternating Harmonic Series.

At x = 1/3; Harmonic series.

Convergence Set: [-1,1/3).

(b) $\left| \dfrac{(2x-3)^{n+1}}{4^{n+1}\sqrt{n+1}} \dfrac{4^n\sqrt{n}}{(2x-3)^n} \right| = \dfrac{|2x-3|\sqrt{n}}{4\sqrt{n+1}} \to \dfrac{|2x-3|}{4} < 1$ if $-1/2 < x < 7/2$.

At x =-1/2: Divergence since it is the p=1/2 series.

At x=7/2: Convergence by Alternating Series Test.

Convergence Set: (-1/2, 7/2].

29. See the write-up of Problem 30 and let p=3.

Problem Set 11.7 Operations on Power Series

1. This is the sum of the geometric series with a=1, r=-x. The series is $1-x+x^2-x^3+\cdots$. Convergence if $|-x| < 1$. Radius of convergence is 1.

3. $(1-x)^{-2} = 1 + 2x + 3x^2 + 4x^3 + \cdots$, $-1 < x < 1$ [1st part of Example 1]

Then $(1-x)^{-3} = \dfrac{1}{2}\dfrac{d}{dx}[(1-x)^{-2}] = \dfrac{1}{2}[(1)(2) + (2)(3)x + (3)(4)x^2 + \cdots]$

$= 1 + 3x + 6x^2 + \cdots + \dfrac{(n+1)(n+2)}{2}x^n + \cdots, -1 < x < 1.$

5. This is the sum of the geometric series with $a=1/2$, $r=3x/2$. The series is $(1/2) + (3/4)x + (9/8)x^2 + \cdots$ which converges iff $|3x/2| < 1$; i.e., x is in $(-2/3, 2/3)$. The radius of convergence is $2/3$.

7. This is the sum of the geometric series with $a=x^2$, $r=x^4$. The series is $x^2 + x^6 + x^{10} + \cdots$ and converges iff $|x^4| < 1$. Radius of convergence is 1.

9. The f used here will always be the f of Problem 9. Be careful not to confuse it with the f of Problem 1.

$$f'(x) = \ln(1+x); \quad f''(x) = (1+x)^{-1}.$$
so f'' is the function of Problem 1.

$$f''(x) = 1 - x + x^2 - x^3 + \cdots. \quad \text{[from Problem 1]}$$

$$f'(x) = \int_0^x f''(t)\,dt = x - \frac{x^2}{2} + \frac{x^3}{3} - \frac{x^4}{4} + \cdots.$$

$$f(x) = \int_0^x f'(t)\,dt = \frac{x^2}{1\cdot 2} - \frac{x^3}{2\cdot 3} + \frac{x^4}{3\cdot 4} - \frac{x^5}{4\cdot 5} + \cdots = \sum \frac{(-1)^n x^{n+2}}{(n+1)(n+2)}$$

for x in $(-1,1)$. The radius of convergence is 1.

11. $\ln(1+x) - \ln(1-x) = (x - \frac{x^2}{2} + \frac{x^3}{3} - \frac{x^4}{4} + \cdots) - (-x - \frac{x^2}{2} - \frac{x^3}{3} - \cdots)$

$$= 2(x + \frac{x^3}{3} + \frac{x^5}{5} + \cdots). \quad \text{The radius of convergence is } 1.$$

13. $e^{-x} = 1 - x + \frac{x^2}{2!} - \frac{x^3}{3!} + \cdots$ for all x in $\mathbf{R}$.

15. Add the series from Example 3 and the one from Problem 13.

$$(1 + x + \frac{x^2}{2!} + \frac{x^3}{3!} + \cdots) + (1 - x + \frac{x^2}{2!} - \frac{x^3}{3!} + \cdots)$$

$$= 2 + \frac{2x^2}{2!} + \frac{2x^4}{4!} + \cdots = \sum \frac{2x^{2n}}{(2n)!} \text{ for all } x \text{ in } \mathbf{R}.$$

17. $\left(1 - x + \dfrac{x^2}{2!} - \dfrac{x^3}{3!} + \cdots\right)\left(1 + x + x^2 + x^3 + \cdots\right)$

$\qquad = 1 + \dfrac{x^2}{2} + \dfrac{x^3}{3} + \dfrac{3x^4}{8} + \cdots$, for x in $(-1,1)$.

19. $e^{-x}\tan^{-1}x = \left(1 - x + \dfrac{x^2}{2!} - \dfrac{x^3}{3!} + \cdots\right)\left(x - \dfrac{x^3}{3} + \dfrac{x^5}{5} - \dfrac{x^7}{7} + \cdots\right)$

$\qquad = x - x^2 + \dfrac{x^3}{6} + \dfrac{x^4}{6} + \dfrac{3x^5}{40} + \cdots$, x in $(-1,1)$.

21. See Example 2 for the series for $\tan^{-1}x$. Then for x in $(-1,1)$.

$(\tan^{-1}x)(1+x^2+x^4) = \left(x - \dfrac{x^3}{3} + \dfrac{x^5}{5} - \dfrac{x^7}{7} + \cdots\right)(1+x^2+x^4)$

$\qquad = x + \left(1 - \dfrac{1}{3}\right)x^3 + \left(1 - \dfrac{1}{3} + \dfrac{1}{5}\right)x^5 + \left(-\dfrac{1}{3} + \dfrac{1}{5} - \dfrac{1}{7}\right)x^7 + \cdots$

$\qquad = x + \dfrac{2x^3}{3} + \displaystyle\sum_{n=0}^{\infty}(-1)^n\left(\dfrac{1}{2n+1} - \dfrac{1}{2n+3} + \dfrac{1}{2n+5}\right)x^{2n+5}$.

23. $\dfrac{e^t}{1+t} = \left(1 + t + \dfrac{t^2}{2} + \dfrac{t^3}{6} + \cdots\right)(1 - t + t^2 - t^3 + \cdots)$

$\qquad = 1 + \dfrac{t^2}{2} - \dfrac{t^4}{3} + \dfrac{3t^5}{8} - \cdots$.

Therefore, $\displaystyle\int_0^x \dfrac{e^t}{t+1}\, dt = x + \dfrac{x^3}{6} - \dfrac{x^5}{15} + \dfrac{3x^6}{48} - \cdots$, x in $(-1,1)$.

25. (a) $\dfrac{x}{1-(-x)} = \dfrac{x}{1+x}$ [It is the geometric series with a=x, r=-x.]

(b) $\dfrac{e^x - 1 - x}{x^2}$, $x \neq 0$, since $\left(1 + x + \dfrac{x^2}{2!} + \dfrac{x^3}{3!} + \cdots\right) - 1 - x = e^x - 1 - x$.

(c) $(2x) + \dfrac{(2x)^2}{2} + \dfrac{(2x)^3}{3} + \cdots = -\ln(1-2x)$,

$\qquad$ for 2x in $(-1,1)$, so x in $(-1/2,1/2)$. [See Example 1.]

27. $x + 2x^2 + 3x^3 + 4x^4 + \cdots = x(1 + 2x + 3x^2 + 4x^3 + \cdots)$

$= x\, D_x(x + x^2 + x^3 + x^4 + \cdots) = x\, D_x\left[\dfrac{x}{1-x}\right] = x\,\dfrac{1}{(1-x)^2} = \dfrac{x}{(1-x)^2}\,,$

for x in (-1,1).

29. (a) $\tan^{-1}(e^x-1) = (x + \dfrac{x^2}{2!} + \dfrac{x^3}{3!} + \cdots) - (1/3)\,(x + \dfrac{x^2}{2!} + \dfrac{x^3}{3!} + \cdots)^3 + \cdots$

$= x + \dfrac{x^2}{2} - \dfrac{x^3}{6} + \cdots,\ x$ in $(-\infty,\ \ln2)$.

(b) $1+(x+ \dfrac{x^2}{2!} + \dfrac{x^3}{3!} +\cdots)+ \dfrac{1}{2!}(x+ \dfrac{x^2}{2!} + \dfrac{x^3}{3!} +\cdots)^2+ \dfrac{1}{3!}\,(x+ \dfrac{x^2}{2!} + \dfrac{x^3}{3!}+\cdots)^3+\cdots$

$= 1 + x + x^2 + \dfrac{5x^3}{6} + \cdots,\ x$ in $\mathbb{R}$.

31. $\dfrac{x}{x^2-3x+2} = \dfrac{1}{1-x} - \dfrac{2}{2-x} = \dfrac{1}{1-x} - \dfrac{1}{1-x/2}$

$= (1 + x + x^2 + x\cdots) - (1 + \dfrac{x}{2} + \dfrac{x^2}{4} + \cdots),\ -1 < x < 1,$

$= \dfrac{x}{2} + \dfrac{3x^2}{4} + \dfrac{7x^3}{8} +\cdots,\ -1 < x < 1.$

33.

$$F(x) = f_0 \qquad\quad + f_1 x \qquad\quad + f_2 x^2 \qquad\quad + f_3 x^3+\cdots$$
$$- xF(x) = \qquad\quad - f_0 x \qquad\quad - f_1 x^2 \qquad\quad - f_2 x^3-\cdots$$
$$- x^2 F(x) = \qquad\qquad\qquad\quad - f_0 x^2 \qquad\quad - f_1 x^3-\cdots$$

$(1-x-x^2)F(x) = f_0 + (f_1-f_0)x + (f_2-f_1-f_0)x^2 + (f_3-f_2-f_1)x^3+\cdots$

$= 0\ + (1-0)x +(0)x^2 + (0)x^3 + \cdots$ [since $f_{n+2} = f_n+f_{n+1}$]

$= x$

Therefore, $F(x) = \dfrac{x}{1-x-x^2}$.

35. $\tan^{-1}x = x - \dfrac{x^3}{3} + \dfrac{x^5}{5} - \dfrac{x^7}{7} + \dfrac{x^9}{9} - \cdots, \ -1 < x < 1.$

$\tan^{-1}(0.2) \approx 0.2 - \dfrac{0.008}{3} + \dfrac{0.00032}{5} - \dfrac{0.0000128}{7} + \dfrac{0.000000512}{9}.$

$\tan^{-1}(1/239) \approx 1/239.$

Thus, $\pi = 16\tan^{-1}(0.2) - 4\tan^{-1}(1/239) \approx 3.14159$ (to first 6 digits).

Problem Set 11.8   Taylor and Maclaurin Series

1. $\tan x = \dfrac{\sin x}{\cos x} = x + \dfrac{x^3}{3} + \dfrac{2x^5}{15} + \cdots, \ x$ in $(-\pi/2, \pi/2).$ [Used division.]

3. $e^x \sin x = \left(1 + x + \dfrac{x^2}{2!} + \dfrac{x^3}{3!} + \dfrac{x^4}{4!} + \dfrac{x^5}{5!} + \cdots\right)\left(x - \dfrac{x^3}{3!} + \dfrac{x^5}{5!} + \cdots\right)$

$= x + x^2 + \left(\dfrac{-1}{6} + \dfrac{1}{2}\right)x^3 + \left(\dfrac{-1}{6} + \dfrac{1}{6}\right)x^4 + \left(\dfrac{1}{120} - \dfrac{1}{12} + \dfrac{1}{24}\right)x^5 + \cdots$

$= x + x^2 + \dfrac{x^3}{3} - \dfrac{x^5}{30} + \cdots,$ for all x in $\mathbf{R}.$

5. $\cos x \ \ln(1+x) = \left(1 - \dfrac{x^2}{2!} + \dfrac{x^4}{4!} - \dfrac{x^6}{6!} + \cdots\right)\left(x - \dfrac{x^2}{2} + \dfrac{x^3}{3} - \dfrac{x^4}{4} + \cdots\right)$

$= x - \dfrac{x^2}{2} - \dfrac{x^3}{6} + \dfrac{3x^5}{40} + \cdots,$ for all x in $(-1,1).$

7. $e^x + x + \sin x = \left(1 + x + \dfrac{x^2}{2!} + \dfrac{x^3}{3!} + \cdots\right) + x + \left(x - \dfrac{x^3}{3!} + \dfrac{x^5}{5!} - \cdots\right)$

$= 1 + 3x + \dfrac{x^2}{2} + \dfrac{x^4}{24} + \dfrac{x^5}{60} + \cdots,$ for all x in $\mathbf{R}.$

9. $\frac{1}{1-x}$ coshx = $(1 + x + x^2 + x^3 + x^4 + x^5 + \cdots)(1 + \frac{x^2}{2!} + \frac{x^4}{4!} + \cdots)$

$$= 1 + x + (\frac{1}{2} + 1)x^2 + (\frac{1}{2} + 1)x^3 + (\frac{1}{24} + \frac{1}{2} + 1)x^4 + (\frac{1}{24} + \frac{1}{2} + 1)x^5 + \cdots$$

$$= 1 + x + \frac{3x^2}{2} + \frac{3x^3}{2} + \frac{37x^4}{24} + \frac{37x^5}{24} + \cdots, \text{ for all x in } (-1,1).$$

11. $1 - x + x^3 - x^4 + \cdots$, x in $(-1,1)$. [by long division]

13. $(\sin x)^3 = (x - \frac{x^3}{3!} + \frac{x^5}{5!} + \cdots)(x - \frac{x^3}{3!} + \frac{x^5}{5!} + \cdots)(x - \frac{x^3}{3!} + \frac{x^5}{5!} + \cdots)$

$$= x^3 - \frac{x^5}{2} + \cdots, \text{ for all x in } \mathbb{R}.$$

15. $x\sec(x^2) + \sin x = \frac{x}{\cos(x^2)} + \sin x$

$$= \frac{x}{1 - (x^4/2!) + (x^8/4!) - \cdots} + (x - \frac{x^3}{3!} + \frac{x^5}{5!} - \cdots)$$

$$= (x + \frac{x^5}{2} + \cdots) + (x - \frac{x^3}{6} + \frac{x^5}{120} - \cdots) \text{ [long division for first part]}$$

$$= 2x - \frac{x^3}{6} + \frac{61x^5}{120} + \cdots, \text{ for all x in } (-\pi/2, \pi/2)$$

17. $(1+x)^{3/2} = 1 + (3/2)x + \frac{(3/2)(1/2)x^2}{2!} + \frac{(3/2)(1/2)(-1/2)x^3}{3!} + \cdots$

$$= 1 + \frac{3x}{2} + \frac{3x^2}{8} - \frac{x^3}{16} + \frac{3x^4}{128} - \frac{3x^5}{256} + \cdots, \text{ for x in } (-1,1).$$

19. $e^x = e + e(x-1) + (e/2)(x-1)^2 + (e/3!)(x-1)^3 + \cdots, x \text{ in } \mathbb{R}.$

21. $f(x) = \cos x$, $f'(x) = -\sin x$, $f''(x) = -\cos x$, $f^{(3)}(x) = \sin x$.

$f(\pi/3) = 1/2$, $f'(\pi/3) = -\sqrt{3}/2$, $f''(\pi/3) = -1/2$, $f^{(3)}(x) = \sqrt{3}/2$.

$\cos x = \dfrac{1}{2} + (-\sqrt{3}/2)(x-\pi/3) + \dfrac{(-1/2)(x-\pi/3)^2}{2!} + \dfrac{(\sqrt{3}/2)(x-\pi/3)^3}{3!} + \cdots$

$= \dfrac{1}{2} - \dfrac{\sqrt{3}(x-\pi/3)}{2} - \dfrac{(x-\pi/3)^2}{4} + \dfrac{\sqrt{3}(x-\pi/3)^3}{12} + \cdots$, x in **R**.

23. $1+x^2+x^3 = 3 + 5(x-1) + 4(x-1)^2 + (x-1)^3$.

25. $f(x) = f(-x) \Rightarrow \Sigma[a_n x^n] = \Sigma[a_n(-x)^n] = \Sigma[(-1)^n(a_n x^n)]$

$\Rightarrow a_n = (-1)^n(a_n) \Rightarrow a_n = -a_n$ (if n odd) $\Rightarrow a_n = 0$ (if n odd).

27. $(1-t^2)^{-1/2} = 1 + (-1/2)(-t^2) + \dfrac{(-1/2)(-3/2)(-t^2)^2}{2!} + \dfrac{(-1/2)(-3/2)(-5/2)(-t^2)^3}{3!}$

$+\cdots = 1 + \dfrac{t^2}{2} + \dfrac{3t^4}{8} + \dfrac{5t^6}{16} + \cdots$

$\sin^{-1}x = \displaystyle\int_0^x \left(1 + \dfrac{t^2}{2} + \dfrac{3t^4}{8} + \dfrac{5t^6}{16} + \cdots\right) dt = x + \dfrac{x^3}{6} + \dfrac{3x^5}{40} + \dfrac{5x^7}{112} + \cdots.$

29. $\cos(x^2) = 1 - (1/2)x^4 + (1/4!)x^8 - (1/6!)x^{12} + \cdots$

$\displaystyle\int_0^1 \cos(x^2)\,dx = \left[x - \dfrac{x^5}{2!5} + \dfrac{x^9}{4!9} - \dfrac{x^{13}}{6!13} + \dfrac{x^{17}}{8!17} - \dfrac{x^{21}}{10!21} + \cdots\right]_0^1$

$= 1 - \dfrac{1}{2!5} + \dfrac{1}{4!9} - \dfrac{1}{6!13} + \dfrac{1}{8!17} - \dfrac{1}{10!21} + \cdots$, which satisfies the

Alternating-Series Test. $\dfrac{1}{8!17} < 0.00005$ but $\dfrac{1}{6!13} > 0.0005$, so

$1 - \dfrac{1}{2!5} + \dfrac{1}{4!9} - \dfrac{1}{6!13} \approx 0.9045$ has the desired accuracy.

31. $\dfrac{1}{x} = \dfrac{1}{1-(1-x)} = 1 + (1-x) + (1-x)^2 + \cdots = 1 - (x-1) + (x-1)^2 = \cdots.$

33. Only terms through the 4th degree need to be expressed.

(a) $f(x) = e^{x+x^2} = 1 + (x+x^2) + \dfrac{(x+x^2)^2}{2!} + \dfrac{(x+x^2)^3}{3!} + \dfrac{(x+x^2)^4}{4!} + \cdots$

$= 1 + (x+x^2) + \dfrac{x^2+2x^3+x^4}{2!} + \dfrac{x^3+3x^4+\cdots}{3!} + \dfrac{x^4+\cdots}{4!} + \cdots$

$= 1 + x + \dfrac{3x^2}{2} + \dfrac{7x^3}{6} + \dfrac{25x^4}{24} + \cdots \ .$

Therefore, $f^{(4)}(0) = 4!\left(\dfrac{25}{24}\right) = 25.$

(b) $f(x) = e^{\sin x} = 1 + \sin x + \dfrac{\sin^2 x}{2!} + \dfrac{\sin^3 x}{3!} + \dfrac{\sin^4 x}{4!} + \cdots$

$= 1 - \left(x - \dfrac{x^3}{3!} + \cdots\right) + \dfrac{1}{2}\left(x - \dfrac{x^3}{3!} + \cdots\right)^2 + \dfrac{1}{6}(x - \cdots)^3 + \dfrac{1}{24}(x - \cdots)^4 + \cdots$

$= 1 - \left(x - \dfrac{x^3}{6} + \cdots\right) + \dfrac{1}{2}\left(x^2 - \dfrac{x^4}{3} + \cdots\right)^2 + \dfrac{1}{6}(x^3 - \cdots) + \dfrac{1}{24}(x^4 - \cdots) + \cdots$

$= 1 + x + \dfrac{x^2}{2} - \dfrac{x^4}{8} + \cdots \ .$

Therefore, $f^{(4)}(0) = 4!\left(\dfrac{-1}{8}\right) = -3.$

(c) $f(x) = \displaystyle\int_0^x \dfrac{e^{t^2}-1}{t^2}\, dt = \int_0^x \dfrac{1}{t^2}\left[\left[1 + t^2 + \dfrac{t^4}{2!} + \dfrac{t^6}{3!} + \cdots\right] - 1\right] dt$

$= \displaystyle\int_0^x \left[1 + \dfrac{t^2}{2} + \dfrac{t^4}{6} + \cdots\right] dt = \left[t + \dfrac{t^3}{6} + \dfrac{t^5}{30} + \cdots\right]_0^x$

$= x + \dfrac{x^3}{6} + \dfrac{x^5}{30} + \cdots,\ \text{so } f^{(4)}(0) = 0.$

(d) $f(x) = e^{\cos x} = (e)e^{\cos x - 1} = (e)\exp\left[\dfrac{-x^2}{2!} + \dfrac{x^4}{4!} \cdots\right]$

$$= e\left[1 + \dfrac{-x^2}{2!} + \dfrac{x^4}{4!} - \cdots\right] + \dfrac{\left[\dfrac{-x^2}{2!} + \cdots\right]^2}{2!} + \cdots$$

$$= e\left[1 - \dfrac{x^2}{2} + \dfrac{x^4}{24} + \dfrac{x^4}{8} + \cdots\right] + e\left[1 - \dfrac{x^2}{2} + \dfrac{x^4}{6} + \cdots\right]$$

Therefore, $f^{(4)}(0) = e[4!(\tfrac{1}{6})] = 4e \approx 10.8731$.

(e) $f(x) = \ln(\cos^2 x) = \ln(1 - \sin^2 x) = \ln([1 + (-\sin^2 x)]$

$$= (-\sin^2 x) - \dfrac{(-\sin^2 x)^2}{2} + \dfrac{(-\sin^2 x)^3}{3} - \cdots$$

$$= -\left[x - \dfrac{x^3}{3!} + \cdots\right]^2 - \dfrac{1}{2}\left[x - \dfrac{x^3}{3!} + \cdots\right]^4 + \cdots$$

$$= -\left[x^2 - \dfrac{x^4}{3} + \cdots\right] - \dfrac{1}{2}\left[x^4 + \cdots\right] = -x^2 - \dfrac{x^4}{6} + \cdots \; .$$

Therefore $f^{(4)}(0) = 4!\left(\dfrac{-1}{6}\right) = -4$.

35. $\tanh x = \dfrac{\sinh x}{\cosh x} = a_0 + a_1 x + a_2 x^2 + a_3 x^3 + a_4 x^4 + a_5 x^5 + \cdots$

Thus, $x + \dfrac{x^3}{3!} + \dfrac{x^5}{5!} + \cdots$

$$= (a_0 + a_1 x + a_2 x^2 + a_3 x^3 + a_4 x^4 + a_5 x^5 + \cdots)(1 + \dfrac{x^2}{2!} + \dfrac{x^4}{4!} + \cdots)$$

$$= a_0 + a_1 x + (\dfrac{a_0}{2} + a_2)x^2 + (\dfrac{a_1}{2} + a_3)x^3 + (\dfrac{a_0}{24} + \dfrac{a_2}{2} + a_4)x^4 +$$

$$(\dfrac{a_1}{24} + \dfrac{a_3}{2} + a_5)x^5 + \cdots .$$

Therefore, $a_0 = 0$, $a_1 = 1$, $a_2 = -a_0/2 = 0$, $a_3 = 1/6 - a_1/2 = -1/3$, $a_4 = -a_0/24 - a_2/2 = 0$, and $a_5 = 1/120 - a_1/24 - a_3/2 = 2/15$, so $\tanh x = x - (1/3)x^3 + (2/15)x^5 + \cdots .$

37. $f^{(4)}(0) = \begin{cases} 0, & \text{if } t < 0 \\ \text{undefined}, & \text{if } t = 0 \\ 24, & \text{if } t > 0 \end{cases}$, so a Maclaurin series doesn't exist.

Even if a Maclaurin series did exist, as is conceivable in the case of the car, $f^{(n)}(0) = 0$ for all n due to $f(x)$ for $x < 0$, so the Maclaurin series would have to be identically 0, which would not represent the function for $x > 0$.

39. $x - \frac{1}{6}x^3 + \frac{1}{120}x^5 - \frac{1}{5040}x^7$

41. $-2 + x - x^2 - \frac{5}{6}x^3$

43. $x + \frac{1}{2}x^2 - \frac{5}{24}x^4 - \frac{23}{120}x^5$

45. $x + x^2 + \frac{1}{3}x^3 - \frac{1}{30}x^5$

Problem Set 11.9   Chapter Review

True-False Quiz

1. False C^{ex}: $a_n = 2 + (-1)^n$, $b_n = 5 + 1/n$.

3. True.   If a sequence converges to L, every subsequence converges to L.

5. False. C^{ex}: Let $a_n = n$ for n prime; $a_n = 0$ for n not prime.

7. False. C^{ex}: Let a_n be the nth partial sum of the harmonic series.

9. True. $a_n \rightarrow L \Rightarrow L-1 < a_n < L+1$ for all $n > N$ (for some N)

$$\Rightarrow \frac{L}{n} - \frac{1}{n} < \frac{a_n}{n} < \frac{L}{n} + \frac{1}{n}. \quad \text{Now apply Squeeze Theorem.}$$

11. True.   The first assertion by the Alternating-Series Test; the second was proven in the "A Convergence Test" section (page 477).

13. False. C^{ex}: $\Sigma(-1)^n$.

15. True. $\rho = 1$, so the test is inconclusive.

17. False.   nth-Term Test. [nth term converges to 1/e]

19. True.   Use linearity. First series converges by the Integral Test; second converges by Ordinary Comparison Test using p=2 series.

21. True.   These conditions imply the convergence of $\sum\limits_{n=101}^{\infty} a_n$ by Ordinary

Comparison Test. Then $a_1 + \cdots + a_{100} + \sum\limits_{n=101}^{\infty} a_n$ converges.

23. True.   The left-hand side is less than the sum of the geometric series with a=r=1/3, and that sum is 1/2.

25. True. Σb_n converges $\Rightarrow \Sigma(-b_n)$ converges   [Linearity Theorem]

                                  $\Rightarrow \Sigma(-a_n)$ converges   [Ord. Comparison Test]

                                  $\Rightarrow \Sigma a_n$ converges   [Linearity Theorem]

27. True.   The left-hand side is the error in using the first 99 terms of the alternating harmonic series to approximate its sum. That error is less than the 100th term, 1/100, or 0.01.

29. True.   Radius of convergence is at least 3-(-1.1) = 4.1, so the convergence set contains (-1.1,7.1).

31. True.   If the series converges at x = 1.5, then it converges at x=1.

$$\int_0^x f(t)\,dt = \int_0^x \left[\sum_{n=0}^{\omega} a_n t^n \right] dt = \sum_{n=0}^{\infty} \frac{a_n x^{n+1}}{n+1}. \quad \text{Now let } x=1.$$

33. False. C^{ex}; See Problems 37 ad 38, Section 11.8.

35. True. $f(x) = e^{-x}$, $f'(x) = -e^{-x}$, so $f'(x) + f(x) = 0$ on R.

Sample Test Problems

1. Converges. $\lim\limits_{n\to\infty} \dfrac{9}{(9 + 1/n^2)^{1/2}} = 3.$

3. Converges. $\lim\limits_{n\to\infty} \left[1 + \dfrac{4}{n}\right]^n = \lim\limits_{m\to\infty} \left[1 + \dfrac{1}{m}\right]^{4m}$ [letting n = 4m]

$$= \lim_{m\to\infty} \left[\left[1 + \frac{1}{m}\right]^m \right]^4 = \left[\lim_{m\to\infty} \left[1 + \frac{1}{m}\right]^m \right]^4 = e^4 \approx 54.5982.$$

5. Converges to 1. [See write-up of Sample Test Problem 14, Section 9.5]

7. Converges by the Squeeze Theorem. $[0 \leq (\sin^2 n)/(n^{1/2}) \leq 1/(n^{1/2}).]$

9. Converges. Partial sums, S_n, are collapsing. $S_n = 1 - \dfrac{1}{\sqrt{n+1}} \rightarrow 1$, so the sum is 1.

11. Diverges. [Partial sums are collapsing, $s_n = -\ln(n+1) \rightarrow \infty.]$

13. Converges to $\dfrac{1}{1-e^{-2}} = \dfrac{e^2}{e^2-1}$. [Geometric — $a=1, r=e^{-2}$]

15. Converges to $\dfrac{0.91}{1-0.01} = \dfrac{91}{99}$. [Geometric — $a=0.91$, $r=0.01$]

17. Converges to $\cos 2 \approx -0.4164$. [Power series for $\cos x$ evaluated at $x=2$.]

19. Diverges. [Use Limit Comparison test with Harmonic Series.]

21. Converges. [Use Absolute Convergence Test with the $p=(3/2)$ series.]

23. Converges. [Use linearity; two converging geometric series.]

25. Diverges. [Use Divergence Test.]

27. Converges by Ratio Test. $\dfrac{(n+1)^2}{(n+1)!}\dfrac{n!}{n^2} = \dfrac{n+1}{n^2} = \dfrac{1}{n} + \dfrac{1}{n^2} \rightarrow 0 < 1.$

29. Diverges. [Use Ratio Test.]

31. Converges. [Use Ratio Test.]

33. Conditionally convergent. The series converges by the Alternating Series Test, but the corresponding positive-term series diverges by the Limit Comparison Test, using the Harmonic Series.

35. Diverges. [Geometric — $r=-3/2.]$

37. $\left| \dfrac{x^{n+1}}{(n+1)^3+1} \dfrac{n^3+1}{x^n} \right| = \dfrac{(n^3+1)|x|}{(n+1)^3+1} \to |x| < 1$, if $-1 < x < 1$.

At $x = 1$: Convergence by Ordinary Comparison Test using the p=3 series.

At $x = -1$: Convergence by Absolute Convergence Test using result of x=1.

Convergence Set: $[-1,1]$.

39. $\left| \dfrac{(x-4)^{n+1}}{n+2} \dfrac{n+1}{(x-4)^n} \right| = \dfrac{(n+1)|x-4|}{n+2} \to |x-4| < 1$ if $3 < x < 5$. [Ratio Test]

At $x = 3$: Divergence [Harmonic Series].

At $x = 5$: Convergence [Alternating Harmonic Series].

Convergence Set: $(3,5]$.

41. $\left| \dfrac{(x-3)^{n+1}}{2^{n+1}+1} \dfrac{2^n+1}{(x-3)^n} \right| = \dfrac{(2^n+1)|x-3|}{2^{n+1}+1} \to \dfrac{|x-3|}{2} < 1$, if x is in $(1,5)$.

At $x=1$, $x=-1$: Divergence by nth-Term Test. Convergence Set: $(1,5)$.

43. $D(1+x)^{-1} = -(1+x)^{-2}$. $D(1-x+x^2-x^3+\cdots) = -1+2x-3x^2+4x^3-\cdots$.

Therefore, $(1+x)^{-2} = 1-2x+3x^2-4x^3+\cdots$ on $(-1,1)$.

45.

$f(x)$	$= \sin^2 x$.	$f(0)$	$= 0$.
$f'(x)$	$= 2\sin x \cos x = \sin 2x$.	$f'(0)$	$= 0$.
$f''(x)$	$= 2\cos 2x$.	$f''(0)$	$= 2$.
$f^{(3)}(x)$	$= -4\sin 2x$.	$f^{(3)}(0)$	$= 0$.
$f^{(4)}(x)$	$= -8\cos 2x$.	$f^{(4)}(0)$	$= -8$.
$f^{(5)}(x)$	$= 16\sin 2x$.	$f^{(5)}(0)$	$= 0$.
$f^{(6)}(x)$	$= 32\cos 2x$.	$f^{(6)}(0)$	$= 32$.

Therefore, $\sin^2 x = \dfrac{2x^2}{2!} - \dfrac{8x^4}{4!} + \dfrac{32x^6}{6!} - \dfrac{128x^8}{8!} + \cdots$ for all x in $\mathbf{R}$.

47. $\sin x + \cos x = (x - \frac{x^3}{3!} + \frac{x^5}{5!} - \cdots) + (1 - \frac{x^2}{2!} + \frac{x^4}{4!} - \cdots)$

$$1 + x - \frac{x^2}{2!} - \frac{x^3}{3!} + \frac{x^4}{4!} + \frac{x^5}{5!} - \frac{x^6}{6!} - \frac{x^7}{7!} + \cdots \text{ on } \mathbf{R}.$$

49. $\dfrac{e^x - 1}{x} = 1 + \dfrac{x}{2!} + \dfrac{x^2}{3!} + \dfrac{x^3}{4!} + \cdots .$

$$\int_0^x \frac{e^t - 1}{t}\, dt = x + \frac{x^2}{2!2} + \frac{x^3}{3!3} + \frac{x^4}{4!4} + \frac{x^5}{5!5} + \cdots .$$

Now let $x = 0.2$. The integral approximately equals 0.21046.

51. $\cos x = 1 - \dfrac{x^2}{2!2} + \dfrac{x^4}{4!} - \cdots .$

The error will be less than $\dfrac{x^4}{4!} < \dfrac{(0.1)^4}{4!} < 0.00000417.$

Problem Set 12.1   The Parabola

1. $y^2 = 4(4)x$; p=4.
    Focus: (4,0).
    Directrix: x=-4.

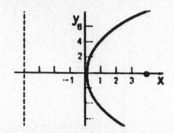

3. $x^2 = -4(4)y$; p=4.
    Focus: (0,-4).
    Directrix: y=4.

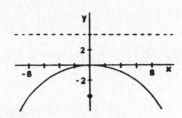

5. $y^2 = -4(1/2)x$; p=1/2.
    Focus: (1,2, 0).
    Directrix: x=-1/2.

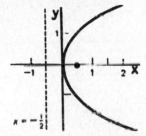

7. $x^2 = 4(3/2)y$; p=3/2.
    Focus: (0,3/2).
    Directrix: y=-3/2.

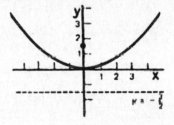

9. p = 3. Directrix is
    x=-3. Equation is
    $y^2 = 4(3)x$ or $y^2 = 12x$.

11. p=4. Graph opens upward.
     $x^2 = 4(4)y$ or $x^2 = 16y$.

13. p=5. Graph opens to left.
     $y^2 = -4(5)x$ or $y^2 = -20x$.

15. $y^2 = 4px$.

$(-1)^2 = 4p(3) \Rightarrow p = 1/12$.

$y^2 = 4(1/12)x; \quad y^2 = (1/3)x$.

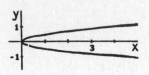

17. $x^2 = -4py$.

$(6)^2 = -4p(-5) \Rightarrow p = 9/5$.

$x^2 = (-36/5)y$.

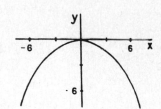

19. $y^2 = 16x, \quad 2yy' = 16, \quad y' = 8/y$.

At $(1,-4)$:   (1) $y' = -2$.

(2) Tangent is $y = -2x-2$.

(3) Normal is $y = (1/2)x - 9/2$.

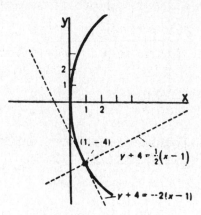

21. $x^2 = 2y, \quad 2x = 2y', \quad y' = x$.

At $(4,8)$:   (1) $y' = 4$.

(2) Tangent is $y-8 = 4(x-4)$.
$y = 4x - 8$.

(3) Normal is $y-8 = (-1/4)(x-4)$.
$y = (-1/4)x + 9$.

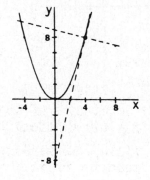

23. $y^2 = -15x, \quad 2yy' = -15, \quad y' = -15/2y$.

At $(-3, -3\sqrt{5})$:   (1) $y' = \sqrt{5}/2$.

(2) Tangent is $y = (\sqrt{5}/2)x - (3/2)\sqrt{5}$.

(3) Normal is $y = (-2/\sqrt{5})x - 21/\sqrt{5}$.

25. $x^2 = -6y, \quad 2x = -6y', \quad y' = (-1/3)x$.

At $(3\sqrt{2}, -3)$: $y' = -\sqrt{2}$.

Tangent is $y = -\sqrt{2}x + 3$.

Normal is $y = (1/\sqrt{2})x - 6$.

27. $y^2 = 5x$, $2yy' = 5$, $y' = 5/2y$.

The slope, y', is $\sqrt{5}/4$, so $5/2y = \sqrt{5}/4$.

Therefore, $y = 2\sqrt{5}$; then $x = 4$.

The point we seek is $(4, 2\sqrt{5})$.

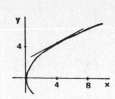

29. $y^2 = -18x$, $2yy' = -18$, $y' = -9/y$.

$3x - 2y + 4 = 0$ has slope $3/2$, so $-9/y = 3/2$.   Therefore, $y = -6$, $x = -2$.

Tangent is $y = (3/2)x - 3$.

31. A focal chord [nonvertical line through $(0,p)$] has form $y = mx + p$.
Simultaneously solve $y = mx + p$ and $x^2 = 4py$ to obtain that the endpoints
of the focal chord with slope m occur where $x = 2p[m \pm (m^2 + 1)^{1/2}]$.   The
slope of a tangent to the parabola is $x/2p$, so the slopes of the two
tangents are $[m + (m^2 + 1)^{1/2}]$ and $[m - (m^2 + 1)^{1/2}]$.   The product of these
slopes is -1, so the tangents are perpendicular.

33. Minimize the square of the distance, z, between focus $(0,p)$ and an
arbitrary point of the parabola $(x, x^2/4p)$, x any real number.

$$z(x) = x^2 + \left[\frac{x^2}{4p} - p\right]^2 = \frac{x^2}{2} + \frac{x^4}{16p^2} + p^2 \Rightarrow z'(x) = x + \frac{x^3}{4p^2} = \frac{x(4p^2 + x^2)}{4p^2}.$$

$z'(x) = 0$ iff $x = 0$ (and the corresponding value of y is 0).

$z''(x) = 1 + \dfrac{3x^2}{4p^2}$; then $z''(0) = 1 > 0$, so z is minimum at $(0,0)$

[vertex].

35. The y-coordinate of the spaceship can be represented:

(1) $p + (40 \text{ million}) \sin 15°$.

(2) $(1/4p)[(40 \text{ million})\cos 15°]^2$, using the x-coordinate and the
equation of the curve.

Set these equal, and then solve for p.

$p = (20 \text{ million})(1 - \sin 15°) \approx 14.8236$ million miles.

37. Let $y^2 = 4px$. $|FR| + |RG| = [(p - x_0)^2 + y_0^2]^{1/2} + (p - x_0)$

$$= [(p - x_0)^2 + 4px_0]^{1/2} + (p - x_0) = 2p.$$

---

39. $|FP| + |FQ| = \sqrt{(p-x_1)^2 + 4px_1} + \sqrt{(p-x_2)^2 + 4px_2}$

$$= \sqrt{(p^2 + 2px_1 + x_1^2)} + \sqrt{(p^2 + 2px_2 + x_2^2)}$$

$$= \sqrt{(p + x_1)^2} + \sqrt{(p + x_2)^2}$$

$$= (p + x_1) + (p + x_2) = x_1 + x_2 + 2p.$$

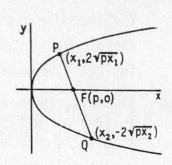

Therefore L = p + p + 2p = 4p. [Could also use the result of Problem 37 to get this.]

41. $\frac{dy}{dx} = \frac{\delta x}{H}$, so $y = \frac{\delta x^2}{2H} + C$. $y(0) = 0 \Rightarrow C = 0$. Therefore, $y = \frac{\delta x^2}{2H}$. Thus, the loaded hanging cable hangs in the shape of a parabola. The unloaded hanging cable (of Problem 29, Section 7.8) hangs in the shape of a catenary.

43. $f(x) = .25*x^2 + 2$

$f(x) = 1$

$f(x) = 3$

$f(x) = x+1$

$f(x) = -x+1$

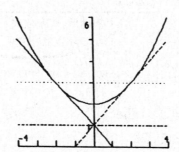

Problem Set 12.2   Ellipses and Hyperbolas

1. Vertical Ellipse.

3. Vertical hyperbola. The plus sign is associated with the y^2 term.

5. $x^2 = (4/9)y$. Vertical parabola opening upward.

7. $\frac{x^2}{1} + \frac{y^2}{(4/9)} = 1$. Horizontal ellipse.

9. Vertical ellipse since the divisor of
    y^2 is larger than the divisor of x^2.

    $a = 4$, $b = 3$, $c = \sqrt{16-9} \approx 2.65$.
    Vertices: $(0,\pm 4)$.
    Foci $(0,\pm 2.65)$.

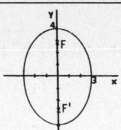

11. Vertical hyperbola.
    $a = 3$, $b = 4$, $c = 5$.
    Vertices: $(0,\pm 3)$.
    Foci: $(0,\pm 5)$.
    Asymptotes: $y = \pm(3/4)x$.

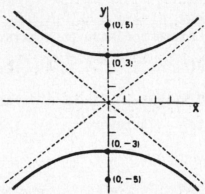

13. $\dfrac{x^2}{36} + \dfrac{y^2}{9} = 1$. Horizontal Ellipse.

    $a = 6$, $b = 3$, $c = \sqrt{27} \approx 5.20$.
    Vertices: $(\pm 6,0)$.
    Foci: $(\pm 5.20,0)$.

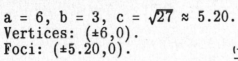

15. $\dfrac{x^2}{25} - \dfrac{y^2}{4} = 1$.   Horizontal hyperbola

    since the plus sign is with the x^2

    term. $a=5$, $b=2$, $c=\sqrt{25+4} \approx 5.39$.
    Vertices: $(\pm 5,0)$.
    Foci: $(\pm 5.39,0)$.
    Asymptotes: $y = \pm(2/5)x$.

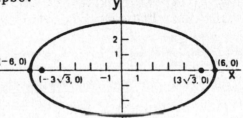

17. $a = 6$ and $c=3$, so $b^2 = 27$. $\dfrac{x^2}{36} + \dfrac{y^2}{27} = 1$.

19. $c=5$ and $e = 1/3$, so $a = 15$ and then $b^2 = 200$. $\dfrac{x^2}{200} + \dfrac{y^2}{225} = 1$.

21. Horizontal ellipse since the vertices are on the x-axis. $a=5$. Then the

    equation has the form $\dfrac{x^2}{25} + \dfrac{y^2}{b^2} = 1$ so $\dfrac{4}{25} + \dfrac{9}{b^2} = 1$. Thus, $b^2 = \dfrac{225}{21}$. $\dfrac{x^2}{25} +$

    $\dfrac{y^2}{225/21} = 1$ or $\dfrac{x^2}{25} + \dfrac{21y^2}{225} = 1$.

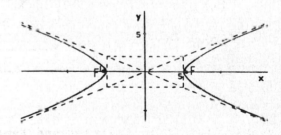

317

23. a=4 and c=5, so $b^2 = 9$. (Vertical hyperbola) $\dfrac{-x^2}{9} + \dfrac{y^2}{16} = 1$.

25. a=8 and $y = \pm(1/2)x$, so b = 4. (Horizontal hyperbola) $\dfrac{x^2}{64} - \dfrac{y^2}{16} = 1$.

27. Horizontal ellipse since foci are on x-axis. $\dfrac{x^2}{16} + \dfrac{y^2}{12} = 1$.

c = 2, $a^2 = 8c = 16$, $b^2 = a^2 - c^2 = 16 - 4 = 12$.

29. Asymptotes are $y = \pm(1/2)x$. It is a vertical asymptote since (4,3) is above both asymptotes. Therefore, a/b = 1/2, so b = 2a or $b^2 = 4a^2$.

Equation then has the form $\dfrac{-x^2}{4a^2} + \dfrac{y^2}{a^2} = 1$.

At (4,3): $\dfrac{-16}{4a^2} + \dfrac{9}{a^2} = 1$, so $a^2 = 5$. Then $b^2 = 20$. $\dfrac{-x^2}{20} + \dfrac{y^2}{5} = 1$.

31. a=5 and b=4, so $\dfrac{x^2}{25} + \dfrac{y^2}{16} = 1$. If y = 2, $x = \sqrt{75}/2$.

Therefore, the box can be $\sqrt{75} \approx 8.66$ ft wide.

33. $\dfrac{x^2}{a^2} + \dfrac{y^2}{b^2} = 1$, so $y^2 = \dfrac{b^2(a^2 - x^2)}{a^2}$.

At the focus, x=c, so $y^2 = \dfrac{b^2(a^2 - c^2)}{a^2} = \dfrac{b^2(b^2)}{a^2}$, $y = \dfrac{b^2}{a}$, $2y = \dfrac{2b^2}{a}$.

35. a = 18.09 AU, b = 4.56 AU, so $a - c = a - (a^2 - b^2)^{1/2} \approx 0.584$ AU.

37. a-c = 4132 and a+c = 4583, so a = 4357.5 and c = 225.5.
Therefore, e = c/a = 2255/43575 $\approx$ 0.05175.

39. $\sqrt{pq} = \sqrt{|A'F||FA|}$
$= \sqrt{(a-c)(a+c)}$
$= \sqrt{a^2 - c^2} = \sqrt{b^2} = b$.

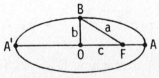

41. In terms of θ, the coordinates of P are $(a\cos\theta, b\sin\theta)$, so $x = a\cos\theta$, $y = b\sin\theta$. Therefore, $\dfrac{x^2}{a^2} + \dfrac{y^2}{b^2} = \cos^2\theta + \sin^2\theta = 1$.

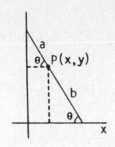

43. Let $\dfrac{x^2}{a^2} - \dfrac{y^2}{b^2} = 1$ and $\dfrac{y^2}{A^2} - \dfrac{x^2}{B^2} = 1$ be the hyperbolas. The asymptotes are

$y = \pm(b/a)x$ and $y = \pm(A/B)x$, so $b/a = A/B$. Thus, $\dfrac{a\sqrt{1-e^2}}{a} = \dfrac{A}{A\sqrt{1-E^2}}$;

$(1-e^2)(1-E^2) = 1$; $e^2 + E^2 = e^2 E^2$; $E^{-2} + e^{-2} = 1$.

45. ellipse: $a < 0$

parallel lines: $a = 0$

hyperbola: $a > 0$

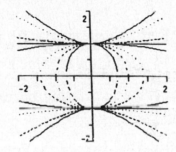

Problem Set 12.3 More on Ellipses and Hyperbolas

1. Vertical ellipse. $a = 10/2 = 5$ and $c = 4$, so $b^2 = 9$. $\dfrac{x^2}{9} + \dfrac{y^2}{25} = 1$.

3. Horizontal hyperbola. $a = 8/2 = 4$, $c = 5$, $b^2 = c^2 - a^2 = 25 - 16 = 9$. $\dfrac{x^2}{16} - \dfrac{y^2}{9} = 1$.

5. $\dfrac{3y}{27} + \dfrac{\sqrt{6}x}{9} = 1$. [Then simplify.]

7. $\dfrac{-4x}{8} - \dfrac{2y}{4} = 1.$

9. $\dfrac{x^2}{25} + \dfrac{y^2}{25} = 1.$ Circle. [Can be regarded as ellipse with a = b.]

   $a^2 = b^2 = 25, \quad x_0 = 3, \quad y_0 = 4. \quad \dfrac{3x}{25} + \dfrac{4y}{25} = 1; \quad y = \dfrac{-3}{4}x + \dfrac{25}{4}.$

11. Since the point is on the y-axis, it must be a vertex, so the tangent is horizontal and its equation is y = 5.

13. Since (0,6) is on the tangent $9(0)x_0 + 4(6)y_0 = 36$; then $y_0 = 3/2$. The points of the ellipse at which y = 3/2 are where $x = \pm\sqrt{3}$, so the points of tangency are $(\pm\sqrt{3}, 3/2)$.

15. $\dfrac{x^2}{35/2} - \dfrac{y^2}{5} = 1.$ Slope is $\dfrac{-2}{3} = \dfrac{5x_0}{(35/2)y_0}$, so $y_0 = (-3/7)x_0$.

    Then substituting into the equation of the hyperbola and solving for x_0,

    x_0, $x_0 = \pm 7$; $y_0 = (-3/7)(\pm 7) = \mp 3$. The points are (7,-3) and (-7,3).

17. In the first octant part, $y = (b/a)(a^2 - x^2)^{1/2}$. Therefore the area is

    $4\displaystyle\int_0^a (b/a)(a^2 - x^2)^{1/2}dx = 4(b/a)\left[(1/4)\pi a^2\right] = \pi ab.$

19. $y = (b/a)(x^2 - a^2)^{1/2}$ and $c = (a^2 + b^2)^{1/2}$. Then (by disks) the volume is

    $\displaystyle\int_a^{(a^2+b^2)^{1/2}} \pi\left[(b/a)(x^2-a^2)^{1/2}\right]^2 dx = \dfrac{\pi b^2}{a^2}\left[\dfrac{x^3}{3} - a^3 x\right]_a^{(a^2+b^2)^{1/2}}$

    $= \dfrac{\pi b^2\left[(a^2+b^2)^{1/2}(b^2 - 2a^2) + 2a^3\right]}{3a^2}.$

21. Let z denote the square of the area of the rectangle, and (x,y) denote its vertex in the first quadrant. The area will be maximum where z is maximum. Maximize $z = [(2x)(2y)]^2 = 16x^2y^2$.

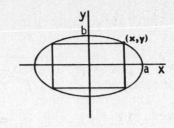

$$z(x) = 16x^2 \frac{b^2(a^2-x^2)}{a^2} = \frac{16b^2(a^2x^2-x^4)}{a^2} \text{ where x is in } [0,a].$$

$$z'(x) = \frac{32b^2x(a^2-2x^2)}{a^2} \text{ and } z''(x) = \frac{32b^2(a^2-6x^2)}{a^2}.$$

$z'(x) = 0$ if $x = a/\sqrt{2}$ and $z''(a/\sqrt{2}) < 0$, so z is maximum at $x = a/\sqrt{2}$. The corresponding value of y is $b/\sqrt{2}$, so the dimensions of the largest rectangle is $2a/\sqrt{2}$ by $2b/\sqrt{2}$ or $\sqrt{2}a$ by $\sqrt{2}b$.

23. Adding corresponding sides of the equations, we get $9y^2 = 675$. Then we obtain $x^2 = 36$. Therefore, the first quadrant point is $(6, 5\sqrt{3})$.

25.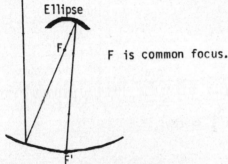

F is common focus.

27. It crosses the major axis between a focus and the far vertex.

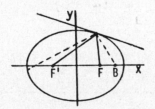

29. Tie a knot in a long piece of string so that the difference of the lengths of the two parts is 2a. Choose points F and F' such that the distance between F and F' is 2c where c > a. Attach the ends of the string at F and F'; slip the knot through a small ring and place a pencil point through the ring. Keep the point on the paper and keep the string taut by holding the knot above the surface. Move the pencil to trace out part of one branch of the hyperbola. Interchange ends of the string to draw part of the other branch.

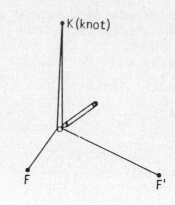

$$2a = (|F'P| + |PK|) - (|FP| + |PK|) = |F'P| - |FP|.$$

Since B and C heard the explosion at the same time, E is on the perpendicular bisector of BC. Thus, y = 5.

$t_A = t_B + 12$, so $\dfrac{d_A}{1/3} = \dfrac{d_B}{1/3} + 12$. Therefore, $d_A - d_B = 4$, so E is on a hyperbola with c = 8 and 2a = 4. Then $b^2 = (8)^2 - (2)^2 = 60$. Equation of the hyperbola is $\dfrac{x^2}{4} - \dfrac{y^2}{60} = 1$. When y = 5, $x = \sqrt{17/3}$; explosion occurred at $(\sqrt{17/3}, 5) \approx (2.38, 5)$.

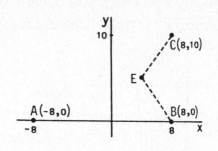

Problem Set 12.4   Translation of Axes

1. $(x-1)^2 + (y+2)^2 = 1$. Circle with center (1, -2) and radius 1.

3. $4(x^2 - 4x + 4) + 9(y^2 + 8y + 16) = -124 + 16 + 144.$

$4(x-2)^2 + 9(y+4)^2 = 36$ or $\dfrac{(x-2)^2}{9} + \dfrac{(y+4)^2}{4} = 1$. Horizontal ellipse; Center is at (2, -4).

5. $4(x-2)^2 + 9(y+4)^2 = 0$. Point (2, -4).

7. $(y-4)^2 = 10(x+3)$. Horizontal parabola opening to the right with vertex $(-3,4)$.

9. $(x^2-2x+1) + (y^2+4y+4) = -20+1+4$; $(x-1)^2 + (y+2)^2 = -15$. Empty set.

11. $(x+2)^2 - (y + 5/2)^2 = 0$. Intersecting lines $y + 5/2 = \pm (x+2)$.

13. $(2x-3)(2x-5) = 0$. Parallel lines $x = 3/2$ and $x = 5/2$.

15. $25(x^2+6x+9) - 4(y^2+2y+1) = -129 + 225 - 4$; $25(x+3)^2 - 4(y+1)^2 = 92$; $\dfrac{(x+3)^2}{95/25} - \dfrac{(y+1)^2}{23} = 1$. Horizontal hyperbola with center at $(-3,-1)$.

17.

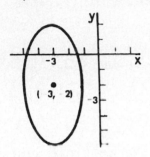

19.

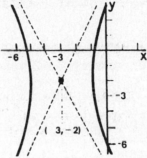

21.

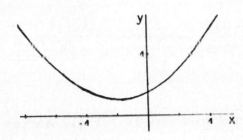

23. $y-1 = \pm4$.
     $y=5$, $y=-3$.

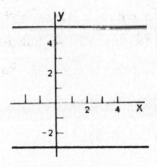

25. $\dfrac{(x-1)^2}{16} + \dfrac{(y+2)^2}{4} = 1$.

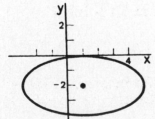

27. $\dfrac{(x+3)^2}{16} - \dfrac{(y-2)^2}{9} = 1$.

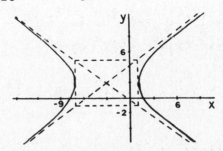

29. $(x+2)^2 = 4(y-1)$.

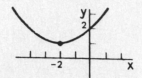

31. $(y-1)^2 = 4(5/4)(x + 1/5)$. Vertex is $(-1/5,1)$ and $p = 5/4$.
Focus is $(-1/5 + 5/4,1) = (21/20,1)$
Directrix is $x = -1/5 - 5/4$, $x = -29/20$.

33. $\dfrac{(x-1)^2}{25} + \dfrac{(y+2)^2}{16} = 1$ is a horizontal ellipse with center at $(1,-2)$.

$a = 5$, $b = 4$, $c^2 = a^2-b^2 = 25-16 = 9$, so $c = 3$.
Then the foci are $(1+3,-2)$ and $(1-3,-2)$; i.e., $(4,-2)$ and $(-2,-2)$.

35. $\dfrac{(x-5)^2}{25} + \dfrac{(y-1)^2}{16} = 1$.

37. Vertical parabola (since vertex and focus are on the vertical line $x=2$)
opening upward (since focus is above vertex), with $p = 5-3 = 2$.
$(x-2)^2 = 4(2)(y-3)$.

39. Vertical hyperbola since vertices and focus are on vertical line $x = 0$.
Center is $(0,3)$, $a = 6-3=3$, $c = 8-3 = 5$, so $b^2 = c^2-a^2 = 25-9 = 16$.
$\dfrac{-x^2}{16} + \dfrac{(y+3)^2}{9} = 1$.

41. Parabola is horizontal opening to the left and $p = (10-2)/2 = 4$. Vertex
is $(2+4,5) = (6,5)$. $(y-5)^2 = -4(4)(x-6)$.

43. Horizontal ellipse with center $(0,2)$ and $c=2$. $\dfrac{x^2}{a^2} + \dfrac{(y-2)^2}{b^2} = 1$.

At $(0,4)$: $0 + 4/b^2 = 1$, so $b^2 = 4$. Then $a^2 = b^2+c^2 = 4+4 = 8$.
$\dfrac{x^2}{8} + \dfrac{(y-2)^2}{4} = 1$.

45. (a) $y = ax^2 + bx + c$

$(0,0)$ on curve $\Rightarrow c = 0$

$(-1,2)$ on curve $\Rightarrow 2 = a - b$ ⎤
$\Rightarrow a = 1, b = -1$
$(3,6)$ on curve $\Rightarrow 6 = 9a + 3b$ ⎦

Equation of vertical parabola: $y = x^2 - x$

(b) $x = ay^2 + by + c$

$(0,0)$ on curve $\Rightarrow c = 0$

$(-1,2)$ on curve $\Rightarrow -1 = 4a + 2b$ ⎤
$\Rightarrow a = 1/4, b = -1$
$(3,6)$ on curve $\Rightarrow 3 = 36a + 6b$ ⎦

Equation of horizontal parabola: $x = \frac{1}{4}y^2 - y$

(c) $x^2 + y^2 + ax + by + c = 0$

$(0,0)$ on curve $\Rightarrow c = 0$

$(-1,2)$ on curve $\Rightarrow 5 - a + 2b = 0$ ⎤
$\Rightarrow a = -5, b = -5$
$(3,6)$ on curve $\Rightarrow 45 + 3a + 6b = 0$ ⎦

Equation of circle: $x^2 + y^2 - 5x - 5y = 0$

47. $K = 0$: $y^2 = Lx$, parabola. Latus rectum is $|L|$ [Section 12.1, No. 33]

$K > 0$: $y^2 + Kx^2 - Lx = 0$, ellipse.

The standard form is $\dfrac{(x - L/2K)^2}{L^2/4K^2} + \dfrac{y^2}{+L^2/4K} = 1$.

Latus rectum is $\dfrac{2b^2}{a} = \dfrac{-2L^2/4K}{|L/2K|} = |L|$ [Section 12.2, No. 33].

$K < 0$: $y^2 + Kx^2 - Lx = 0$, hyperbola.

The standard form is $\dfrac{(x + L/2K)^2}{L^2/4K^2} + \dfrac{y^2}{L^2/4K} = 1$.

Latus rectum is $\dfrac{2b^2}{a} = \dfrac{-2L^2/4K}{|L/2K|} = |L|$ [Section 12.2, No. 34].

Problem Set 12.5   Rotation of Axes

1. $\cot 2\theta = 0$, so let $\theta = \pi/4$.
     $x = (\sqrt{2}/2)(u-v)$, $y = (\sqrt{2}/2)(u+v)$.
     $\dfrac{-u^2}{1} + \dfrac{v^2}{4} = 1$.

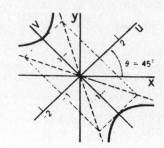

3. $\cot 2\theta = \dfrac{A-C}{B} = \dfrac{4-0}{-3} = \dfrac{-4}{3}$.

     $\cos 2\theta = \dfrac{-4}{5} = -0.8$; $\sin^2\theta = \dfrac{1-\cos 2\theta}{2} = 0.9$; $\cos^2\theta = 1 - \sin^2\theta = 0.1$.

     $\tan\theta = \dfrac{\sin\theta}{\cos\theta} = 3$.

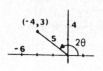

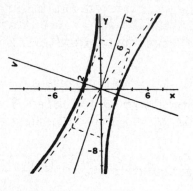

     $x = 0.1u - 0.9v = 0.1(u-3v)$.
     $y = 0.9u + 0.1v = 0.1(3u+v)$.
     $4(0.1)(u-3v)^2 - 3(0.1)(u-3v)(3u+v) = 18$.
     $-5u^2 + 45v^2 = 180$.
     $\dfrac{-u^2}{36} + \dfrac{v^2}{4} = 1$.

5. $\cot 2\theta = \sqrt{3}/3$, so let $\theta = \pi/6$.
     $x = (1/2)(\sqrt{3}u - v)$, $y = (1/2)(u + \sqrt{3}v)$.
     $v^2 = 6u$.

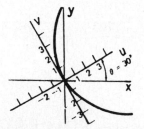

7. $\cot 2\theta = \sqrt{3}/3$.

     See write-up of Problem 5 for
     angle and equations of rotation.

     $\dfrac{u^2}{2} + \dfrac{v^2}{8} = 1$.

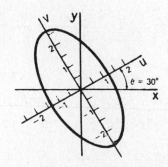

9. $\cot 2\theta = \dfrac{9-16}{-24} = \dfrac{7}{24}$.

$\cos 2\theta = 7/25 = 0.28$.

$\sin^2\theta = \dfrac{1-\cos 2\theta}{2} = 0.36$, $\cos^2\theta = 1-\sin^2\theta = 0.64$, $\tan\theta = \dfrac{\sin\theta}{\cos\theta} = \dfrac{3}{4}$.

$x = 0.8u - 0.6v = 0.2(4u-3v)$.
$y = 0.6u + 0.8v = 0.2(3u+4v)$.

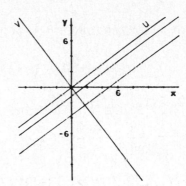

Substituting x and y into the given equation and simplifying yields $v^2+4v+3 = 0$; or $(v+1)(v+2) = 0$, or $v = -1$, $v = -3$ (parallel lines).

[We could have saved ourselves some work if we had realized this at the beginning since it means that the original equation can be expressed in a factored form. The factorization is $(3x-4y-5)(3x-4y-15) = 0$.]

11. $\cot 2\theta = -4/3$.

See write-up of Problem 3 for θ and equations of rotation.

$$\dfrac{(u-\sqrt{10})^2}{44} + \dfrac{(v-2\sqrt{10})^2}{4} = 1.$$

$$\dfrac{U^2}{44} + \dfrac{v^2}{2} = 1. \quad \text{[After translation]}$$

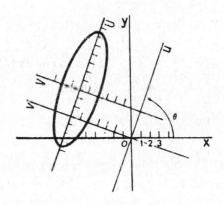

13. $x = (\cos a)u - (\sin a)v$, $y = (\sin a)u + (\cos a)v$.

$[(\cos^2 a)u - (\sin a \cos a)v] + [(\sin^2 a)u + (\sin a \cos a)v] = d$. $u = d$. The equation of the line in the uv-system is $u=d$, so its distance from the origin is $|d|$.

15. $x = (\cos\theta)u - (\sin\theta)v$; $y = (\sin\theta)u + (\cos\theta)v$. Therefore, $(\cos\theta)x = (\cos^2\theta)u - (\sin\theta\cos\theta)v$; $(\sin\theta)y = (\sin^2\theta)u + (\sin\theta\cos\theta)v$. $(\cos\theta)x + (\sin\theta)y = (\cos^2\theta+\sin^2\theta)u = u$ [adding corresponding sides].

Similarly, obtain $(-\sin\theta)x + (\cos\theta)y = v$.
That is, $u = (\cos\theta)x + (\sin\theta)y$, $v = (-\sin\theta)x + (\cos\theta)y$.

17. $\cot 2\theta = -24/7$. Let $\cos 2\theta = -24/25$.

Then $\sin\theta = (7\sqrt{2})/10$, $\cos\theta = \sqrt{2}/10$, $\tan\theta = 7$.

$x = (\sqrt{2}/10)(u-7v)$, $y = (\sqrt{2}/10)(7u+v)$. $u^2 = 2$. (parallel lines).

The points closest to the origin are $(\pm\sqrt{2},0)$ [uv-system]; $(\pm 1/5, \pm 7/5)$ [xy-system].

19. $b^2 - 4ac = [-2A\sin\theta\cos\theta + B\cos^2\theta - B\sin^2\theta + 2C\sin\theta\cos\theta]^2$

$$-4[A\cos^2\theta + B\sin\theta\cos\theta + C\sin^2\theta][A\sin^2\theta - B\sin\theta\cos\theta + C\cos^2\theta]$$

$$= 4A^2\sin^2\theta\cos^2\theta - 2AB\sin\theta\cos^3\theta + 2AB\sin^3\theta\cos\theta - 4AC\sin^2\theta\cos^2\theta$$

$$-2AB\sin\theta\cos^3\theta + B^2\cos^4\theta - B^2\sin^2\theta\cos^2\theta + 2BC\sin\theta\cos^3\theta$$

$$+2AB\sin^3\theta\cos\theta - B^2\sin^2\theta\cos^2\theta + B^2\sin^4\theta - 2BC\sin^3\theta\cos\theta$$

$$-4AC\sin^2\theta\cos^2\theta + 2BC\sin\theta\cos^3\theta - 2BC\sin^3\theta\cos\theta + 4C^2\sin^2\theta\cos^2\theta$$

$$= 4A^2\sin^2\theta\cos^2\theta - 4AB\sin\theta\cos^3\theta - 4AC\cos^4\theta$$
$$+ 4AB\sin^3\theta\cos\theta + 4B^2\sin^2\theta\cos^2\theta + 4BC\sin\theta\cos^3\theta$$
$$- 4AC\sin^4\theta + 4BC\sin^3\theta\cos\theta - 4C^2\sin^2\theta\cos^2\theta$$

$$= B^2\cos^4\theta - 4AC\cos^4\theta + 2B^2\sin^2\theta\cos^2\theta - 8AC\sin^2\theta\cos^2\theta + B^2\sin^4\theta - 4AC\sin^4\theta$$

$$= (B^2-4AC)\cos^4\theta + (B^2-4AC)\,2\sin^2\theta\cos^2\theta + (B^2-4AC)\sin^4\theta$$
$$= (B^2-4AC)(\cos^4\theta + 2\sin^2\theta\cos^2\theta + \sin^4\theta)$$
$$= (B^2-4AC)(\cos^2\theta + \sin^2\theta)^2 = B^2-4AC.$$

21. (a) From Problem 19, $B^2-4AC = 0^2-4ac \Rightarrow -\Delta = -4ac \Rightarrow \dfrac{1}{ac} = \dfrac{4}{\Delta}$.

(b) $\dfrac{1}{a} + \dfrac{1}{c} = \dfrac{a+c}{ac} = \dfrac{1}{ac}(a+c) = \dfrac{4}{\Delta}(A+C)$. [From Problem 18 and part (a)]

(c) $\dfrac{1}{a} + \dfrac{1}{c} = \dfrac{4(A+C)}{\Delta}$. [From (b)]

Therefore, $\dfrac{1}{a^2} + \dfrac{1}{ac} = \dfrac{4(A+C)}{a\Delta}$.   [Multiplying through by 1/a]

$$\left[\dfrac{1}{a}\right]^2 - \dfrac{4(A+C)}{\Delta}\left[\dfrac{1}{a}\right] + \dfrac{1}{ac} = 0.$$

$$\left[\dfrac{1}{a}\right]^2 - \dfrac{4(A+C)}{\Delta}\left[\dfrac{1}{a}\right] + \dfrac{4}{\Delta} = 0. \qquad \text{[From (a)]}$$

This is a quadratic equation in $\dfrac{1}{a}$. The same can be obtained for $\dfrac{1}{c}$.

Using the quadratic formula to solve the equation yields the result we are to obtain.

Let V^+ and V^- be the roots of the quadratic equation. Then there are three possibilities:

(1) 1/a and 1/c each equals V^+,
(2) 1/a and 1/c each equals V^-,
(3) 1/a is one of the values and 1/c is the other.

(1) and (2) would each require that a=c which is not a restriction in this problem. Therefore, (3) is the answer that is valid for the general case.

23. $B^2 - 4AC - B^2 \; 4$.

(a) Ellipse if $B^2 - 4 < 0$ and $B \ne 0$; i.e., if $|B| < 2$ and $B \ne 0$.

(b) Circle if $B = 0$.

(c) Hyperbola if $B^2 - 4 > 0$; i.e., if $|B| > 2$.

(d) Two parallel lines if $B^2 - 4 = 0$; i.e. if $|B| = 2$.

Problem Set 12.6   The Polar Coordinate System

1. A$(4, \pi/3)$
    B$(2, \pi/2)$
    C$(5, \pi/6)$
    D$(0, 11\pi)$
    E$(3, 3\pi/2)$
    F$(7/2, 4\pi/3)$
    G$(4, 0)$

3. A$(-5, \pi/4)$
    B$(5, -3\pi/4)$
    C$(2, -\pi/3)$
    D$(-2, 2\pi/3)$
    E$(-6, 0)$
    F$(3, -\pi)$
    G$(-4, -2\pi/3)$
    H$(-3, \pi)$

5. (a) $(4,7\pi/3)$, $(4,-5\pi/3)$, $(-4,4\pi/3)$, $(-4,-2\pi/3)$
    (b) $(-3,13\pi/4)$, $(-3,-3\pi/4)$, $(3,9\pi/4)$, $(3,\pi/4)$
    (c) $(5,13\pi/6)$, $(-5,-11\pi/6)$, $(5,7\pi/6)$, $(5,-5\pi/6)$
    (d) $(7,4\pi/3)$, $(7,-8\pi/3)$, $(-7,\pi/3)$, $(-7,-5\pi/3)$

[Points plotted below left.]

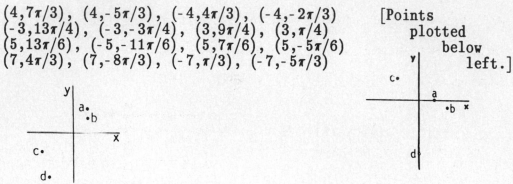

7. (a) $(2,2\sqrt{3})$ (b) $(3\sqrt{2}/2, 3\sqrt{2}/2)$ (c) $(-5\sqrt{3}/2, -2.5)$ (d) $(-3.5, -7\sqrt{3}/2)$

9. (a) The point is in the 3rd quadrant. $r^2 = 12+4 = 16$; $\tan\theta = 1\sqrt{3}$. $(-4,\pi/6)$ and $(4,7\pi/6)$ are two of the infinitely many possibilities.

    (b) The point is in the 1st quadrant. $r^2 = 1+3 = 4$; $\tan\theta = \sqrt{3}$. $(2,\pi/3)$ or $(-2,4\pi/3)$ will do.

    (c) The point is in the 4th quadrant. $r^2 = 2+2 = 4$; $\tan\theta = -1$. $(2,-\pi/4)$, $(-2,3\pi/4)$.

    (d) $(0,0)$, or more generally, $(0,k)$ for any k in $\mathbb{R}$.

11. $r\cos\theta - 4r\sin\theta + 2 = 0$. $r = 2/[4\sin\theta - \cos\theta]$.

13. $r\sin\theta = -5$. $r = -5\csc\theta$.

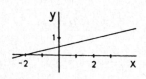

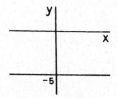

15. $r = 4$.

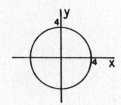

17. $y/x = \tan\theta = \tan(\pi/3) = \sqrt{3}$. $y = \sqrt{3}x$. (line)

19. x+6 = 0; x = -6. (line)

21. $r\sin\theta$ - 4 = 0; y - 4 = 0; y = 4.   (line)

23. Circle.

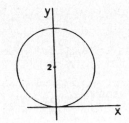

25. Line.

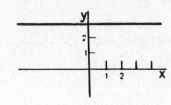

27. Circle.

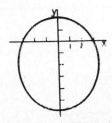

29. Parabola; e=1, θ_0=0, d=4.

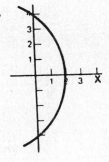

31. Ellipse; e=1/2, θ_0=0, d=6.

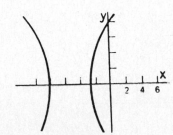

33. Parabola; e=1, θ_0=0, d=2.

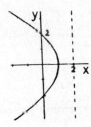

35. Hyperbola; e=2, θ_0=π, d=4.

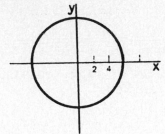

37. $(x - c \cos a)^2 + (y - c \sin a)^2 = a^2$.

$(x^2 + y^2) + c^2(\cos^2 a + \sin^2 a) - 2c(x\cos a + y\sin a) = a^2$.

$r^2 + c^2(1) - 2c(r\cos\theta \cos a + r\sin\theta \sin a) = a^2$.

$r^2 + c^2 - 2cr \cos(\theta - a) = a^2$.

39. Rotate to obtain $r = \dfrac{ed}{1 + e\cos\theta}$. Then the focus is at the pole and the directrix is along the y-axis, so calculate r for $\theta = \pi/2$, obtaining r = ed. Therefore, the length of the latus rectum is 2ed.

41. $\dfrac{ed}{1+e} = 17$ and $\dfrac{ed}{1-e} = 183$. [Problem 40]

Thus, ed = 17(1+e) and ed = 18(1-e), so 17(1+e) = 183(1-e); e = 0.83.

43. e=1 for parabola,

so polar equation is $r = \dfrac{d}{1+\cos\theta}$.

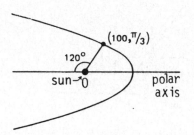

At $(100, \pi/3)$ $100 = \dfrac{d}{1 + 1/2}$, so d = 150.
Then when $\theta = 0$, $r = \dfrac{150}{1+1} = 75$, so the
closest the comet comes to the sun is
75 million miles.

45. (a) e = 0.1   (b) e = 0.5

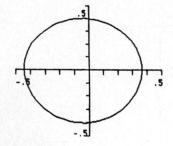

(c) e = 0.9   (d) e = 1

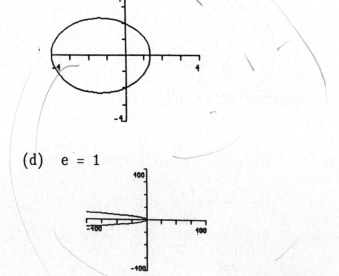

332

$r =$

(e)   e = 1.1   (f)   e = 1.3

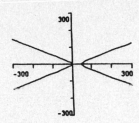   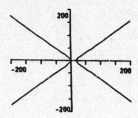

Problem Set 12.7   Graphs of Polar Equations

1. $\theta = \pm 1.$

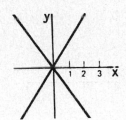

3. $r\sin\theta+6 = 0;\ y+6 = 0;\ y =-6.$

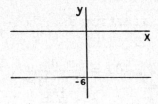

5. $r = 6\sin\theta.$

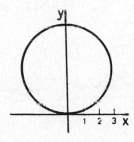

7. Parabola; e=1, d=4.

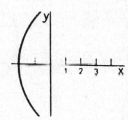

9. $r = 5-5\sin\theta.$ y-axis symmetry since $= 5-5\sin(\pi-\theta) = 5-5\sin\theta.$

r	θ
10	-90°
9.33	-60°
7.50	-30°
6.29	-15°
5	0°
3.71	15°
2.5	30°
0.67	60°
0	90°

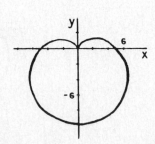

333

11. $r = 3 - 3\cos\theta$.

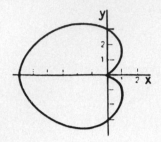

13. $r = 2 - 4\cos\theta$.

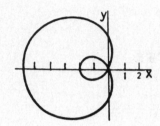

15. $r = 4 - 3\sin\theta$.   y-axis symmetry since $4 - 3 \times \sin(\pi - \theta) = 4 - 3\sin\theta$.

r	θ
7	-90°
6.60	-60°
5.50	-30°
4.77	-15°
4	0°
3.22	15°
2.50	30°
1.40	60°
1	90°

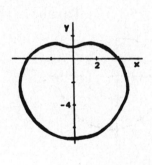

17. $r^2 = 9\sin2\theta$.

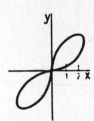

19. $r^2 = -16\cos2\theta$.

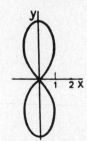

21. r=5cos3θ. x-axis symmetry since 5cos3(-θ) = 5cos(-3θ) = 5cos3θ.

r	θ
5	0°
4.83	5°
4.33	10°
3.54	15°
2.50	20°
1.29	25°
0	30°

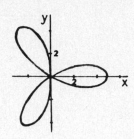

23. r = 6sin2θ.

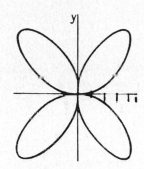

25. r = 7cos5θ.

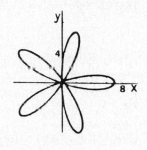

27. r = (1/2)θ, $\theta \geq 0$. [Note that θ is in radians.]

r	θ
0	0
$\pi/4$	$\pi/2$
$\pi/2$	π
$3\pi/4$	$3\pi/2$
π	2π
$5\pi/4$	$5\pi/2$
$3\pi/2$	3π
$7\pi/4$	$7\pi/2$
2π	4π

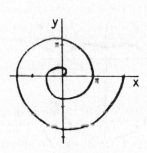

29. r = e$^{\theta}$, $\theta \geq 0$.

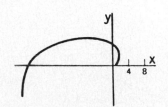

31. r = 2/θ, $\theta > 0$.

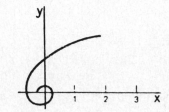

33. r = 6[circle]; r = 4+4cosθ[cardioid].

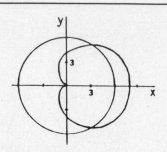

6 = 4+4cosθ.
2 = 4cosθ.
cosθ = 1/2.
θ = $\pm\pi/3$ and r = 6.

Points: (6,π/3) and (6,-π/3).

35. r = 3$\sqrt{3}$cosθ; r = 3sinθ [circles].

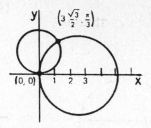

(π/3,3 3/2) is obtained
by simultaneous solutions
method. Another point of
intersection.

(0,0), $(\frac{3\sqrt{3}}{2}, \frac{\pi}{3})$

37. r = 6[circle]; r = 6/(1+2sinθ) [hyperbola].

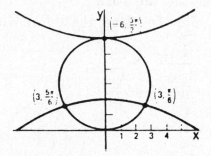

(6,0) and (6,π) are
obtained by simultaneous
solutions method. Another
point of intersection is
(6,π/2).

39. (*) r= cos$(\frac{2\theta}{3})$.

(1/2,π/2) satisfies (*) but (-1/2,-π/2) does not
(1,0) satisfies (*) but (1,π-0) does not.
However, there is symmetry with respect to the y-axis since, for all
(r,θ), replacing (r,θ) by(r,3π- θ) yields an equivalent equation.

41. $a^4 = |PF|^2|PF'|^2$

$= (a^2 + r^2 -2ar\cos\theta)(a^2 + r^2 - 2ar\cos\theta)$
[by Law of Cosines]

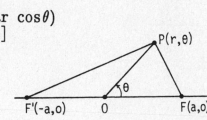

$= a^4 + 2a^2r^2 + r^4 - 4a^2r^2\cos^2\theta$

$= a^4 + r^4 + 2a^2r^2(1 - 2\cos^2\theta)$

$= a^4 + r^4 + 2a^2r^2\cos2\theta$

Thus, $r^2 = 2a^2\cos2\theta$, a lemniscate.

43. (a) b = 2   (b) b = 2.5

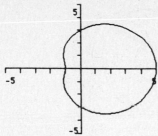   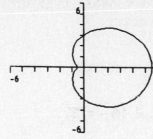

(c) b = 2.9   (d) b = 3
Does not have a sharp
corner.

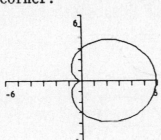   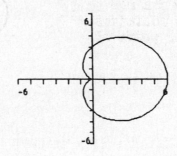

(e) b = 3.1   (f) b = 4
Has loop at the pole

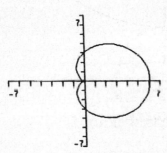   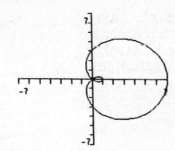

Problem Set 12.8   Calculus in Polar Coordinates

1. Circle with radius a, so area is πa^2.

3. $r = +3\cos\theta$ [limacon, a>b case].

Making use of symmetry,

$$\text{Area} = 2\int_0^\pi (1/2)(3+\cos\theta)^2 d\theta$$

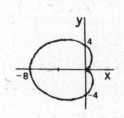

$$= \int_0^\pi \left(9 + 6\cos\theta + \frac{1+\cos2\theta}{2}\right)d\theta$$

$$= \left[9\theta + 6\sin\theta + \frac{\theta}{2} + \frac{\sin2\theta}{4}\right]_0^\pi = 9\pi + \frac{\pi}{2} = \frac{19\pi}{2} \approx 29.8451$$

5. $\text{Area} = 2\int_0^\pi (1/2)(4-4\cos\theta)^2 d\theta$

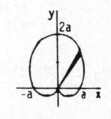

$$= \left[16\theta - 32\sin\theta + 8\theta + 4\sin2\theta\right]_0^\pi$$

$$= 24\pi \approx 75.3982.$$

7. $\text{Area} = 2\int_0^\pi (1/2)[a(1+\sin\theta)]^2 d\theta$

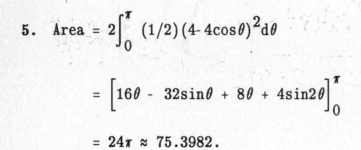

$$= a^2\left[\theta - 2\cos\theta + \theta/2 - (1/4)\sin2\theta\right]_0^\pi$$

$$= (3/2)\pi a^2.$$

9. $r^2 = 4\sin2\theta$ lemniscate].

Making use of symmetry,

$$\text{Area} = 4\int_0^{\pi/4} (1/2)4\sin2\theta \; d\theta$$

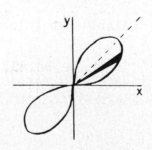

$$= 4\left[-\cos2\theta\right]_0^{\pi/4} = 4[0-(-1)] = 4.$$

11. Tangents at the pole are $\theta = \pm\pi/3$.

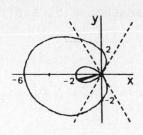

$$\text{Area} = 2\int_0^{\pi/3} (1/2)(2-4\cos\theta)^2 d\theta$$

$$= \left[4\theta - 16\sin\theta + 8\theta + 4\sin2\theta\right]_0^{\pi/3}$$

$$= 4\pi - 6\sqrt{3} \approx 2.1741.$$

13. Tangents at the pole are $\theta = \pm\pi/6$.

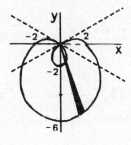

$$\text{Area} = 2\int_{-\pi/2}^{\pi/6} (1/2)(2-4\sin\theta)^2 d\theta$$

$$= \left[4\theta + 16\cos\theta + 8\theta - 4\sin2\theta\right]_{-\pi/2}^{\pi/6}$$

$$= 8\pi + 6\sqrt{3} \approx 35.5250.$$

15. $r = 4\cos3\theta$.
Tangents at pole: Let $r=0$. Then $\cos3\theta = 0$.
$3\theta = \pm\pi/2, \pm3\pi/2\cdots\cdot$
$\theta = \pm\pi/6, \pm\pi/2\cdots\cdot$

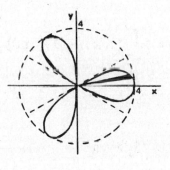

$$\text{Area} = 6\int_0^{\pi/6} (1/2)(4\cos3\theta)^2 d\theta$$

$$= 3\int_0^{\pi/6} (8 + 8\cos6\theta) d\theta$$

$$= 3\left[8\theta + \frac{8\sin6\theta}{6}\right]_0^{\pi/6} = 4\pi \approx 12.5664.$$

17. $\pi(10)^2 - \pi(7)^2 = 51\pi \approx 160.2212.$

19. One point of intersection is $(2,\pi/6)$.

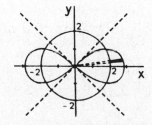

$$\text{Area} = 4\int_0^{\pi/6} (1/2)[(8\cos2\theta) - (2)^2] d\theta$$

$$= 2\left[4\sin2\theta - 4\theta\right]_0^{\pi/6} = 4\sqrt{3} - 4\pi/3 \approx 2.7394.$$

21. Point of intersection is where $\theta = \pi/4$.

$$\text{Area} = \int_0^{\pi/4} (1/2)\left[(3+3\cos\theta)^2 - (3+3\sin\theta)^2\right]d\theta$$

$$= \int_0^{\pi/4}\left[9\cos\theta - 9\sin\theta + \frac{9(1+\cos2\theta)}{4} - \frac{9(1-\cos2\theta)}{4}\right]d\theta$$

$$= \int_0^{\pi/4}\left(9\cos\theta - 9\sin\theta + \frac{9\cos2\theta}{2}\right)d\theta = \left[9\sin\theta + 9\cos\theta + \frac{9\sin2\theta}{4}\right]_0^{\pi/4}$$

$$= 9\sqrt{2} + \frac{9}{4} - 9 \approx 5.9779.$$

23. (a) Let $r = f(\theta) = 2\cos\theta$. Then $f'(\theta) = -2\sin\theta$.

$f(\pi/3) = 1$, $f'(\pi/3) = -\sqrt{3}$, $\sin(\pi/3) = \sqrt{3}/2$, $\cos(\pi/3) = 1/2$.

Slope at $\theta = \pi/3$ is $\dfrac{(1)(1/2) + (-\sqrt{3})(\sqrt{3}/2)}{-(1)(\sqrt{3}/2) + (-\sqrt{3})(1/2)} = \sqrt{3}/3 \approx 0.5774.$

(b) Let $r = f(\theta) = 1 + \sin\theta$. Then $f'(\theta) = \cos\theta$.

$f(\pi/3) = 1 + \sqrt{3}/2$, $f'(\pi/3) = 1/2$, $\sin(\pi/3) = \sqrt{3}/2$, $\cos(\pi/3) = 1/2$.

Slope at $\theta = \pi/3$ is $\dfrac{(1 + \sqrt{3}/2)(1/2) + (1/2)(\sqrt{3}/2)}{-(1 + \sqrt{3}/2)(\sqrt{3}/2) + (1/2)(1/2)} = -1.$

(c) Let $r = f(\theta) = \sin2\theta$. Then $f'(\theta) = 2\cos2\theta$.

$f(\pi/3) = \sqrt{3}/2$, $f'(\pi/3) = -1$, $\sin(\pi/3) = \sqrt{3}/2$, $\cos(\pi/3) = 1/2$.

Slope at $\theta = \pi/3$ is $\dfrac{(\sqrt{3}/2)(1/2) + (-1)(\sqrt{3}/2)}{-(\sqrt{3}/2)(\sqrt{3}/2) + (-1)(1/2)} = \sqrt{3}/5 \approx 0.3464.$

(d) Let $r = f(\theta) = 4 - 3\cos\theta$. Then $f'(\theta) = 3\sin\theta$.

$f(\pi/3) = 5/2$, $f'(\pi/3) = 3\sqrt{3}/2$, $\sin(\pi/3) = \sqrt{3}/2$, $\cos(\pi/3) = 1/2$.

Slope at $\theta = \pi/3$ is $\dfrac{(5/2)(1/2) + (3\sqrt{3}/2)(\sqrt{3}/2)}{-(5/2)(\sqrt{3}/2) + (3\sqrt{3}/2)(1/2)} = -7\sqrt{3}/3 \approx -4.0415.$

25. Let $f(\theta) = 1 - 2\sin\theta$. Then $f'(\theta) = -2\cos\theta$.

Slope at θ is $\dfrac{(1-2\sin\theta)\cos\theta + (-2\cos\theta)\sin\theta}{-(1-2\sin\theta)\sin\theta + (-2\cos\theta)\sin\theta} = \dfrac{(1-4\sin\theta)\cos\theta}{4\sin^2\theta - \sin\theta - 2}$.

Slope is 0 if $\sin\theta = 1/4$ or $\cos\theta = 0$, but denominator is not 0; i.e., if θ is $\sin^{-1}(1/4) \approx 0.2527$ [14.5°], $\pi - \sin^{-1}(1/4) \approx 2.8889$ [165.5°]

The points then are $(1/2, 0.2527)$, $(1/2, 2.8889)$, $(-1, \pi/2)$, $(3, 3\pi/2)$.

27. Let $f(\theta) = a(1+\cos\theta)$. Then $f'(\theta) = -a\sin\theta$. Making use of symmetry,

$$\text{Perimeter} = 2\int_0^\pi [a^2(1+\cos\theta)^2 + a^2\sin^2\theta]^{1/2}d\theta$$

$$= 2a\int_0^\pi (1 + 2\cos\theta + \cos^2\theta + \sin^2\theta)^{1/2}d\theta = 2a\int_0^\pi [2(1+\cos\theta)^{1/2}d\theta$$

$$= 2a\int_0^\pi [4\cos^2(\theta/2)]^{1/2}d\theta = 4a\int_0^\pi \cos(\theta/2)d\theta = 4a\Big[2\sin(\theta/2)\Big]_0^\pi = 8a$$

29. $\text{Area(1 petal)} = (1/2)\int_0^{\pi/n} a^2\cos^2(n\theta)d\theta = \pi a^2/4n$.

$$\text{Total area} = \begin{bmatrix} (1/4)\pi a^2, & \text{if } n \text{ is odd} \\ (1/2)\pi a^2, & \text{if } n \text{ is even} \end{bmatrix}.$$

31. (a) At Q, $\theta = \tan^{-1}(b/a)$.

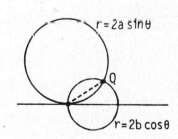

$$\text{Area} = (1/2)\int_0^{\tan-1(b/a)} 4a^2\sin^2\theta\, d\theta +$$

$$+ (1/2)\int_{\tan-1(b/a)}^{\pi/2} 4b^2\cos^2\theta\, d\theta$$

$$= (1/2)\pi b^2 - (a^2-b^2)\tan^{-1}(b/a) - ab.$$

(b) For r = f(θ) = 2asinθ , | For r = g(θ) = 2bcosθ,

$$\frac{dy}{dx} = \frac{2a\sin\theta\cos\theta + 2a\cos\theta\sin\theta}{-2a\sin^2\theta + 2a\cos^2\theta}$$ | $$\frac{dy}{dx} = \frac{2b\cos^2\theta - 2b\sin^2\theta}{-2b\cos\theta\sin\theta - 2b\sin\theta\cos\theta}$$

$$= \frac{2a\sin2\theta}{2a\cos2\theta} = \tan2\theta \ .$$ | $$= \frac{2b\cos2\theta}{-2b\sin2\theta} = -\cot2\theta.$$

Therefore, for each θ, the respective tangent lines are perpendicular. Hence, they are perpendicular at the point of intersection.

33.

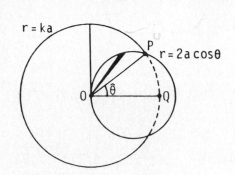   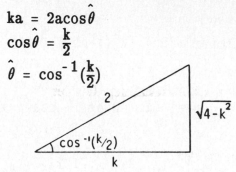

$$ka = 2a\cos\hat\theta$$
$$\cos\hat\theta = \frac{k}{2}$$
$$\hat\theta = \cos^{-1}\left(\frac{k}{2}\right)$$

$$2\text{Area (Shaded region)} = 2\int_{\cos^{-1}\left(\frac{k}{2}\right)}^{\pi/2} \frac{1}{2}(2a\cos\theta)^2 d\theta$$

$$= 4a^2\int_{\cos^{-1}\left(\frac{k}{2}\right)}^{\pi/2} \cos^2\theta\ d\theta = 2a^2\left[\theta + \sin\theta\ \cos\theta\right]_{\cos^{-1}\left(\frac{k}{2}\right)}^{\pi/2}$$

$$= 2a^2\left[\left[\frac{\pi}{2}\right] - \left[\cos^{-1}\left[\frac{k}{2}\right] + \frac{\sqrt{4-k^2}}{2}\ \frac{k}{2}\right]\right] = a^2\pi - 2a^2\cos^{-1}\left[\frac{k}{2}\right] - \frac{k\sqrt{4-k^2}}{2}\ .$$

Grazing area
  = Area(Circle of radius ka) - 2Area(Sector OPQ) - 2Area(Shaded region).

$$= \pi(ka)^2 - 2\left[\left[\frac{1}{2}\right](ka)\cos^{-1}\left[\frac{k}{2}\right]\right] - a^2\pi - 2a^2\cos^{-1}\left[\frac{k}{2}\right] - \frac{k\sqrt{4-k^2}}{2}$$

$$= a^2\left[\frac{\pi}{2}k^2 + (k^2-2)\sin^{-1}\left[\frac{k}{2}\right] + \frac{k\sqrt{4-k^2}}{2}\right]\ .$$

35. Set $a^2[(1/2)\pi k^2 + (1/3)k^3] = \pi a^2$; solve for k. $2k^3 + 3\pi k^2 - 6\pi = 0$. Let
$f(k) = 2k^3 + 3\pi k^2 - 6\pi$. Use Newton's Method to obtain $k \approx 1.2566$. Thus,
length of rope is ka $\approx (1.2566)a$.

37. 12.5664, 26.7298   **39.** L = 63.4618

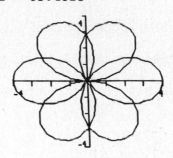

Problem Set 12.9   Chapter Review

True-False Quiz

1. False.   If a=0, it is not a parabola.

3. False. $e = \dfrac{\text{distance to focus}}{\text{distance to directrix}} < 1$, so focus is closer.

5. True. $y = \pm(b/a)x$.

7. True.

9. False.   It represents the two intersecting lines, $y = \pm x$.

11. True.   Horizontal hyperbola if k > 0; vertical hyperbola if k < 0.

13. True.   It is 2c and $c = (a^2 - b^2)^{1/2}$.

15. True.   See write-up of Problem 27, Section 12.3.

17. True.   Since the vertex is the point nearest the sun, the planet has to
travel farther in a given time period.

19. True.   The translation that eliminates the x and y terms yields $u^2 + v^2$
+ f = 0.   [Point for f = 0, circle for f < 0, empty set for
f > 0.]

21. **False.** C^{ex}: For B=C=E=0, A=1, D=-3, F=2 $[x^2-3x+2 = 0]$, the graph is two parallel lines, which can not be obtained from such an intersection. For A=B=C=D=E=F=0, it is the plane.

23. **False.** C^{ex}: The graph of xy=1 doesn't enter the 2nd or 4th.

25. **True.** See second box on page 553.

27. **False.** See Example 5, Section 12.6.

29. **True.** Apply test for symmetry given in the text. $f(-\theta) = f(\theta) = r$.

Sample Test Problems

1. a-5, b-9, c-4, d-3, e-2, f-8, g-8, h-1, i-7, j-6.

3. $\dfrac{x^2}{4} + \dfrac{y^2}{9} - 1$.

   Vertical ellipse with vertices $(0,\pm3)$.

   $c^2 = a^2-b^2 = 9-4 = 5$. Foci are $(0,\pm2.24)$.

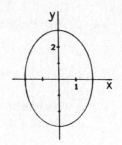

5. $x^2 = -4(9/4)y$.

   Vertical parabola opening down.

   Vertex is $(0,0)$.

   Focus is $(0,-9/4)$.

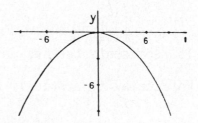

7. $\dfrac{x^2}{25} + \dfrac{y^2}{9} = 1$.

   Horizontal ellipse with vertices $(\pm5,0)$.

   $c^2 = a^2-b^2$. Foci are $(\pm4,0)$.

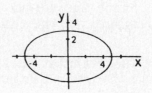

9. $r = \dfrac{(5/2)(1)}{1+1\sin\theta}$; $e = 1$, $d = \dfrac{5}{2}$.

   Vertical parabola opening downward
   with vertex $(5/4, \pi/2)$ and focus $(0,0)$.

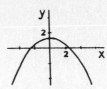

11. Horizontal ellipse. $a = 4$, $c = ae = 2$, $b^2 = a^2 - c^2 = 12$. $\dfrac{x^2}{16} + \dfrac{y^2}{12} = 1$.

13. Horizontal parabola opening left. $y^2 = -4px$. $(3)^2 = -4p(-1)$, $p = 9/4$.
    $y^2 = -4(9/4)x$ or $y^2 = -9x$.

15. Hyperbola (since there are asymptotes) that is horizontal (since the
    vertices are on the horizontal line $y = 0$); $a = 2$.

    $\dfrac{b}{a} = \dfrac{1}{2}$ (from asymptotes), so $b = 1$. An equation is $\dfrac{x^2}{4} - \dfrac{y^2}{1} = 1$.

17. Horizontal ellipse. $c = 4-1 = 3$. $2a = 10$, so $a = 5$, $b^2 = a^2 - c^2 = 16$.
    $\dfrac{(x-1)^2}{25} + \dfrac{(y-2)^2}{16} = 1$.

19. $(x-3)^2 + (y + 9/2)^2 = 0$.
    Circle.

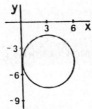

21. $(x^2 + 8x + 16) = -6y - 28 + 16$.

    $(x+4)^2 = -6(y+2)$.

    Vertical parabola opening downward.

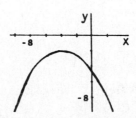

23. $x = (\sqrt{2}/2)(u-v)$, $y = (\sqrt{2}/2)(u+v)$

$(5/2)u^2 - (1/2)v^2 = 10$, so $r = 5/2$ and $s = -1/2$.

$$\frac{u^2}{4} - \frac{v^2}{20} = 1. \quad c^2 = a^2 + b^2 = 20 + 4 = 24.$$

Hyperbola

Distance between foci is $2c = 4\sqrt{6} \approx 9.7980$.

25. Circle.

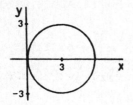

27. $r = \cos 2\theta$ is a four-leaved rose.

For tangents at the pole, let $r = 0$.
Then $2\theta = \pm\pi/2, \pm 3\pi/2, \cdots$, so
$\theta = \pm\pi/4, \pm 3\pi/4, \cdots$.

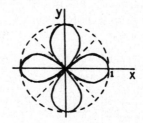

29. Circle.

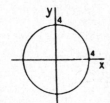

31. Horizontal limaçon
(a > b case).

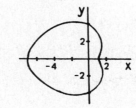

33. Line.

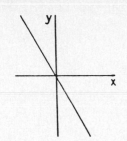

35. Lemniscate.

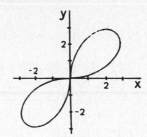

37. $r^2 - 6r\cos\theta - 6r\sin\theta + 9 = 0.$
$(x^2+y^2) - 6x - 6y + 9 = 0.$
$(x-3)^2 + (y-3)^2 = 9.$

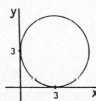

39. Let $r = f(\theta) = 3+3\cos\theta.$ Then $f'(\theta) = -3\sin\theta.$
$f(\pi/6) = 3+3(\sqrt{3}/2)$ and $f'(\pi/6) = -3(1/2) = -3/2.$ Then the slope of the
tangent where $\theta = \pi/6$ is $\dfrac{(3+3\sqrt{3}/2)(\sqrt{3}/2) + (-3/2)(1/2)}{-(3+3\sqrt{3}/2)(1/2) + (-3/2)(\sqrt{3}/2)}$

$$= \frac{6\sqrt{3} + 9 - 3}{6 \quad 3\sqrt{3} \quad 3\sqrt{3}} = -1.$$

41. Area $= 2\displaystyle\int_0^\pi (1/2)(5-5\cos\theta)^2 d\theta$

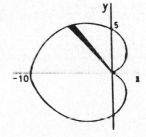

$$= 25\left[\theta - 2\sin\theta + \theta/2 + (1/4)\sin 2\theta\right]_0^\pi$$

$$= 75\pi/2 \approx 117.8097.$$

43. $y' = -x/4y.$

At $(16,6)$ on ellipse, $y' = -2/3.$

Equation of tangent is $(y-6) = (-2/3)(x-16),$
$y = (-2/3)x + (50/3).$

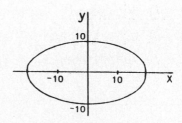

At $(14,k)$ on the tangent line,
$k = (-2/3)(14) + (50/3) = 22/3.$

Problem Set 13.1   Plane Curves: Parametric Representation

1. (b) Simple; not closed.
   (c) $t = x/2$, so $y = (3/2)x$.

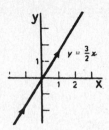

3. $x = t\text{-}4$, $y = t^{1/2}$, t in $[0,4]$.

   (a)

t	x	y
0	-4	0
1	-3	1
2	-2	1.41
3	-1	1.73
4	0	2

   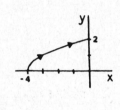

   (b) Simple, not closed.

   (c) $t = y^2$

   $x = y^2\text{-}4$, y in $[0,2]$.

5. (b) Simple, not closed.
   (c) $t = y^{1/3}$, so $x = y^{2/3}$,
   y in $[-1,8]$.

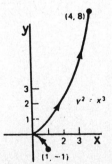

7. (b) Not simple; not closed.
   (c) $t^2 = y+4$ and $x^2 = t^2(t^2\text{-}4)^2$.
   Therefore, $x^2 = (y+4)y^2$.

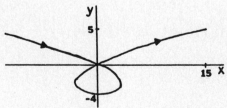

9. $x = 3\sin t$, $y = 5\cos t$, t in $[0, 2\pi]$.

(a)

t	x	y
0	0	5
$\pi/6$	1.50	4.33
$\pi/4$	2.12	3.54
$\pi/3$	2.60	2.50
$\pi/2$	3	0

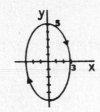

(b) Simple; closed.

(c) $x/3 = \sin t$, $y/3 = \cos t$. Therefore,

$$(x/3)^2 + (y/5)^2 = 1.$$

11. (b) Simple; not closed.

(c) $\sqrt{x} + \sqrt{y} = 2(\sin^2 t + \cos^2 t)$.

$\sqrt{x} + \sqrt{y} = 2$.

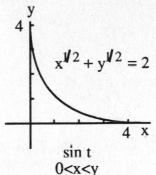

$x^{1/2} + y^{1/2} = 2$

$\sin t$
$0 < x < y$
$0 < y < 4$

13. $\dfrac{dy}{dx} = \dfrac{dy/dt}{dx/dt} = \dfrac{6t^2}{6t} = t$. $\dfrac{dy'}{dx} = \dfrac{dy'/dt}{dx/dt} = \dfrac{1}{6t}$.

15. $\dfrac{dy}{dx} = \dfrac{dy/dt}{dx/dt} = \dfrac{2 - 3t^{-2}}{2 + 3t^{-2}} = \dfrac{2t^2 - 3}{2t^2 + 3}$.

$\dfrac{d^2y}{dx^2} = \dfrac{dy'}{dx} = \dfrac{dy'/dt}{dx/dt} = \dfrac{[(2t^2 + 3)4t - (2t^2 - 3)4t]}{(2t^2 + 3)^2} \dfrac{t^2}{2t^2 + 3} = \dfrac{24t^3}{(2t^2 + 3)^3}$.

17. $\dfrac{dy}{dx} = \dfrac{5\sec t \tan t}{3\sec^2 t} = (5/3)\sin t$. $\dfrac{dy'}{dx} = \dfrac{(5/3)\cos t}{3\sec^2 t} = (5/9)\cos^3 t$.

19. $\dfrac{dy}{dx} = \dfrac{3t^2}{2t} = (3/2)t$.

If $t = 2$, then $x = 4$, $y = 8$, and $y' = 3$.
Tangent is $y - 8 = 3(x - 4)$ or $y = 3x - 4$.

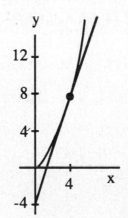

21. $\dfrac{dy}{dx} = \dfrac{dy/dt}{dx/dt} = \dfrac{2\sec^2 t}{2\sec t \, \tan t} = \csc t.$

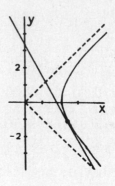

At $t = \dfrac{-\pi}{6}$ $x = \dfrac{4\sqrt{3}}{3}$, $y = \dfrac{-2\sqrt{3}}{3}$, $y' = -2$.

Equation of Tangent: $y + \dfrac{-2\sqrt{3}}{3} = -2(x - \dfrac{4\sqrt{3}}{3})$

or $y = -2x + 2\sqrt{3}$.

23. $\displaystyle\int_0^1 (x^2 - 4y)\,dx = \int_{-1}^0 [(t+1)^2 - 4(t^2 + 4)]\,dt$

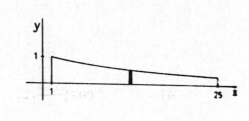

$$= \left[-t^4 + (1/3)t^3 + t^2 - 15t\right]_{-1}^0 = -44/3.$$

25. Area $= \displaystyle\int_1^{25} y\,dx$ $\qquad\qquad [dx = 2e^{2t}dy; \; x=25 \Rightarrow t=\ln 5; \; x=1 \Rightarrow t=0]$

$$= \int_0^{\ln 5} e^{-t} 2e^{2t}\,dt = \int_0^{\ln 5} 2e^t\,dt$$

$$= \left[2e^t\right]_0^{\ln 5} = 2e^{\ln 5} - 2e^0 = 10 - 2 = 8.$$

27. Add the points Q and T (as indicated to the right) to Figure 6 on text page 554. The coordinates of Q are:

$x = |ON| - |QT| = at - b\sin t.$

$y = |NC| + |CT| = a - b\cos t.$

For $x = 8t - 4\sin t$, $y = 8 - 4\cos t$:

t	x	y
0	0	4
$\pi/2$	$4\pi - 4$	8
π	8π	12
$3\pi/2$	$12\pi + 4$	8
2π	16π	4

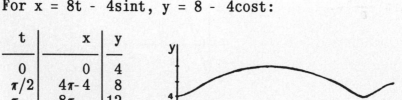

29. (a) $y = -16(v_0 \cos\alpha)^{-2}x^2 + (v_0 \sin\alpha)(v_0 \cos\alpha)^{-1}x$, which has the form of a

vertical parabola, $y = ax^2 + bx$, where a and b are constants. Note
that $a \neq 0$ unless $v_0 = 0$ (projectile doesn't move) or $\cos\alpha = 0$
(projectile goes straight up or down).

(b) Letting $y = 0$ and solving for t, $t = 0$ or $t = (1/16)v_0 \sin\alpha$, so the
time of the flight is $(1/16)v_0 \sin\alpha$.

(c) When $t = (1/16)v_0 \sin\alpha$, $x = (v_0 \cos\alpha)(1/16)v_0 \sin\alpha = (1/32)v_0^2 \sin 2\alpha$.

(d) Range is maximum if $\sin 2\alpha$ is maximum; i.e., if $2\alpha = \pi/2$ or $\alpha = \pi/4$.

31. See Figure at the right. Note:
   (1) $b\beta$ (length of arc BP) = at (length of arc AB),
   (2) Hypotenuse of triangle ONC is a-b.

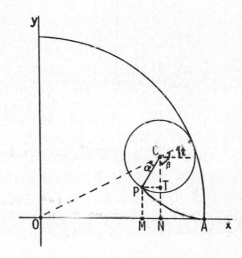

$x = |OM| = |ON| - |TP|$

$= (a-b)\cos t - b\sin(\beta - \pi/2 - t)$

$= (a-b)\cos t + b\cos(\beta - t)$

$= (a-b)\cos t + b\cos\left(\dfrac{at}{b} - t\right)$

$= (a-b)\cos t + b\cos\left(\dfrac{a-b}{b}\right)t.$

$y = |MP| = |NC| - |CT|$

$= (a-b)\sin t - b\cos(\beta - \pi/2 - t)$

$= (a-b)\sin t - b\sin(\beta - t)$

$= (a-b)\sin t - b\sin\left(\dfrac{at}{b} - t\right)$

$= (a-b)\sin t - b\sin\left(\dfrac{a-b}{b}\right)t.$

33. Use the figure below to help obtain the coordinates of P. Note that:

      (1) $b\beta$ (length of arc BP) = at (length of arc AB), so β = at/b.

      (2) Therefore, $\beta + t = \dfrac{at}{b} + t = (\dfrac{a+b}{b})t$.

      (3) The hypotenuse of triangle ONC is a+b.

$$x = |OM| = |ON| + |TP| \qquad\qquad y = |MP| = |NC| - |CT|$$
$$= (a+b)\cos t + b\sin[\beta - (\tfrac{\pi}{2} - t)] \qquad = (a+b)\sin t - b\cos[\beta - (\tfrac{\pi}{2} - t)]$$
$$= (a+b)\cos t - b\cos(\beta + t) \qquad\qquad = (a+b)\sin t - b\sin(\beta + t)$$
$$= (a+b)\cos t - b\cos(\tfrac{a+b}{b})t. \qquad\qquad = (a+b)\sin t - b\cos(\tfrac{a+b}{b})t.$$

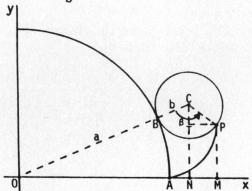

35. (a) $x = |ON| = 2a\cot\theta$ (since $\cot\theta = \dfrac{|ON|}{|AN|} = \dfrac{|ON|}{2a}$).

Now solve the equation of the circle and the equation of the line OA simultaneously to obtain the nonzero y-coordinate of a point of intersection:

$$x^2 + (y-a)^2 = a^2 \text{ [circle] and } y = (\tan\theta)x \text{ [line OA].}$$
$$x^2 = (\cot^2\theta)y^2.$$
$$(\cot^2\theta)y^2 + (y-a)^2 = a^2, \quad (\cot^2\theta + 1)y^2 - 2ay = 0,$$
$$(\csc^2\theta)y^2 - 2ay = 0, \quad y[(\csc^2\theta)y - 2a] = 0. \text{ Finally, } y = 2a\sin^2\theta.$$

(b) $\begin{bmatrix} x = 2a\cot\theta \\ y = 2a\sin^2\theta \end{bmatrix}$ so $\begin{bmatrix} x^2 = 4a^2\cot^2\theta \\ \csc^2\theta = 2a/y \end{bmatrix}$.

Then, $x^2 = 4a^2\cot^2\theta = 4a^2(\csc^2\theta - 1) = 4a^2(2a/y - 1) = 8a^3/y - 4a^2$.

Therefore, $y = 8a^3/(x^2 + 4a^2)$.

37. $x'(t) = -(a+b)\sin t + (a+b)\sin(\frac{a+b}{b})t$.

$y'(t) = (a+b)\cos t - (a+b)\cos(\frac{a+b}{b})t$.

$[x'(t)]^2 + [y'(t)]^2 = 4(a+b)^2\sin^2(a/2b)t$. One loop is completed when $ta = 2\pi b$ (circumference of the circle), i.e., when $t = 2\pi b/a$. There will be $2\pi/(2\pi b/a) = a/b$ loops.

Thus, $L = (a/b)\int_0^{2\pi b/a} 2(a+b)\sin(a/2b)t\ dt = 8(a+b)$.

39. (a)
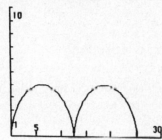

(b)
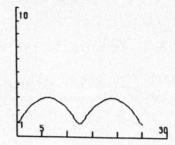

(c)
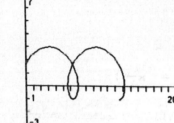

41. Conjectures: Same as in Problem 40.

I. $(0,\infty)$
II. $(-1,0)$
III.   None
IV. $(-\infty,-1)$

(a) a=2
     b=2
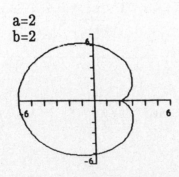

(b) a=3
     b=1
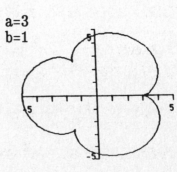

(c) a=5
     b=2

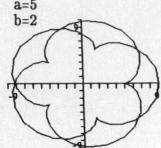

(d) a=7
     b=4

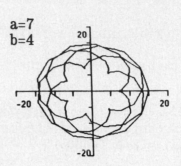

Problem Set 13.2 Vectors in the Plane: Geometric Approach

1.

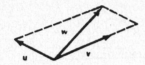

3.

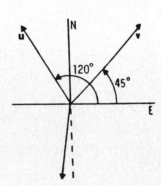

5. $\mathbf{w} = (1/2)\,(\mathbf{u} + \mathbf{v})$

7. Consider vertical components.
    $|\mathbf{w}| = |\mathbf{u}|\sin 150° + |\mathbf{v}|\sin 30° = 1/2 + 1/2 = 1.$

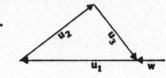

9. $\mathbf{u}$ and $\mathbf{v}$ together have a N-S force of
    $|\mathbf{v}|\sin(45°) + |\mathbf{u}|\sin(120°)$

    $= 10\sqrt{2}/2 + 10\sqrt{3}/2 \approx 15.73;$
    E-W force of $|\mathbf{v}|\cos(45°) + |\mathbf{u}|\cos(120°)$

    $= 10\sqrt{2}/2 + 10(-1/2) \approx 2.07.$

    $a = \tan^{-1}(2.07/15.73) \approx 7.50\text{o}.$
    $[(2.07)^2 + (15.73)^2]^{1/2} \approx 15.87$

    Thus, $\mathbf{w}$ needs to have direction S7.50°W
    and a magnitude of 15.87 lbs.

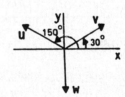

11. $|\mathbf{v}| = (1/2)|\mathbf{w}| = 125$ newtons.

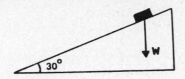

13. Need $58\cos110° + 425\cos a = 0$; $\cos a = \dfrac{\text{-}58\cos110°}{425}$ Therefore $a \approx 87.32°$, so the plane should fly N2.68°E. Ground speed will be $58\sin110° + 425\sin a \approx 479.04$ mph.

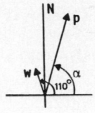

15. Magnitude of vertical components are equal. Therefore, $|\mathbf{p}|\sin(30°) = |\mathbf{w}| = 80$, so $|\mathbf{p}| = 80/\sin(30°) = 160$ mph.

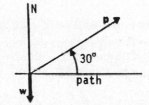

17. $\mathbf{u}+\mathbf{v} = \mathbf{w}$.

$\mathbf{z} = (1/2)\mathbf{u} + (1/2)\mathbf{v} = (1/2)(\mathbf{u}+\mathbf{v}) = (1/2)\mathbf{w}$.

Therefore, DE is parallel to AC.

(This also showed that the length of DE is half the length of AC.)

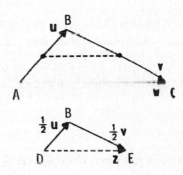

19. $\mathbf{v}_1 + \mathbf{v}_2 = \cdots + \mathbf{v}_n = P_1P_2 + P_2P_3 + \cdots + P_nP_1$ [Tail of $\mathbf{v}_1$ is p_i.]
$= P_1P_1 = 0.$

21. Coordinitize the line containing PA so that P is the origin. Do the same for the lines containing PB and PC. Then for each line the components in the direction of that line must balance. That is:

$$(1) \quad w\cos\alpha + w\cos\beta = w,$$
$$(2) \quad w\cos\beta + w\cos\gamma = w,$$
$$(3) \quad w\cos\alpha + w\cos\gamma = w.$$

Thus, $(1')$ $\cos\alpha + \cos\beta = 1$, $(2')$ $\cos\beta + \cos\gamma = 1$, $(3')$ $\cos\alpha + \cos\gamma = 1$.

Hence, $\cos\alpha = \cos\beta = \cos\gamma = \frac{1}{2}$; and so $\alpha = \beta = \gamma$, since $\alpha+\beta+\gamma = \pi$.

23. [See write up for Problem 21.]
$$(1) \quad 5w\cos\alpha + 4w\cos\beta = 3w,$$
$$(2) \quad 3w\cos\beta + 5w\cos\gamma = 4w,$$
$$(3) \quad 3w\cos\alpha + 4w\cos\gamma = 5w.$$

Thus, $5\cos\alpha + 4\cos\beta = 3$, $3\cos\beta + 5\cos\gamma = 4$, and $3\cos\alpha + 4\cos\gamma = 5$.

Therefore, $\cos\alpha = 3/5$, $\cos\beta = 0$, $\cos\gamma = 4/5$, from which it follows that $\sin\alpha = 4/5$, $\sin\beta = 1$, $\sin\gamma = 3/5$.

Therefore, $\cos(\alpha+\beta) = -4/5$, $\cos(\alpha+\gamma) = 0$, $\cos(\beta+\gamma) = -3/5$, so $\alpha+\beta = \cos^{-1}(-4/5) \approx 143.13°$, $\alpha+\gamma = 90°$, $\beta+\gamma = \cos^{-1}(-3/5) \approx 126.87°$.

This problem can be modeled with three strings going through A, four strings through B, and five strings through C, with equal weights attached to the twelve strings. Then the quantity to be minimized is $3|AP| + 4|BP| + 5|CP|$.

Problem Set 13.3 Vectors in the Plane: Algebraic Approach

1. (a) $-14i+20j$. (b) -18. (c) -38.
   (d) 425. (e) -100. (f) $13 - \sqrt{13}$.

3. (a) $\cos\theta = \dfrac{\langle 2,-3\rangle \cdot \langle -1,4\rangle}{|\langle 2,-3\rangle||\langle -1,4\rangle|} = \dfrac{-2-12}{\sqrt{4+9}\sqrt{1+16}} = \dfrac{-14}{\sqrt{221}} \quad -0.9417.$

   (b) $\cos\theta = \dfrac{\langle -5,-2\rangle \cdot \langle 6,0\rangle}{|\langle -5,-2\rangle||\langle 6,0\rangle|} = \dfrac{-30+0}{\sqrt{25+4}\sqrt{36+0}} = \dfrac{-5}{\sqrt{29}} \quad -0.9285.$

   (c) $\cos\theta = \dfrac{\langle -3,-1\rangle \cdot \langle -2,-4\rangle}{|\langle -3,-1\rangle||\langle -2,-4\rangle|} = \dfrac{6+4}{\sqrt{9+1}\sqrt{4+16}} = \dfrac{10}{200} \quad 0.7071.$

(d) $\cos\theta = \dfrac{\langle 4,-5\rangle \cdot \langle -8,10\rangle}{|\langle 4,-5\rangle||\langle -8,10\rangle|} = \dfrac{-32-50}{\sqrt{16+25}\,\sqrt{64+100}} = -1.$

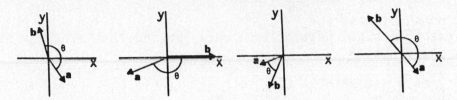

5. (a) $-5\mathbf{i} + 5\mathbf{j}.$ (b) $-6\mathbf{i} - 5\mathbf{j}.$

    (c) $-\sqrt{2}\mathbf{i} + e\mathbf{j}.$ (d) $3\mathbf{i} - (5/3)\mathbf{j}.$

7. Interpret $\mathbf{u}+\mathbf{v}$ as a diagonal of a parallelogram. Then $\mathbf{u}-\mathbf{v}$ is the other diagonal. The parallelogram is a rhombus since the diagonals are perpendicular. Therefore, the sides are of equal length, so $|\mathbf{u}| = |\mathbf{v}|$.

9. $a\mathbf{u}+b\mathbf{u} = a\langle u_1,u_2\rangle + b\langle u_1,u_2\rangle = \langle au_1,au_2\rangle + \langle bu_1,bu_2\rangle$

    $= \langle au_1+bu_1, au_2+bu_2\rangle = \langle (a+b)u_1,(a+b)u_2\rangle = (a+b)\langle u_1,u_2\rangle = (a+b)\mathbf{u}.$

11. $c(\mathbf{u}\cdot\mathbf{v}) = c(\langle u_1,u_2\rangle \cdot \langle v_1,v_2\rangle) = c(u_1 v_1 + u_2 v_2) = c(u_1 v_1) + c(u_2 v_2)$

    $= (cu_1)v_1 + (cu_2)v_2 = \langle cu_1,cu_2\rangle \cdot \langle v_1,v_2\rangle$

    $= (c\langle u_1,u_2\rangle) \cdot \langle v_1,v_2\rangle = (c\mathbf{u})\cdot\mathbf{v}.$

13. $3(3\mathbf{i}-4\mathbf{j}) = 9\mathbf{i}-12\mathbf{j}.$

15. $\langle 6,3\rangle \cdot \langle -1,2\rangle = -6+6 = 0$, so they are perpendicular.   [Theorem C]

17. $0 = \langle c,3\rangle \cdot \langle c,-4\rangle = c^2 - 12$ iff $c = \pm 2\sqrt{3}.$

19. $7\mathbf{i}-8\mathbf{j} = k(3\mathbf{i}-2\mathbf{j}) + m(-\mathbf{i}+4\mathbf{j}) = (3k-m)\mathbf{i} + (-2k+4m)\mathbf{j}$

    iff $7 = 3k-m$ and $-8 = -2k+4m$ iff $k=2, m=-1.$

21. $r_1 i + r_2 j = k(a_1 i + a_2 j) + m(b_1 i + b_2 j) = (ka_1 + mb_1)i + (ka_2 + mb_2)j$

$\Leftrightarrow r_1 = ka_1 + mb_1$ and $r_2 = ka_2 + mb_2$.

Solve these two equations simultaneously [noting that $a_1 b_2 - a_2 b_1 \neq 0$ since a and b are noncollinear] and obtain

$$k = \frac{b_2 r_1 - b_1 r_2}{a_1 b_2 - a_2 b_1} \quad \text{and} \quad m = \frac{a_1 r_2 - a_2 r_1}{a_1 b_2 - a_2 b_1}.$$

23. Work $= \mathbf{F} \cdot \mathbf{D} = (3i+10j) \cdot (10j) = 0 + 100 = 100$ ft-lb.

25. $\mathbf{D} = \langle 6-1, 8-0 \rangle = \langle 5,8 \rangle$, $\mathbf{F} = \langle 3,4 \rangle$.

Work $= \mathbf{D} \cdot \mathbf{F} = \langle 5,8 \rangle \cdot \langle 3,4 \rangle = 15+32 = 47$ ft-lb.

27. $|\mathbf{u} \cdot \mathbf{v}| = |\cos\theta||\mathbf{u}||\mathbf{v}| \leq |\mathbf{u}||\mathbf{v}|$ since $|\cos\theta| \leq 1$.

29. Assume $\mathbf{u} \neq 0$ and $\mathbf{v} \neq 0$. Then:

(a) $|\mathbf{u}|^2 + |\mathbf{v}|^2 = 2\mathbf{u} \cdot \mathbf{v} \Rightarrow \mathbf{u} \cdot \mathbf{u} - 2\mathbf{u} \cdot \mathbf{v} + \mathbf{v} \cdot \mathbf{v} = 0 \Rightarrow (\mathbf{u}-\mathbf{v}) \cdot (\mathbf{u}-\mathbf{v}) = 0$
$\Rightarrow |\mathbf{u}-\mathbf{v}|^2 = 0 \Rightarrow \mathbf{u}-\mathbf{v}\, 0 \Rightarrow \mathbf{u} = \mathbf{v}$.

(b) $|\mathbf{u}|^2 + |\mathbf{v}|^2 = 2\mathbf{u} \cdot \mathbf{v} \Rightarrow |\mathbf{u}| = |\mathbf{v}|$ and $|\mathbf{u}|^2 = 2|\mathbf{u}||\mathbf{v}|\cos\theta$
$\Rightarrow |\mathbf{u}|^2 = 2|\mathbf{u}|^2\cos\theta \Rightarrow \cos\theta = 1/2 \Rightarrow \theta = \pi/3$.

Thus, u and v have the same length; the angle between them is $\pi/3$.

31. Let $x^2+y^2 = r^2$ be the equation of the circle. Then

$\mathbf{AC} \cdot \mathbf{BC} = \langle x+r,y \rangle \cdot \langle x-r,y \rangle = (x+r)(x-r) + y^2 = (x^2+y^2) - r^2 = r^2 - r^2 = 0.$

33. $\text{pr}_\mathbf{v}\mathbf{u} = (\text{scalar projection of } \mathbf{u} \text{ on } \mathbf{v})(\text{unit vector in direction of } \mathbf{v})$

$$= (|\mathbf{u}|\cos\theta)\frac{\mathbf{v}}{|\mathbf{v}|} \quad \frac{|\mathbf{u}||\mathbf{v}|\cos\theta}{|\mathbf{v}|}\frac{\mathbf{v}}{|\mathbf{v}|} = \frac{\mathbf{u}\cdot\mathbf{v}}{|\mathbf{v}|^2}\mathbf{v} = \frac{\mathbf{u}\cdot\mathbf{v}}{\mathbf{v}\cdot\mathbf{v}}\mathbf{v}$$

(a) $\text{pr}_{\langle 3,4 \rangle}\langle 0,5 \rangle = \frac{\langle 0,5 \rangle \cdot \langle 3,4 \rangle}{\langle 3,4 \rangle \cdot \langle 3,4 \rangle}\langle 3,4 \rangle \cdot = \frac{20}{25}\langle 3,4 \rangle = \frac{4}{5}\langle 3,4 \rangle.$

(b) $\text{pr}_{\langle 3,4 \rangle}\langle -3,2 \rangle = \frac{\langle -3,2 \rangle \cdot \langle 3,4 \rangle}{\langle 3,4 \rangle \cdot \langle 3,4 \rangle}\langle 3,4 \rangle \cdot = \frac{-1}{25}\langle 3,4 \rangle.$

Problem Set 13.4 Vector-Valued Functions and Curvilinear Motion

1. $3\mathbf{i} - 1\mathbf{j} = 3\mathbf{i} - \mathbf{j}$.

3. $\lim\limits_{t \to 2} \dfrac{t-2}{t^2-4}\,\mathbf{i} + \lim\limits_{t \to 2} \dfrac{t^2+t-6}{t-2}\,\mathbf{j} \;\overset{\textcircled{L}}{=}\; \lim\limits_{t \to 2} \dfrac{1}{2t}\,\mathbf{i} + \lim\limits_{t \to 2} \dfrac{2t+1}{1}\,\mathbf{j} = (1/4)\mathbf{i} + 5\mathbf{j}$.

5. $1\mathbf{i} + 0\mathbf{j} = \mathbf{i}$.

7. Doesn't exist since $\lim\limits_{t \to 0^+} [\ln(t^2)] = -\infty$.

9. (a) Require that $t-2 \neq 0$ and $4+t \geq 0$, so $t \neq 2$ and $t \geq -4$. The domain is $[-4, 2) \cup (2, \infty)$.

    (b) Require that $t^2+1 \geq 0$. The domain is $\mathbf{R}$.

11. (a) $1/(t-2)$ is continuous at each t except at $t=2$ and $\sqrt{4+t}$ is continuous for each $t > -4$, so $\mathbf{r}$ is continuous for t in $(-4, 2) \cup (2, \infty)$.

    (b) $[t]$ is continuous except at each integer and $(t^2+1)^{1/2}$ is continuous for all t, so $\mathbf{r}$ is continuous except at each integer.

13. (a) $D_t\mathbf{r}(t) = 4(2t+3)\mathbf{i} - 2e^{2t}\mathbf{j}$. $D_t^2\mathbf{r}(t) = 8\mathbf{i} - 4e^{2t}\mathbf{j}$.

    (b) $D_t\mathbf{r}(t) = (-2\sin 2t)\mathbf{i} - (3\sin^2 t \cos t)\mathbf{j}$.

    $D_t^2\mathbf{r}(t) = (-4\cos 2t)\mathbf{i} - [(3\sin^2 t)(-\sin t) + (\cos t)(6\sin t \cos t)]\mathbf{j}$

    $= (-4\cos 2t)\mathbf{i} + (3\sin^3 t - 6\sin t \cos^2 t)\mathbf{j}$.

15. $\mathbf{r}(t) = \langle e^{2t}, 3\ln t \rangle$; $\mathbf{r}'(t) = \langle 2e^{2t}, 3t^{-1} \rangle$; $\mathbf{r}''(t) = \langle 4e^{2t}, -3t^{-2} \rangle$.

    $D_t[\mathbf{r}(t) \cdot \mathbf{r}'(t)] = \mathbf{r}(t) \cdot \mathbf{r}''(t) + \mathbf{r}'(t) \cdot \mathbf{r}'(t)$

    $= [(e^{2t})(4e^{2t}) + (3\ln t)(-3t^{-2})] + [(2e^{2t})^2 + (3t^{-1})^2]$

    $= 8e^{4t} + 9t^{-2}(1 - \ln t)$.

17. $h(t)\mathbf{r}(t) = e^{-3t}\langle(t-1)^{1/2}, \ln(2t^2)\rangle$.

$D_t[h(t)\mathbf{r}(t)] = e^{-3t}\langle(1/2)(t-1)^{-1/2}, 4t/2t^2\rangle + (-3e^{-3t})\langle(t-1)^{1/2}, \ln(2t^2)\rangle$

$= e^{-3t}\langle(1/2)(7-6t)(t-1)^{-1/2}, 2/t - 3\ln(2t^2)\rangle$.

19. $\mathbf{F}(t) = \mathbf{f}(g(t)) = \langle\cos(3t^2-4), \exp(9t^2-12)\rangle$.

$\mathbf{F}'(t) = \langle-6t\sin(3t^2-4), 18t\exp(9t^2-12)\rangle$.

21. $\displaystyle\int_0^1 \langle e^t, e^{-t}\rangle dt = \left[\langle e^t, -e^{-t}\rangle\right]_0^1 = \langle e, -e^{-1}\rangle - \langle1, -1\rangle = \langle e-1, 1-e^{-1}\rangle$

$\langle1.7182, 0.6321\rangle$.

23. $\mathbf{r}(t) = \langle e^{-t}, e^t\rangle$; $\mathbf{v}(t) = \langle-e^{-t}, e^t\rangle$;

$\mathbf{a}(t) = \langle e^{-t}, e^t\rangle$.

$\mathbf{r}(1) = \langle1/e, e\rangle$; $\mathbf{v}(1) = \langle-1/e, e\rangle$;

$\mathbf{a}(1) = \langle1/e, e\rangle$;

$|\mathbf{v}(1)| = [(1/e)^2+e^2]^{1/2} \approx 2.7431$.

Cartesian equation of the curve is $y = 1/x$, $x>0$.

25. $\mathbf{r}(t) = \langle2\cos t, -3\sin^2 t\rangle$; $\mathbf{r}(\pi/3) = \langle1, -9/4\rangle$.

$\mathbf{v}(t) = \langle-2\sin t, -6\sin t\cos t\rangle = \langle-2\sin t, -3\sin2t\rangle$;

$\mathbf{v}(\pi/3) = \langle-\sqrt{3}, -3\sqrt{3}/2\rangle$;

$|\mathbf{v}(t)| = \sqrt{39}/2 \approx 3.1225$.

$\mathbf{a}(t) = \langle-2\cos t, -6\cos2t\rangle$; $\mathbf{a}(\pi/3) = \langle-1, 3\rangle$.

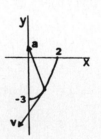

Cartesian equation of the curve is
$x^2 = (4/3)(y+3)$, x in $[-2,2]$.

27. $\mathbf{r}(t) = \langle 3t^2, t^3 \rangle$; $\mathbf{v}(t) = \langle 6t, 3t^2 \rangle$;

   $\mathbf{a}(t) = \langle 6, 6t \rangle$.

   $\mathbf{r}(2) = \langle 12, 8 \rangle$.
   $\mathbf{v}(2) = \langle 12, 12 \rangle$.
   $|\mathbf{v}(2)| = \sqrt{144+144} \approx 16.97$.
   $\mathbf{a}(2) = \langle 6, 12 \rangle$.

   $x = 3t^2$, $y = t^3$

   $\Rightarrow x^3 = 27t^6$, $y^2 = t^6$

   $\Rightarrow x^3 = 27y^2$.

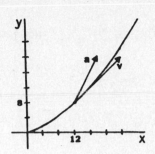

29. $\mathbf{r}(t) = \langle \cos t, -2\tan t \rangle$; $\mathbf{v}(t) = \langle -\sin t, -2\sec^2 t \rangle$;

   $\mathbf{a}(t) - \langle -\cos t, -4\sec^2 t \tan t \rangle$.

   $\mathbf{r}(-\pi/4) = \langle \sqrt{2}/2, 2 \rangle$; $\mathbf{v}(-\pi/4) = \langle \sqrt{2}/2 \rangle -4 \rangle$;
     $|\mathbf{v}(-\pi/4)| = \sqrt{33/2} \approx 4.0620$;
     $\mathbf{a}(-\pi/4) = \langle \sqrt{2}/2, 8 \rangle$.

Cartesian equation of the curve is

   $$y^2 = 4x^{-2} - 4, \ x \text{ in } [-1,0) \cup (0,1].$$

31. $\mathbf{a}(t) = \langle 0, -32 \rangle$, so $\mathbf{v}(t) = \langle k_1, -32t + k_2 \rangle$.

   $\mathbf{v}(0) = \langle 0, 0 \rangle$ and $\mathbf{v}(0) = \langle k_1, k_2 \rangle$, so $k_1 = k_2 - 0$.

   $\mathbf{v}(t) = \langle 0, -32t \rangle$, so $\mathbf{r}(t) = \langle C_1, -16t^2 + C_2 \rangle$.
     $\mathbf{r}(0) = \langle 0, 0 \rangle$ and $\mathbf{r}(0) = \langle C_1, C_2 \rangle$, so $C_1 = C_2 = 0$.

   $\mathbf{r}(t) = \langle 0, -16t^2 \rangle$.

33. $\mathbf{a}(t) = \langle 1, e^{-t} \rangle$.

$\mathbf{v}(t) = \langle t + C_1, -e^{-t} + C_2 \rangle$. $\mathbf{v}(0) = \langle 2,1 \rangle$, so $C_1 = C_2 = 2$.

$\mathbf{v}(t) = \langle t+2, -e^{-t}+2 \rangle$.

$\mathbf{r}(t) = \langle \frac{t^2}{2} + 2t + k_1, e^{-t} + 2t + k_2 \rangle$. $\mathbf{r}(0) = \langle 1,1 \rangle$, so $k_1 = 1$, $k_2 = 0$.

$\mathbf{r}(t) = \langle \frac{t^2}{2} + 2t + 1, e^{-t} + 2t \rangle$.

35. $\mathbf{r}(t) = 5\langle \cos 6t, \sin 6t \rangle$.

$\mathbf{v}(t) = 30\langle -\sin 6t, \cos 6t \rangle$.

$\mathbf{a}(t) = -180\langle \cos 6t, \sin 6t \rangle$.

$|\mathbf{v}(t)| = 30(\sin^2 6t + \cos^2 6t)^{1/2} = 30$.

37. Substituting $\theta = 30°$ and $v_0 = 96$ ft/sec into equations in Example 5,

$\mathbf{r}(t) = (48\sqrt{3}t)\mathbf{i} + (48t - 16t^2)\mathbf{j}$ and $\mathbf{v}(t) = 48\sqrt{3}\mathbf{i} + (48 - 32t)\mathbf{j}$.

The projectile is at ground level when $48t - 16t^2 = 0$ (y-component of r); i.e. when t=0 (when it starts) or t=3. It hits the ground after 3 seconds. Its speed then is $|48\sqrt{3}\mathbf{i} + (-48)\mathbf{j}| = 96$ ft/sec. At that moment it will be $48\sqrt{3}(3) = 144\sqrt{3} \approx 249.42$ ft from the origin.

39. $\mathbf{r}(t) = \langle \cosh \omega t, \sinh \omega t \rangle$; $\mathbf{v}(t) = \langle \omega \sinh \omega t, \omega \cosh \omega t \rangle$;

$\mathbf{a}(t) = \langle \omega^2 \cosh \omega t, \omega^2 \sinh \omega t \rangle = \omega^2 \langle \cosh \omega t, \sinh \omega t \rangle = \omega^2 \mathbf{r}(t)$. $c = \omega^2$.

41. Let $\theta = 45°$ in Example 5.

$\mathbf{r}(t) = [tv_0(\sqrt{2}/2)]\mathbf{i} + [tv_0(\sqrt{2}/2) - 16t^2]\mathbf{j}$.

$\mathbf{v}(t) = [v_0(\sqrt{2}/2)]\mathbf{i} + [v_0(\sqrt{2}/2) - 32t]\mathbf{j}$.

It is given that $300 = tv_0(\sqrt{2}/2)$ amd $0 = tv_0(\sqrt{2}/2) - 16t^2$.

Solve for t in the second equation to obtain $t = (\sqrt{2}/32)v_0$.

Then substitute that value of t in the first equation and solve for v_0 to obtain $v_0 = \sqrt{9600} \approx 97.98$ ft/sec.

43. [See Example 3.] The acceleration due to the rotation must at least balance the acceleration due to gravity. Therefore,

$$980 \leq |\mathbf{a}(t)| = |-\omega^2 \mathbf{r}(t)| = \omega^2|<60\cos\omega t, 60\sin\omega t>| = 60\omega^2.$$

Thus, $\omega^2 \geq 980/60$; $\omega \geq 4.0415$.

Angular speed must be at least 4.0416 rad/sec   [about 0.65 rev/sec].

45. (a) Use formula $v^2 = \dfrac{k}{r} = \dfrac{gR^2}{r}$.   [From 44(a),(c)]

$$v^2 = \frac{(32.17)[(3960)(5280)]^2}{(4160)(5280)}   \text{[Changed miles to feet]}$$

≈ 640297822, so $v \approx 25,304$ ft/sec   [or 17,253 mph].

(b) [We will express units in miles and seconds for this part.]

$$T^2 = \frac{4\pi^2 r^3}{k}, \text{ so } r^3 = \frac{kT^2}{4\pi^2} = \frac{gR^2T^2}{4\pi^2}.   \text{[From 44(b),(c)]}$$

$$g = \frac{32.17 \text{ ft}}{1 \text{ sec}^2}\frac{1 \text{ mi}}{5280 \text{ ft}} = \frac{32.17 \text{ mi}}{5280 \text{ sec}^2};$$

$R = 3960$ mi; $T = 1$ day $= (24)(60)(60)$sec $= 86400$ sec.

Therefore, $r^3 = \dfrac{(32.17/5280)(3960)^2(86400)^2}{4\pi^2}$, so $r \approx 26,240$ miles.

47. (a) $\mathbf{r}(t)$

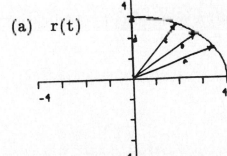

(b) $\mathbf{r}'(t)$

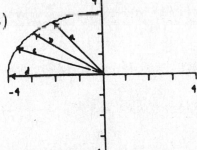

(c) $\mathbf{r}''(t)$

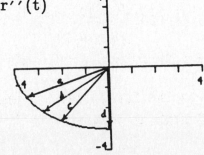

49. (a) r(t)

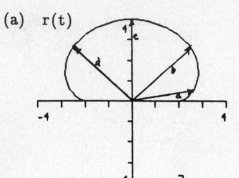

(b) r'(t)

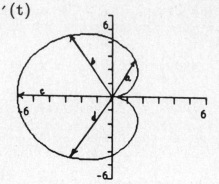

(c) r''(t)

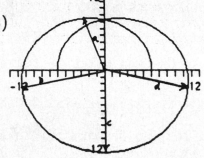

Problem Set 13.5   Curvature and Acceleration

1. [Use Theorem A and $\mathbf{v}(t) = \langle x', y' \rangle$.]

$x = 4t^2$, $x' = 8t$, $x'' = 8$, $y = 4t$, $y' = 4$, $y'' = 0$.

At t=1/2: $x'=4$, $x''=8$, $y'=4$, $y'' = 0$.

$$T(1/2) = \frac{\mathbf{v}(1/2)}{|\mathbf{v}(1/2)|} = \frac{\langle 4,4 \rangle}{|\langle 4,4 \rangle|} = (4/4)\langle 1,1 \rangle.$$

$$\kappa(1/2) = \frac{|(4)(0) - (4)(8)|}{[(4)^2 + (4)^2]^{3/2}} = \sqrt{2}/8 \approx 0.1768.$$

3. $\mathbf{r}(t) = \langle 4\cos t, 3\sin t \rangle$; $\mathbf{v}(t) = \langle -4\sin t, 3\cos t \rangle$; $\mathbf{v}(\pi/4) = \langle -2\sqrt{2}, 3\sqrt{2}/2 \rangle$.

Therefore, $T(\pi/4) = \dfrac{\mathbf{v}(\pi/4)}{|\mathbf{v}(\pi/4)|} = \langle -0.8, 0.6 \rangle$.

$x = 4\cos t$, $y = 3\sin t$, $x' = -4\sin t$, $y' = 3\cos t$, $x''= -4\cos t$, $y''= -3\sin t$.

$$\kappa(t) = \frac{|12\sin^2 t + 12\cos^2 t|}{(16\sin^2 t + 9\cos^2 t)^{3/2}}, \text{ so } (\pi/4) = \frac{12}{(8 + 9/2)^{3/2}} \approx 0.2715.$$

5. $y' = 2x$, $\quad y'' = 2$.

At $(1,1)$: $\quad y' = 2$.

$\qquad\qquad y'' = 2$.

$\qquad\qquad \kappa = 2/(1+4)3/2 = 2/5\sqrt{5}$.

$\qquad\qquad R = 5\sqrt{5}/2 \approx 5.59$.

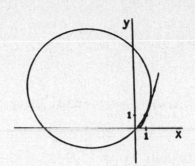

7. A translation $(X = x+4, \; Y = y)$ followed by a reflection with respect to the line, $y=x$, yields the same parabola and the same point as in Problem 5, so the curvature and radius of curvature are the same as for that problem.

9. $y = e^{x}-x$; $\; y' = e^{x}-1$; $\; y'' = e^{x}$.

$y'(0) = 0$; $\; y''(0) = 1$.

$\kappa(0) = \dfrac{|1|}{(1+0)^{3/2}} = 1$; $\; R(0) = 1$.

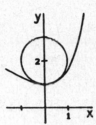

11. $y' = (1/2)\sinh(x/2)$, $\quad y'' = (1/4)\cosh(x/2)$.

At $(0,1)$: $\quad y' = 0$.

$\qquad\qquad y'' = 1/4$.

$\qquad\qquad \kappa = 1/4$.

$\qquad\qquad R = 4$.

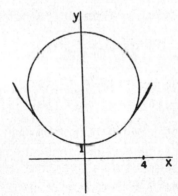

13. $y' = \cos x$, $\quad y'' = -\sin x$.

At $(\pi/4, \sqrt{2}/2)$: $\quad y' = \sqrt{2}/2$.

$\qquad\qquad\qquad y'' = -\sqrt{2}/2$.

$\qquad\qquad\qquad \kappa = 2/3\sqrt{3}$.

$\qquad\qquad\qquad R = 3\sqrt{3}/2 \approx 2.60$.

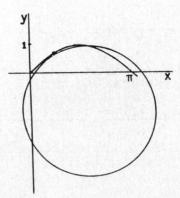

15. $y = \ln x$; $y' = x^{-1}$; $y'' = -x^{-2}$; $\kappa(x) = \dfrac{|-x^{-2}|}{(1+x^{-2})^{3/2}} = \dfrac{x}{(x^2+1)^{3/2}}$.

$\kappa'(x) = \dfrac{1-2x^2}{(x^2+1)^{5/2}}$; Aux. axis for κ':

$$\begin{array}{ccccc} & (+) & (0) & (-) & \\ \hline 0 & & \sqrt{2}/2 & & x \end{array}$$

is maximum at $x = \sqrt{2}/2$, $y = \ln(\sqrt{2}/2) = -\ln\sqrt{2}$; i.e. at $(\sqrt{2}/2, -\ln\sqrt{2})$.

17. Due to symmetry we can restrict our search to $[0, \pi]$.

$y' = \cos x$, $y'' = -\sin x$, $\kappa(x) = \dfrac{\sin x}{(1+\cos^2 x)^{3/2}}$, $\kappa'(x) = \dfrac{2(\cos x)(2-\cos^2 x)}{(1+\cos^2 x)^{5/2}}$.

is maximum where $x = \pi/2$. The points are $(\pi/2, 1)$ and $(-\pi/2, -1)$.

19. $\mathbf{v}(t) = \langle 2t, 1\rangle$, $\mathbf{a}(t) = \langle 2, 0\rangle$, $|\mathbf{v}(t)| = (4t^2+1)^{1/2}$, $|\mathbf{a}(t)| = 2$.

$a_T = D_t[(4t^2+1)^{1/2}] = 4t(4t^2+1)^{-1/2}$, $a_N^2 = |\mathbf{a}|^2 - a_T^2 = 4(4t^2+1)^{-1}$.

At $t=1$: $a_T = 4/\sqrt{5}$ 1.79. $a_N = 2/\sqrt{5} \approx 0.89$.

21. $\mathbf{r}(t) = \langle -a\cos t, a\sin t\rangle$; $\mathbf{v}(t) = \langle -a\sin t, -a\cos t\rangle$; $\mathbf{a}(t) = \langle -a\cos t, -a\sin t\rangle$.

$|\mathbf{v}(t)| = (a^2\cos^2 t + a^2\sin^2 t)^{1/2} = |a|$; similarly, $|\mathbf{a}(t)| = |a|$.

$a_T = \dfrac{d^2 s}{dt^2} = \dfrac{d}{dt}|\mathbf{v}(t)| = \dfrac{d}{dt}|a| = 0$; $a_N^2 = |\mathbf{a}(t)|^2 - a_T^2 = a^2$, so $a_N = |a|$.

23. See the discussion dealing with s and a parameter t (page 578 of the text). The important ingredients are that s increases as t increases and t can be expressed as a function of s. Use the chain rule.

$\mathbf{N} = \dfrac{d\mathbf{T}/ds}{|d\mathbf{T}/ds|} = \dfrac{\mathbf{T}'(t)/|\mathbf{v}(t)|}{|\mathbf{T}'(t)/|\mathbf{v}(t)||} = \dfrac{\mathbf{T}'(t)}{|\mathbf{T}'(t)|} = \dfrac{d\mathbf{T}/dt}{|d\mathbf{T}/dt|}$.

25. $\mathbf{v}(t) = \langle \cos t, 2\cos 2t\rangle$, $\mathbf{a}(t) = \langle -\sin t, -4\sin 2t\rangle$. $\mathbf{a}(t) = 0$ iff $-\sin t = 0$ and $-4\sin 2t = 0$, which occurs iff $t = 0$, π, 2π, so it occurs only at the origin.

$\mathbf{a}(t)$ points to the origin iff $\mathbf{a}(t) = -k\mathbf{r}(t)$ for some k and $\mathbf{r}(t)$ is not $\langle 0, 0\rangle$. This occurs iff $t = \pi/2$, $3\pi/2$, so it occurs only at $(1, 0)$ and $(-1, 0)$.

27. $s''(t) = a_T = 0 \Rightarrow$ speed $= s'(t) = c$ (a constant).

$\kappa(ds/dt)^2 = a_N = 0 \Rightarrow \kappa = 0$ or $\dfrac{ds}{dt} = 0$

$$\Rightarrow \kappa = 0.$$

29. $\mathbf{T \cdot T} = 1 \Rightarrow \mathbf{T} \cdot (d\mathbf{T}/ds) + \mathbf{T} \cdot (d\mathbf{T}/ds) = 0 \Rightarrow \mathbf{T} \cdot (d\mathbf{T}/ds) = 0 = \mathbf{T} \cdot \kappa \mathbf{N} = 0.$

Therefore, $\mathbf{T}$ is perpendicular to $\kappa \mathbf{N}$, and hence to $\mathbf{N}$.

31. It is given that at $(-12,16)$, $s'(t) = 10$ ft/sec and $s''(t) = 5$ ft/sec^2. $\kappa = 1/20$ [from Example 2].

Therefore, $a_T = 5$ and $a_N = (1/20)(10)^2 = 5$, so $\mathbf{a} = 5\mathbf{T} = 5\mathbf{N}$.

Let $\mathbf{r}(t) = \langle 20\cos t, 20\sin t \rangle$ describe the circle.
$\mathbf{r}(t) = \langle -12,16 \rangle \Rightarrow \cos t = -3/5$ and $\sin t = 4/5$.

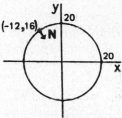

$\mathbf{v}(t) = \langle -20\sin t, 20\cos t \rangle$, so $|\mathbf{v}(t)| = 20$.

Then $\mathbf{T}(t) = \dfrac{\mathbf{v}(t)}{|\mathbf{v}(t)|} = \langle -\sin t, \cos t \rangle$.

Thus, at $(-12,16)$, $\mathbf{T} = \langle -4/5, -3/5 \rangle$ and $\mathbf{N} = \langle 3/5, -4/5 \rangle$ [since $\mathbf{N}$ is a unit vector perpendicular to $\mathbf{T}$ and pointing to the concave side of the curve.

Therefore, $\mathbf{a} = 5\langle -4/5, -3/5 \rangle + 5\langle 3/5, -4/5 \rangle = \langle -1, -7 \rangle = \mathbf{i} \cdot 7\mathbf{j}$.

33. Let $\mu mg = \dfrac{mv_R^2}{R}$. Then $v_R = \sqrt{\mu gR}$.

At the values given, $v_R = \sqrt{(0.4)(32)(400)} = \sqrt{5120} \approx 71.55$ ft/sec

[about 48.79 mph]

35. $x' = r' \cos\theta - r\sin\theta$; $y' = r'\sin\theta + r\cos\theta$.

$x'' = (r''-r)\cos\theta - 2r'\sin\theta$; $y'' = (r''-r)\sin\theta + 2r'\cos\theta$

Substitute these into $\kappa = \dfrac{|x'y''- y'x''|}{[(x')^2 + (y')^2]^{3/2}}$ and simplify to obtain the desired formula.

(a) $r(\theta) = 4\cos\theta$, $r'(\theta) = -4\sin\theta$; $r''(\theta) = -4\cos\theta$.

$$\kappa = \frac{|32(\cos^2\theta + \sin^2\theta)|}{|16(\cos^2\theta + \sin^2\theta)|^{3/2}} = \frac{32}{64} = \frac{1}{2}.$$

(b) $r(\theta) = 1+\cos\theta$, $r'(\theta) = -\sin\theta$; $r''(\theta) = -\cos\theta$,
   so $r(0) = 2$, $r'(0) = 0$, $r''(0) = -1$.

$$\kappa = \frac{|4 + 0 - (-2)|}{(4 + 0)^{3/2}} = \frac{3}{4}.$$

37. max k = 0.7606
     min k = 0.1248

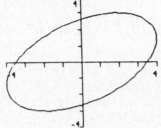

39. Conjecture: If a > b, the graph is a central ellipse with major diameter 2a, minor diameter 2b, with major axis making an angle θ with the positive x-axis. Proof.

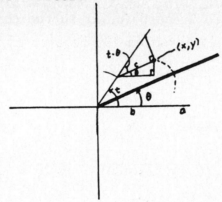

When $\theta = 0$, we get the ellipse x=acost, y=bsint as in Problem 24 of Section 12.2. Now the construction above clearly gives the same ellipse rotated through angle θ. Moreover

$x = b\cos t + c \cos\theta = b\cos t + (a-b)\cos(t-\theta)\cos\theta$

$\quad = b\cos t + \dfrac{a-b}{2}\,[\cos t + \cos(t-2\theta)]$

$\quad = \dfrac{a+b}{2}\cos t + \dfrac{a-b}{2}\cos(t-2\theta)$

Similarly

$y = \dfrac{a+b}{2}\sin t - \dfrac{a-b}{2}\sin(t-2\theta)$

Problem Set 13.6   Chapter Review

True-False Quiz

1. False.   See Example 3, Section 13.1.

3. False.   But we can if f^{-1} exists. See Example 3, Section 13.1.

5. False. $x = f(t) = t^4$, $y = g(t) = t^8$ provides a counterexample.

7. True. $\langle 2,-3\rangle \cdot \langle 6,4\rangle = 0$.

9. False. $(\mathbf{a}\cdot\mathbf{b})\cdot\mathbf{c}$ is not defined since $\mathbf{a}\cdot\mathbf{b}$ is a scalar.

11. True. $|\mathbf{u}\cdot\mathbf{v}| = |\mathbf{u}||\mathbf{v}||\cos\theta|$, so $|\mathbf{u}\cdot\mathbf{v}| = |\mathbf{u}||\mathbf{v}|$ implies $|\cos\theta| = 1$. Then θ is 0 or π.

13. True.   See Problem 7, Section 13.3. Here is an alternate approach. $\mathbf{u}+\mathbf{v}$ perpendicular to $\mathbf{u}-\mathbf{v} \Rightarrow 0 = (\mathbf{u}+\mathbf{v})\cdot(\mathbf{u}-\mathbf{v}) = \mathbf{u}\cdot\mathbf{u} - \mathbf{v}\cdot\mathbf{v} \Rightarrow \mathbf{u}\cdot\mathbf{u} = \mathbf{v}\cdot\mathbf{v}$ $=\cdot |\mathbf{u}|^2 = |\mathbf{v}|^2 \Rightarrow |\mathbf{u}| = |\mathbf{v}|$.

15. True.   See definitions of limit and continuity, Section 13.4.

17. True.   The curve is the line $y = (2/3)x - (11/3)$.

19. True.   See the work done for Problem 29, Section 13.5.

Sample Test Problems

1. $x = 3y+2$ or $y = (1/3)x - (2/3)$,
   x in $\mathbf{R}$.

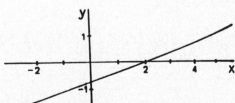

3. $\dfrac{x+2}{4} = \sin t$ and $\dfrac{y-1}{3} = \cos t$.

$\dfrac{(x+2)^2}{16} + \dfrac{(y-1)^2}{9} = 1$.
[Since $\sin^2 t + \cos^2 t = 1$.]

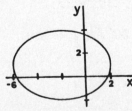

5. $\dfrac{dy}{dx} = \dfrac{dy/dt}{dx/dt} = \dfrac{1+(t+1)^{-1}}{6t^2-4}$. At t=0:   x=7, y=0, y'=-1/2.

Tangent is y = (-1/2)x + (7/2).

Normal is y = 2x-14.

7. dx/dt = tcost. dy/dt = tsint.

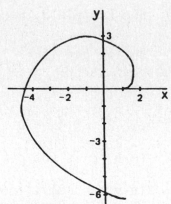

Length $= \displaystyle\int_0^{2\pi} [t^2\cos^2 t + t^2\sin^2 t]^{1/2}dt$

$= \displaystyle\int_0^{2\pi} t\,dt = \Big[(1/2)t^2\Big]_0^{2\pi}$

$= 2\pi^2 \approx 19.7392$.

9. (a) 3<2,-5> - 2<1,1> = <6,-15> - <2,2> = <4,-17>.

(b) <2,-5>·<1,1> = 2 + (-5) = -3.

(c) <2,-5>·(<1,1> + <-6,0>) = <2,-5>·<-5,1> = -10 + (-5) = -15.

(d) (4<2,-5> + 5<1,1>)·3<-6,0> = <13,-15>·<-18,0> = -234 + 0 = -234.

(e) $\sqrt{36+0}$<-6,0>·<1-1> = 6(-6 + 0) = -36.

(f) <-6,0>·<-6,0> - $\sqrt{36+0}$ = (36+0) - 6 = 30.

11. 5i-4j = k(-2i) + m(3i-2j) = (-2k+3m)i - (2m)j, so 5 = -2k+3m and 4 =2m. Therefore, m=2 and k=1/2.

13. One vector that makes such an angle is <cos150°,sin150°>. The required vector is then $10\,\dfrac{<\cos150°,\sin150°>}{|<\cos150°,\sin150°>|} = 5<-\sqrt{3},1>$.

15. Let the wind vector be **w** = <100cos30°,100sin30°> = <50 $\sqrt{3}$,50>. Let **p** = <p_1,p_2> be the plane's air velocity vector. We want **w** + **p** = 450**j** = <0,450>.

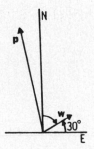

<50 $\sqrt{3}$,50> + <p_1,p_2> = <0,450>

⇒ 50 $\sqrt{3}$ + p_1 = 0, 50 + p_2 = 450

⇒ p_1 = -50 $\sqrt{3}$, p_2 = 400.

Therefore, p = <-50 $\sqrt{3}$,400>. The angle

β formed with the vertical satisfies

$\cos\beta = \dfrac{\mathbf{p} \cdot \mathbf{j}}{|\mathbf{p}||\mathbf{j}|} = \dfrac{400}{\sqrt{167500}}$; $\beta \approx 12.22°$. Thus, the heading is N12.22°W.

The air speed is $|p| = \sqrt{167500} \approx 409.27$ mph.

17. (a) $\mathbf{r}'(t) = \langle 1/t, -6t\rangle$; $\mathbf{r}''(t) = \langle -t^{-2}, -6\rangle$.

(b) $\mathbf{r}'(t) = \langle \cos t, -2\sin 2t\rangle$; $\mathbf{r}''(t) = \langle -\sin t, -4\cos 2t\rangle$.

(c) $\mathbf{r}'(t) = \langle \sec^2 t, -4t^3\rangle$; $\mathbf{r}''(t) = \langle 2\sec^2 t \tan t, -12t^2\rangle$.

19. $\mathbf{v}(t) = \langle 4t, 4\rangle$; $\mathbf{a}(t) = \langle 4,0\rangle$; $\mathbf{v}(-1) = \langle -4,4\rangle$; $|\mathbf{v}(-1)| = 4\,2$; $\mathbf{a}(-1) = \langle 4,0\rangle$.

21. (a) $y = x^2 - x$; $y' = 2x-1$; $y'' = 2$; $y'(1) = 1$; $y''(1) = 2$;

$\kappa(1) = \dfrac{|2|}{(1+1)^{3/2}} = \dfrac{1}{\sqrt{2}} \approx 0.7071$

(b) Let $x = t+t^3$; $x' = 1+3t^2$; $x'' = 6t$; $y = t+t^2$; $y' = 1+2t$; $y'' - 2$.

At $(2,2)$, $t = 1$, so $x' = 4$, $x'' = 6$, $y' = 3$, $y'' = 2$.

Therefore, $\kappa = \dfrac{|(4)(2)-(3)(6)|}{(16+9)^{3/2}} = 0.08$.

(c) $y = a\cosh(x/a)$; $y' = \sinh(x/a)$; $y'' = (1/a)\cosh(x/a)$.

At $(a, a\cosh 1)$, $y' = \sinh 1$, $y'' = (1/a)(\cosh 1)$.

Therefore, $\kappa = \dfrac{(1/a)(\cosh 1)}{(1 + \sinh^2 1)^{3/2}} = \dfrac{\cosh 1}{a(\cosh^2 1)^{3/2}} = \dfrac{1}{a\cosh^2 1} = \dfrac{0.4120}{a}$.

23. $\mathbf{v}(t) = \langle -2t, 2 \rangle$; $|\mathbf{v}(t)| = 2(t^2+1)^{1/2}$; $\mathbf{a}(t) = \langle -2, 0 \rangle$; $|\mathbf{a}(t)| = 2$.

$a_\mathbf{T} = d^2s/dt^2 = (d/dt)[2(t^2+1)^{1/2}] = 2t(t^2+1)^{-1/2}$.

At $(0,2)$: $t=1$.

$a_\mathbf{T} = \sqrt{2}$.

$a_\mathbf{N}^2 = |a|^2 - a_\mathbf{T}^2 = 4-2 = 2$, so $a_\mathbf{N} = \sqrt{2}$.

Problem Set 14.1   Cartesian Coordinates in Three-Space

1.
A(3,2,4).
B(2,0,3).
C(-3,4,5).
D(0,4,0).
E(-2,-6,-1).

3. The y-coordinate is 0 in the xz-plane.
The x- and z-coordinates are 0 on the y-axis.

5. (a) $(16+25+49)^{1/2} = 3\sqrt{10} \approx 9.4868.$ (b) $(25+9+49)^{1/2} = \sqrt{83} \approx 9.1104.$
(c) $(2+4+3)^{1/2} = 3.$

7. The square of the lengths of the sides are: 4+36+9 = 49 (1st and 2nd),
36+16+144 = 196 (1st and 3rd), and 16+4+225 = 245.   245 = 49+196, so it
is a right triangle.

9. Coordinates of lid: (2,-2,4), (2,3,4),
(5,-2,4), (5,3,4).

Coordinates of base: (2,-2,0), (2,3,0),
(5,-2,0), (5,3,0).

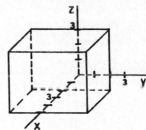

11. (a) $(x-3)^2+(y-1)^2+(z-4)^2 = 25.$ (b) $(x+6)^2+(y-2)^2+(z+3)^2 = 4.$

(c) $(x+1)^2+y^2+(z-4)^2 = 6.$

13. $(x-6)^2+(y+7)^2+(z-4)^2 = 100.$ Center is (6,-7,4) and radius is 10.

15. $4[x^2-x+(1/4)] + 4[y^2+2y+1] + 4[z^2+4z+4] = 13+1+4+16$
$[x-(1/2)]^2 + [y+1]^2 + [z+2]^2 = 8.5$
Center: (1/2,-1,-2)   Radius: $\sqrt{8.5} \approx 2.92$

17.   2x+6y+3z = 12.

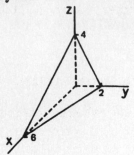

19.   x+3y-z = 6.

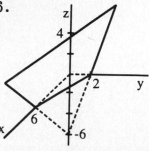

21.   x+3y=8.   Since no z-term appears, changes in the value of z do not effect changes in x and y, so the graph is parallel to the z-axis. This is analogous to the 2-dimensional situation where, for example, x=5 (no y-term appears) is parallel to the y-axis.

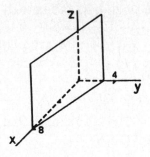

23.   Sphere with origin for center and radius 3.

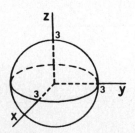

25.   The center is $(1,1,11/2)$.
      Radius is $(1/2)(36+16+1)^{1/2}$
          $= \sqrt{53}/22$.
      Equation is
      $(x-1)^2+(y-1)^2+(z-11/2)^2 = 53/4$.

27. The center is in the plane parallel to the xy-plane and 6 units above it, so z = 6. Similarly, x = 6 and y = 6, so the center is (6,6,6).

Equation: $(x-6)^2 + (y-6)^2 + (z-6)^2 = 36$.

29. (a) Plane that is parallel to and two units above the xy-plane.

(b) Plane that is perpendicular to the xy-plane and whose trace in the xy-plane is the line x=y.

(c) Union of the yz-plane (x=0) and the xz-plane (y=0).

(d) Union of all three coordinate planes.

(e) Cylinder of radius 2 and z-axis for axis.

(f) Hemisphere above the xy-plane with radius 3 and origin as the center of the base.

31. If P(x,y,z) denotes the moving point,

$$[(x-1)^2+(y-2)^2+(z+3)^2]^{1/2} = 2[(x-1)^2+(y-2)^2+(z-3)^2]^{1/2},$$

which simplifies to $(x-1)^2+(y-2)^2+(z-5)^2 = 16$, a sphere of radius 4 and center (1,2,5).

33. $11\pi/12$.

Problem Set 14.2   Vectors in Three-Space

1.   (a) 3i+3j-2k.

(b) 6i+2j+4k.

3.   (a) Length is $\sqrt{21}$; $\cos\alpha = 4/\sqrt{21}$; $\cos\beta = 1/\sqrt{21}$; $\cos\gamma = 2/\sqrt{21}$.

(b) Length is $\sqrt{50}$; $\cos\alpha = -3/\sqrt{50}$; $\cos\beta\ -4/\sqrt{50}$; $\cos\gamma = 5/\sqrt{50}$.

5.   Unit vector is $\dfrac{\langle 3,1,-7\rangle}{(9+1+49)^{1/2}} = \dfrac{\langle 3,1,-7\rangle}{(59)^{1/2}}$. Other vector is $\dfrac{-5\langle 3,1,-7\rangle}{(59)^{1/2}}$.

7. $\cos\theta = \dfrac{-4-6-10}{(29)^{1/2}(30)^{1/2}} = \dfrac{-20}{(870)^{1/2}}$. $\theta \approx 2.3159$ rad. [132.69°].

9. Let $\mathbf{v} = \langle a,b,c\rangle$ be a vector perpendicular to $\langle 4,3,6\rangle$ and $\langle -2,-3,-2\rangle$. Then $\langle a,b,c\rangle \cdot \langle 4,3,6\rangle = 0$ and $\langle a,b,c\rangle \cdot \langle -2,-3,-2\rangle = 0$. Therefore, $4a+3b+6c = 0$ and $-2a-3b-2c = 0$. This pair of equations has infinitely many solutions. Let c be any nonzero constant and solve for a and b.

Let $c = 3$. Then $a = -6$ and $b = 2$. Thus, $\langle -6,2,3\rangle$, which has length 7, is perpendicular to $\langle 4,3,6\rangle$ and $\langle -2,-3,-2\rangle$. Then the vectors required are: $\dfrac{10\langle -6,2,3\rangle}{7}$ and $\dfrac{-10\langle -6,2,3\rangle}{7}$.

11. $\mathbf{BA} = \langle 4,-9,3\rangle$ and $\mathbf{BC} = \langle 4,-6,1\rangle$. Then the cosine of angle ABC is

$$\frac{16+54+3}{(106)^{1/2}(53)^{1/2}} = \frac{73}{(5618)^{1/2}}, \text{ so angle ABC} \approx 0.2288 \text{ rad. } [13.11°]$$

13. Let θ be the angle between $\mathbf{u}$ and $\mathbf{v}$. The scalar projection of $\mathbf{u}$ on $\mathbf{v}$ is

$$\frac{\mathbf{u}\cdot\mathbf{v}}{|\mathbf{v}|} = \frac{-4+3+6}{(19)^{1/2}} \approx 1.1471$$

15. $\mathbf{m} = \text{(scaler projection)} \dfrac{\mathbf{v}}{|\mathbf{v}|} = \dfrac{\mathbf{u}\cdot\mathbf{v}}{|\mathbf{v}|}\dfrac{\mathbf{v}}{|\mathbf{v}|} = \dfrac{\mathbf{u}\cdot\mathbf{v}}{\mathbf{v}\cdot\mathbf{v}}\mathbf{v}$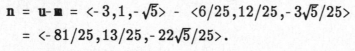

$$= \frac{-6+4+5}{4+16+5}\langle 2,4,-\sqrt{5}\rangle = \langle 6/25,12/25,-3\sqrt{5}/25\rangle.$$

$\mathbf{n} = \mathbf{u}-\mathbf{m} = \langle -3,1,-\sqrt{5}\rangle - \langle 6/25,12/25,-3\sqrt{5}/25\rangle$

$\quad = \langle -81/25,13/25,-22\sqrt{5}/25\rangle.$

17. (a) $|\mathbf{u}| = \sqrt{14}$. Direction cosines are $3/\sqrt{14}$, $-2/\sqrt{14}$, $1/\sqrt{14}$.

(b) $|\mathbf{u}| = 2\sqrt{14}$. Direction cosines are $-1/\sqrt{14}$, $2/\sqrt{14}$, $-3/\sqrt{14}$.

19. $(4+9+z^2)^{1/2} = 5$ and $z > 0 \Rightarrow z = 2\sqrt{3} \approx 3.4641.$

21. There are infinitely many such pairs. By inspection, note that $\langle 1,2,0\rangle$ is perpendicular to $\langle -4,2,5\rangle$ (dot product is 0). And $\langle -2,1,c\rangle$ is perpendicular to $\langle 1,2,0\rangle$ for every real number c (dot product is 0). Now find c so that $\langle -2,1,c\rangle$ is perpendicular to $\langle -4,2,5\rangle$. The dot product is $8+2+5c$ which will be 0 if c is -2. Hence, a pair of perpendicular vectors, each perpendicular to $\langle -4,2,5\rangle$ are $\langle 1,2,0\rangle$ and $\langle -2,1,-2\rangle$.

23. (a) Since $\mathbf{v}\cdot\mathbf{w}$ is not a vector.   (b) Since $\mathbf{u}\cdot\mathbf{w}$ is not a vector.
    (f) Since $|\mathbf{u}|$ is not a vector.

25. (a) $2(x-1) - 4(y-2) + 3(z+3) = 0$ or $2x-4y+3z = -15.$

(b) $3(x+2) - 2(y+3) - 1(z-4) = 0$ or $3x-2y-z = -4.$

27. Let Ω be the smaller angle. Then $\cos\Omega = \dfrac{|\langle 2,-4,3\rangle \cdot \langle 3,-2,-1\rangle|}{\sqrt{4+16+9}\ \sqrt{9+4+1}} = \dfrac{11}{\sqrt{406}}$.

Therefore, $\Omega = \cos^{-1}(11/\sqrt{406}) \approx 0.9933$ rad. [56.91°]

29. (a) Planes parallel to the xy-plane may be expressed as z=D, so z=2 is an equation of the plane.
(b) An equation of the plane is $2(x+4) - 3(y+1) - 4(z-2) = 0$ or $2x-3y-4z = -13$.

31. Distance is 0 since the point is in the plane.

33. $(1,0,0)$ is on $5x-3y-2z = 5$ (by inspection). The distance from $(1,0,0)$ to $-5x+3y+2z = 7$ is $\dfrac{|-5(1)+3(0)+2(0)-7|}{(25+9+4)^{1/2}} = \dfrac{12}{\sqrt{38}} \approx 1.9467$.

35. $|\mathbf{u+v}|^2 + |\mathbf{u-v}|^2 = (\mathbf{u+v})\cdot(\mathbf{u+v}) + (\mathbf{u-v})\cdot(\mathbf{u-v})$
$= [\mathbf{u\cdot u} + 2(\mathbf{u\cdot v}) + \mathbf{v\cdot v}] + [\mathbf{u\cdot u} - 2(\mathbf{u\cdot v}) + \mathbf{v\cdot v}] = 2|\mathbf{u}|^2 + 2|\mathbf{v}|^2$.

37. Let Ω be the angle. Place the cube in the "corner" of the first octant. A vector along main diagonal through the origin is $\langle 1,1,1\rangle$. The projection of that vector onto the face in the xy-plane is $\langle 1,1,0\rangle$.

Then $\cos\Omega = \dfrac{1+1+0}{(3)^{1/2}(2)^{1/2}} = \dfrac{2}{\sqrt{6}}$ so $\Omega \approx 0.6155$ rad. [35.26°]

39. Place the box in the "corner" of the 1st octant so that its vertices are $(0,0,0)$, $(4,0,0)$, $(0,6,0)$, $(4,6,0)$, $(0,0,10)$, $(4,0,10)$, $(0,6,10)$, and $(4,6,10)$.

The main diagonals are $(0,0,0)$ to $(4,6,10)$, $(4,0,0)$ to $(0,6,10)$, $(0,6,0)$ to $(4,0,10)$, $(4,6,0)$ to $(0,0,10)$.

Corresponding vectors are $\langle 4,6,10\rangle$, $\langle -4,6,10\rangle$, $\langle 4,-6,10\rangle$, $\langle -4,-6,10\rangle$.

The smallest angle Ω between main diagonals is obtained if the numerator in the formula $\cos\Omega = \dfrac{|\mathbf{u\cdot v}|}{|\mathbf{u}||\mathbf{v}|}$ is largest.

There are six ways of pairing up the four main diagonals. The largest value of $|\mathbf{u\cdot v}|$ is 120, using $\langle 4,6,10\rangle$ and $\langle -4,6,10\rangle$ [or use $\langle 4,-6,10\rangle$ and $\langle -4,-6,10\rangle$].

That is, $\cos\Omega = \dfrac{|\langle 4,6,10\rangle\cdot\langle -4,6,10\rangle|}{\sqrt{16+36+100}\ \sqrt{16+36+100}} = \dfrac{15}{19}$, so $\Omega \approx 0.6608$ rad. [37.86°]

41. $W = \left[5\ \dfrac{\langle 2,2,-1\rangle}{|\langle 2,2,-1\rangle|}\right]\cdot\langle 3-0,5-1,7-2\rangle = 15$ newton meters [15 joules].

43. $\langle x,y,z \rangle = \langle 2,3,-1 \rangle + (1/5)\langle 7-2,-2-3,9-(-1) \rangle = \langle 3,2,1 \rangle$, so the point is $(3,2,1)$.

45. $(x^2+2x+1) + (y^2+6y+9) + (z^2-8z+16) = 1+9+16$. [Completing the squares]
$(x+1)^2 = (y+3)^2+(z-4)^2 = 26$; center is $(-1,-3,4)$ and radius is $\sqrt{26}$.
d(sphere,plane) = d(sphere's center,plane) - radius of sphere.

$$= \frac{|3(-1) + 4(-3) + 1(4) - 15|}{\sqrt{(3)^2 + (4)^2 + (1)^2}} - \sqrt{26} = \sqrt{26} - \sqrt{26} = 0.$$

47. $(\mathbf{x-a}) \cdot (\mathbf{x-b}) = 0$
$\mathbf{x \cdot x} - (\mathbf{a+b}) \cdot \mathbf{x} + \mathbf{a \cdot b} = 0$

$$\mathbf{x \cdot x} - (\mathbf{a+b}) \cdot \mathbf{x} + \frac{(\mathbf{a+b}) \cdot (\mathbf{a+b})}{4} = \frac{(\mathbf{a+b}) \cdot (\mathbf{a+b})}{4} - \mathbf{a \cdot b}$$

$$\left[\mathbf{x} - \frac{\mathbf{a+b}}{2}\right] \cdot \left[\mathbf{x} - \frac{\mathbf{a+b}}{2}\right] = \frac{(\mathbf{a+b}) \cdot (\mathbf{a+b}) - 4\mathbf{a \cdot b}}{4}$$

$$\left|\mathbf{x} - \frac{\mathbf{a+b}}{2}\right|^2 = \frac{\mathbf{a \cdot a} + 2\mathbf{a \cdot b} + \mathbf{b \cdot b} - 4\mathbf{a \cdot b}}{4}; \quad \left|\mathbf{x} - \frac{\mathbf{a+b}}{2}\right|^2 = \frac{\mathbf{a \cdot a} + 2\mathbf{a \cdot b} + \mathbf{b \cdot b}}{4}$$

$$\left|\mathbf{x} - \frac{\mathbf{a+b}}{2}\right|^2 = \frac{(\mathbf{a-b}) \cdot (\mathbf{a-b})}{4}; \quad \left|\mathbf{x} - \frac{\mathbf{a+b}}{2}\right|^2 = \frac{|\mathbf{a-b}|^2}{4}$$

This is a vector equation of the sphere with center $\left[\frac{a_1+b_1}{2}, \frac{a_2+b_2}{2}, \frac{a_3+c_3}{2}\right]$

and radius $\frac{|\mathbf{a-b}|}{2} = \frac{\sqrt{(a_1-b_1)^2 + (a_2-b_2)^2 + (a_3-b_3)^2}}{2}$.

49. $\mathbf{OP} = \mathbf{OA} + \mathbf{AP} = \mathbf{OA} + (2/3)\mathbf{AQ}$

$= \mathbf{a} + (2/3) [(\mathbf{b-a}) + (1/2)(\mathbf{c-b})]$

$= (\mathbf{a+b+c})/3$.

Thus, for the triangle given,
$\mathbf{OP} = [\langle 2,6,5 \rangle + \langle 4,-1,2 \rangle + \langle 6,1,2 \rangle]/3 = \langle 12,6,9 \rangle/3 = \langle 4,2,3 \rangle$.
Therefore, P is $(4,2,3)$.

Problem Set 14.3 The Cross Product

1. (a) $\langle -2,7,10 \rangle$.
   (b) $\langle 4,15,9 \rangle$.
   (c) $\langle -3,2-2 \rangle \cdot \langle -13,2,7 \rangle = 29$.
   (d) $\langle 3,2,-2 \rangle \times \langle -13,2,7 \rangle$
   $= \langle 18,47,20 \rangle$.

3. $\langle -2,1,-4 \rangle \times \langle 3,-4,5 \rangle = \langle 5-16,-12+10,8-3 \rangle = \langle -11,2,5 \rangle$ is perpendicular to both. Therefore, every vector perpendicular to both is of the form $k\langle -11,2,5 \rangle$, k in $\mathbf{R}$.

5. Two vectors in the plane are $\langle 6,-3,3 \rangle = 3\langle 2,-1,1 \rangle$ and $\langle -1,1,-5 \rangle$. Then $\langle 2,-1,1 \rangle \times \langle -1,1,-5 \rangle = \langle 4,9,1 \rangle$ is a normal of the plane. The required vectors are $\dfrac{\pm\langle 4,9,1 \rangle}{(98)^{1/2}}$.

7. Area $= |\langle -2,1,4 \rangle \times \langle 4,-2,-5 \rangle| = |\langle 3,6,0 \rangle| = 3\sqrt{5} \approx 6.7082$.

9. The area of the triangle is half the area of the corresponding parallelogram. Adjacent sides of the triangle can be represented by the vectors $\langle -1,2,7 \rangle$ and $\langle -4,0,8 \rangle = 4\langle -1,0,2 \rangle$ using $\langle 3,2,-1 \rangle$ as the vertex.

   Then Area $= (1/2)|\langle -1,2,7 \rangle \times 4\langle -1,0,2 \rangle| = 2|\langle -1,2,7 \rangle \times \langle -1,0,2 \rangle|$
   $= 2|\langle 4-0,-7+2,0+2 \rangle| = 2|\langle 4,-5,2 \rangle| = 2\sqrt{45} \approx 13.4164$.

11. A normal is $\langle 1,6,4 \rangle \times \langle 3,1,4 \rangle = \langle 20,8,-17 \rangle$.
    An equation is $20(x-4) + 8(y-0) - 17(z-6) = 0$ or $20x+8y-17z = -22$.

13. The plane's normals will be perpendicular to the normals of the other two planes. Then a normal is $\langle 1,-3,2 \rangle \times \langle 2,-2,-1 \rangle = \langle 7,5,4 \rangle$. An equation of the plane is $7(x+1) + 5(y+2) + 4(z-3) = 0$ or $7x+5y+4z = -5$.

15. Each vector normal to the plane is parallel to the line of intersection of the given planes. Also, the cross product of vectors normal to those planes is parallel to each of those plane, and therefore is parallel to the line of intersection of the planes. Thus, a normal to the plane we seek is $\langle 4,-3,2 \rangle \times \langle 3,2,-1 \rangle = \langle 3-4,6+4,8+9 \rangle = \langle -1,10,17 \rangle$. Equation of the plane is $-1(x-6)+10(y-2)+17(x+1) = 0$ or $x-10y-17z = 3$.

17. Volume $= |\langle 2,3,4 \rangle \cdot \langle 0,4,-1 \rangle \times \langle 5,1,3 \rangle| = |\langle 2,3,4 \rangle \cdot \langle 13,-5,-20 \rangle| = -69$.

19. (a) $|\langle 3,-4,2 \rangle \cdot \langle -1,2,1 \rangle \times \langle 3,-2,5 \rangle| = |\langle 3,-4,2 \rangle \cdot \langle -3,-1,2 \rangle| = 9$.

    (b) $|\langle 3,2,1 \rangle \times \langle 1,1,2 \rangle| = |\langle 3,-5,1 \rangle| = \sqrt{35} \approx 5.9161$.

(c) Let Ω be the angle. Then Ω is the complement of the smaller angle between **u** and **vxw**. Therefore

$$\sin\Omega = \frac{|\mathbf{u} \cdot (\mathbf{vxw})|}{|\mathbf{u}||\mathbf{vxw}|} = \frac{9}{(14)^{1/2}(14)^{1/2}} = \frac{9}{14} \text{ so } \Omega \approx 0.6982 \text{ rad. } [40.01°]$$

21. Volume = (1/3)(Area of triangular base)(height)
        = (1/3)[(1/2)(Area of corresponding parallelogram base)](height)
        = (1/6)(Volume of corresponding parallelopiped)
        = (1/6) $|\mathbf{a} \cdot \mathbf{bxc}|$.

23. Use Theorems B (text page 567) and C (text page 603), or expand each side in terms of components and compare components.

$$|\mathbf{uxv}|^2 = (\mathbf{uxv}) \cdot (\mathbf{uxv}) \quad \text{(by B5)}$$

$$= \mathbf{u} \cdot [\mathbf{vx}(\mathbf{uxv})] \quad \text{(by C5)}$$

$$= \mathbf{u} \cdot [(\mathbf{v} \cdot \mathbf{v})\mathbf{u} - (\mathbf{v} \cdot \mathbf{u})\mathbf{v}] \quad \text{(by C6)}$$

$$= (\mathbf{v} \cdot \mathbf{v})(\mathbf{u} \cdot \mathbf{u}) - (\mathbf{v} \cdot \mathbf{u})(\mathbf{u} \cdot \mathbf{v}) \quad \text{(by B2 and B3)}$$

$$= |\mathbf{v}|^2|\mathbf{u}|^2 - (\mathbf{u} \cdot \mathbf{v})(\mathbf{u} \cdot \mathbf{v}) \quad \text{(by B5 and B1)}$$

$$= |\mathbf{v}|^2|\mathbf{u}|^2 - |\mathbf{u} \cdot \mathbf{v}|^2 \quad \text{(by B5)}.$$

25. $(\mathbf{v} - \mathbf{w})\mathbf{xu} = -\mathbf{ux}(\mathbf{v} + \mathbf{w}) = -[\mathbf{uxv} + \mathbf{uxw}] = \mathbf{uxv} - \mathbf{uxw} = \mathbf{vxu} + \mathbf{wxu}$.

27. **PQ** = $\langle -a,b,0 \rangle$ and **PR** = $\langle -a,0,c \rangle$ are adjacent sides of the triangle.

$$\text{Area} = \frac{1}{2}|\langle -a,b,0 \rangle \times \langle -a,0,c \rangle| = \frac{1}{2}|\langle bc,ac,ab \rangle| = \sqrt{b^2c^2 + a^2c^2 + a^2b^2}.$$

29. $D^2 = (1/4)(b^2c^2 + a^2c^2 + a^2b^2)$ [Problem 27]
       $= [(1/2)bc]^2 + [(1/2)ac]^2 + [(1/2)ab]^2 = A^2 + B^2 + C^2$.

31. $A^2 = (1/4)|\mathbf{axb}|^2 = (1/4)[|\mathbf{a}|^2|\mathbf{b}|^2 - (\mathbf{a} \cdot \mathbf{b})^2]$ [See Problem 23.]

       $= (1/4)\left[a^2b^2 - [(1/2)(|\mathbf{a}|^2 + |\mathbf{b}|^2 - |\mathbf{a} - \mathbf{b}|^2)]\right]$ [The given identity]

       $= (1/16)[4a^2b^2 - (a^2 + b^2 - c^2)^2]$

       $= (1/16)(2a^2b^2 + 2a^2c^2 + 2b^2c^2 - a^4 - b^4 - c^4)$ [After simplifying]

The same expression is obtained when s is replaced by $(1/2)$ $(a+b+c)$ in $s(s-a)(s-b)(s-c)$, and then multiplying and simplifying is done.

Therefore, A $\sqrt{s(s-a)(s-b)(s-c)}$.

Problem Set 14.4   Lines and Curves in Three-Space

1. $x = 1+3t$, $y = -2 + 7t$, $z = 3+3t$, since $(1,-2,3)$ is a point of the line and $\langle 3,7,3 \rangle$ is a vector in the direction of the line.

3. $\langle 6-4, 2-2, -1-3 \rangle = \langle 2,0,-4 \rangle = 2\langle 1,0,-2 \rangle$ so $\langle 1,0,-2 \rangle$ is a vector in the direction of the line. Therefore, using point $(4,2,3)$, parametric equations of the line are $x = 4+1t$, $y = 2+0t$, $z = 3-2t$, or more simply $x = 4+t$, $y = 2$, $z = 3-2t$.

5. Parametric: $x = 4-2t$, $y = -6+t$, $z = 3+5t$. Symmetric: $\dfrac{x-4}{-2} = \dfrac{y+6}{1} = \dfrac{z-3}{5}$.

7. Parametric: $x = 2-3t$, $y = 5+4t$, $z = -4+2t$. Symmetric: $\dfrac{x-2}{-3} = \dfrac{y-5}{4} = \dfrac{z+4}{2}$.

9. A vector in the direction of the line is $\langle 5,2,-5 \rangle$ x $\langle 10, 6, -5 \rangle$ which is $\langle -10+30, -50+25, 30-20 \rangle = \langle 20,-25,10 \rangle = 5\langle 4,-5,2 \rangle$.

    To find a point on the intersection, let $x = 0$ and solve $2y-5z = 5$ and $6y-5z=35$ simultaneously, obtaining $y = 5$, $z = 1$. Therefore, $(0,5,1)$ is a point on the line. Then symmetric equations of the line are

    $$\dfrac{x-0}{4} = \dfrac{y-5}{-5} = \dfrac{z-1}{2}.$$

11. $\langle 1,4,2 \rangle$ x $\langle 2,-1,-2 \rangle = \langle -6,6,-9 \rangle = -3\langle 2,-2,3 \rangle$ is in the direction of the line. Let $y = 0$; solve $x+2z = 13$ and $2x-2z = 10$; conclude that a point of the line is $(23/3, 0, 8/3)$. $\dfrac{x-23/3}{2} = \dfrac{y}{-2} = \dfrac{z-8/3}{3}$.

13. $\langle 1,-5,2 \rangle$ is a vector in the direction of the line. $\dfrac{x-4}{1} = \dfrac{y}{-5} = \dfrac{z-6}{2}$.

15. The point of intersection on the z-axis is $(0,0,4)$. A vector in the direction of the line is $\langle 5-0, -3-0, 4-4 \rangle = \langle 5,-3,0 \rangle$. Parametric equations are $x = 0+5t$, $y = 0-3t$, $z = 4+0t$; i.e., $x = 5t$, $y = -3t$, $z = 4$.

17. Two points on the first line are $(-2,1,2)$ and $(0,5,1)$ [for t=0 and t=1]. A point on the second line is $(2,3,1)$ [for t=0]. Then two nonparallel vectors in the plane are $\langle 2,4,-1 \rangle$ and $\langle 2,-2,0 \rangle = 2\langle 1,-1,0 \rangle$. Therefore, $\langle 2,4,-1 \rangle$ x $\langle 1,-1,0 \rangle = -\langle 1,1,6 \rangle$ is a normal of the plane. An equation of the plane is $1(x-0) + 1(y-5) + 6(z-1) = 0$, or $x+y+6z = 11$.

19. $(1,-1,4)$ is a point of the line [for $t=0$] and $\langle 2,3,1\rangle$ is a vector in the direction of the line. Another vector in the plane is $\langle 0,0,1\rangle$ [between the two points]. Then $\langle 2,3,1\rangle \times \langle 0,0,1\rangle = \langle 3,-2,0\rangle$ is normal to the plane. An equation of the plane is $3(x-1) - 2(y+1) + 0(z-5) = 0$ or $3x-2y = 5$.

21. (b) Vectors in the direction of the lines are $\langle -1,4,2\rangle$ and $\langle 1,0,2\rangle$, so $\langle -1,4,2\rangle \times \langle 1,0,2\rangle = \langle 8-0,2+2,0-4\rangle = \langle 8,4-4\rangle = 4\langle 2,1,-1\rangle$ is perpendicular to both lines, so is normal to π. Then an equation of π is $2(x-2)+1(y-3)-1(z-0) = 0$, or $2x+y-z = 7$.

   (c) Let $t = 0$. Then $x = -1$, $y = 2$, $z = -1$, so let $Q = (-1,2,-1)$.

   (d) $d(Q,\pi) = \dfrac{|2(-1)+(2)-(-1)-7|}{\sqrt{4+1+1}} = \sqrt{6} \approx 2.4495$.

23. $r(t) = \langle 2\cos t, 6\sin t, t\rangle$ so $r(\pi/3) = \langle 1, 3\sqrt{3}, \pi/3\rangle$.

   $r'(t) = \langle -2\sin t, 6\cos t, 1\rangle$ so $r'(\pi/3) = \langle -\sqrt{3}, 3, 1\rangle$. Therefore, symmetric

   equations of tangent at $t = \pi/3$ are $\dfrac{x-1}{\sqrt{3}} = \dfrac{y-3\sqrt{3}}{3} = \dfrac{z-\pi/3}{1}$.

25. Let $r(t) = \langle 3t, 2t^2, t^5\rangle$; $r'(t) = \langle 3, 4t, 5t^4\rangle$. $(-3,2,-1)$ is the point of the curve when $t = -1$, and $r'(-1) = \langle 3,-4,5\rangle$ is tangent to the curve, and hence perpendicular to the plane, at $(-3,2,-1)$. Then an equation of the plane is $3(x+3) - 4(y-2) + 5(z+1) = 0$ or $3x-4y+5z = -22$.

27. (a) $x^2+y^2+z^2 = (\sin t\cos t)^2 + (\sin^2 t)^2 + (\cos t)^2$
   $$= \sin^2 t\cos^2 t + \sin^4 t + \cos^2 t = (\sin^2 t)(\cos^2 t + \sin^2 t) + \cos^2 t$$
   $$= (\sin^2 t)(1) + \cos^2 t = \sin^2 t + \cos^2 t = 1.$$
   Therefore, the curve lies on the sphere $x^2+y^2+z^2 = 1$.

   (b) $r(t) = \langle \sin t\cos t, \sin^2 t, \cos t\rangle = \langle \frac{1}{2}\sin 2t, \sin^2 t, \cos t\rangle$.

   $r'(t) = \langle \cos 2t, 2\sin t\cos t, -\sin t\rangle = \langle \cos 2t, \sin 2t, -\sin t\rangle$.

   $r(\pi/6) = \langle \frac{\sqrt{3}}{4}, \frac{1}{4}, \frac{\sqrt{3}}{2}\rangle = \frac{1}{4}\langle \sqrt{3}, 1, 2, \sqrt{3}\rangle$ is a vector to the tangent line.

   $r'(\pi/6) = \langle \frac{1}{2}, \frac{\sqrt{3}}{2}, \frac{-1}{2}\rangle$, and hence $\langle 1, \sqrt{3}, -1\rangle$, is a vector in the direction of the tangent line.

   $R(t) = \frac{1}{4}\langle \sqrt{3}, 1, 2\sqrt{3}\rangle + T\langle 1, \sqrt{3}, -1\rangle$ is an equation of the tangent. $z=0$ at the point where the line intersects the xy-plane.

Thus, $\frac{1}{4}(2\sqrt{3}) + (-1)T = 0$, so $T = \frac{\sqrt{3}}{2}$.

Then $x = \frac{1}{4}(\sqrt{3}) + (\frac{\sqrt{3}}{2})(1) = \frac{3\sqrt{3}}{4}$; $y = \frac{1}{4}(1) + (\frac{\sqrt{3}}{2})(\sqrt{3}) = \frac{7}{4}$.

Therefore, the tangent line intersects the xy-plane at $(\frac{3\sqrt{3}}{4}, \frac{7}{4}, 0)$.

29. $d^2 = |PQ|^2 - [\text{scalar projection of } PQ \text{ on } n]^2$

$= |PQ|^2 - \left(\frac{PQ \cdot n}{|n|}\right)^2 = \frac{|PQ|^2|n|^2 - (PQ \cdot n)^2}{|n|^2} = \frac{|PQ \times n|^2}{|n|^2}$ [LaGrange]

Thus, $d = \frac{|PQ \times n|}{|n|}$.

(a) $P(3,-2,1)$ is on the line so $PQ = \langle-2,2,-5\rangle$ and $n = \langle2,-2,1\rangle$.

Then $d = \frac{|\langle-8,-8,0\rangle|}{\sqrt{4+4+1}} = \frac{8\sqrt{2}}{3} \approx 3.7712$.

(b) $P(1,-1,0)$ is on the line so $PQ = \langle1,0,3\rangle$ and $n = \langle2,3,-6\rangle$.

Then $d = \frac{|\langle-9,12,3\rangle|}{\sqrt{4+9+36}} = \frac{3\sqrt{26}}{7} \approx 2.1853$.

Problem Set 14.5 Velocity, Acceleration, and Curvature

1. $v(t) = \langle3,8t,6t^2\rangle$; $a(t) = \langle0,8,12t\rangle$.
   $v(t) = \langle3,8,6\rangle$; $s(1) = \sqrt{109} \approx 10.4403$; $a(1) = \langle0,8,12\rangle$.

3. $r(t) = \langle2t-t^2,3t,t^3+1\rangle$; $v(t) = \langle2-2t,3,3t^2\rangle$; $a(t) = \langle-2,0,6t\rangle$.
   $v(1) = \langle0,3,3\rangle$; $s(1) = |v(1)| = \sqrt{0+9+9} \approx 4.2426$; $a(1) = \langle-2,0,6\rangle$.

5. $v(t) = \langle-2\sin2t,-2e^{-t},3\cos t\rangle$; $a(t) = \langle-4\cos2t,2e^{-t},-3\sin t\rangle$.
   $v(0) = \langle0,-2,3\rangle$; $s(0) = \sqrt{13} \approx 3.6056$; $a(0) = \langle-4,2,0\rangle$.

7. $|v| = k \Rightarrow |v|^2 = v \cdot v = k^2 \Rightarrow D_t(v \cdot v) = 0 \Rightarrow 2v \cdot v = 0$
   $\Rightarrow v \cdot a = 0 \Rightarrow v$ and a are perpendicular.

9. $\mathbf{v}(t) = \langle t\cos t + \sin t, -t\sin t + \cos t, \sqrt{8} \rangle$.

$|\mathbf{v}(t)| = [t^2\cos^2 t + 2t\sin t\cos t + \sin^2 t) + (t^2\sin^2 t - 2t\sin t\cos t + \cos^2 t) + (8)]^{1/2}$

$= [t^2(\cos^2 t + \sin^2 t) + (\sin^2 t + \cos^2 t) + 8]^{1/2} = (t^2 + 9)^{1/2}$.

Then length $= \displaystyle\int_0^4 \sqrt{t^2 + 9}\, dt = \left[\frac{t}{2}\sqrt{t^2 + 9} + \frac{9}{2}\ln(t + \sqrt{t^2 + 9})\right]_0^4$ [Formula 44]

$= [2(5) + (9/2)\ln|4 + 5|] - [0 + 9/2)\ln|0 + 3|] \approx 14.9438$.

11. $\mathbf{v}(t) = \langle 3t^{1/2}, 1, 3 \rangle$; $|\mathbf{v}(t)| = (9t + 10)^{1/2}$.

Length $= \displaystyle\int_6^{10} (9t + 10)^{1/2} dt = \left[(2/27)(9t + 10)^{3/2}\right]_6^{10} = 976/27 \approx 36.1481$.

13. $\mathbf{v}(t) = \langle 2\cosh 2t, 2\sinh 2t, 2 \rangle$.

$|\mathbf{v}(t)| = (4\cosh^2 2t + 4\sinh^2 2t + 4)^{1/2} = 2(2\cosh^2 2t)^{1/2} = 2\sqrt{2}\cosh 2t$.

$L = \displaystyle\int_{-12}^{1} 2\sqrt{2}\cosh 2t\, dt = \left[\sqrt{2}\sinh 2t\right]_{-1/2}^{1} = \sqrt{2}(\sinh 2 + \sinh 1) \approx 6.7911$.

15. $\mathbf{r}(t) = \langle t^2 - 1, 2t + 3, t^2 - 4 \rangle$; $\mathbf{v}(t) = \langle 2t, 2, 2t - 4 \rangle$; $\mathbf{a}(t) = \langle 2, 0, 2 \rangle$.

$\mathbf{r}(2) = \langle 3, 7, -4 \rangle$; $\mathbf{v}(2) = \langle 4, 2, 0 \rangle$; $|\mathbf{v}(2)| = 2\sqrt{5}$; $\mathbf{a}(2) = \langle 2, 0, 2 \rangle$.

$\kappa = \dfrac{|\mathbf{v} \times \mathbf{a}|}{|\mathbf{v}|^3} = \dfrac{|2\langle 2,1,0 \rangle \times 2\langle 1,0,1 \rangle|}{(2\sqrt{5})^3} = \dfrac{4|\langle 1,-2,-1 \rangle|}{40\sqrt{5}} = \dfrac{\sqrt{6}}{10\sqrt{5}} \approx 0.1095$.

$T = \dfrac{\mathbf{v}}{|\mathbf{v}|} = \dfrac{2\langle 2,1,0 \rangle}{2\sqrt{5}} = \dfrac{\langle 2,1,0 \rangle}{\sqrt{5}}$.

$a_N = \kappa|\mathbf{v}|^2 = \dfrac{\sqrt{6}}{10\sqrt{5}}(2\sqrt{5})^2 = \dfrac{2\sqrt{30}}{5}$; $a_T = \dfrac{\mathbf{v} \cdot \mathbf{a}}{|\mathbf{v}|} = \dfrac{2\langle 2,1,0 \rangle \cdot 2\langle 1,0,1 \rangle}{2\sqrt{5}} = \dfrac{4}{\sqrt{5}}$.

$N = \dfrac{\mathbf{a} - a_T T}{a_N} = \dfrac{\langle 2,0,2 \rangle - \langle 8/5, 4/5, 0 \rangle}{2\sqrt{30}/5} = \dfrac{\langle -3,-4,5 \rangle}{\sqrt{30}}$.

$B = T \times N = \dfrac{\langle 2,1,0 \rangle \times \langle -3,-4,5 \rangle}{\sqrt{5}\ \sqrt{30}} = \dfrac{\langle 1,-2,-1 \rangle}{\sqrt{6}}$.

17. $\mathbf{v}(t) = \langle 3\cos 3t, -3\sin 3t, 1\rangle$; $\mathbf{a}(t) = \langle -9\sin 3t, -9\cos 3t, 0\rangle$.

$\mathbf{v}(\pi/9) = (1/2)\langle 3, -3\sqrt{3}, 2\rangle$; $|v(\pi/9)| = \sqrt{10}$; $\mathbf{a}(\pi/9) = (-9/2)\langle\sqrt{3}, 1, 0\rangle$.

$a_N = \dfrac{|\mathbf{v}\mathbf{x}\mathbf{a}|}{|\mathbf{v}|} = 9$. $a_T = \dfrac{\mathbf{v}\cdot\mathbf{a}}{|\mathbf{v}|} = 0$. $\kappa = \dfrac{a_N}{|\mathbf{v}|^2} = \dfrac{9}{10}$. $\mathbf{T} = \dfrac{\mathbf{v}}{|\mathbf{v}|} = \dfrac{\langle 3, -3\sqrt{3}, 2\rangle}{2\sqrt{10}}$.

$\mathbf{N} = \dfrac{\mathbf{a} - a_T\mathbf{T}}{a_N} = \dfrac{\langle\sqrt{3}, 1, 0\rangle}{-2}$. $\mathbf{B} = \mathbf{T}\mathbf{x}\mathbf{N} = \dfrac{\langle 1, -\sqrt{3}, -6\rangle}{2\sqrt{10}}$.

19. $\mathbf{v}(t) = e^t\langle\cos t + \sin t, -\sin t + \cos t, 1\rangle$; $\mathbf{a}(t) = e^t\langle 2\cos t, -2\sin t, 1\rangle$.

$\mathbf{v}(\pi/2) = e^{\pi/2}\langle 1, -1, 1\rangle$; $|v(\pi/2)| = e^{\pi/2}\sqrt{3}$; $\mathbf{a}(\pi/2) = e^{\pi/2}\langle 0, -2, 1\rangle$.

$a_N = \dfrac{|\mathbf{v}\mathbf{x}\mathbf{a}|}{|\mathbf{v}|} = e^{\pi/2}\sqrt{2}$. $a_T = \dfrac{|\mathbf{v}\cdot\mathbf{a}|}{|\mathbf{v}|} e^{\pi/2}\sqrt{3}$. $\kappa = \dfrac{a_N}{|\mathbf{v}|^2} = \sqrt{2}/3\, e^{-\pi/2} \approx 0.098$.

$\mathbf{T} = \dfrac{\mathbf{v}}{|\mathbf{v}|} = \dfrac{\langle 1, -1, 1\rangle}{\sqrt{3}}$. $\mathbf{N} = \dfrac{\mathbf{a} - a_T\mathbf{T}}{a_N} = \dfrac{\langle 1, 1, 0\rangle}{-\sqrt{2}}$. $\mathbf{B} = \mathbf{T}\mathbf{x}\mathbf{N} - \dfrac{\langle 1, -1, -2\rangle}{\sqrt{6}}$.

21. $\mathbf{v}(t) = \langle 1, 2t, 3t^2\rangle$; $|v(t)| = (1+4t^2+9t^4)^{1/2}$; $\mathbf{a}(t) = \langle 0, 2, 6t\rangle$.

$a_T = \dfrac{\mathbf{v}\cdot\mathbf{a}}{|\mathbf{v}|} = \dfrac{4t+18t^3}{(1+4t^2+9t^4)^{1/2}}$. $a_N = \dfrac{|\mathbf{v}\mathbf{x}\mathbf{a}|}{|\mathbf{v}|} = \left[\dfrac{4(9t^4+9t^2+1)}{(9t^4+4t^2+1)}\right]^{1/2}$.

23. $\mathbf{v}(t) = \langle 1, t^2, -t^{-2}\rangle$; $|v(t)| = t^{-2}(t^8+t^4+1)^{1/2}$; $\mathbf{a}(t) = \langle 0, 2t, 2t^{-3}\rangle$.

$a_T = \dfrac{\mathbf{v}\cdot\mathbf{a}}{|\mathbf{v}|} = \dfrac{2(t^8-1)}{t^3(t^8+t^4+1)^{1/2}}$. $a_N = \dfrac{|\mathbf{v}\mathbf{x}\mathbf{a}|}{|\mathbf{v}|} = \dfrac{2(t^8+4t^4+1)^{1/2}}{t(t^8+t^4+1)^{1/2}}$.

25. $\mathbf{r}(t) = \langle 10\cos t, 10\sin t, (34/2\pi)t\rangle$.

Using the result of Example 1 with $a = 10$ and $c = 34/2\pi$, the length of one complete turn is $2\pi\sqrt{(10)^2 + (34/2\pi)^2}$ angstroms $= 10^{-8}\sqrt{400\pi^2+34^2}$ cm. Therefore, the total length of the helix is $(2.9)(10^8)(10^{-8})\sqrt{400\pi^2+34^2}$ ≈ 207.1794 cm.

27. $r(t) = 100e^{-t} \langle \cos t, \sin t, 1 \rangle$. Consider the $\lim_{t \to \infty}$ of each component of r.

$$\lim_{t \to \infty} (100e^{-t}\cos t) = \lim_{t \to \infty} \frac{100 \cos t}{e^t} = 0.$$

Similarly for the second and third components of $r(t)$, so the bee's final resting place is the origin.

$$r'(t) = -100e^{-t} \langle \cos t, \sin t, 1 \rangle + 100e^{-t} \langle -\sin t, \cos t, 0 \rangle$$
$$= 100e^{-t} \langle -\cos t - \sin t, -\sin t + \cos t, -1 \rangle$$

$$|r'(t)| = 100e^{-t} \sqrt{[-(\cos t - \sin t)^2 + (-\sin t + \cos t)^2 + (-1)^2]}$$

$$= 100e^{-t} \sqrt{[2(\cos^2 t + \sin^2 t) + 1]} = 100e^{-t}\sqrt{[2(1)+1)]} = 100\sqrt{3}e^{-t}.$$

The bee traveled $\int_0^\infty 100\sqrt{3}e^{-t}dt = \lim_{b \to \infty} \int_0^b 100\sqrt{3}e^{-t}dt = \lim_{b \to \infty}\left[-100\sqrt{3}e^{-t}\right]_0^b$

$$= \lim_{b \to \infty}\left[-100\sqrt{3}e^{-t}\right]_0^b = \lim_{b \to \infty}\left[\frac{-100\sqrt{3}}{e^b} + 100\sqrt{3}\right] = 100\sqrt{3} \approx 173.2051$$

29. $r(t) = \langle \cos t, \sin t, 16t \rangle$. Death occurred at $r(12) = \langle \cos 12, \sin 12, 192 \rangle$. $r'(t) = \langle -\sin t, \cos t, 16 \rangle$. $r'(12) = \langle -\sin 12, \cos 12, 16 \rangle$. Now let T=0 when the bee died, and let $s(T)$ describe the path of the bee from the instant of death.

Then $s_0 = s(0) = \langle \cos 12, \sin 12, 192 \rangle$ and $v_0 = v(0) = \langle -\sin 12, \cos 12, 16 \rangle$.

We know that $d^2s/dT^2 = \langle 0, 0, -32 \rangle$, so using the result of Problem 28(c),
   $s(T) = \langle 0, 0, -32 \rangle [(1/2)T^2] + \langle -\sin 12, \cos 12, 16 \rangle T + \langle \cos 12, \sin 12, 192 \rangle$.
It landed when z=0; $-32[(1/2)T^2] + 16T + 192 = 0$; solve and get T=4.

$s(4) = \langle -4\sin 12 + \cos 12, 4\cos 12 + \sin 12, 0 \rangle$, so it landed on the xy-plane approximately at the point (2.99, 2.84, 0).

31. $(F \times G)' = [\langle f_2 g_3 - f_3 g_2, f_3 g_1 - f_1 g_3, f_1 g_2 - f_2 g_1 \rangle]'$

$$= \langle (f_2 g_3' + f_2' g_3) - f_3 g_2' + f_3' g_2), (f_3 g_1' + f_3' g_1) - (f_1 g_3' + f_1' g_3) (f_1 g_2' + f_1' g_2)$$

$$- (f_2 g_1' + f_2' g_1) \rangle.$$

$$FxG' + F'xG = \langle f_2 g_3' - f_3 g_2', f_3 g_1' - f_1 g_3', f_1 g_2' - f_2 g_1' \rangle$$

$$+ \langle f_2' g_3 - f_3' g_2 \, f_3' g_1 - f_1' g_3, f_1' g_2 - f_2' g_1 \rangle$$

$$= \langle (f_2 g_3' + f_2' g_3) - (f_3 g_2' + f_3' g_2), (f_3 g_1' + f_3' g_1) - (f_1 g_3' + f_1' g_3),$$

$$(f_1 g_2' + f_1' g_2) - (f_2 g_1' + f_2' g_1) \rangle.$$

Therefore, $(d/dt)\,(\mathbf{r} \times \mathbf{r}') = \mathbf{r} \times \mathbf{r}'' + \mathbf{r}' \times \mathbf{r}' = \mathbf{r} \times \mathbf{r}'' + \mathbf{0} = \mathbf{r}' \times \mathbf{r}''$.

33. (a) $\mathbf{L}'(t) = m\mathbf{r}(t) \times \mathbf{a}(t) = \tau(t)$. [Using result of Problem 31]

(b) $\tau(t) = \mathbf{0}$ for all $t \Rightarrow \mathbf{L}'(t) = \mathbf{0}$ for all t
$\Rightarrow \mathbf{L}(t) = \mathbf{k}$ for all t [$\mathbf{k}$ is some constant.]

(c) A particle moving under the influence of a central force satisfies $\mathbf{r}''(t) = c\mathbf{r}(t)$, or $\mathbf{a}(t) = c\mathbf{r}(t)$. Thus, $\tau(t) = m\mathbf{r}(t) \times c\mathbf{r}(t) = \mathbf{0}$, so from (b) we can conclude that $\mathbf{L}(t)$ is a constant.

35. $\mathbf{B} = \mathbf{T} \times \mathbf{N} = (1/\sqrt{6})\langle 1,2,1 \rangle \times (1/\sqrt{2})\langle -1,0,1 \rangle = (1/\sqrt{3})\langle 1,-1,1 \rangle$, so $\langle 1,-1,1 \rangle$ is perpendicular to the osculating plane at $(1,1,1/3)$. Therefore, an equation of the osculating plane at $(1,1,1/3)$ is

$1(x-1) - 1(y-1) + 1(z-1/3) = 0$, or $3x - 3y + 3z = 1$.

Problem Set 14.6 Surfaces in Three-Space

1. $\dfrac{x^2}{16} + \dfrac{y^2}{25} = 1$ (Elliptic cylinder).   3. $2x + 5z - 12 = 0$ is a plane (cylinder) parallel to the y-axis. The xz-trace is a line.

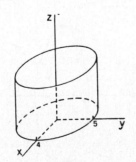
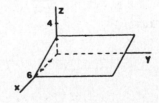

5. $(x-4)^2+(y+1)^2 = 4$
    (Circular cylinder).

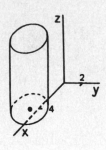

7. $\dfrac{x^2}{16} + \dfrac{y^2}{4} + \dfrac{z^2}{9} = 1$ (Ellipsoid).

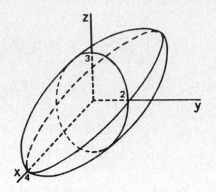

9. $z = \dfrac{x^2}{3} + \dfrac{y^2}{4/3}$
    is an ellliptic paraboloid
    with z-axis for the axis of
    symmetry.

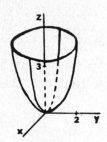

11. (Cylindrical surface).

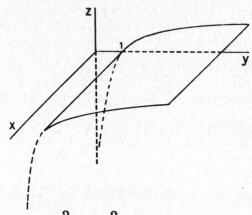

13. (Hyperbolic paraboloid).

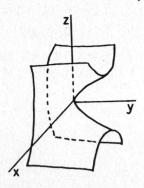

15. $y = \dfrac{x^2}{4} + \dfrac{z^2}{9}$ is an elliptic
    paraboloid with y-axis for
    axis of symmetry.

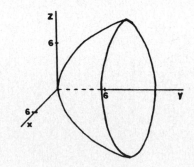

17. (Plane)   **19.** (Hemisphere).

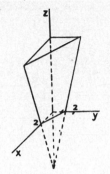   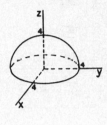

21. (a) Replacing x by -x results in an equivalent equation.

　　　(b) Replacing x by -x and y by -y results in an equivalent equation.

　　　(c) Replacing y by -y and z by -z results in an equivalent equation.

　　　(d) Replacing x by -x, y by -y, and z by -z, results in an equivalent equation.

23. For each value of y=k (a constant) a circle whose radius squared is $x^2 = y/2 = k/2$ is generated, so in the plane y=k an equation of the circle is $x^2 + z^2 = k/2$ or $k = 2x^2 + 2z^2$. Hence, an equation of the surface is $y = 2x^2 + 2z^2$.

25. For each value of y=k (a constant) a circle whose radius squared is $x^2 = (1/4)(12 - 3k^2)$ is generated, so in the plane y=k an equation of the circle is $x^2 + z^2 = (1/4)(12 - 3k^2)$ or $4x^2 + 3k^2 + 4z^2 = 12$. Hence, an equation of the surface is $4x^2 + 3y^2 + 4z^2 - 12$.

27. Substituting z = 4 into the given equation results in $\dfrac{x^2}{16} + \dfrac{y^2}{36} = 1$. $a^2 = 36$, $b^2 = 16$; then $c^2 = a^2 - b^2 = 36 - 16 = 20$, so $c = 2\sqrt{5}$. Therefore, the foci are $(0, \pm 2\sqrt{5}, 4)$.

29. $\dfrac{x^2}{a^2} + \dfrac{y^2}{b^2} + \dfrac{h^2}{c^2} = 1$ can be expressed as $\dfrac{x^2}{\dfrac{a^2(c^2-h^2)}{c^2}} + \dfrac{y^2}{\dfrac{b^2(c^2-h^2)}{c^2}} = 1$.

Hence, area of the cross section is $\pi \left[\dfrac{a\sqrt{c^2-h^2}}{c}\right]\left[\dfrac{b\sqrt{c^2-h^2}}{c}\right] =$ $\dfrac{\pi ab(c^2-h^2)}{c^2}$.

31. $x^2+z^2 = 4-x^2$, or $2x^2+z^2 = 4$, or $\dfrac{x^2}{2} + \dfrac{z^2}{4} = 1$.

Major diameter is $2\sqrt{4} = 4$; minor diameter is $2\sqrt{2}$.

33. $x^2+y^2-z^2 = (t\cos t)^2 + (t\sin t)^2 - (t)^2 = t^2(\cos^2 t+\sin^2 t)-t^2 = t^2(1)-t^2 = 0$, so each point of the spiral $\mathbf{r}(t) = \langle t\cos t, t\sin t, t\rangle$ lies on the circular cone $x^2+y^2-z^2 = 0$.

Each point of $\mathbf{r}(t) = \langle 3t\cos t, t\sin t, t\rangle$ satisfies $x^2+9y^2-9z^2 = 0$ since $x^2+9y^2-9z^2 = (3t\cos t)^2+9(t\sin t)^2-9(t)^2 = 9t^2(\cos^2 t+\sin^2 t)-9t^2 = 0$, so the spiral lies on the elliptic cone.

Problem Set 14.7   Cylindrical and Spherical Coordinates

1. (a) $x = 6\cos(\pi/6) = 3\sqrt{3}$.   (b) $x = 4\cos(4\pi/3) = -2$.

     $y = 6\sin(\pi/6) = 3$. $y = 4\sin(4\pi/3) = 2\sqrt{3}$.

     $z = -2$. $z = -8$.

3. (a) $\rho = \sqrt{x^2+y^2+z^2} = \sqrt{4+12+16} = 4\sqrt{2}$.

     $\tan\theta = \dfrac{y}{x} = \dfrac{-2\sqrt{3}}{2} = -\sqrt{3}$ and (x,y) is in the 4th quadrant so $\theta = 5\pi/3$.

     $\cos\phi = \dfrac{z}{\rho} = \dfrac{4}{4\sqrt{2}} = \dfrac{\sqrt{2}}{2}$ so $\phi = \pi/4$. Spherical: $(4\sqrt{2},\ 5\pi/3,\ \pi/4)$

   (b) $\rho = \sqrt{2+2+12} = 4$.

     $\tan\theta = \dfrac{\sqrt{2}}{-\sqrt{2}} = -1$ and (x,y) is in 2nd quadrant so $\theta = \dfrac{3\pi}{4}$.

     $\cos\phi = \dfrac{2\sqrt{3}}{4} = \dfrac{\sqrt{3}}{2}$ so $\phi = \pi/6$. Spherical: $(4, 3\pi/4, \pi/6)$

5. r-5 (Cylinder).

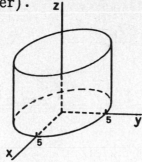

7. $\phi = \pi/6$ (Cone).

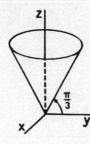

9. r = 3cosθ (Circular cylinder).

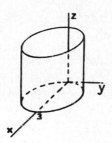

11. $\rho = 3\cos\phi$
$x^2 + y^2 + (z - 3/2)^2 = 9/4$ (Sphere)

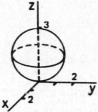

13. $r^2 + z^2 = 9$.
$x^2 + y^2 + z^2 = 9$ (Sphere).

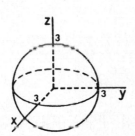

15. $x^2 + y^2 = 9$; $r^2 = 9$; r = 3.

17. $r^2 + 4z^2 = 10$.

19. $(x^2+y^2+z^2)-3z^2 = 0$; $\rho^2 - 3\rho^2\cos^2\phi = 0$; $\cos^2\phi = 1/3$ (pole is not lost); $\cos^2\phi = 1/3$ [or $\sin^2\phi = 2/3$ or $\tan^2\phi = 2$].

21. $(r^2+z^2) + z^2 = 4$; $\rho^2 + \rho^2\cos^2\phi = 4$; $\rho^2 = \dfrac{4}{1+\cos^2\phi}$.

23. $r\cos\theta + r\sin\theta = 4$; $r = 4(\sin\theta + \cos\theta)^{-1}$.

25. $(x^2+y^2+z^2)-z^2=9$; $\rho^2-\rho^2\cos^2\phi = 9$; $\rho^2(1-\cos^2\phi) = 9$; $\rho^2\sin^2\phi = 9$, $\rho = \dfrac{3}{\sin\phi}$.

27. $r^2\cos2\theta = z$; $r^2(\cos^2\theta-\sin^2\theta) = z$; $(r\cos\theta)^2-(r\sin\theta)^2 = z$; $x^2-y^2=z$.

29. $z = 2x^2+2y^2 = 2(x^2+y^2)$ (Cartesian); $z = 2r^2$ (cylindrical).

31. Use results of Example 7 for St. Paul: $P_1(-151.4,-2796,2800)$.

Oslo: $\rho = 3960$, $\theta = 10.5°$, $\phi = 90° - 59.6° = 30.4°$.
 $x = (3960)\sin(30.4°)\cos(10.5°) \approx 1970$.
 $y = (3960)\sin(30.4°)\sin(10.5°) \approx 365.2$.
 $z = (3960)\cos(30.4°) \approx 3416$.

$\cos a \approx \dfrac{(-151.4)(1970) + (-2796)(365.2) + (2800)(3416)}{(3960)(3960)} \approx 0.5258$.

$a \approx 1.0170$ rad, so $d \approx 3960(1.0171) \approx 4028$ miles.

33. Using the results of Example 7, the Cartesian coordinates of St. Paul area $(-151.4, -2796, 2800)$.

Turin: $\rho = 3960$, $\theta = 7.4°$, $\phi = 90°-45° = 45°$. Therefore,
 $x = (3960)(\sin45°)(\cos7.4°) \approx 2777$.
 $y = (3960)(\sin45°)(\sin7.4°) \approx 360.6$
 $z = (3960)(\cos45°) \approx 2800$.

Let Ω be the angle formed by the Earth radius to St. Paul with the Earth radius to Turin.

Then $\cos\Omega = \dfrac{\langle-151.4, -2796,2800\rangle\cdot\langle2777,360.6,2800\rangle}{(3960)(3960)} \approx 0.0488$; $\Omega \approx 1.1496$.

Therefore, the great-circle distance between St. Paul and Turin in $3960(1.1496) \approx 4552$ miles.

35. Let β be the angle formed by the plane of 0-P_1-P_2 and the z-axis. It is the complement of the smaller angle between $OP_1 \times OP_2$ and the z-axis.

Thus, $\beta = \dfrac{\pi}{2} - \cos^{-1}\left[\dfrac{|(OP_1 \times OP_2) \cdot \mathbf{k}|}{|OP_1 \times OP_2||\mathbf{k}|}\right] \approx 0.5690$.

Therefore, the distance between the North Pole and the NY-Turin great circle is $(3960)(0.5690) \approx 2253$ miles.

37. The great circle distance is at $a\gamma$ if γ is the central angle.

By Problem 36, $d^2 = (a-a)^2 + 2a^2[1 - \cos(\theta_1 - \theta_2)\sin\phi_1\sin\phi_2 - \cos\phi_1\cos\phi_2]$

$= 2a^2[1 - \cos(\theta_1 - \theta_2)\sin\phi_1\sin\phi_2 - \cos\phi_1\cos\phi_2]$.

On the other hand, if the law of cosines is used,

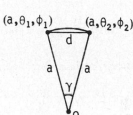

$d^2 = a^2 + a^2 - 2a^2\cos\gamma = 2a^2(1 - \cos\gamma)$.

Thus, $\cos\gamma = \cos(\theta_1 - \theta_2)\sin\phi_1\sin\phi_2 + \cos\phi_1\cos\phi_2$.

39. [Express γ in radians since the $d = a\gamma$ formula is based on γ being in radians.]

(a) NY$(-74°, 40.4°)$; Greenwich $(0°, 51.3°)$

$\cos\gamma = \cos(-74° - 0°)\cos(40.40°)\cos(51.3°) + \sin(40.4°)\sin(51.3°)$
≈ 0.637
Then $\gamma \approx 0.880$, so $d \approx 3960(0.880) \approx 3480$ mi.

(b) St. Paul $(-93.1°, 45°)$; Turin $(7.4°, 45°)$
$\cos\gamma = \cos(-93.1° - 7.4°)\cos(45°)\cos(45°) + \sin(45°)\sin(45°) \approx 0.409$
Then $\gamma \approx 1.150$, so $d \approx 3960(1.14957) \approx 4550$ mi.

(c) South Pole $(7.4°, -90°)$; Turin $(7.4°, 45°)$
Then $\gamma = 135° = 3\pi/4$ radians is clear, so $d = 3960(3\pi/4) \approx 9330$ mi.

(d) NY$(-74°, 40.4°)$; Cape Town$(18.4°, -33.9°)$
$\cos\gamma = \cos(-74° - 18.4°)\cos(40.4°)\cos(-33.9°) + \sin(40.4°)\sin(-33.9°)$
≈ -0.388
Then $\gamma \approx 1.969$, so $d \approx 3960(1.969) \approx 7800$ mi.

(e) It is clear the $\gamma = 180° = \pi$ radians, so $d = 3960\pi \approx 12,440$ mi.

Problem Set 14.8   Chapter Review

1. True. The coordinates are defined in terms of distances from the coordinate planes in such a way that they are unique.

3. True. See PLANES, Section 14.2.

5. False. The distance between $(0,0,3)$ and $(0,0,-3)$ (a point from each plane) is 6, so the distance between the planes is less than or equal to 6.

7. True. At $t = 1/2$.

9. True. $||u|u| = |u||u| = |u|^2$.   [LENGTH AND DOT PRODUCT, Section 1.3.3]

11. True. $|uxv| = |-vxu| = |-1||vxu| = |vxu|$.

13. False. Obviously not true if $u = v$. [More generally, it is only true when u and v are also perpendicular.]

15. True. $\dfrac{|uxv|}{(u \cdot v)} = \dfrac{|u||v|\sin\theta}{|u||v|\cos\theta} = \tan\theta$.

17. True. $|(2ix2j) \cdot (jxi)| = 4|(k) \cdot (-k)| = 4(k \cdot k) = 4$.

19. True. Since $\langle b_1, b_2, b_3 \rangle$ is normal to the plane.

21. False. C^{ex}: Let $r(t) = \langle 0,1,t \rangle$. Then $|r'(t)| = |\langle 0,0,1 \rangle| = 1$; but $D_t|r(t)| = D_t(1+t^2)^{1/2} = \dfrac{t}{(1+t^2)^{1/2}}$ which is never 1.   In general, $D_t|r(t)| = \dfrac{r(t) \cdot r'(t)}{|r(t)|}$.

23. False. It is the nonnegative part of the z-axis.

25. True.

Sample Test Problems

1. Center is the midpoint $(1,2,4)$. Square of the radius is $9+1+1 = 11$. The equation is $(x-1)^2 + (y-2)^2 + (z-4)^2 = 11$.

3. (a) $|\mathbf{a}| = \sqrt{4+1+4} = 3$; $|\mathbf{b}| = \sqrt{25+1+9} = \sqrt{35}$.

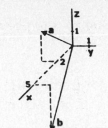

(b) $2/3, -1/3, 2/3$ for $\mathbf{a}$;

$5/\sqrt{35}, 1/\sqrt{35}, -3/\sqrt{35}$ for $\mathbf{b}$.

(c) $\frac{\mathbf{a}}{3} = \langle 2/3, -1/2, 2/3 \rangle$.

(d) $\cos\theta = \dfrac{\mathbf{a}\cdot\mathbf{b}}{|\mathbf{a}||\mathbf{b}|} = \dfrac{10-1-6}{3\sqrt{35}} = \dfrac{1}{\sqrt{35}}$. $\theta \approx 1.4010$ rad. $[80.27°]$

5. $k\langle 3,3,-1\rangle \times \langle -1,-2,4\rangle = k\langle 10,-11,-3\rangle$ for each k in $\mathbb{R}$.

7. (a) $y = 7$ (since y must be a constant).

(b) $x = -5$ (since it is parallel to the yz-plane).

(c) $z = -2$ (since it is parallel to the xy-plane).

(d) $3x-4y+z = -45$ [since it can be expressed as

$3x-4y+z = D$ and then D must satisfy

$3(-5)-4(7)+(-2) = D$, so $D = -45$].

9. Vectors normal to the planes are perpendicular so $\langle 1,5,C\rangle \cdot \langle 4,-1,1\rangle = 0$. Therefore, $4-5+C = 0$, $C=1$.

11. A vector in the direction of the line is $\langle 8,1,-8\rangle$. Then parametric equations are $x = -2+8t$, $y = 1+t$, $z = 5-8t$.

13. $\langle 50,25,0\rangle = 25\langle 2,1,0\rangle$ is in direction of the line. Parametric equations are $x = 0+2t$, $y = 25+1t$, $z = 16+0t$ or $x = 2t$, $y = 25+t$, $x = 16$.

15. $\langle 5,-4,-3\rangle$ is a vector in the direction of the line, and $\langle 2,-2,1\rangle$ is a position vector to the line. Then a vector equation of the line is $\mathbf{r}(t) = \langle 2,-2,1\rangle + t\langle 5,-4,-3\rangle$.

17. $\mathbf{r}'(t) = \langle 1,t,t^2\rangle$, $\mathbf{r}'(2) = \langle 1,2,4\rangle$ and $\mathbf{r}(2) = \langle 2,2,8/3\rangle$. Symmetric equations for tangent lines are $\dfrac{x-2}{1} = \dfrac{y-2}{2} = \dfrac{z-8/3}{4}$. Normal plane is $1(x-2) + 2(y-2) + 4(x-8/3) = 0$ or $3x+6y+12z = 50$.

19. $\mathbf{r}'(t) = e^t\langle\cos t + \sin t, -\sin t, +\cos t, 1\rangle$ $|\mathbf{r}'(t)| = \sqrt{3}e^t$

Length is $\int_1^5 \sqrt{3}e^t \, dt = \left[\sqrt{3} \, e^t\right]_1^5 = \sqrt{3}(e^5 - e) \approx 252.3509.$

21. $\mathbf{v}(t) = \langle 1, 2t, 3t^2\rangle$; $\mathbf{a}(t) = \langle 0, 2, 6t\rangle.$

$\mathbf{v}(1) = \langle 1, 2, 3\rangle$; $|\mathbf{v}(1)| = \sqrt{14}$; $\mathbf{a}(1) = \langle 0, 2, 6\rangle.$

$a_T = \dfrac{\mathbf{v}\cdot\mathbf{a}}{|\mathbf{v}|} = \dfrac{0+4+18}{\sqrt{14}} = \dfrac{22}{\sqrt{14}} \approx 5.880$; $\quad a_N = \dfrac{|\mathbf{v}\times\mathbf{a}|}{|\mathbf{v}|} = \dfrac{|\langle 6, -6, 2\rangle|}{\sqrt{14}} = \dfrac{2\sqrt{19}}{\sqrt{14}} \approx 2.330.$

23. (Sphere).

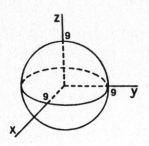

25. (Circular paraboloid).

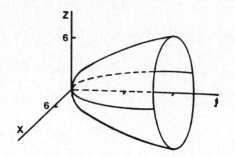

27. $3x+3y-6z = 12$ is a plane.

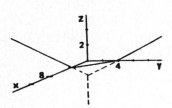

29. $\dfrac{x^2}{12} + \dfrac{y^2}{9} + \dfrac{z^2}{4} = 1.$ (Ellipsoid).

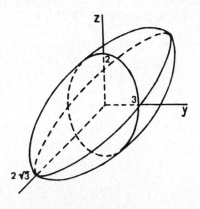

31. (a) $r^2 = 9$; $r = 3.$

(b) $(x^2+y^2) + 3y^2 = 16.$ $r^2+3r^2\sin^2\theta = 16$, $r^2 = \dfrac{16}{1+3\sin^2\theta}$.

(c) $r^2 = 9z.$

(d) $r^2+4z^2 = 10.$

33. (a) $\rho^2 = 4$; $\rho = 2$.

(b) $x^2 + y^2 + z^2 - 2z^2 = 0$; $\rho^2 - 2\rho^2 \cos^2\phi = 0$; $\rho^2(1 - 2\cos^2\phi) = 0$;
$1 - 2\cos^2\phi = 0$; $\cos^2\phi = 1/2$; $\phi = \pi/4$ or $\phi = 3\pi/4$.

Any of the following (as well as others) would be acceptable:

$(\phi - \pi/4)(\phi - 3\pi/4) = 0$,

$\cos^2\phi = 1/2$,

$\sec^2\phi - 2$,

$\tan^2\phi = 1$.

(c) $2x^2 - (x^2 + y^2 + z^2) = 1$; $2\rho^2 \sin^2\phi \cos^2\theta - \rho^2 = 1$; $\rho^2 = \dfrac{1}{2\sin^2\phi \cos^2 - 1}$.

(d) $x^2 + y^2 = z$; $\rho^2 \sin^2\phi \cos^2\theta + \rho^2 \sin^2\phi \sin^2\theta = \rho\cos\phi$;

$\rho^2 \sin^2\phi (\cos^2\theta + \sin^2\theta) = \rho\cos\phi$; $\rho\sin^2\phi = \cos\phi$; $\rho = \cot\phi\csc\phi$.

[Note that when we divided through by ρ in (c) and (d) we did not lose
the pole since it is also a solution of the resulting equations.]

35. $(2,0,0)$ is a point of the first plane. The distance between the planes
is then $\dfrac{|2(2) - 3(0) + 3(0) - 9|}{(4+9+3)^{1/2}} = 1.25$.

37. See Problem 7, Section 14.5.

Problem Set 15.1   Functions of Two or More Variables

1.   (a) 5.   (b) 0.   (c) 6.   (d) $a^6 \pm a^2$.   (e) $x^2 \pm x^2$, $x \neq 0$.   (f) Undefined.

The natural domain is the set of all (x,y) such that y is nonnegative.

3.   (a) $\sin(2\pi) = 0$.   (b) $4\sin(\pi/6) = 2$.
(c) $16\sin(\pi/2) = 16$.   (d) $\pi^2 \sin(\pi^2) \approx -4.2469$.

(e) $1.44\sin[(3.1)(4.2)] \approx 0.6311$.

5. $F(t\cos t, \sec^2 t) = t^2 \cos^2 t \, \sec^2 t = t^2$, $\cos t \neq 0$.

7. $z = 6$ (Plane).   9. $x+2y+z = 6$ is a plane.

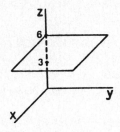   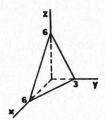

11. $x^2+y^2+z^2 = 16$, $z \geq 0$,   13. $z = 3-x^2-y^2$ (paraboloid).
(hemisphere).

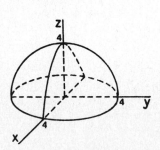   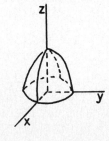

15. $z = \exp[-(x^2+y^2)]$.

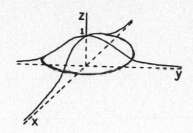

17. $x^2+y^2 = 2z$; $x^2+y^2 = 2k$.

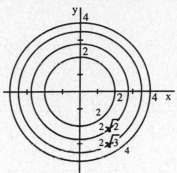

k=0	Point	
k=2	$x^2+y^2 = 4$	(Circle $r = 2$)
k=4	$x^2+y^2 = 8$	(Circle $r = 2\sqrt{2}$)
k=6	$x^2+y^2 = 12$	(Circle $r = 2\sqrt{3}$)
k=8	$x^2+y^2 = 16$	(Circle $r = 4$)

19. $x^2 = zy$, $y \neq 0$; $x^2 = ky$, $y \neq 0$.

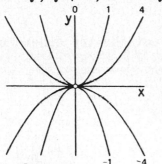

21. $z = \dfrac{x^2+y}{x+y^2}$, $x \neq -y^2$.

k=0: $y = -x^2$.
 [parabola except $(0,0)$ and
                                    $(-1,-1)$]

k=1: $x^2+y = x+y^2$.
 $(x-1/2)^2 - (y-1/2)^2 = 0$.
 $y = x$ or $y = -x+1$.
 [intersecting lines except
      $(0,0)$ and $(-1,-1)$].

k-2: $x^2+y - 2x+2y^2$.

$\dfrac{(x-1)^2}{7/8} - \dfrac{(y-1/4)^2}{7/16} = 1$.
 [hyperbola except $(0,0)$ and
                                    $(-1,-1)$]

k=4: $x^2+y = 4x+4y^2$.

$\dfrac{(x-2)^2}{63/16} - \dfrac{(y-1/8)^2}{63/64} = 1$.
 [hyperbola except $(0,0)$ and
                                    $(-1,-1)$].

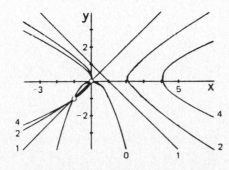

23. $x = 0$, if $T = 0$:

$y^2 = (1/T - 1)x^2$, if $y \neq 0$.

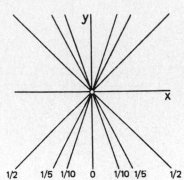

1/2   1/5 1/10   0   1/10 1/5   1/2

25. $x^2+y^2+z^2 \geq 16$. The set of all points on and outside the sphere of radius 4 that is centered at the origin.

27. $\frac{x^2}{9} + \frac{y^2}{16} + \frac{z^2}{1} \leq 1$. Points inside and on the ellipsoid.

29. $x^2+y^2+z^2 = k$, $k > 0$. Set of all spheres centered at the origin.

31. $\frac{x^2}{1/16} + \frac{y^2}{1/16} - \frac{z^2}{1/9} = k$. The circular cone $\frac{x^2}{9} + \frac{y^2}{9} = \frac{z^2}{16}$

and all hyperboloids (one and two sheets) with z-axis for axis such that a:b:c is $(1/4):(1/4):(1/3)$ or 3:3:4.

33. $4x^2-9y^2 = k$, k in R; $\frac{x^2}{k/4} - \frac{y^2}{k/9} = 1$, if $k \neq 0$.

Planes $y = \pm 2x/3$ [for k=0] and all hyperbolic cylinders parallel to the z-axis such that the ratio a:b is $(1/2):(1/3)$ or 3:2 [where a is associated with the x-term].

35. (a) AC is the least steep path and BC is the most steep path between A and C since the level curves are farthest apart along AC and closest together along BC.

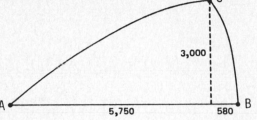

(b) $|AC| \approx \sqrt{(5750)^2 + (3000)^2} \approx 6490$ ft.

$|BC| \approx \sqrt{(580)^2 + (3000)^2} \approx 3060$ ft.

37.

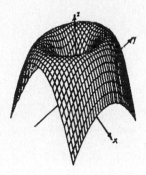 $\sin(\sqrt{2*x^2+y^2})$

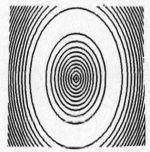

39. $(2*x-y^2)*\exp(-x^2-y^2)$

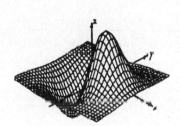

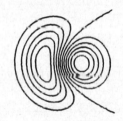

Problem Set 15.2   Partial Derivatives

1. $f_x(x,y) = 8(2x-y)^3$. $f_y(x,y) = -4(2x-y)^3$.

3. $f_x(x,y) = \dfrac{(xy)(2x) - (x^2-y^2)(y)}{(xy)^2} = \dfrac{x^2+y^2}{x^2 y}$.

$f_y(x,y) = \dfrac{(xy)(-2y) - (x^2-y^2)(x)}{(xy)^2} = \dfrac{-(x^2+y^2)}{xy^2}$.

5. $f_x(x,y) = e^y \cos x$. $f_y(x,y) = e^y \sin x$.

7. $f_x(x,y) = x(x^2-y^2)^{-1/2}$. $f_y(x,y) = -y(x^2-y^2)^{-1/2}$.

9. $g_x(x,y) = -ye^{-xy}$. $g_y(x,y) = -xe^{-xy}$.

11. $f_x(x,y) = 4[1+(4x-7y)^2]^{-1}$. $f_y(x,y) = -7[1+(4x-7y)^2]^{-1}$.

13. $f_x(x,y) = -2xy\sin(x^2+y^2)$. $f_y(x,y) = -2y^2\sin(x^2+y^2) + \cos(x^2+y^2)$.

15. $F_x(x,y) = 2\cos x \cos y$. $F_y(x,y) = -2\sin x \sin y$.

17. $f_x(x,y) = 4xy^3 - 3x^2y^5$; $f_{xy}(x,y) = 12xy^2 - 15x^2y^4$.

   $f_y(x,y) = 6x^2y^2 - 5x^3y^4$; $f_{yx}(x,y) = 12xy^2 - 15x^2y^4$.

19. $f_x(x,y) = 6e^{2x}\cos y$; $f_{xy}(x,y) = -6e^{2x}\sin y$.

   $f_y(x,y) = -3e^{2x}\sin y$; $f_{yx}(x,y) = -6e^{2x}\sin y$.

21. $F_x(x,y) = \dfrac{(xy)(2)-(2x-y)(y)}{(xy)^2} = \dfrac{y^2}{x^2y^2} = \dfrac{1}{x^2}$; $F_x(3,-2) = \dfrac{1}{9}$ $[y \neq 0]$.

   $F_y(x,y) = \dfrac{(xy)(-1)-(2x-y)(x)}{(xy)^2} = \dfrac{-2x^2}{x^2y^2} = \dfrac{-2}{y^2}$; $F_y(3,-2) = \dfrac{-1}{2}$ $[x \neq 0]$.

23. $f_x(x,y) = -y^2(x^2+y^4)^{-1}$; $f_x(\sqrt{5},-2) = -4/21 \approx -0.1905$.
   $f_y(x,y) = 2xy(x^2+y^4)^{-1}$; $f_y(\sqrt{5},-2) = -4\sqrt{5}/21 \approx -0.4259$.

25. Let $z = f(x,y) = x^2/9 + y^2/4$. $f_y(x,y) = y/2$. Slope is $f_y(3,2) = 1$.

27. Let $z = f(x,y) = (1/2)(9x^2+9y^2-36)^{1/2}$. $f_x(x,y) = \dfrac{9x}{2(9x^2+9y^2-36)^{1/2}}$.
   $f_x(2,1) = 3$.

29. $V_r(r,h) = 2\pi rh$; $V_r(6,10) = 120\pi \approx 376.99$ in^3/in.

31. $P(V,T) = kT/V$. $P_T(V,T) = k/V$; $P_T(100,40) = k/100$ lb/in^2 per degree.

33. $f_x(x,y) = 3x^2y - y^3$; $f_{xx}(x,y) = 6xy$; $f_y(x,y) = x^3 - 3xy^2$; $f_{yy}(x,y) = -6xy$.
Therefore, $f_{xx}(x,y) + f_{yy}(x,y) = 0$.

35. $F_y(x,y) = 15x^4y^4 - 6x^2y^2$; $F_{yy}(x,y) = 60x^4y^3 - 12x^2y$;
$F_{yyy}(x,y) = 180x^4y^2 - 12x^2$.

37. (a) $\dfrac{\partial^3 f}{\partial y^3}$.   (b) $\dfrac{\partial^3 f}{\partial y \partial x^2}$.   (c) $\dfrac{\partial^4 f}{\partial y^3 \partial x}$.

39. (a) $f_x(x,y,z) = 6xy - yz$.   (b) $f_y(x,y,z) = 3x^2 - xz + 2yz^2$; $f_y(0,1,2) = 8$.

(c) Using the result in (a), $f_{xy}(x,y,z) = 6x - z$.

41. $-yze^{-xyz} - y(xy - z^2)^{-1}$.

43. If $f(x,y) = x^4 + xy^3 + 12$, $f_y(x,y) = 3xy^2$; $f_y(1,-2) = 12$.
Therefore, along the tangent line $\Delta y = 1 \Rightarrow \Delta z = 12$, so $\langle 0,1,12 \rangle$ is a tangent vector (since $\Delta x = 0$). Then parametric equations of the

tangent line are $\begin{cases} x = 1 \\ y = -2+t \\ z = 5+12t \end{cases}$. Then the point of the xz-plane at which

the bee hits is $(1,0,29)$ [since $y=0 \Rightarrow t=2 \Rightarrow x=1$, $z=29$].

45. Domain: (Case $x < y$)

The lengths of the sides are then x, $y-x$, and $1-y$. The sum of the lengths of any two sides must be greater than the length of the remaining side, leading to three inequalities:

$x + (y-x) > 1-y \qquad \Rightarrow y > \dfrac{1}{2}$

$(y-x) + (1-y) > x \quad \Rightarrow x < \dfrac{1}{2}$

$x + (1-y) > y-x \qquad \Rightarrow y < x + \dfrac{1}{2}$

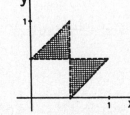

The case for $y < x$ yields similar inequalities (x and y interchanged). The graph of D_A, the domain of A is given above. In set notation it is

$$D_A = \{(x,y): x < \tfrac{1}{2}, y > \tfrac{1}{2}, y < x + \tfrac{1}{2}\} \cup \{(x,y): y < \tfrac{1}{2}, x > \tfrac{1}{2}, x < y + \tfrac{1}{2}\}.$$

Range:
The area is greater than zero but can be arbitrarily close to zero since one side can be arbitrarily small and the other two sides are bounded above. It seems that the area would be largest when the triangle is equilateral. An equilateral triangle with sides equal to 1/3 has area $\sqrt{3}/36$. Hence the range of A is $(0, \sqrt{3}/36)$. [In Sections 8 and 9 of this chapter, methods will be presented which will make it easy to prove that the largest value of A will occur when the triangle is equilateral.]

47. (a) Moving parallel to the y-axis from the point (1,1) to the nearest level curve and approximating $\Delta z/\Delta y$, we obtain

$$f_y(1,1) = \frac{4 - 5}{1.25 - 1} = -4.$$

(b) Moving parallel to the x-axis from the point (-4,2) to the nearest level curve and approximating $\Delta z/\Delta x$, we obtain

$$f_x(-4,2) \approx \frac{1 - 0}{-2.5 - (-4)} = \frac{2}{3}.$$

(c) Moving parallel to the x-axis from the point (-5,-2) to the nearest level curve and approximately $\Delta z/\Delta x$, we obtain

$$f_x(-4,-5) \approx \frac{1 - 0}{-2.5 - (-5)} = \frac{2}{5}.$$

(d) Moving parallel to the y-axis from the point (0,-2) to the nearest level curve and approximating $\Delta z/\Delta y$, we obtain

$$f_y(0,2) \approx \frac{0 - 1}{-19/8 - (-2)} = \frac{8}{3}.$$

Problem Set 15.3 Limits and Continuity

1. -18.

3. $\displaystyle\lim_{(x,y)\to(2,\pi)} [x\cos^2 xy - \sin(xy/3)] = 2\cos^2 2\pi - \sin(2\pi/3) = 2 - \frac{\sqrt{3}}{2} \approx 1.1340.$

5. 1/3.

7. The limit doesn't exist since the function is not defined anywhere along the line y=x. That is, there is no neighborhood of the origin in which the function is defined everywhere except possibly at the origin.

9. The entire plane since x^2+y^2+1 is never zero.

11. Require $y-x^2 \neq 0$. S is the entire plane except the parabola $y = x^2$.

13. Require $x-y+1 \geq 0$; $y \leq x+1$. S is region below and on the line $y = x+1$.

15. Along x-axis (y=0): $\displaystyle\lim_{(x,y)\to(0,0)} \frac{0}{x^2+0} = 0$.

    Along y=x: $\displaystyle\lim_{(x,y)\to(0,0)} \frac{x^2}{2x^2} = \lim_{(x,y)\to(0,0)} \frac{1}{2} = \frac{1}{2}$.

    Hence, the limit does not exist because for some points near the origin $f(x,y)$ is getting closer to 0, but for others it is getting closer to 1/2.

17. (a) $\displaystyle\lim_{x\to0} \frac{x^2(mx)}{x^4+(mx)^2} = \lim_{x\to0} \frac{mx^3}{x^4+m^2x^2} = \lim_{x\to0} \frac{mx}{x^2+m^2} - 0$.

    (b) $\displaystyle\lim_{x\to0} \frac{x^2(x^2)}{x^4+(x^2)^2} = \lim_{x\to0} \frac{x^4}{2x^4} = \lim_{x\to0} \frac{1}{2} = \frac{1}{2}$.

    (c) $\displaystyle\lim_{(x,y)\to(0,0)} \frac{x^2y}{x^4+y^2}$ doesn't exist.

19. The boundary consists of the points of the rectangle. The set is closed.

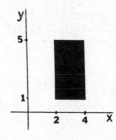

21. The boundary consists of the circle and the origin. The set is neither open [since, for example, (2,0) is not an interior point], nor closed [since (0,0) is not in the set].

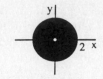

23. The boundary consists of the graph of $y = \sin(1/x)$ along with the part of the y-axis for which $y \leq 1$. The set is open.

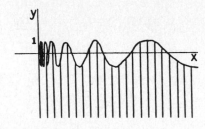

25. $\dfrac{x^2 - 4y^2}{x - 2y} = \dfrac{(x+2y)(x-2y)}{x-2y} = x+2y$ [if $x \neq 2y$].

We want $g(x) = x + 2(x/2)$ [if $x = 2y$, or $y = x/2$] $= 2x$.

27. Note: $(x,y) \to (0,0)$ is equivalent to $r \to 0$.

(a) $f(x,y) = \dfrac{(r\cos\theta)(r\sin\theta)}{\sqrt{r^2}} = |r|\sin\theta\cos\theta = \dfrac{|r|\sin 2\theta}{2} \to 0$ as $r \to 0$.

(b) $f(x,y) = \dfrac{(r\cos\theta)(r\sin\theta)}{r^2} = \dfrac{\sin 2\theta}{2}$ which does not approach 0 as $r \to 0$.

(c) $f(x,y) = \dfrac{r^{7/3}\cos^{7/3}\theta}{r^2} = r^{1/3}\cos^{7/3}\theta \to 0$ as $r \to 0$.

(d) $f(x,y) = (r\cos\theta)(r\sin\theta)(\cos 2\theta)$ [See introduction to this problem for third factor.]

$= \dfrac{r^2\sin 2\theta\cos 2\theta}{2} = \dfrac{r^2\sin 4\theta}{4} \to 0$ as $r \to 0$.

(e) $f(x,y) = \dfrac{r^4\cos^2\theta\sin^2\theta}{r^2\cos^2\theta + r^4\sin^4\theta} = r^2\left[\dfrac{\cos^2\theta\sin^2\theta}{\cos^2\theta + r^2\sin^4\theta}\right] = r^2\left[\dfrac{\sin^2\theta}{1 + r^2\sin^2\theta\tan^2\theta}\right]$

if $\theta \neq \pm\pi/2$. This converges to 0 as r→0 since the fraction is bounded (the numerator is in $[0,1]$ and the denominator is greater than or equal to 1). If $\theta = \pm\pi/2$, $f(x,y) = 0$.

(f)   This one is not easier in polar coordinates. Here is a Cartesian coordinates solution.

Along curve $x=y^2$: $\dfrac{xy^2}{x^2+y^4} = \dfrac{(y^2)y^2}{(y^2)^2+y^4} = \dfrac{1}{2}$ which does not approach 0.

Conclusion: The functions of parts (a), (c), (d), and (e) are continuous at the origin. Those of parts (b) and (f) are discontinuous at the origin.

29.   (a) $\{(x,y,z):x^2+y^2 = 1,\ z\ \text{in}\ [1,2]\}$. [For $x^2+y^2 < 1$, the particle hits the hemisphere and then slides to the origin (or bounces toward the origin); for $x^2+y^2 = 1$, it bounces up; for $x^2+y^2 > 1$, it falls straight down.]

(b) $\{(x,y,z):x^2+y^2 = 1,\ z = 1\}$. [As one moves at a level of z=1 from the rim of the bowl toward any position away from the bowl there is a change from seeing all of the interior of the bowl to seeing none of it.

(c) $\{(x,y,z):z=1\}$. [$f(x,y,z)$ is undefined (infinite) at $(x,y,1)$.]

(d) $\emptyset$ [Small changes in points of the domain result in small changes in the shortest path from the points to the origin.]

31.   (a) $f(x,y) = \begin{cases} (x^2+y^2)^{1/2}+1, & \text{if } y\neq 0 \\ |x-1|, & \text{if } y=0 \end{cases}$.   Check discontinuities where y=0.

As y=0, $(x^2+y^2)^{1/2}+1\ |x|+1$, so f is continuous if $|x|+1 = |x-1|$. Squaring each side and simplifying yields $|x| = -x$, so f is continuous for $x \leq 0$. That is, f is discontinuous along the positive x-axis.

(b)   Let $P = (u,v)$ and $Q = (x,y)$.

$$f(u,v,x,y) = \begin{cases} |OP|+|OQ|, & \text{if } P \text{ and } Q \text{ are not on same ray from the origin and neither is the origin} \\ \\ |PQ|, & \text{otherwise} \end{cases}.$$

This means that in the first case one travels from P to the origin and then to Q; in the second case one travels directly from P to Q without passing through the origin, so f is discontinuous on the set $\{(u,v,x,y): \langle u,v \rangle = k\langle x,y \rangle$ for some $k > 0$, $\langle u,v \rangle \neq 0$, $\langle x,y \rangle \neq 0\}$.

33.

35.

Problem Set 15.4 Differentiability

1. $\langle 2xy+3y, x^2+3x \rangle$.

3. $\nabla f(x,y) = \langle (x)(e^{xy}y)+(e^{xy})(1), xe^{xy}x \rangle = e^{xy} \langle xy+1, x^2 \rangle$.

5. $x(x+y)^{-2} \langle y(x+2y), x^2 \rangle$.

7. $(x^2+y^2+z^2)^{-1/2} \langle x,y,z \rangle$.

9. $\nabla f(x,y) = \langle (x^2y)(e^{x-z})+(e^{x-z})(2xy), x^2e^{x-z}, x^2ye^{x-z}(-1) \rangle$
    $= xe^{x-z} \langle y(x+2), x, -xy \rangle$.

11. $\nabla f(x,y) = <2xy-y^2, x^2-2xy>$; $\nabla f(-2,3) = <-21,16>$.

$z = f(-2,3) + <-21,16> \cdot <x+2, y-3> = 30 + (-21x-42+16y-48)$;
$z = -21x+16y-60$.

13. $\nabla f(x,y) = <-\pi\sin(\pi x)\sin(\pi y), \pi\cos(\pi x)\cos(\pi y) + 2\pi\cos(2\pi y)>$
$\nabla f(-1,1/2) = <0,-2\pi>$

$z = f(-1,1/2) + <0,-2\pi> \cdot <x+1, y-1/2> = -1 + (0-2\pi y+\pi)$;
$z = -2\pi y + (\pi-1)$.

15. $\nabla f(x,y,z) = <6x+z^2, -4y, 2xz>$, so $\nabla f(1,2,-1) = <7,-8,-2>$.
Tangent hyperplane: $w = f(1,2,-1) + \nabla f(1,2,-1) \cdot <x-1, y-2, z+1>$
$$= -4 + <7,-8,-2> \cdot <x-1, y-2, z+1>$$
$$= -4 + (7x-7-8y+16-2z-2).$$
$$w = 7x-8y-2z+3.$$

17. $\nabla f(f/g) = \dfrac{<gf_x-fg_x, gf_y-fg_y, gf_z-fg_z>}{g^2} = \dfrac{g<f_x,f_y,f_z> - f<g_x,g_y,g_z>}{g^2}$

$$= \dfrac{g\nabla f - f\nabla g}{g^2}.$$

19. [In the context of the definition of differentiability.]

$0 = \mathbf{q}_1 \cdot \mathbf{h} + |\mathbf{h}|\epsilon_1(\mathbf{h}) - \mathbf{q}_2 \cdot \mathbf{h} + |\mathbf{h}|\epsilon_2(\mathbf{h}) = (\mathbf{q}_1-\mathbf{q}_2) \cdot \mathbf{h} + |\mathbf{h}|[\epsilon_1(\mathbf{h}) - \epsilon_2(\mathbf{h})]$

$= |\mathbf{q}_1-\mathbf{q}_2||\mathbf{h}|\cos\theta + |\mathbf{h}|[\epsilon_1(\mathbf{h}) - \epsilon_2(\mathbf{h})]$

$= |\mathbf{h}|(|\mathbf{q}_1-\mathbf{q}_2|\cos\theta + [\epsilon_1(\mathbf{h}) - \epsilon_2(\mathbf{h})])$.

Therefore, $0 = |\mathbf{q}_1-\mathbf{q}_2|\cos\theta + [\epsilon_1(\mathbf{h}) - \epsilon_2(\mathbf{h})]$.

Therefore, $0 = |\mathbf{q}_1-\mathbf{q}_2| + [\epsilon_1(\mathbf{h}) - \epsilon_2(\mathbf{h})]$ [since true when $\theta = 0$].
Therefore, $0 = |\mathbf{q}_1-\mathbf{q}_2|$ [since the 1st term is a constant and the 2nd term can be made arbitrarily small]. Therefore, $\mathbf{q}_1 = \mathbf{q}_2$.

21. $\mathbf{\nabla}f(x,y) = \left\langle -10\left[\dfrac{1}{2\sqrt{|xy|}}\,\dfrac{|xy|}{xy}\,y\right],\ -10\left[\dfrac{1}{2\sqrt{|xy|}}\,\dfrac{|xy|}{xy}\,x\right]\right\rangle$

$\qquad\qquad = \dfrac{-5xy}{|xy|^{3/2}}\,\langle y,x\rangle.\quad \left[\text{Note that } \dfrac{|a|}{a} = \dfrac{a}{|a|}.\right]$

$\mathbf{\nabla}f(1,-1) = \langle -5,5\rangle.$

Tangent plane: $z = f(1,-1) + \mathbf{\nabla}f(1,-1)\cdot\langle x-1,y+1\rangle$
$\qquad\qquad\qquad = -10 + \langle -5,5\rangle\cdot\langle x-1,y+1\rangle = -10 + (-5x+5+5y+5)$
$\qquad\qquad z = -5x+5y.$

23. $\mathbf{\nabla}f(\mathbf{p}) = \mathbf{\nabla}g(\mathbf{p}) \Rightarrow \mathbf{\nabla}[f(\mathbf{p})-g(\mathbf{p})] = 0 \Rightarrow f(\mathbf{p})-g(\mathbf{p})$ is a constant.

25. The hint gives precisely what is needed to show f is differentiable at 0 with $\mathbf{q} = \mathbf{\nabla}f(0) = 0$, leaving only to show that

$\qquad \epsilon(\mathbf{h}) = \epsilon(h,k) = \dfrac{\sin(h^2+k^2) - (h^2+k^2)}{(h^2+k^2)^{3/2}}$ approaches 0 as $\mathbf{h}\to 0.$

Let $u = h^2+k^2$, so $u\to 0$ as $\mathbf{h}\to 0.$

Then $\displaystyle\lim_{h\to 0} \dfrac{\sin(h^2+k^2) - (h^2+k^2)}{(h^2+k^2)^{3/2}} = \lim_{u\to 0}\dfrac{\sin u - u}{u^{3/2}} = 0$ [Use l'Hopital's Rule twice.]

27.

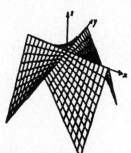
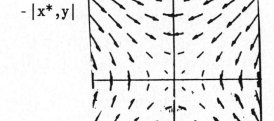

$-|x^*,y|$

(a) The gradient points in the direction of greatest increase of the function.

(b) No. If it were, $|0+h|-|0| = 0 + |h|\,\delta(h)$ where $\delta(h)\to 0$ as $h\to 0$, which is impossible.

Problem Set 15.5 Directional Derivatives and Gradients

1. $D_u f(x,y) = \langle 2xy, x^2 \rangle \cdot \langle 3/5, -4/5 \rangle$; $D_u f(1,2) = 8/5$.

3. $D_u f(x,y) = f(x,y) \cdot u$ [where $u = a/|a|$]

    $= \langle 4x+y, x-2y \rangle \cdot \dfrac{\langle 1,-1 \rangle}{\sqrt{2}}$; $D_u f(3,-2) = \langle 10,7 \rangle \cdot \dfrac{\langle 1,-1 \rangle}{\sqrt{2}} = \dfrac{3}{\sqrt{2}} \approx 2.1213$.

5. $D_u f(x,y) = e^x \langle \sin y, \cos y \rangle \cdot [(1/2)\langle 1, \sqrt{3} \rangle]$; $D_u f(0, \pi/4) = (\sqrt{2}+\sqrt{6})/4 \approx 0.9659$.

7. $D_u f(x,y,z) = \langle 3x^2 y, x^3 - 2yz^2, -2y^2 z \rangle \cdot [(1/3)\langle 1,-2,2 \rangle]$; $D_u f(-2,1,3) = 52/3$.

9. f increases most rapidly in the direction of the gradient.

    $\nabla f(x,y) = \langle 3x^2, -5y^4 \rangle$; $\nabla f(2,-1) = \langle 12,-5 \rangle$. $\dfrac{\langle 12,-5 \rangle}{13}$ is the unit vector in that direction. The rate of change of f(x,y) in that direction at that point is the magnitude of the gradient. $|\langle 12,-5 \rangle| = 13$.

11. $\nabla f(x,y,z) = \langle 2xyz, x^2 z, x^2 y \rangle$; $\nabla f(1,-1,2) = \langle -4,2,-1 \rangle$. A unit vector in that direction is $(1/\sqrt{21})\langle -4,2,-1 \rangle$. The rate of change in that direction is $\sqrt{21} \approx 4.5826$.

13. $-\nabla f(x,y) = 2\langle x,y \rangle$; $-\nabla f(-1,2) = 2\langle -1,2 \rangle$ is the direction of most rapid decrease. A unit vector in that direction is $u = (1/\sqrt{5})\langle -1,2 \rangle$.

15. The level curves are $y/x^2 = k$. For p = (1,2), $k = 2$, so the level curve through (1,2) is $y/x^2 = 2$ or $y = 2x^2$ $[x \neq 0]$.
    $\nabla f(x,y) = \langle -2yx^{-3}, x^{-2} \rangle$. $\nabla f(1,2) = \langle -4,1 \rangle$,

    which is perpendicular to the parabola at (1,2).

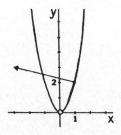

17. $u = \langle 2/3, -2/3, 1/3 \rangle$. $D_u f(x,y,z) = \langle y,x,2z \rangle \cdot \langle 2/3,-2/3,1/3 \rangle$. $D_u f(1,1,1) = 2/3$.

19. (a) Hottest if denominator is smallest; i.e., at the origin.

(b) $\nabla T(x,y,z) = \dfrac{-200\langle 2x,2y,2z\rangle}{(5+x^2+y^2+z^2)^2}$; $\nabla T(1,-1,1) = (-25/4)\langle 1,-1,1\rangle$.

-$\langle 1,-1,1\rangle$ is one vector in the direction of great increase.

(c) Yes.

21. He should move in the direction of
$-\nabla f(\mathbf{p}) = -\langle f_x(\mathbf{p}), f_y(\mathbf{p})\rangle = -\langle -1/2, -1/4\rangle$
$= (1/4)\langle 2,1\rangle$. Or use $\langle 2,1\rangle$.
The angle a formed with the East is
$\tan^{-1}(1/2) \approx 26.57°$ [N63.43°E].

23. The climber is moving in the direction of $\mathbf{u} = (1/\sqrt{2})\langle -1,1\rangle$.

Let $f(x,y) = 3000e^{-(x^2+2y^2)/100}$.

$\nabla f(x,y) = 3000e^{-(x^2+2y^2)/100}\langle -x/50, -y/25\rangle$; $\nabla f(10,10) = -600e^{-3}\langle 1,2\rangle$.

She will move at a slope of $D_{\mathbf{u}}(10,10) = -600e^{-3}\langle 1,2\rangle \cdot (1/\sqrt{2})\langle -1,1\rangle$

$$= (-300\sqrt{2})e^{-3} \approx -21.1229.$$

She will descend. Slope is about -21.

25. $\nabla T(x,y) = \langle -4x, -2y\rangle$. $dx/dt = -4x$, $dy/dt = -2y$. $\dfrac{dx/dt}{-4x} = \dfrac{dy/dt}{-2y}$ has
solution $|x| = 2y^2$. Since the particle starts at $(-2,1)$, this
simplifies to $x = -2y^2$.

27. (a) $\nabla T(x,y,z) = \left\langle \dfrac{-10(2x)}{(x^2+y^2+z^2)^2}, \dfrac{-10(2y)}{(x^2+y^2+z^2)^2}, \dfrac{-10(2z)}{(x^2+y^2+z^2)^2}\right\rangle$

$$= \dfrac{-20}{(x^2+y^2+z^2)^2}\langle x,y,z\rangle.$$

$\mathbf{r}(t) = \langle t\cos\pi t, t\sin\pi t, t\rangle$, so $\mathbf{r}(1) = \langle -1,0,1\rangle$. Therefore, when t=1,
the bee is at $(-1,0,1)$, and $\nabla T(-1,0,1) = -5\langle -1,0,1\rangle$.

$r'(t) = \langle \cos\pi t - \pi t \sin\pi t, \sin\pi t + \pi t \cos\pi t, 1 \rangle$, so $r'(1) = \langle -1, -\pi, 1 \rangle$.

$u = \dfrac{r'(1)}{|r'(1)|} = \dfrac{\langle -1, -\pi, 1 \rangle}{\sqrt{2+\pi^2}}$ is the unit tangent vector at $(-1,0,1)$.

$D_u T(-1,0,1) = u \cdot \nabla T(-1,0,1) = \dfrac{\langle -1,-\pi,1 \rangle \cdot \langle 5,0,-5 \rangle}{\sqrt{2+\pi^2}} = \dfrac{-10}{\sqrt{2+\pi^2}} \approx -2.9026.$

Therefore, the temperature is decreasing at about $2.9°C$ per meter traveled when the bee is at $(-1,0,1)$; i.e., when $t = 1$ sec.

(b)   Method 1:   (First express T in terms of t)

$$T = \frac{10}{x^2+y^2+z^2} = \frac{10}{(t\cos\pi t)^2+(t\sin\pi t)^2+(t)^2} = \frac{10}{2t^2} = \frac{5}{t^2}.$$

$T(t) = 5t^{-2}$; $T'(t) = -10t^{-3}$; $t'(1) = -10$.

Method 2:   (Use chain rule)

$D_t T(t) = \dfrac{dT}{ds}\dfrac{ds}{dt} = (D_u T)(|r(t)|)$, so

$D_t T(t) = [D_u T(-1,0,1)](|r(1)|) = \dfrac{-10}{\sqrt{2+\pi^2}} \left(\sqrt{2+\pi^2}\right) = -10.$

Therefore, the temperature is decreasing at about $10°C$ per second when the bee is at $(-1,0,1)$; i.e., when $t = 1$ sec.

29.

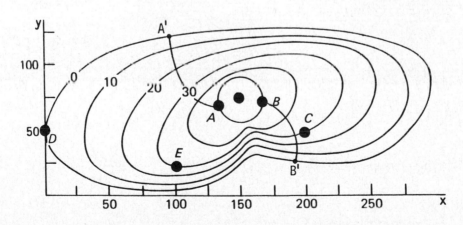

(a) $A'(100,120)$ (b) $B'(190,25)$

(c) $f_x(C) \approx \dfrac{20-30}{230-200} = \dfrac{-1}{3}$; $f_y(D) = 0$; $D_u f(E) \approx \dfrac{40-30}{25} = \dfrac{2}{5}.$

31. leave: (-0.1), -5) x^2-y^2

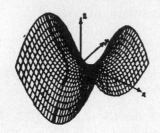

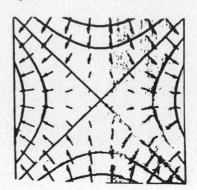

33. leave: (3,5) $z = x^3 - 3xy^2$

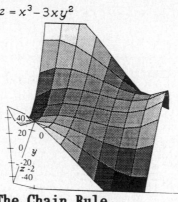

Problem Set 15.6   The Chain Rule

1. $dw/dt = (2xy^3)(3t^2) + (3x^2y^2)(2t) = (2t^9)(3t^2) + (3t^{10})(2t) = 12t^{11}$.

3. $\dfrac{dw}{dt} = (e^x\sin y + e^y\cos x)(3) + (e^x\cos y + e^y\sin x)(2)$

$\qquad = 3e^{3t}\sin 2t + 3e^{2t}\cos 3t + 2e^{3t}\cos 2t + 2e^{2t}\sin 3t.$

5. $dw/dt = [yz^2(\cos(xyz^2)](3t^2) + [xz^2\cos(xyz^2)](2t) + [2xyz\cos(xyz^2)](1)$

$\qquad = (3yz^2t^2 + 2xz^2t + 2xyz)\cos(xyz^2) = (3t^6 + 2t^6 + 2t^6)\cos(t^7) = 7t^6\cos(t^7).$

7. $\partial w/\partial t = (2xy)(s) + (x^2)(-1) = 2st(s-t)s - s^2t^2 = s^2t(2s-3t).$

9. $\dfrac{\partial w}{\partial t} = e^{x^2+y^2}(2x)(s\cos t) + e^{x^2+y^2}(2y)(\sin s)$

$\qquad = 2e^{x^2+y^2}(xs\cos t + y\sin s)$

$\qquad = 2(s^2\sin t\cos t + t\sin^2 s)\exp(s^2\sin^2 t + t^2\sin^2 s).$

11. $\dfrac{\partial w}{\partial t} = \dfrac{x(-s\ \sin st)}{(x^2+y^2+z^2)^{1/2}} + \dfrac{y(s\cos st)}{(x^2+y^2+z^2)^{1/2}} + \dfrac{z(s^2)}{(x^2+y^2+z^2)^{1/2}} = s^4t(1+s^4t^2)^{-1/2}.$

13. $\partial z/\partial t = (2xy)(2) + (x^2)(-2st) = 4(2t+s)(1-st^2) - 2st(2t+s)^2;$

$(\partial z/\partial t)\big|_{(1,-2)} = 72.$

15. $\dfrac{dw}{dx} = (2u-\tan v)(1) + (-u\sec^2 v)(\pi)$

$= 2x - \tan\pi x - \pi x\sec^2\pi x.$

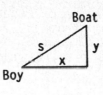

$\dfrac{dw}{dx}\bigg|_{x=1/4} = (1/2)-1-(\pi/2) = \dfrac{1+\pi}{-2} \approx -2.0708.$

17. $V(r,h) = \pi r^2 h,\ dr/dt = 0.5$ in/yr, $dh/dt = 8$ in/yr.

$dV/dt = (2\pi rh)(dr/dt) + (\pi r^2)(dh/dt);\ (dV/dt)\big|_{(20,400)} = 11200\pi$ in^3/yr

$= \dfrac{11200\pi\ \text{in}^3}{1\ \text{yr}} \times \dfrac{1\ \text{board ft}}{144\ \text{in}^3} \approx 244.35$ board ft per year.

19. The stream carries the boat along at 2 ft/sec with respect to the boy.

$dx/dt = 2,\ dy/dt = 4,\ s^2 = x^2+y^2.$
$2s(ds/dt)=2x(dx/dt)+2y(dy/dt).$
$ds/dt = (2x+4y)/s.$

When $t=3$, $x=6$, $y=12$, $s=3\sqrt{20}$. Thus,
$(ds/dt)\big|_{t=3} = \sqrt{20} \approx 4.47$ ft/sec.

21. Let $F(x,y) = x^3+2x^2y-y^3 = 0.$

Then $\dfrac{dy}{dx} = \dfrac{-\partial F/\partial x}{\partial F/\partial y} = \dfrac{-(3x^2+4xy)}{2x^2-3y^2} = \dfrac{3x^2+4xy}{3y^2-2x^2}.$

23. Let $F(x,y) = x\sin y + y\cos x = 0.$ $\dfrac{dy}{dx} = \dfrac{-(\sin y - y\sin x)}{x\cos y + \cos x} = \dfrac{y\sin x - \sin y}{x\cos y + \cos x}.$

25. Let $F(x,y,z) = = 3x^2z+y^3-xyz^3 = 0.$ $\dfrac{\partial z}{\partial x} = \dfrac{-(6xz-yz^3)}{3x^2-3xyz^2} = \dfrac{yz^3-6xz}{3x^2-3xyz^2}.$

415

27. $\dfrac{\partial T}{\partial s} = \dfrac{\partial T}{\partial x}\dfrac{\partial x}{\partial s} + \dfrac{\partial T}{\partial y}\dfrac{\partial y}{\partial s} + \dfrac{\partial T}{\partial z}\dfrac{\partial z}{\partial s} + \dfrac{\partial T}{\partial w}\dfrac{\partial w}{\partial s}.$

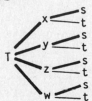

29. $y = (1/2)[f(u) + f(v)]$, where $u = x-ct$, $v = x+ct$.

$y_x = (1/2)[f'(u)(1) + f'(v)(1)] = (1/2)[f'(u) + f'(v)].$

$y_{xx} = (1/2)[f''(u)(1) + f''(v)(1)] = (1/2)[f''(u) + f''(v)].$

$y_t = (1/2)[f'(u)(-c) + f'(v)(c)] = (-c/2)[f'(u) - f'(v)].$

$y_{tt} = (-c/2)[f''(u)(-c) - f''(v)(c)] = (c^2/2)[f''(u) + f''(v)] = c^2 y_{xx}.$

31. Let $w = \displaystyle\int_x^y f(u)\,du = -\int_y^x f(u)\,du$, where $x = g(t)$, $y = h(t)$.

Then $\dfrac{dw}{dt} = \dfrac{dw}{dx}\dfrac{dx}{dt} + \dfrac{dw}{dy}\dfrac{dy}{dt} = -f(x)g'(t) + f(y)h'(t)$
$\qquad\qquad = f(h(t))h'(t) - f(g(t))g'(t).$

Thus, for the particular function given

$$F'(t) = \sqrt{9+(t^2)^4}\,(2t) - \sqrt{9+(\sin\sqrt{2}\pi t)^4}\,(\sqrt{2}\pi\cos\sqrt{2}\pi t);$$

$$F'(\sqrt{2}) = (5)(2\sqrt{2}) - (3)(\sqrt{2}\pi) = 10\sqrt{2} - 3\sqrt{2}\pi \approx 0.8135.$$

33. $c^2 = a^2 + b^2 - 2ab\cos 40°$ [Law of Cosines] where a, b, and c are functions of t.
$2cc' = 2aa' + 2bb' - 2(a'b + ab')\cos 40°$. So

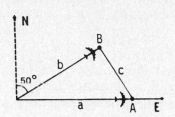

$$c' = \frac{aa' + bb' - (a'b + ab')\cos 40°}{c}$$

When a=200 and b=150,

$$c^2 = (200)^2 + (150)^2 - 2(200)(150)\cos 40° = 62500 - 60000\cos 40°.$$

It is given that a'= 450 and b'= 400, so at that instant,

$$c' = \frac{(200)(450) + (150)(400) - [(450)(150) + (200)(400)]\cos 40°}{\sqrt{62500 - 60000\cos 40°}} \approx 288.$$

Thus, the distance between the planes is increasing at about 288 mph.

Problem Set 15.7 Tangent Planes, Approximations

1. $\nabla F(x,y,z) = 2\langle x,y,z \rangle$; $\nabla F(2,3,\sqrt{3}) = 2\langle 2,3,\sqrt{3} \rangle$.
Tangent Plane: $2(x-2) + 3(y-3) + \sqrt{3}(z-\sqrt{3}) = 0$, or $2x + 3y + \sqrt{3}z = 16$.

3. Let $F(x,y,z) = x^2 - y^2 + z^2 + 1 = 0$. $\nabla F(x,y,z) = \langle 2x, -2y, 2z \rangle = 2\langle x, -y, z \rangle$.
$\nabla F(1,3,\sqrt{7}) = 2\langle 1, -3, \sqrt{7} \rangle$, so $\langle 1, -3, \sqrt{7} \rangle$ is normal to the surface at the point. Then the tangent plane is $1(x-1) - 3(y-3) + \sqrt{7}(z-\sqrt{7}) = 0$, or more simply, $x - 3y + \sqrt{7}z = -1$.

5. $\nabla f(x,y) = (1/2)\langle x,y \rangle$; $\nabla f(2,2) = \langle 1,1 \rangle$.

Tangent Plane: $z-2 = 1(x-2) + 1(y-2)$, or $x + y - z = 2$.

7. $\nabla f(x,y) = \langle -4e^{3y}\sin 2x, 6e^{3y}\cos 2x \rangle$; $\nabla f(\pi/3, 0) = \langle -2\sqrt{3}, -3 \rangle$.

Tangent Plane: $z+1 = -2\sqrt{3}(x - \pi/3) - 3(y-0)$, or $2\sqrt{3}x + 3y + z = (2\sqrt{3}\pi - 3)/3$.

9. Let $z = f(x,y) = 2x^2y^3$; $dz = 4xy^3 dx + 6x^2y^2 dy$. For the points given,
$dx = -0.01$, $dy = 0.02$, $dz = 4(-0.01) + 6(0.02) = 0.08$.
$\Delta z = f(0.99, 1.02) - f(1,1) = 2(0.99)^2(1.02)^3 - 2(1)^2(1)^3 \approx 0.080179921$.

11. $dz = 2x^{-1}dx + y^{-1}dy = (-1)(0.02) + (1/4)(-0.04) = -0.03.$

$\Delta z = f(-1.98, 3.96) - f(-2,4) = \ln[(-1.98)^2(3.96)] - \ln16 \approx -0.030151.$

13. Let $F(x,y,z) = x^2 - 2xy - y^2 - 8x + 4y - z = 0;$ $\nabla F(x,y,z) = \langle 2x - 2y - 8, -2x - 2y + 4, -1 \rangle.$
Tangent plane is horizontal if $\nabla F = \langle 0,0,k \rangle$ for any $k \neq 0.$
$2x - 2y - 8 = 0$ and $-2x - 2y + 4 = 0$ if $x=3$ and $y=-1.$ Then $z=-14.$ There is a
horizontal tangent plane at $(3,-1,-14).$

15. For $F(x,y,z) = x^2 + 4y + z^2 = 0,$ $F(x,y,z) = \langle 2x, 4, 2z \rangle = 2\langle x, 2, z \rangle.$

$F(0,-1,2) = 0,$ and $\nabla F(0,-1,2) = 2\langle 0,2,2 \rangle = 4\langle 0,1,1 \rangle.$

For $G(x,y,z) = x^2 + y^2 + z^2 - 6z + 7 = 0,$ $\nabla G(x,y,z) = \langle 2x, 2y, 2z - 6 \rangle = 2\langle x, y, z - 3 \rangle.$

$G(0,-1,2) = 0,$ and $\nabla G(0,-1,2) = 2\langle 0,-1,-1 \rangle = -2\langle 0,1,1 \rangle.$

$\langle 0,1,1 \rangle$ is normal to both surfaces at $(0,-1,2)$ so the surfaces have the
same tangent plane; hence, they are tangent to each other at $(0,-1,2).$

17. Let $F(x,y,z) = x^2 + 2y^2 + 3z^2 - 12 = 0;$ $\nabla F(x,y,z) = 2\langle x, 2y, 3z \rangle$ is normal to the
plane.

A vector in the direction of the line, $\langle 2,8,-6 \rangle = 2\langle 1,4,-3 \rangle,$ is normal to
the plane.

$\langle x, 2y, 3z \rangle = k\langle 1,4,-3 \rangle$ and (x,y,z) is on the surface for points $(1,2,-1)$
[when $k=1$] and $(-1,-2,1)$ [when $k=-1$].

19. $\nabla f(x,y,z) = 2\langle 9x, 4y, 4z \rangle;$ $\nabla f(1,2,2) = 2\langle 9,8,8 \rangle.$

$\nabla g(x,y,z) = 2\langle 2x, -y, 3z \rangle;$ $\nabla f(1,2,2) = 4\langle 1,-1,3 \rangle.$

$\langle 9,8,8 \rangle \times \langle 1,-1,3 \rangle = \langle 32,-19,-17 \rangle.$ Line: $x = 1 + 32t,$ $y = 2 - 19t,$ $z = 2 - 17t.$

21. $dS = S_A dA + S_W dW = \dfrac{-W}{(A-W)^2}\, dA + \dfrac{A}{(A-W)^2}\, dW = \dfrac{-WdA + AdW}{(A-W)^2}.$

At $W = 20,$ $A = 36:$ $dS = \dfrac{-20dA + 36dW}{256} = \dfrac{-5dA + 9dW}{64}.$

Thus, $|dS| \leq \dfrac{5|dA| + 9|dW|}{64} \leq \dfrac{5(0.02) + 9(0.02)}{64} = 0.004375.$

23. $V = \pi r^2 h$, $dV = 2\pi r h \, dr + \pi r^2 \, dh$.

$|dV| \leq 2\pi r h |dr| + \pi r^2 |dh| \leq 2\pi r h (0.02 r) + \pi r^2 (0.03 h)$
$= 0.04 \pi r^2 h + 0.03 \pi r^2 h = 0.07 V$. **Maximum error in V is 7%.**

25. Solving for R, $R = \dfrac{R_1 R_2}{R_1 + R_2}$, so $\dfrac{\partial R}{\partial R_1} = \dfrac{R_2^2}{(R_1 + R_2)^2}$ and $\dfrac{\partial R}{\partial R_2} = \dfrac{R_1^2}{(R_1 + R_2)^2}$

Therefore, $dR = \dfrac{R_2^2 \, dR_1 + R_1^2 \, dR_2}{(R_1 + R_2)^2}$; $|dR| \leq \dfrac{R_2^2 |dR_1| + R_1^2 |dR_2|}{(R_1 + R_2)^2}$.

Then at $R_1 = 25$, $R_2 = 100$, $dR_1 = dR_2 = 0.5$,

$R = \dfrac{(25)(100)}{25 + 100} = 20$ and $|dR| \leq \dfrac{(100)^2 (0.5) + (25)^2 (0.5)}{(125)^2} = 0.34$.

27. Let $F(x,y,z) = xyz = k$; let (a,b,c) be any point on the surface of F.
$\nabla F(x,y,z) = \langle yz, xz, xy \rangle = \langle k/x, k/y, k/z \rangle = k\langle 1/x, 1/y, 1/z \rangle$.
$\nabla F(a,b,c) = k\langle 1/a, 1/b, 1/c \rangle$. An equation of the tangent plane at the
point is $(1/a)(x-a) + (1/b)(x-b) + (1/c)(x-c) = 0$, or $\dfrac{x}{a} + \dfrac{y}{b} + \dfrac{z}{c} = 3$.

Points of intersection of the tangent plane on the coordinate axes are
$(3a,0,0)$, $(0,3b,0)$, and $(0,0,3c)$.

The volume of the tetrahedron is $(1/3)(\text{area of base})(\text{altitude})$

$= \dfrac{1}{3} \left[\dfrac{1}{2} |3a| |3b| \right] \left[|3c| \right] = \dfrac{9|abc|}{2} = \dfrac{9|k|}{2}$ (a constant).

29. $f(x,y) = (x^2 + y^2)^{1/2}$; $f(3,4) = 5$.

$f_x(x,y) = x(x^2 + y^2)^{-1/2}$; $f_x(3,4) = 3/5 = 0.6$.

$f_y(x,y) = y(x^2 + y^2)^{-1/2}$; $f_y(3,4) = 4/5 = 0.8$.

$$f_{xx}(x,y) = y^2(x^2+y^2)^{-3/2}; \quad f_{xx}(3,4) = 16/125 = 0.128.$$

$$f_{xy}(x,y) = -xy(x^2+y^2)^{-3/2}; \quad f_{xy}(3,4) = -12/125 = -0.096$$

$$f_{yy}(x,y) = x^2(x^2+y^2)^{-3/2}; \quad f_{yy}(3,4) = 9/125 = 0.072$$

Therefore, the second order Taylor approximation is

$$f(x,y) = 5 + 0.6(x-3) + 0.8(y-4) +$$
$$+ 0.5 \ [0.128(x-3)^2 + 2(-0.096)(x-3)(y-4) + 0.072(y-4)^2]$$

(a) First order Taylor approximation: $f(x,y) = 5 + 0.6(x-3) + 0.8(y-4)$.

   Thus, $f(3.1,3.9) \approx 5 + 0.6(0.1) + (0.8)(-0.1) = 4.98$

(b) $f(3.1,3.9) \approx 5 + 0.6(0.1) + 0.8(-0.1) +$
$$+ 0.5 \ [0.128(0.1)^2 + 2(-0.096)(0.1)(-0.1) + 0.072(-0.1)^2]$$
$$= 4.98196.$$

(c) $f(3.1,3.9) \approx 4.981967483$.

Problem Set 15.8 Maxima and Minima

1. $\nabla f(x,y) = \langle 2x-4, 8y \rangle = \langle 0,0 \rangle$ at $(2,0)$, a stationary point.
    $D = f_{xx}f_{yy} - f_{xy}^2 = (2)(8) - (0)^2 = 16 > 0$ and $f_{xx} = 2 > 0$.
    Local minimum at $(2,0)$.

3. $\nabla f(x,y) = \langle 8x^3-2x, 6y \rangle = \langle 2x(4x^2-1), 6y \rangle = \langle 0,0 \rangle$, at $(0,0)$, $(0.5,0)$, $(-0.5,0)$ all stationary points.
    $f_{xx} = 24x^2-2; \quad D = f_{xx}f_{yy} - f_{xy}^2 = (24x^2-2)(6) - (0)^2 = 12(12x^2-1)$.
    At $(0,0)$: $D = -12$, so $(0,0)$ is a saddle point.
    At $(0.5,0)$ and $(-0.5,0)$: $D = 24$ and $f_{xx} = 4$, so local minima occur at these points.

5. $\nabla f(x,y) = \langle y,x \rangle = \langle 0,0 \rangle$ at $(0,0)$, a stationary point.
    $D = f_{xx}f_{yy} - f_{xy}^2 = (0)(0) - (1)^2 = -1$, so $(0,0)$ is a saddle point.

7. $\nabla f(x,y) = \langle y-2x^{-2}, x-4y^{-2}\rangle = \langle 0,0\rangle$ at $(1,2)$.
    $D = f_{xx}f_{yy} - f_{xy}^2 = (4x^{-3})(8y^{-3}) - (1)^2 = 32x^{-3}y^{-3} - 1$, $f_{xx} = 4x^{-3}$.
    At $(1,2)$: $D > 0$, and $f_{xx} > 0$, so a local minimum at $(1,2)$.

9. Let $\nabla f(x,y) = \langle -\sin x - \sin(x+y), -\sin y - \sin(x+y)\rangle = \langle 0,0\rangle$.

    Then $\begin{bmatrix} -\sin x - \sin(x+y) = 0 \\ \sin y + \sin(x+y) = 0 \end{bmatrix}$. Therefore, $\sin x = \sin y$, so $x = y = \pi/4$.

    However, these values satisfy neither equation. Therefore, the gradient is defined but never zero in its domain, and the boundary of the domain is outside the domain, so there are no critical points.

11. We do not need to use calculus for this one. $3x$ is minimum at 0 and $4y$ is minimum at -1. $(0,-1)$ is in S, so $3x+4y$ is minimum at $(0,-1)$; the minimum value is -4. Similarly, $3x$ and $4y$ are each maximum at 1. $(1,1)$ is in S, so $3x+4y$ is maximum at $(1,1)$; the maximum value is 7. [Use calculus techniques and compare.]

13. $\nabla f(x,y) = \langle 2x, -2y\rangle = \langle 0,0\rangle$ at $(0,0)$.

    $D = f_{xx}f_{yy} - f_{xy}^2 = (2)(-2) - (0)^2 < 0$, so $(0,0)$ is a saddle point. A parametric representation of the boundary of S is $x = \cos t$, $y = \sin t$, t in $[0,2\pi]$.
    $f(x,y) = f(x(t),y(t)) = \cos^2 t - \sin^2 t + 1 = \cos 2t - 1$. $\cos 2t - 1$ is maximum if $\cos 2t = 1$, which occurs for $t = 0,\pi,2\pi$. The points of the curve are $(\pm 1,0)$. $f(\pm 1,0) = 2$.

    $f(x,y) = \cos 2t - 1$ is minimum if $\cos 2t = -1$, which occurs for $t = \pi/2$, $3\pi/2$. The points of the curve are $(0,\pm 1)$. $f(0,\pm 1) = 0$.
    Global minimum of 0 at $(0,\pm 1)$; global maximum of 2 at $(\pm 1,0)$.

15. Let x,y,z denote the numbers, so $x+y+z = N$.
    Maximize $P = xyz = xy(N-x-y) = Nxy - x^2 y - xy^2$.
    Let $\nabla P(x,y) = \langle Ny-2xy-y^2, Nx-x^2-2xy\rangle = \langle 0,0\rangle$.
    Then $\begin{bmatrix} Ny-2xy-y^2 = 0 \\ Nx-x^2-2xy = 0 \end{bmatrix}$. $N(x-y) = x^2-y^2 = (x+y)(x-y)$. $x=y$ or $N = x+y$.

    Therefore, $x = y$ [since $N = x+y$ would mean that $P = 0$, certainly not a maximum value].

Then, substituting into $Nx - x^2 - 2xy = 0$, we obtain $Nx - x^2 - 2x^2 = 0$, from which we obtain $x(N - 3x) = 0$, so $x = N/3$ [since $x = 0 \Rightarrow P = 0$].

$P_{xx} = -2y$; $D = P_{xx}P_{yy} - P_{xy}^2 = (-2y)(-2x) - (N - 2x - 2y)^2 = 4xy - (N - 2x - 2y]^2$.

At $x = y = N/3$: $D = N^2/3 > 0$, $P_{xx} = -2N/3 < 0$ [so local maximum]. If $x = y = N/3$, then $z = N/3$.

Conclusion: Each number is $N/3$. [If the intent is to find three distinct numbers, then there is no maximum value of P that satisfies that condition.]

17. Let S denote the surface area of the box with dimensions x, y, z.

$S = 2xy + 2xz + 2yz$ and $V_o = xyz$, so $S = 2(xy + V_o y^{-1} + V_o x^{-1})$.

Minimize $f(x, y) = xy + V_o y^{-1} + V_o x^{-1}$ subject to $x > 0$, $y > 0$.

$\nabla f(x, y) = \langle y - V_o x^{-2}, x - V_o y^{-2} \rangle = \langle 0, 0 \rangle$ at $(V_o^{1/3}, V_o^{1/3})$.

$D = f_{xx}f_{yy} - f_{xy}^2 = 4V_o^2 x^{-3}y^{-3} - 1$, $f_{xx} = 2V_o x^{-3}$.

At $(V_o^{1/3}, V_o^{1/3})$: $D = 3 > 0$, $f_{xx} = 2 > 0$, so local minimum.

Conclusion: The box is a cube with edge $V_o^{1/3}$.

19. Let S denote the area of the sides and bottom of the tank with base 1 by w and depth h. $S = lw + 2lh + 2wh$ and $lwh = 256$.
$S(l, w) = lw + 2l(256/lw) + 2w(256/lw)$, $w > 0$, $l > 0$.
$\nabla S(l, w) = \langle w - 512l^{-2}, l - 512w^{-2} \rangle = \langle 0, 0 \rangle$ at $(8, 8)$. $h = 4$ when $l = 8$ and $w = 8$.
At $(8, 8)$ $D > 0$ and $S_{ll} > 0$, so $(8, 8)$ is a local minimum. Dimensions are $8' \times 8' \times 4'$.

21. Let $\langle x, y, z \rangle$ denote the vector; let S be the sum of its components.

$x^2 + y^2 + z^2 = 81$, so $z = (81 - x^2 - y^2)^{1/2}$.

Maximize $S(x, y) = x + y + (81 - x^2 - y^2)^{1/2}$, $0 \le x^2 + y^2 \le 9$.

Let $\nabla S(x, y) = \langle 1 - x(81 - x^2 - y^2)^{-1/2}, 1 - y(81 - x^2 - y^2)^{-1/2} \rangle = \langle 0, 0 \rangle$.

Therefore, $x = (81 - x^2 - y^2)^{1/2}$ and $y = (81 - x^2 - y^2)^{1/2}$. We then obtain $x = y = 3\sqrt{3}$ as the only stationary point. For these values of x and y, $z = 3\sqrt{3}$ and $S = 9\sqrt{3} \approx 15.59$.

The boundary needs to be checked. It is fairly easy to check each edge of the boundary separately. The largest value of S at a boundary point occurs at three places and turns out to be $18/\sqrt{2} \approx 12.73$. Conclusion: The vector is $3\sqrt{3}<1,1,1>$.

23. $A = (y + s\, x\sin a)(x\cos a)$ and $2x+y = 12$.

Maximize $A(x,a) = 12x\cos a - 2x^2\cos a + x^2\sin a\,\cos a$, x in $(0,6]$, a in $(0,\pi/2)$.

$\nabla A(x,a) = <12\cos a - 4x\cos a + 2x\sin a\cos a,\ -12x\sin a + 2x^2\sin a + x^2\cos 2a>$
$= <0,0>$ at $(4,\pi/6)$.

At $(4,\pi/6)$, $D > 0$ and $A_{xx} < 0$, so local max., and $A = 12\sqrt{3} \approx 20.78$. At the boundary point of $x=6$, we get $a=\pi/4$, $A = 18$. Thus, maximum occurs for width of turned-up sides = $4''$, and base angle = $\pi/2 + \pi/6 = 2\pi/3$.

25. Let M be the maximum value of $f(x,y)$ on the polygonal region, P. Then $ax+by+(c-M) = 0$ is a line that either contains a vertex of P or divides P into two subregions. In the latter case $ax+by+(c-M)$ is positive in one of the regions and negative in the other. $ax+by+(c-M) > 0$ contradicts that M is the maximum value of $ax+by+c$ on P. [Similar argument for minimum.]

(a)
x	y	2x+3y+4	
-1	2	8	← Max. at (-1,2).
0	1	7	
1	0	6	
-3	0	-2	
0	-4	-8	

(b)
x	y	-3x+2y+1	
-3	0	10	
0	5	11	
2	3	1	
4	0	-11	← Min. at (4,0).
1	-4	-10	

27. $z(x,y) = y^2 - x^2$. $\nabla z(x,y) = <-2x,2y> = <0,0>$ at $(0,0)$.

There are no stationary points and no singular points, so consider boundary points.

On side (1): $y = 2x$, so $z = 4x^2 - x^2 = 3x^2$.
$z'(x) = 6x = 0$ if $x = 0$.
Therefore, $(0,0)$ is a candidate.

On side (2): $y = -4x+6$, so $z = (-4x+6)^2 - x^2$
$$= 15x^2 - 48x + 36.$$

$$z'(x) = 30x - 48 = 0 \text{ if } x = 1.6.$$
Therefore, $(1.6, -0.4)$ is a candidate.

On side (3): $y = -x$, so $z = (-x)^2 - x^2 = 0$.

Also, all vertices are candidates.

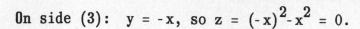

x	y	$z = y^2 - x^2$
0	0	0
1.6	-0.4	-2.4 ← minimum value of -2.4.
2	-2	0
1	2	3 ← maximum value of 3.

29.

x_i	y_i	x_i^2	$x_i y_i$	
3	2	9	6	
4	3	16	12	
5	4	25	20	
6	4	36	24	$m(135) + b(25) = (97)$ and $m(25) + (5)b = (18)$.
7	5	49	35	Solve simultaneously and obtain $m = 0.7$, $b = 0.1$.
$\Sigma = 25$	18	135	97	The least-squares line is $y = 0.7x + 0.1$.

31. $T(x,y) = 2x^2 + y^2 - y$. $\nabla T = \langle 4x, 2y-1 \rangle = 0$
if $x=0$ and $y = 1/2$, so $(0, 1/2)$ is the
only interior critical point.

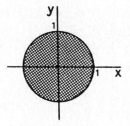

On the boundary $x^2 = 1 - y^2$, so T is a
function of y there.

$T(y) = 2(1-y^2) + y^2 - y = 2 - y - y^2$, $y = [-1, 1]$.

$T'(y) = -1 - 2y = 0$ if $y = -1/2$, so on the boundary, critical points
occur where y is -1, $-1/2$, 1.

Thus, points to consider are $(0, 1/2)$, $(0, -1)$, $(\sqrt{3}/2) - 1/2)$, $(-\sqrt{3}/2, -1/2)$
and $(0, 1)$. Substituting these into $T(x,y)$ yields that the coldest spot
is $(0, 1/2)$ where the temperature is $-1/4$, and there is a tie for the
hottest spot at $(\pm\sqrt{3}/2, -1/2)$ where the temperature is $9/4$.

33. Without loss of generality we will assume that $a \leq \beta \leq \gamma$. We will consider it intuitively clear that for a triangle of maximum area the center of the circle will be inside or on the boundary of the triangle; i.e., a, β, and γ are in the interval $[0,\pi]$. Along with $a+\beta+\gamma = 2\pi$, this implies that $a+\beta \geq \pi$.

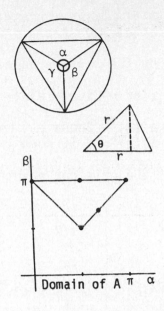

The area of an isosceles triangle with congruent sides of length r and included angle θ is $\frac{1}{2} r^2 \sin\theta$.

$$\text{Area}(\Delta ABC) = \frac{1}{2} r^2 \sin a + \frac{1}{2} r^2 \sin\beta + \frac{1}{2} r^2 \sin\gamma$$

$$= \frac{1}{2} r^2 (\sin a + \sin\beta + \sin[2\pi - (a+\beta)])$$

$$= \frac{1}{2} r^2 [\sin a + \sin\beta - \sin(a+\beta)].$$

Area (ΔABC) will be maximum if
(*) $A(a,\beta) = \sin a + \sin\beta - \sin(a+\beta)$ is maximum.

Restrictions are $0 \leq a \leq \beta \leq \pi$, and $a+\beta \geq \pi$.

Three critical points are the vertices of the triangular domain of A: $(\pi/2, \pi/2)$, $(0,\pi)$, and (π,π). We will now search for others.

$\nabla A(a,\beta) = \langle \cos a - \cos(a+\beta), \cos\beta - \cos(a+\beta) \rangle = 0$ if $\cos a = \cos(a+\beta) = \cos\beta$.

Therefore, $\cos a = \cos\beta$, so $a = \beta$. [Due to the restrictions stated] Then $\cos a = \cos(a+a) = \cos 2a = 2\cos^2 a - 1$, so $\cos a = 2\cos^2 a - 1$.

Solve for a: $2\cos^2 a - \cos a - 1 = 0$; $(2\cos a + 1)(\cos a - 1) = 0$;
$\cos a = -1/2$ or $\cos a = 1$; $a = 2\pi/3$ or $a = 0$.

(We are still in the case where $a = \beta$.) $(2\pi/3, 2\pi/3)$ is a new critical point, but $(0,0)$ is out of the domain of A. There are no critical points in the interior of the domain of A.

On the $\beta = \pi$ edge of the domain of A;

$A(a) = \sin a - \sin(a+\pi) = 2\sin a$ so $A'(a) = 2\cos a$.
$A'(a) = 0$ if $a = \pi/2$. $(\pi/2,\pi)$ is a new critical point.

On the $\beta = \pi - a$ edge of the domain of A:

$A(a) = \sin a + \sin(\pi - a) - \sin(a + \pi - a) = \sin a + \sin a$
$$= 2\sin a$$
$A'(a) = 2\cos a$
$A'(a) = 0$ if $\cos a = 0$ so $a = \pi/2$. $(\pi/2,\pi/2)$ is already a critical point.

(The critical points are indicated on the graph of the domain of A.)

a	β	A
$\pi/2$	$\pi/2$	22
0	π	0
π	π	0
$2\pi/3$	$2\pi/3$	$3\sqrt{3}/2$ ← Maximum value of A. The triangle is isosceles.
$\pi/2$	π	2

35. local max: $f(1.75,0) = 1.15$.
global max: $f(-3.8,0) = 2.30$.

37. global min: $f(0,1) = f(0,-1) = -0.12$.

39. No global max or global min.

41. global max: $f(0.67,0) = 5.06$.
global min: $f(-0.75,0) = -3.54$.

43. global max: $f(2.1,2.1) = 3.5$.
global min: $f(4.2,4.2) = -3.5$.

Problem Set 15.9 Lagrange's Method

1. $\langle 2x,2y \rangle = \lambda \langle y,x \rangle$. $2x=\lambda y$, $2y=\lambda x$, $xy=3$. Critical points are $(\pm\sqrt{3},\pm\sqrt{3})$, $f(\pm\sqrt{3},\pm\sqrt{3}) = 6$. It is not clear whether 6 is the minimum or maximum, so take any other point on $xy=3$, for example $(1,3)$. $f(1,3) = 10$, so 6 is the minimum value.

3. Let $\nabla f(x,y) = \lambda \nabla g(x,y)$, where $g(x,y) = x^2 + y^2 - 1 = 0$.
$\langle 8x - 4y, -4x + 2y \rangle = \lambda \langle 2x, 2y \rangle$.

(1) $4x - 2y = \lambda x$.

(2) $-2x + y = \lambda y$.

(3) $x^2 + y^2 = 1$.

(4) $0 = \lambda x + 2\lambda y$ [From equations (1) and (2)]

(5) $\lambda = 0$ or $x + 2y = 0$.   [(4)]

$\lambda = 0$:   (6) $y = 2x$.   [(1)]

      (7) $x = \pm 1/\sqrt{5}$.   [(6),(3)]

      (8) $y = \pm 2/\sqrt{5}$.   [(7),(6)]

$x + 2y = 0$:   (9) $x = -2y$.

      (10) $y = \pm 1/\sqrt{5}$.   [(9),(3)]

      (11) $x = \pm 2/\sqrt{5}$.   [(10),(9)]

Critical Points: $(1/\sqrt{5}, 2/\sqrt{5})$, $(-1/\sqrt{5}, -2/\sqrt{5})$, $(2/\sqrt{5}, -1/\sqrt{5})$, $(-2/\sqrt{5}, 1/\sqrt{5})$.

$f(x,y)$ is 0 at the first two critical points and 5 at the last two. Therefore, the maximum value of $f(x,y)$ is 5.

5. $\langle 2x, 2y, 2z \rangle = \lambda \langle 1, 3, -2 \rangle$. $2x = \lambda$, $2y = 3\lambda$, $2z = -2\lambda$, $x + 3y - 2z = 12$.
Critical point is $(6/7, 18/7, -12/7)$.
$f(6/7, 18/7, -12/7) = 72/7$ is the minimum   [e.g., $f(12,0,0) = 144$].

7. Let l and w denote the dimensions of the base, h denote the depth.
Maximize $V(l,w,h) = lwh$ subject to $g(l,w,h) = lw + 2lh + 2wh = 48$.

$\langle wh, lh, lw \rangle = \lambda \langle w + 2h, l + 2h, 2l + 2w \rangle$.
$wh = \lambda(w + 2h)$, $lh = \lambda(l + 2h)$, $lw = \lambda(2l + 2w)$, $lw + 2lh + 2wh = 48$.
Critical point is $(4, 4, 2)$.
$V(4,4,2) = 32$ is the maximum.   [$V(11,2,1) = 22$, for example.]

9. Let l and w denote the dimensions of the base, h the depth.
   Maximize $V(l,w,h) = lwh$ subject to $0.60lw + 0.20(lw+2lh+2wh) = 12$ which
   simplifies to $2lw + lh + wh = 30$, or $g(l,w,h) = 2lw + lh + wh - 30$.

   Let $\nabla V(l,w,h) = \lambda \nabla g(l,w,h)$; $\langle wh,lh,lw \rangle = \lambda \langle 2w+h, 2l+h, l+w \rangle$.

       (1) $wh = \lambda(2w+h)$.

       (2) $lh = \lambda(2l+h)$.

       (3) $lw = \lambda(l+w)$.

       (4) $2lw + lh + wh = 30$.

       (5) $(w-l)h = 2\lambda(w-l)$. $[(1),(2)]$

       (6) $w = l$ or $h = 2\lambda$.

$w = l$:   (7) $l = 2\lambda = w$. $[(3)]$ Note: $w \neq 0$, for then $V = 0$.

       (8) $h = 4\lambda$. $[(7),(2)]$

       (9) $\lambda = \sqrt{5}/2$. $[(7),(8),(4)]$

      (10) $l = w = \sqrt{5}$, $h = 2\sqrt{5}$. $[(9),\ (7),(8)]$

$h = 2\lambda$: (11) $\lambda = 0$. $[(2)]$

      (12) $l = w = h = 0$. $[(11),(1)-(3)]$

(Not possible since this does not satisfy (4).)

$(\sqrt{5}, \sqrt{5}, 2\sqrt{5})$ is a critical point and $V(\sqrt{5}, \sqrt{5}, 2\sqrt{5}) = 10\sqrt{5} \approx 22.36$ ft^3 is
the maximum volume [rather than the minimum volume since, for example,
$g(1,1,14) = 30$ and $V(1,1,14) = 14$ which is less than 22.36].

11. Maximize $f(x,y,z) = xyz$,

   subject to $g(x,y,z) = b^2c^2x^2 + a^2c^2y^2 + a^2b^2z^2 - a^2b^2c^2 = 0$.

   $\langle yz, xz, xy \rangle = \lambda \langle 2b^2c^2x, 2a^2c^2y, 2a^2b^2z \rangle$.

   $yz = 2b^2c^2x$, $xz = 2a^2c^2y$, $xy = 2a^2b^2z$, $b^2c^2x^2 + a^2c^2y^2 + a^2b^2z^2 = a^2b^2c^2$.

   Critical point is $(a/\sqrt{3}, b/\sqrt{3}, c/\sqrt{3})$. $V(a/\sqrt{3}, b/\sqrt{3}, c/\sqrt{3}) = 8abc/3\sqrt{3}$, which
   is the maximum.

13. A different hint, which will be used here, is to let Ax+By+Cz = 1 be the plane. [See write-up for Problem 34, Section 15.8.]

Maximize f(A,B,C) = ABC subject to g(A,B,C) = aA+bB+cC-1 = 0.
Let <BC,AC,BC> = λ<a,b,c>.
Then BC = λa, AC = λb, BC = λc, aA+bB+cC = 1.
Therefore, λaA = λbB = λcC [since each equals ABC],
            so aA = bB = cC [since λ=0 implies A=B=C=0 which doesn't satisfy
            the constraint equation].
Then 3aA = 1, so A = 1/3a; similarly B = 1/3b and C = 1/3c.
The rest follows as in the solution for Problem 34, Section 15.8.

15. Let $\alpha+\beta+\gamma = 1$, $\alpha > 0$, $\beta > 0$, and $\gamma > 0$.

Maximize $P(x,y,z) = kx^{\alpha}y^{\beta}z^{\gamma}$, subject to g(x,y,z) = ax+by+cz-d = 0.

Let $\nabla P(x,y,z) = \lambda \nabla g(x,y,z)$. Then

$$\langle k\alpha x^{\alpha-1}y^{\beta}z^{\gamma},\ k\beta x^{\alpha}y^{\beta-1}z^{\gamma},\ k\gamma x^{\alpha}y^{\beta}z^{\gamma-1}\rangle = \lambda\langle a,b,c\rangle.$$

Therefore, $\dfrac{\lambda ax}{\alpha} = \dfrac{\lambda by}{\beta} = \dfrac{\lambda cz}{\gamma}$ [Since each equals $kx^{\alpha}y^{\beta}z^{\gamma}$]

$\lambda \neq 0$ since $\lambda = 0$ would imply x = y = z = 0 which would imply P = 0.

Therefore, $\dfrac{ax}{\alpha} = \dfrac{by}{\beta} = \dfrac{cz}{\gamma}$ (*)

The constraints ax+by+cz = d in the form $\alpha\left(\dfrac{ax}{\alpha}\right) + \beta\left(\dfrac{by}{\beta}\right) + \gamma\left(\dfrac{cz}{\gamma}\right) = d$

becomes $\alpha\left(\dfrac{ax}{\alpha}\right) + \beta\left(\dfrac{ax}{\alpha}\right) + \gamma\left(\dfrac{ax}{\alpha}\right) = d$, using (*).

Then $(\alpha+\beta+\gamma)\left(\dfrac{ax}{\alpha}\right) = d$, or $\dfrac{ax}{\alpha} = d$ [Since $\alpha+\beta+\gamma = 1$.]

$x = \dfrac{\alpha d}{a}$ (**); $y = \dfrac{\beta d}{b}$ and $z = \dfrac{\gamma d}{c}$ then following using (*) and (**).

Since there is only one interior critical point, and since P is 0 on the boundary, P is maximum when $x = \dfrac{\alpha d}{a}$, $y = \dfrac{\beta d}{b}$, $z = \dfrac{\gamma d}{c}$.

17. <-1,2,2> = λ<2x,2y,0> + μ<0,1,2>.
$-1=2\lambda x$, $2=2\lambda y+\mu$, $2=2\mu$, $x^2+y^2=2$, y+2z=1.
Critical points are (-1,1,0) and (1,-1,1).
f(-1,1,0) = 3, the maximum value; f(1,-1,1) = -1, the minimum value.

19. Let $\langle a_1, a_2, \cdots, a_n \rangle = \lambda \langle 2x_1, 2x_2, \cdots, 2x_n \rangle$.

Therefore, $a_i = 2\lambda x_i$, for each $i = 1, 2, \cdots, n$ [since $\lambda = 0$ implies $a_i = 0$, contrary to hypothesis].

$\dfrac{x_i}{a_i} = \dfrac{x_j}{a_j}$ for all i, j. [since each equals $1/2\lambda$].

The constraint equation can be expressed

$$a_1^2 \left[\frac{x_1}{a_1}\right]^2 + a_2^2 \left[\frac{x_2}{a_2}\right]^2 + \cdots + a_n^2 \left[\frac{x_n}{a_n}\right]^2 = 1.$$

Therefore, $\left[a_1^2 + a_2^2 + \cdots + a_n^2\right] \left[\frac{x_1}{a_1}\right]^2 = 1.$

$$x_1^2 = \frac{a_1^2}{a_1^2 + \cdots + a_n^2} ; \text{ similar for each othe } x_i^2.$$

The function to be maximized in a hyperplane with positive coefficients and constant (so intercepts on all axes are positive), and the constraint is a hypersphere of radius 1, so the maximum will occur where each x_i is positive. There is only one such critical point, the one obtained from the above by taking the principal square root to solve for x_i.

Then the maximum value of w is $a_1 \left[\frac{a_1}{\sqrt{A}}\right] + a_2 \left[\frac{a_2}{\sqrt{A}}\right] + \cdots + a_n \left[\frac{a_n}{\sqrt{A}}\right] = \frac{A}{\sqrt{A}} = \sqrt{A}$

where $A = a_1^2 + a_2^2 + \cdots + a_n^2$.

21. Min: $f(4,0) = -4$

23. Min: $f(0,3) = f(0,-3) = -0.99$

Problem Set 15.10   Chapter Review

True-False Quiz

1. True.   Except for the trivial case of $z=0$, which gives a point.

3. True.   Since $g'(0) = f_x(0,0)$.

5. True.   Use "Continuity of a Product" Theorem.

7. False. See Problem 23, Section 15.4.

9. True.   Since <0,0,-1> is normal to the tangent plane.

11. True.   It will point in the direction of greatest increase of heat, and at the origin, $\nabla T(0,0) = \langle 1,0 \rangle$ is that direction.

13. True.   Along the x-axis, $f(x,0) \to \pm\infty$ as $x \to \pm\infty$.

15. True. $-D_{\mathbf{u}}f(x,y) = -[\nabla f(x,y)\cdot\mathbf{u}] = \nabla f(x,y)\cdot(-\mathbf{u}) = D_{-\mathbf{u}}f(x,y)$.

17. True.   By the Min-Max Existence Theorem.

19. False. $f(\pi/2,1) = \sin(\pi/2) = 1$, the maximum value of f, and $(\pi/2,1)$ is in the set.

Sample Test Problems

1. (a) $x^2 + 4y^2 - 100 \geq 0$.

        $\dfrac{x^2}{100} + \dfrac{y^2}{25} \geq 1$.

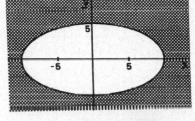

    (b) $2x - y - 1 \geq 0$.

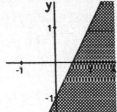

3. $f_x(x,y) = 12x^3y^2+14xy^7$.

   $f_{xx}(x,y) = 36x^2y^2+14y^7$.

   $f_{xy}(x,y) = 24x^3y+98xy^6$.

5. $f_x(x,y) = e^{-y}\sec^2 x$.

   $f_{xx}(x,y) = 2e^{-y}\sec^2 x \tan x$.

   $f_{xy}(x,y) = -e^{-y}\sec^2 x$.

7. $F_y(x,y) = 30x^3y^5-7xy^6$; $F_{yy}(x,y) = 150x^3y^4-42xy^5$;

   $F_{yyx}(x,y) = 450x^2y^4-42y^5$.

9. $z_y(x,y) = y/2$; $z_y(2,2) = 2/2 = 1$.

11. No. On the path $y=x$, $\lim_{x\to 0}\frac{x-x}{x+x} = 0$. On the path $y=0$, $\lim_{x\to 0}\frac{x-0}{x+0} = 1$.

13. (a) $\nabla f(x,y,z) = \langle 2xyz^3, x^2z^3, 3x^2yz^2\rangle$; $\nabla f(1,2,-1) = \langle -4,-1,6\rangle$.

    (b) $\nabla f(x,y,z) = \langle y^2 z\cos xz, 2y\sin xz, xy^2\cos xz\rangle$;
    $\nabla f(1,2,-1) = 4\langle\cos(1),\sin(1),-\cos(1)\rangle \approx \langle -2.1612,-3.3659,2.1612\rangle$.

15. $z = f(x,y) = x^2+y^2$.

    $\langle 1,-\sqrt{3},0\rangle$ is horizontal and is normal to the vertical plane that is given. By inspection, $\langle\sqrt{3},1,0\rangle$ is also a horizontal vector and is perpendicular to $\langle 1,-\sqrt{3},0\rangle$ and therefore is parallel to the vertical plane. Then $\mathbf{u} = \langle\sqrt{3}/2,1/2\rangle$ is the corresponding 2-dimensional unit vector.

    $D_{\mathbf{u}}f(x,y) = \nabla f(x,y)\cdot\mathbf{u} = \langle 2x,2y\rangle\cdot\langle\sqrt{3}/2,1/2\rangle = \sqrt{3}x+y$.

    $D_{\mathbf{u}}f(1,2) = \sqrt{3}+2 \approx 3.7321$ is the slope of the tangent to the curve.

17. (a) $f(4,1) = 9$, so $\frac{x^2}{2}+y^2 = 9$, or $\frac{x^2}{18}+\frac{y^2}{9} = 1$. (c)

    (b) $\nabla f(x,y) = \langle x,2y\rangle$, so $\nabla f(4,1) = \langle 4,2\rangle$.

19. $f_x = f_u u_x + f_v u_y = (1/v)(2x) + (-u/v^2)(yz) = x^{-2}y^{-1}z^{-1}(x^2+3y-4z).$

$f_y = f_u u_y + f_v v_y = (1/v)(-3) + (-u/v^2)(xz) = x^{-1}y^{-2}z^{-1}(-x^2-4z).$

$f_z = f_u u_z + f_v v_z = (1/v)(4) + (-u/v^2)(xy) = x^{-1}y^{-1}z^{-2}(3y-x^2).$

21. $F_t = F_x x_t + F_y y_t + F_z z_t$

$$= \left[\frac{10xy}{z^3}\right]\left[\frac{3t^{1/2}}{2}\right] + \left[\frac{5x^2}{z^3}\right]\left[\frac{1}{t}\right] + \left[\frac{-15x^2y}{z^4}\right]\left[3e^{3t}\right]$$

$$= \frac{15xy\sqrt{t}}{z^3} + \frac{5x^2}{z^3 t} - \frac{45x^2 y e^{3t}}{z^4}.$$

23. Let $F(x,y,z) = 9x^2+4y^2+9z^2-34 = 0.$
$\nabla F(x,y,z) = \langle 18x, 8y, 18z \rangle$, so $\nabla F(1,2,-1) = 2\langle 9,8,-9 \rangle.$
Tangent plane is $9(x-1)+8(y-2)-9(z+1) = 0$, or $9x+8y-9z = 34.$

25. $df = y^2(1+z^2)^{-1}dx + 2xy(1+z^2)^{-1}dy - 2xy^2z(1+z^2)^{-2}dz.$
If $x=1$, $y=2$, $z=2$, $dx=0.01$, $dy=-0.02$, $dz=0.03$, then $df = -0.0272.$
Therefore, $f(1.01,1.98,2.03) \approx f(1,2,2)+df = 0.8 - 0.0272 = 0.7728.$

27. Let (x,y,z) denote the coordinates of the 1st octant vertex of the box.

Maximize $f(x,y,z) = xyz$ subject to $g(x,y,z) = 36x^2+4y^2+9z^2-36 = 0.$

[Where $x,y,z > 0$ and the box's volume is $V(x,y,z) = 8f(x,y,z)$]

Let $\nabla f(x,y,z) = \lambda \nabla g(x,y,z).$
$\langle yz, xz, xy \rangle = \lambda \langle 72x, 8y, 18z \rangle.$

(1) $yz = 72\lambda x.$

(2) $xz = 8\lambda y.$

(3) $xy = 18\lambda z.$

(4) $36x^2+4y^2+9z^2 = 36.$

(5) $\dfrac{yz}{xz} = \dfrac{72\lambda x}{8\lambda y}$, so $y^2 = 9x^2.$ $\qquad\qquad$ [(1),(2)]

(6) $\dfrac{yz}{xy} = \dfrac{72\lambda x}{18\lambda z}$, so $z^2 = 4x^2.$ $\qquad\qquad$ [(1),(3)]

(7) $36x^2+36x^2+36x^2 = 36$, so $x = 1/\sqrt{3}$. $[(5),(6),(4)]$

(8) $y = 3/\sqrt{3}$, $z = 2/\sqrt{3}$. $[(7),(5),(6)]$

$V(1/\sqrt{3},3/\sqrt{3},2/\sqrt{3}) = 8(1/\sqrt{3})(3/\sqrt{3})(2\sqrt{3}) = 16/\sqrt{3} \approx 9.2376$.

The nature of the problem indicates that the critical point yields a maximum value rather than a minimum value.

[For a generalization of this problem, see Problem 11 of Section 15.9.]

29. Maximize $V(r,h) = \pi r^2 h$, subject to $S(r,h) = 2\pi r^2 + 2\pi rh - 24\pi = 0$.

$\langle 2\pi rh, \pi r^2 \rangle = \lambda \langle 4\pi r + 2\pi h, 2\pi r \rangle$. $rh = \lambda(2r+h)$, $r = 2\lambda$, $r^2 + rh = 12$.
Critical point is $(2,4)$. The nature of the problem indicates that the critical point yields a maximum value rather than a minimum value.

Conclusion: The dimensions are radius of 2 and height of 4.

Problem Set 16.1   Double Integrals over Rectangles

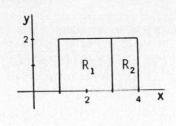

1. $\displaystyle\iint_{R_1} 2dA + \iint_{R_2} 3dA = 2A(R_1) + 3A(R_2)$

$$= 2(4) + 3(2)$$

$$= 14.$$

3. $\displaystyle\iint_R f(x,y)dA = \iint_{R_1} 2dA + \iint_{R_2} 1dA + \iint_{R_3} 3dA$

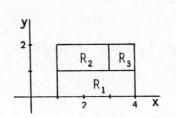

$$= 2A(R_1) + 1A(R_2) + 3A(R_3)$$

$$= 2(2) + 1(2) + 3(2) = 12.$$

5. $\displaystyle 3\iint_R f(x,y)dA - \iint_R g(x,y)dA = 3(3) - (5) = 4.$

7. $\displaystyle\iint_R g(x,y)dA - \iint_{R_1} g(x,y)dA = (5) - (2) = 3.$

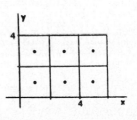

9. $[f(1,1)+f(3,1)+f(5,1)+f(1,3)+f(3,3)+f(5,3)](4)$

$$= [(10) + (8) + (6) + (8) + (6) + (4)](4)$$

$$= 168.$$

11. $4(3+11+27+19+27+43) = 520.$

13. $4(\sqrt{2}+\sqrt{4}+\sqrt{6}+\sqrt{4}+\sqrt{6}+\sqrt{8}) \approx 52.5665.$

15. $z = 6-y$ is a plane parallel to the x-axis. Let T be the area of the front trapezoidal face; let D be the distance between the front and back faces.

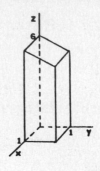

$$\iint_R (6-y)dA = \text{volume of solid}$$
$$= (T)(D) = [(1/2)(6+5)](1) = 5.5.$$

17. $\iint_R 0 dA = 0 \ A(R) = 0.$ The conclusion follows.

19. For c, take the sample point in each square to be the point of the square that is closest to the origin.

Then $c = 2\sqrt{5} + 2\sqrt{2} + 2(2) + 4(1)$
$\approx 15.3006.$

For C, take the sample point in each square to be the point of the square that is farthest from the origin.

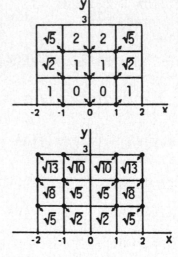

Then $C = 2\sqrt{13} + 2\sqrt{10} + 2\sqrt{8} + 4\sqrt{5} + 2\sqrt{2}$
$\approx 30.9652.$

21. The values of [x] [y] and [x]+[y] are indicated in the various square subregions of R. In each case the value of the integral on R is the sum of the values in the squares since the area of each square is 1.

(a) The integral equals -6.   (b) The integral equals 6

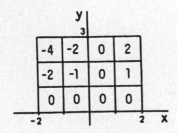

23. Total rainfall in Colorado in 1980; average rainfall in Colorado in 1980.

Problem Set 16.2   Iterated Integrals

1. $\int_0^2 \left[(1/2)x^2 y^2 \right]_{y=1}^3 dx = \int_0^2 4x^2 dx = 32/3.$

3. $\int_1^2 \left[\frac{x^2 y}{2} + xy^2 \right]_{x=0}^3 dx = \int_1^2 \left(\frac{9y}{2} + 3y^2 \right) dy = \left[\frac{9y^2}{4} + y^3 \right]_1^2 = 17 - \frac{13}{4} = 13.75.$

5. $\int_0^\pi \left[(1/2)x^2 \sin y \right]_{x=0}^1 dy = \int_0^\pi (1/2)\sin y \ dy = 1.$

7. $\int_0^{\pi/2} \left[-\cos xy \right]_{y=0}^1 dx = \int_0^{\pi/2} (1-\cos x)dx = \pi/2 - 1 \approx 0.5708.$

9. $\int_0^3 \left[\frac{2(x^2+y)^{3/2}}{3} \right]_{x=0}^1 dy = \int_0^3 \frac{2\left[(1+y)^{3/2} - y^{3/2} \right]}{3} \ dy = \left[\frac{4\left[(1+y)^{5/2} - y^{5/2} \right]}{15} \right]_0^3$

$= \frac{4(32 - 9\sqrt{3}) - 4}{15} = \frac{4(31 - 9\sqrt{3})}{15} \approx 4.1097.$

11. $\int_0^{\ln 3} \left[(1/2)\exp(xy^2) \right]_{y=0}^1 dx = \int_0^{\ln 3} (1/2)(e^x - 1)dx = 1 - (1/2)\ln 3 \approx 0.4507.$

13. $\int_0^1 0 \, dx = 0.$ [Since xy^3 defines an odd function in y.]

15. $\int_0^{\pi/2} \int_0^{\pi/2} \sin(x+y)dxdy = \int_0^{\pi/2} \left[-\cos(x+y) \right]_{x=0}^{\pi/2} dy$

$= \int_0^{\pi/2} \left[-\cos(\pi/2 + y) + \cos y \right] dy = \int_0^{\pi/2} (\sin y + \cos y)dy$

$= \left[-\cos y + \sin y \right]_0^{\pi/2} = (0+1) - (-1+0) = 2.$

437

17. $z = x/2$ is a plane.
$x - 2z = 0$.

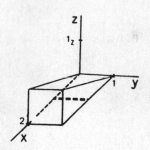

19. $z = x^2 + y^2$ is a paraboloid opening upward with z-axis for axis.

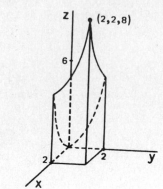

21. $\int_1^3 \int_0^1 (x+y+1)\,dx\,dy = \int_1^3 \left[(1/2)x^2 + yx + x \right]_{x=0}^1 dy = \int_1^3 (y + 3/2)\,dy = 7.$

23. $x^2 + y^2 + 2 > 1.$ $\int_{-1}^1 \int_0^1 [(x^2+y^2+2) - 1]\,dy\,dx = \int_{-1}^1 \left[x^2 y + (1/3)y^3 + y \right]_{y=0}^1 dx$

$$= \int_{-1}^1 (x^2 + 4/3)\,dx = 10/3.$$

25. $\int_a^b \int_c^d g(x)h(y)\,dy\,dx = \int_a^b g(x) \int_c^d h(y)\,dy \; dx = \int_c^d h(y)\,dy \int_a^b g(x)\,dx.$

[First step, used linearity of integration with respect to y; second step, linearity of integration with respect to x; now commute.]

27. $\int_0^1 \int_0^1 xy e^{x^2} e^{y^2}\,dy\,dx = \left[\int_0^1 x e^{x^2}\,dx \right] \left[\int_0^1 y e^{y^2}\,dy \right] = \left[\int_0^1 x e^{x^2}\,dx \right]^2$ [Changed the dummy variable y to the dummy variable x]

$$= \left[\left[\frac{e^{x^2}}{2} \right]_0^1 \right]^2 = \left(\frac{e-1}{2} \right)^2 \approx 0.7381.$$

29. (a) $\int_{-2}^{2} x^2 dx \int_{-1}^{1} |y^3| dy = 2\int_{0}^{2} x^2 dx \; 2\int_{0}^{1} y^3 dy = 2(8/3) \; 2(1/4) = 8/3.$

(b) $\int_{-2}^{2} [x^2] dx \int_{-1}^{1} y^3 dy = 0$ [since second integral equals 0].

(c) $\int_{-2}^{2} [x^2] dx \int_{-1}^{1} |y^3| dy = 2\int_{0}^{2} [x^2] dx \; 2\int_{0}^{1} y^3 dy$

$$= 2\left[\int_{0}^{1} 0 dx + \int_{1}^{\sqrt{2}} 1 dx + \int_{\sqrt{2}}^{\sqrt{3}} 2 dx + \int_{\sqrt{3}}^{2} 3 dx\right]\left[2(1/4)\right]$$

$$= 2[0 + (\sqrt{2}-1) + 2(\sqrt{3} - \sqrt{2}) + 3(2-\sqrt{3})][1/2] = 5-\sqrt{3}-\sqrt{2} \approx 1.8537.$$

31. $0 \leq \int_{a}^{b}\int_{a}^{b} [f(x)g(y) - f(y)g(x)]^2 dxdy$

$$= \int_{a}^{b}\int_{a}^{b} [f^2(x)g^2(y) - 2f(x)g(x)f(y)g(y) + f^2(y)g^2(x)] dxdy$$

$$= \int_{a}^{b} f^2(x) dx \int_{a}^{b} g^2(y) dy - 2\int_{a}^{b} f(x)g(x) dy \int_{a}^{b} f(y)g(y) dy +$$

$$+ \int_{a}^{b} f^2(y) dy \int_{a}^{b} g^2(x) dx$$

$$= 2\int_{a}^{b} f^2(x) dx \int_{a}^{b} g^2(x) dx - 2\left[\int_{a}^{b} f(x)g(x) dx\right]^2.$$

Therefore, $\left[\int_{a}^{b} f(x)g(x) dx\right]^2 \leq \int_{a}^{b} f^2(x) dx \int_{a}^{b} g^2(x) dx.$

Problem Set 16.3 Double Integrals over Nonrectangular Regions

1. $\int_0^1 \left[x^2 y\right]_{y=0}^{3x} dx = \int_0^1 3x^3 dx = 3/4.$

3. $\int_{-1}^3 \left[\frac{x^3}{3} + y^2 x\right]_{x=0}^{3y} dy = \int_{-1}^3 (9y^3 + 3y^3) dy = \left[3y^4\right]_{-1}^3 = 243 - 3 = 240.$

5. $\int_1^3 \left[(1/2)x^2 \exp(y^3)\right]_{x=-y}^{2y} dy = \int_1^3 (3/2)y^2 \exp(y^3)\ dy = (1/2)(e^{27} - e)$

    $\approx 2.660 \times 10^{11}.$

7. $\int_{1/2}^1 \left[y\cos(\pi x^2)\right]_{y=0}^{2x} dx = \int_{1/2}^1 2x\cos(\pi x^2) dx = -\sqrt{2}/2\pi \approx -0.2251.$

9. $\int_0^{\pi/9} \left[\tan\theta\right]_{\theta=\pi/4}^{3r} dr = \int_0^{\pi/9} (\tan 3r - 1) dr = \left[\frac{-\ln|\cos 3r|}{3} - r\right]_0^{\pi/9}$

    $= \left[\frac{\ln(1/2)}{3} - \frac{\pi}{9}\right] - \left[\frac{-\ln(1)}{3} - 0\right] = \frac{3\ln 2 - \pi}{9} \approx -0.1180.$

11. $\int_0^2 \left[xy + (1/2)y^2\right]_{y=0}^{\sqrt{4-x^2}} dx = \int_0^2 \left[x(4-x^2)^{1/2} + 2 - (1/2)x^2\right] dx = 16/3.$

13. $\int_{-1}^1 \int_{x^2}^1 xy\, dy\, dx = 0.$

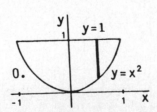

15. $\int_0^1 \int_{x^2}^{\sqrt{x}} (x^2+2y)\,dy\,dx = \int_0^1 \left[x^2y+y^2\right]_{y=x^2}^{\sqrt{x}} dx$

$$= \int_0^1 \left[(x^{5/2}+x)-(x^4+x^4)\right]dx = \left[\frac{2x^{7/2}}{7} + \frac{x^2}{2} - \frac{2x^5}{5}\right]_0^1$$

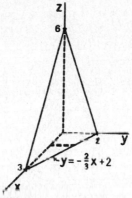

$$= \frac{2}{7} + \frac{1}{2} - \frac{2}{5} = \frac{27}{70} \approx 0.3857.$$

17. $\int_0^2 \int_x^2 2(1+x^2)^{-1}\,dy\,dx$

$$= 4\tan^{-1}2 - \ln 5 \approx 2.8192.$$

19. $\int_0^3 \int_0^{(-2/3)x+2} (6-2x-3y)\,dy\,dx = 6.$

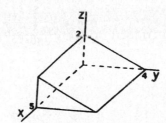

21. $\int_0^5 \int_0^4 \frac{4-y}{2}\,dy\,dx = \left[\int_0^5 1\,dx\right]\left[\int_0^4 \frac{4-y}{2}\,dy\right]$

$$= 5\left[2y - \frac{y^2}{4}\right]_0^4 = 5(8-4) = 20.$$

23. $\int_0^2 \int_0^{(1/2)\sqrt{36-9x^2}} (1/6)(9x+4y)dydx = 10.$

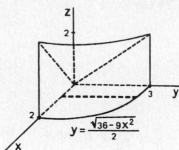

$y = \dfrac{\sqrt{36-9x^2}}{2}$

25. $\int_0^1 \int_0^{\sqrt{y}} (1-y)dxdy = 4/15.$

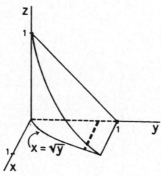

$x = \sqrt{y}$

27. $\int_0^1 \int_0^x \tan x^2 dydx = \int_0^1 \left[y\tan x^2 \right]_{y=0}^x dx$

$= \int_0^1 x\tan x^2 dx = \left[\dfrac{-\ln|\cos x^2|}{2} \right]_0^1$

$= (-1/2)\ln(\cos 1) \approx 0.3078.$

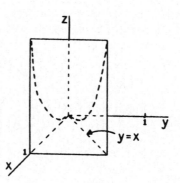

$y = x$

29. $\int_0^2 \int_0^{(3/2)\sqrt{4-x^2}} [4-x^2-(4/9)y^2]dydx = 3\pi \approx 9.4248.$

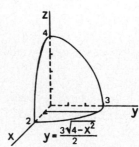

$y = \dfrac{3\sqrt{4-x^2}}{2}$

31. $\int_0^1 \int_y^1 f(x,y)dxdy.$

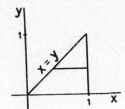

33. $\int_0^1 \int_{y^4}^{\sqrt{y}} f(x,y)dxdy.$

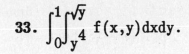

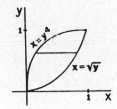

35. $\int_{-1}^0 \int_{-x}^1 f(x,y)dydx + \int_0^1 \int_x^1 f(x,y)dydx.$

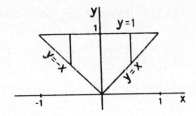

37. $\int_0^2 \int_0^{2-y} xy^2 dxdy + \int_2^4 \int_0^{y-2} xy^2 dxdy = 256/15 \approx 17.0667.$

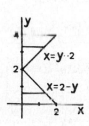

39. $\int_0^2 \int_0^{y^2} \sin(y^3)dxdy = \int_0^2 \left[x\sin(y^3) \right]_{x=0}^{y^2} dy$

$= \int_0^2 y^2 \sin(y^3)dy = \left[\frac{-\cos(y^3)}{3} \right]_0^2$

$= \frac{1-\cos 8}{3} \approx 0.3818.$

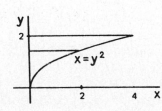

41. The integral over S of x^4y is 0 (see Problem 40). Therefore,

$$\iint_S (x^2+x^4y)\,dA = \iint_S x^2\,dA = 4\left[\iint_{S_1} x^2\,dA + \iint_{S_2} x^2\,dA\right].$$

$$= 4\left[\int_0^1\int_{\sqrt{1-x^2}}^{\sqrt{4-x^2}} x^2\,dy\,dx + \int_1^2\int_0^{\sqrt{4-x^2}} x^2\,dy\,dx\right]$$

$$= 4\left[\int_0^2 x^2\sqrt{4-x^2}\;dx - \int_0^1 x^2\sqrt{1-x^2}\;dx\right]$$

$$= 4\left[16\int_0^{\pi/2}\sin^2\theta\cos^2\theta\;d\theta - \int_0^{\pi/2}\sin^2\phi\cos^2\phi\;d\phi\right]$$

[Using $x = 2\sin\theta$ in 1st integral; $x = \sin\phi$ in 2nd]

$$= 60\int_0^{\pi/2}\sin^2\theta\cos^2\theta\;d\theta = 15\pi/4 \quad \text{[See work in Problem 42.]}$$

$$\approx 11.7810.$$

Problem Set 16.4 Double Integrals in Polar Coordinates

1. $\int_0^{\pi/2}\left[(1/3)r^3\sin\theta\right]_{r=0}^{\cos\theta} d\theta = \int_0^{\pi/2}(1/3)\cos^3\theta\sin\theta\;d\theta = 1/12 \approx 0.0833.$

3. $\int_0^{\pi}\left[\dfrac{r^3}{3}\right]_{r=0}^{\sin\theta} d\theta = \int_0^{\pi}\dfrac{\sin^3\theta}{3}\;d\theta = \int_0^{\pi}\dfrac{(1-\cos^2\theta)\sin\theta}{3}\;d\theta = \left[\dfrac{-\cos\theta}{3} + \dfrac{\cos^3\theta}{9}\right]_0^{\pi}$

$$= \left[\dfrac{1}{3} - \dfrac{1}{9}\right] - \left[\dfrac{-1}{3} + \dfrac{1}{9}\right] = \dfrac{4}{9}.$$

5. $\int_0^{\pi/3}\int_2^{4\cos\theta} r\,dr\,d\theta = 2\left[2\pi/3 + \sqrt{3}\right] \approx 7.6529.$

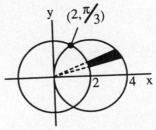

7. $\int_0^{\pi/2} \int_0^{a\sin2\theta} r\,dr\,d\theta = a^2\pi/8.$

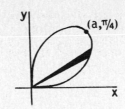

9. $2\int_{5\pi/6}^{3\pi/2} \int_0^{2-4\sin\theta} r\,dr\,d\theta = 2\int_{5\pi/6}^{3\pi/2} \left[\frac{r^2}{2}\right]_0^{2-4\sin\theta} d\theta$

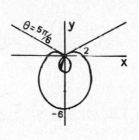

$$= 2\int_{5\pi/6}^{3\pi/2} (6 - 8\sin\theta - 4\cos2\theta)\,d\theta$$

$$= 2\left[6\theta + 8\cos\theta - 2\sin2\theta2\right]_{5\pi/6}^{3\pi/2} = 2(4\pi + 3\sqrt{3})$$

$$\approx 35.525.$$

11. $2\int_0^{2\pi} \int_0^2 e^{r^2} r\,dr\,d\theta = \pi(e^4 - 1) \approx 168.3836.$

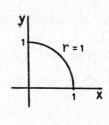

13. $\int_0^{\pi/4} \int_0^2 (4+r^2)^{-1} r\,dr\,d\theta = (\pi/8)\ln2 \approx 0.2722.$

15. $\int_0^{\pi/2} \int_0^1 (4-r^2)^{-1/2} r\,dr\,d\theta = \int_0^{\pi/2} \left[-(4-r^2)^{1/2}\right]_0^1 d\theta$

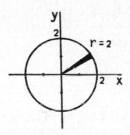

$$= \int_0^{\pi/2} (-\sqrt{3} +2)\,d\theta = (-\sqrt{3} + 2)(\pi/2) \approx 0.4209.$$

17. $\displaystyle\int_{\pi/4}^{\pi/2}\int_{0}^{\csc\theta} r^2\cos^2\theta\ rdrd\theta = 1/12 \approx 0.0833.$

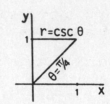

19. $\displaystyle\iint_R (x^2+y^2)dA = \int_0^{\pi/2}\int_0^3 r^2\ rdrd\theta$

$\qquad\qquad = 81\pi/8 \approx 31.8086.$

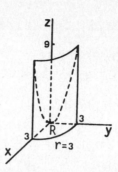

21. $\displaystyle\int_{-5}^0\int_{\sqrt{3}x}^x (y^2)dydx = \int_{-5}^0 \left[\frac{y^3}{3}\right]_{\sqrt{3}x}^{-x} dx$

$\qquad = \displaystyle\int_{-5}^0 \frac{-1-3\sqrt{3}}{3} x^3 dx = \left[\frac{(-1-3\sqrt{3})x^4}{12}\right]_{-5}^0$

$\qquad = \displaystyle\frac{(1+3\sqrt{3})625}{12} \approx 322.7163.$

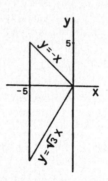

23. This can be done by the methods of this section, but an easier way to do it is to realize that the intersection is the union of two congruent segments (of one base) of the spheres, so (see Problem 20, Section 6.2, with d=h and a=r) the volume is $2[(1/3)\pi d^2(3a-d)] = 2\pi d^2(3a-d)/3.$

25. Volume $= 4\displaystyle\int_0^{\pi/2}\int_0^{a\sin\theta} \sqrt{a^2-r^2}\ rdrd\theta$

$\qquad = \displaystyle\int_0^{\pi/2} [(-1/3)(a^3\cos^3\theta - a^3)]d\theta$

$\qquad = (-4/3)a^3[2/3 - \pi/2] = (2/9)a^3(3\pi-4).$

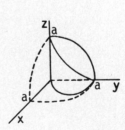

27. Choose a coordinate system so the center of the sphere is the origin and the axis of the part removed is the z-axis.

Volume(Ring) = Volume(Sphere of radius a) - Volume(Part removed)

$$= \frac{4}{3}\pi a^3 - 2\int_0^{2\pi}\int_0^{\sqrt{a^2-b^2}}\sqrt{a^2-r^2}\, r\,dr\,d\theta$$

$$= \frac{4}{3}\pi a^3 - 2(2\pi)\int_0^{\sqrt{a^2-b^2}}(a^2-r^2)^{1/2}r\,dr$$

$$= \frac{4}{3}\pi a^3 - 4\pi\left[\frac{1}{3}(a^2-r^2)^{3/2}\right]_0^{\sqrt{a^2-b^2}} = \frac{4}{3}\pi a^3 + 4\pi\frac{1}{3}\left[(b^3-a^3)\right] = \frac{4}{3}\pi b^3.$$

29. $$\int_0^{\pi/2}\left[\lim_{b\to\infty}\int_0^b (1+r^2)^{-2}r\,dr\right]d\theta = \int_0^{\pi/2}\left[\lim_{b\to\infty}\left[(-1/2)(1+b^2)^{-1} - (-1/2)\right]\right]d\theta$$

$$= \int_0^{\pi/2}(1/2)\,d\theta = \pi/4 \approx 0.7854.$$

Problem Set 10.5 Applications of Double Integrals

1. $m = \int_0^3\int_0^4 (y+1)\,dx\,dy = 30.$ $M_y = \int_0^3\int_0^4 x(y+1)\,dx\,dy = 60.$

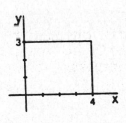

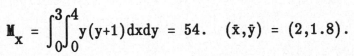

$M_x = \int_0^3\int_0^4 y(y+1)\,dx\,dy = 54.$ $(\bar{x},\bar{y}) = (2,1.8).$

3. $m = \int_0^{\pi}\int_0^{\sin x} y\ dydx = \int_0^{\pi}\left[\frac{y^2}{2}\right]_0^{\sin x} dx$

$\qquad = \int_0^{\pi}\frac{\sin^2 x}{2}\ dx = \int_0^{\pi}\frac{1-\cos 2x}{4}\ dx = \left[\frac{x}{4} - \frac{\sin 2x}{8}\right]_0^{\pi} = \frac{\pi}{4}.$

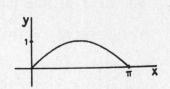

$M_x = \int_0^{\pi}\int_0^{\sin x} yy\ dydx = \int_0^{\pi}\left[\frac{y^3}{3}\right]_0^{\sin x} dx = \int_0^{\pi}\frac{\sin^3 x}{3}dx$

$\qquad = \frac{1}{3}\int_0^{\pi}(1-\cos^2 x)\sin x\ dx = \frac{1}{3}\left[-\cos x + \frac{\cos^3 x}{3}\right]_0^{\pi} = \frac{4}{9}.$

$\bar{y} = \frac{M_x}{m} = \frac{4/9}{\pi/4} = \frac{16}{9\pi} \approx 0.5659; \quad \bar{x} = \pi/2 \text{ (by symmetry)}.$

5. $m = \int_0^1\int_0^{e^{-x}} y^2 dydx = (1/9)(1-e^{-3}).$

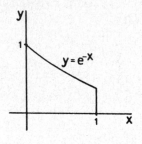

$M_x = \int_0^1\int_0^{e^{-x}} y^3 dydx = (1/16)(1-e^{-4}).$

$M_y = \int_0^1\int_0^{e^{-x}} xy^2 dydx = (1/27)(1-4e^{-3}).$

$(\bar{x},\bar{y}) = ((1/3)(e^3-4)(e^3-1)^{-1}, (9/16)e^{-1}(e^4-1)(e^3-1)^{-1}) \approx (0.2809, 0.5811).$

7. $m = \int_0^{\pi}\int_0^{2\sin\theta} r\ drd\theta = 32/9.$

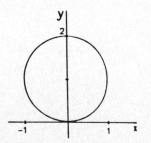

$M_x = \int_0^{\pi}\int_0^{2\sin\theta} (r\sin\theta)r\ drd\theta = 64/15.$

$M_y = 0 \text{ (symmetry)}. \quad (\bar{x},\bar{y}) = (0,1.2).$

9. $I_x = \int_0^3 \int_{y^2}^9 y^2(x+y)\,dx\,dy = \int_0^3 \left[\frac{81y^2}{2} + 9y^3 - \frac{y^6}{2} + y^5\right]dy = \frac{7533}{28} \approx 269.$

$I_y = \int_0^9 \int_0^{\sqrt{x}} x^2(x+y)\,dy\,dx = \int_0^9 \left[x^{7/2} + \frac{x^3}{2}\right]dx = \frac{41553}{8} \approx 5194.$

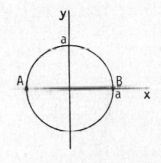

$I_z = I_x + I_y = \frac{305937}{56} \approx 5463.$

11. $I_x = \int_0^a \int_0^a (x+y)y^2\,dx\,dy = (5/12)a^5.$ $I_y = (5/12)a^5.$ $I_z = (5/6)a^5.$

13. $m = \int_0^a \int_0^a (x+y)\,dx\,dy = a^3.$ $\bar{r} = (I_x/m)^{1/2} = (5/12)^{1/2}a \approx 0.6455a.$

15. $m = \delta\pi a^2.$ The moment of inertia about diameter AB is

$I = I_x = \int_0^{2\pi} \int_0^a \delta r^2\sin^2\theta \; r\,dr\,d\theta = \int_0^{2\pi} \frac{\delta a^4 \sin^2\theta}{4}\,d\theta$

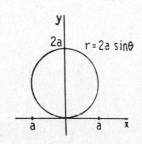

$= \frac{\delta a^4}{8} \int_0^{2\pi} (1-\cos 2\theta)\,d\theta = \frac{\delta a^4}{8}\left[\theta - \frac{\sin 2\theta}{2}\right]_0^{2\pi} = \frac{\delta a^4 \pi}{4}.$

$\bar{r} = (I/m)^{1/2} = \left[\frac{\delta a^4 \pi/4}{\delta \pi a^2}\right]^{1/2} = \frac{a}{2}.$

17. $I_x = \iint_{SS} \delta y^2\,dA = 2\delta \int_0^{\pi/2} \int_0^{2a\sin\theta} (r\sin\theta)^2 r\,dr\,d\theta$

$= 2\delta \int_0^{\pi/2} 4a^4 \sin^6\theta\,d\theta = 8a^4\delta \; \frac{(1)}{(2)}\frac{(3)}{(4)}\frac{(5)}{(6)} \; \frac{\pi}{2} = \frac{5a^4\delta\pi}{4}.$

19. $x = 0$ [by symmetry].

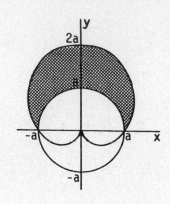

$$M_x = \iint_S ky\,dA = 2k\int_{-0}^{\pi/2}\int_a^{a(1+\sin\theta)}(r\sin\theta)^1 r\,dr\,d\theta$$

$$= 2k\int_0^{\pi/2}(a^3/3)(3\sin^2\theta + 3\sin^3\theta + \sin^4\theta)\,d\theta$$

$$= (2/3)ka^3[(15\pi+32)/16] = (1/24)ka^3(15\pi+32).$$

$$m = \iint_S k\,dA = 2k\int_{-0}^{\pi/2}\int_a^{a(1+\sin\theta)} r\,dr\,d\theta = 2k\int_0^{\pi/2}(1/2)a^2(2\sin\theta + \sin^2\theta)\,d\theta$$

$$= ka^2[(8+\pi)/4] = (1/4)ka^2(\pi+8)$$

Therefore, $\bar{y} = \dfrac{M_x}{m} = \dfrac{(1/24)ka^3(15\pi+32)}{(1/4)ka^2(\pi+8)} = \dfrac{a(15\pi+32)}{6(\pi+8)} \approx 1.1836a.$

21. (a) $m = \iint_S (x+y)\,dA = \int_0^a\int_0^a (x+y)\,dx\,dy$

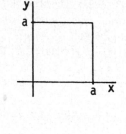

$$= \int_0^a\left[\left[\frac{x^2}{2} + xy\right]_{x=0}^a\right]dy = \int_0^a\left[\frac{a^2}{2} + ay\right]dy$$

$$= \left[\frac{a^2 y}{2} + \frac{ay^2}{2}\right]_0^a = a^3.$$

(b) $M_y = \iint_S x(x+y)\,dA = \int_0^a\int_0^a (x^2+xy)\,dy\,dx$

$$= \int_0^a\left[\left[x^2 y + \frac{xy^2}{2}\right]_{y=0}^a\right]dx = \int_0^a\left[ax^2 + \frac{a^2 x}{2}\right]dx = \left[\frac{ax^3}{3} + \frac{a^2 x^2}{2}\right]_0^a = \frac{7a^4}{12}.$$

Therefore, $\bar{x} = \dfrac{M_y}{m} = \dfrac{7a}{12}.$

(c) $I_y = I_L + d^2 m$, so $\dfrac{5a^5}{12} = I_L + \left[\dfrac{7a}{12}\right]^2[a^3]$; $I_L = \dfrac{11a^5}{144}.$

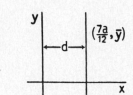

23. $I_x = 2[I_{15}] = 0.5ka^4\pi.$

$I_y = 2[I_{15} + md^2] = 2[0.25a^4\pi + (k\pi a^2)(2a)^2] = 8.5ka^4\pi.$

$I_z = I_x + I_y = 9ka^4\pi.$

25. $M_y = \iint_{S_1 \cup S_2} x\delta(x,y)\,dA = \iint_{S_1} x\delta(x,y)\,dA + \iint_{S_2} x\delta(x,y)\,dA$

$= \dfrac{m_1 \iint_{S_1} x\delta(x,y)\,dA}{m_1} + \dfrac{m_2 \iint_{S_2} x\delta(x,y)\,dA}{m_2} = m_1\bar{x}_1 + m_2\bar{x}_2$

thus, $\bar{x} = \dfrac{M_y}{m} = \dfrac{m_1 x_1 + m_2 x_2}{m_1 + m_2}$ which is equal to what we are to obtain and which is what we would obtain using the center of mass formula for two point masses. [Similar result can be obtained for $\bar{y}$.]

27. $\langle a,b \rangle$ is perpendicular to the line $ax+by = 0$. Therfore, the (signed) distance of (x,y) to L is the scalar projection of $\langle x,y \rangle$ onto $\langle a,b \rangle$, which is $d(x,y) = \dfrac{\langle x,y \rangle \cdot \langle a,b \rangle}{|\langle a,b \rangle|} = \dfrac{ax+by}{|\langle a,b \rangle|}.$

$M_L = \iint_S d(x,y)\delta(x,y)\,dA = \iint_S \dfrac{ax+by}{|\langle a,b \rangle|}\delta(x,y)\,dA$

$= \dfrac{a}{|\langle a,b \rangle|}\iint_S x\delta(x,y)\,dA + \dfrac{a}{|\langle a,b \rangle|}\iint_S y\delta(x,y)\,dA$

$= \dfrac{a}{|\langle a,b \rangle|}(0) + \dfrac{b}{|\langle a,b \rangle|}(0) = 0.$ [Since $(\bar{x},\bar{y}) = (0,0)$].

Problem Set 16.6 Surface Area

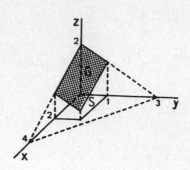

1. $\cos\gamma = \dfrac{\langle 3,4,6\rangle \cdot \langle 0,0,1\rangle}{\sqrt{61}\sqrt{1}}.$

$A(G) = A(S)\sec\gamma = (2)(\sqrt{61}/6)$

$= \sqrt{61}/3 \approx 2.6034.$

3. $z = f(x,y) = (4-y^2)^{1/2}$; $f_x(x,y) = 0$; $f_y(x,y) = -y(4-y^2)^{-1/2}.$

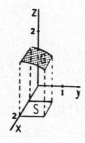

$A(G) = \displaystyle\int_0^1\int_1^2 \sqrt{y^2(4-y^2)^{-1}+1}\; dxdy = \int_0^1\int_1^2 \dfrac{2}{\sqrt{4-y^2}}\; dxdy$

$= \displaystyle\int_0^1 \dfrac{2}{\sqrt{4-y^2}}\; dy = \Big[2\sin^{-1}(y/2)\Big]_0^1$

$= 2(\pi/6) - 2(0) = \pi/3 \approx 1.0472.$

5. Let $z = f(x,y) = (9-x^2)^{1/2}.$

$f_x(x,y) = -x(9-x^2)^{-1/2},\quad f_y(x,y) = 0.$

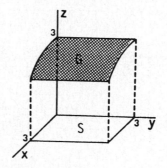

$A(G) = \displaystyle\int_0^2\int_0^3 [x^2(9-x^2)^{-1}+1]\, dydx$

$= \displaystyle\int_0^2\int_0^3 3(9-x^2)^{-1/2}\, dxdy = 9\sin^{-1}(2/3) \approx 6.5675.$

7. $z = f(x,y) = (x^2+y^2)^{1/2}$.

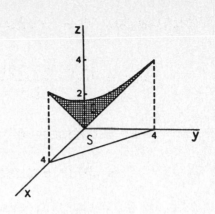

$f_x(x,y) = x(x^2+y^2)^{-1/2}, \quad f_y(x,y) = y(x^2+y^2)^{-1/2}$.

$A(G) = \int_0^4 \int_0^{4-x} [x^2(x^2+y^2)^{-1}+y^2(x^2+y^2)^{-1}+1]^{1/2} dydx$

$\qquad = \int_0^4 \int_0^{4-x} \sqrt{2}\,dydx = 8\sqrt{2}$.

9. Let $F(x,y,z) = x^2+y^2+z^2$.

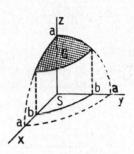

$\sec\gamma = \dfrac{a}{\sqrt{a^2-x^2-y^2}}.$ [See Example 3.]

$A(G) = 8\iint_S \dfrac{a}{\sqrt{a^2-x^2-y^2}}\, dA = 8\int_0^{\pi/2}\int_0^b \dfrac{a}{\sqrt{a^2-r^2}}\, r\, drd\theta$

$\qquad = 8a\left(\dfrac{\pi}{2}\right)\int_0^b (a^2-r^2)^{-1/2} r\,dr = -4a\pi\left[(a^2-r^2)^{1/2}\right]_0^b = 4\pi a(a - \sqrt{a^2-b^2}).$

11. $\sec\gamma = \dfrac{a}{\sqrt{a^2-x^2-y^2}}.$ [See Example 3.]

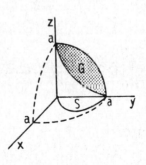

$A(G) = 4\int_0^{\pi/2}\int_0^{a\sin\theta} \dfrac{a}{\sqrt{a^2-r^2}}\, r\, drd\theta$

$\qquad = 4\int_0^{\pi/2} [-a^2(\cos\theta - 1)]d\theta = 2a^2(\pi-2).$

13. Let $F(x,y,z) = x^2 - y^2 - az$.

$$\sec\gamma = \frac{\sqrt{4x^2 + 4y^2 + a^2}}{a}.$$

$$A(G) = \int_0^{2\pi}\int_0^a \frac{\sqrt{4r^2 + a^2}}{a} \; r \; drd\theta = \frac{2\pi}{a}\int_0^a (4r^2 + a^2)^{1/2} r \; dr = \frac{\pi a^2 (5\sqrt{5} - 1)}{6}.$$

15. $\bar{x} = \bar{y} = 0$ [By symmetry]

Let $h = \dfrac{h_1 + h_2}{2}$. Planes $z = h_1$ and $z = h$ cut out the same surface area as planes $z = h$ and $z = h_2$. [See Probelm 20, Section 6.5.]   Therefore, $\bar{z} = h$, the arithmetic average of h_1 and h_2.

17. (a) $A = \pi b^2$.

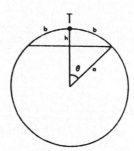

(b) $B = 2\pi a^2 (1 - \cos\phi)$ [Problem 16]

$\quad\quad = 2\pi a^2 [1 - \cos(b/a)]$

$$= 2\pi a^2 \left[\frac{b^2}{2! a^2} - \frac{b^4}{4! a^4} + \frac{b^6}{6! a^4} + \cdots \right]$$

$$= \pi b^2 \left[1 - \frac{b^2}{12a^2} + \frac{b^4}{360a^4} - \cdots \right] \leq \pi b^2.$$

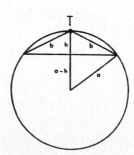

(c) $a^2 - (a-h)^2 = b^2 - h^2$, so $h = b^2/2a$.

$\quad\quad$ Thus, $C = 2\pi ah$ [Problem 20, Section 6.5]

$\quad\quad\quad\quad\quad = 2\pi a(b^2/2a) = \pi b^2.$

(d) $D = 2\pi ah$ [Problem 20, Section 6.5]

$$= 2\pi a(a - \sqrt{a^2-b^2}) = \frac{2\pi a[a^2-(a^2-b^2)]}{a + \sqrt{a^2-b^2}}$$

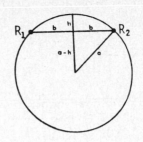

$$= \frac{2\pi ab^2}{a + \sqrt{a^2-b^2}}$$

$$> \pi b^2 .$$

Therefore, $B < A = C < D$.

19. In the following each double integral is over S_{xy}.

$$A(S_{xy})\, f(\bar{x},\bar{y}) = A(S_{xy})\, (a\bar{x}+b\bar{y}+c)$$

$$= \iint dA \left[a\, \frac{\iint x dA}{\iint dA} + b\, \frac{\iint y dA}{\iint dA} + c \right]$$

$$= a\iint x dA + b\iint y dA + c\iint dA$$

$$= \iint (ax+by+c)\, dA$$

$$= \text{Volume of solid cylinder under } S_{xy}.$$

Problem Set 16.7 Triple Integrals (Cartesian Coordinates)

1. $\displaystyle \int_{-3}^{7}\int_{0}^{2x} (x-1-y)\, dy dx = \int_{-3}^{7} -2x\, dx = -40 .$

3. $\int_{1}^{4}\int_{z-1}^{2z}\int_{0}^{y+2z} dxdydz = \int_{1}^{4}\int_{z-1}^{2z}(y+2z)dydz = \int_{1}^{4}\left[\frac{y^2}{2} + 2yz\right]_{y=z-1}^{2z} dz$

$= \int_{1}^{4}\left[\frac{7z^2}{2} + 3z - \frac{1}{2}\right]dz = \left[\frac{7z^3}{6} + \frac{3z^2}{2} - \frac{z}{2}\right]_{1}^{4} = \frac{189}{2} = 94.5.$

5. $\int_{0}^{2}\int_{1}^{z}x^2 dxdz = \int_{0}^{2}(1/3)(z^3-1)dz = 2/3.$

7. $\int_{-2}^{4}\int_{x-1}^{x+1}3y^2 dydx = \int_{-2}^{4}(6x^2+2)dx = 156.$

9. $\int_{0}^{1}\int_{0}^{3}\int_{0}^{(12-3x-2y)/6}f(x,y,z)dzdydx.$ **11.** $\int_{0}^{2}\int_{0}^{4}\int_{0}^{y/2}f(x,y,z)dxdydz.$

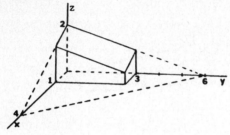
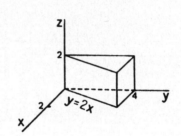

13. $\int_{0}^{2.4}\int_{x/3}^{(4-x)/2}\int_{0}^{4-x-2z}f(x,y,z)dydzdx.$

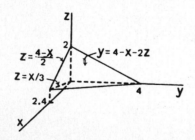

15. Using the cross product of vectors along edges it is easy to show that <2,6,9> is normal to the upward face. Then obtain that its equation is 2x+6y+9z = 18.

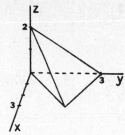

$$\int_0^3 \int_{2x/3}^{(9-x)/3z} \int_0^{(18-2x-6y)/9} f(x,y,z)\,dz\,dy\,dx.$$

17. $$\int_1^4 \int_0^1 \int_0^{\sqrt{1-y^2}} f(x,y,z)\,dz\,dy\,dx.$$

19. $$\int_0^2 \int_{2x^2}^8 \int_0^{2-y/4} 1 \, dz\,dy\,dx = 128/15.$$

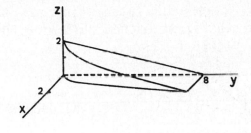

21. $$V = 4\int_0^1 \int_0^{\sqrt{y}} \int_0^{\sqrt{y}} 1 \, dz\,dx\,dy = 4\int_0^1 \int_0^{\sqrt{y}} \sqrt{y} \, dx\,dy$$

$$= 4\int_0^1 \sqrt{y}\,\sqrt{y}\,dy = \left[2y^2\right]_0^1 = 2.$$

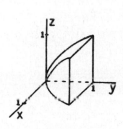

23. Let $\delta(x,y,z) = x+y+z$. [See note with next problem.]

$$m = \int_0^1 \int_0^{1-x} \int_0^{1-x-y} (x+y+z)\,dz\,dy\,dx = 1/8.$$

$$M_{yz} = \int_0^1 \int_0^{1-x} \int_0^{1-x-y} x(x+y+z)\,dz\,dy\,dx = 1/30.$$

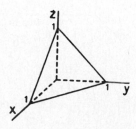

$\bar{x} = 4/15$. The $\bar{y} = \bar{z} = 4/15$ (symmetry).

25. Let $\delta(x,y,z) = 1$. [See note with previous problem.]

$m = (1/8)(\text{volume of sphere}) = (\pi/6)a^3$.

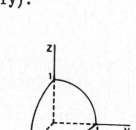

$$M_{xy} = \int_0^a \int_0^{\sqrt{a^2-x^2}} \int_0^{\sqrt{a^2-x^2-y^2}} z\,dz\,dy\,dx$$

$$= \int_0^{\pi/2} \int_0^a \int_0^{\sqrt{a^2-r^2}} z\,r\,dz\,dr\,d\theta = (\pi/16)a^4.$$

$\bar{x} = (3/8)a.$ $\bar{y} = \bar{z} = (3/8)a$ (by symmetry).

27. The limits of integration are those for the first octant part of a sphere of radius 1.

$$\int_0^1 \int_0^{\sqrt{1-x^2}} \int_0^{\sqrt{1-x^2-y^2}} f(x,y,z)\,dz\,dy\,dx$$

29. $\displaystyle\int_0^2 \int_0^{2-z} \int_0^{9-x^2} f(x,y,z)\,dy\,dx\,da.$ Figure is same as for Problem 30 except that the solid doesn't need to be divided into two parts.

31.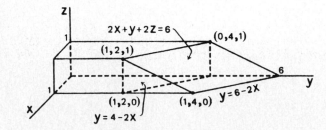

(a) $\displaystyle\int_0^1 \int_0^{4-2x} \int_0^1 1\,dz\,dy\,dx + \int_0^1 \int_{4-2x}^{6-2x} \int_0^{3-x-y/2} 1\,dz\,dy\,dx = 3+1 = 4.$

(b) $\int_0^1\int_0^1\int_0^{6-2x-2z} 1 \, dydxdz = 4.$

(c) $A(S_{xz})f(\bar{x},\bar{z})$ [S_{xz} is the unit square in the corner of xz-plane;

and $(\bar{x},\bar{z}) = (1/2,1/2)$ is the centroid of S_{xz}.]

$= (1)[6 - 2(1/2) - 1(1/2)] = 4.$

33. $\int_0^1\int_0^1\int_0^{6-2x-2z}(30-z)\,dydxdz = \int_0^1\int_0^1 (30-z)(6-2x-2z)\,dxdz$

$= \int_0^1\left[\left[(30-z)(6x-x^2-2xz)\right]_{x=0}^1\right]dz = \int_0^1 (30-z)(5-2z)\,dz$

$= \int_0^1 (150-65z+2z^2)\,dz = \left[150z - \frac{65z^2}{2} + \frac{2z^3}{3}\right]_0^1 = \frac{709}{6}.$

The volume of the solid is 4.   [From Problem 31]

Hence, the average temperature of the solid is $\frac{709/6}{4} = \frac{709}{24} \approx 29.54°.$

$\int_0^1\int_0^1\int_0^{6-2x-2z} z \, dydxdz = \int_0^1\int_0^1 z(6-2x-2z)\,dxdz = \int_0^1\left[\left[z(6x-x^2-2xz)\right]_{x=0}^1\right]dz$

$= \int_0^1 (5z-2z^2)\,dz = \left[\frac{5z^2}{2} \frac{2z^3}{3}\right]_0^1 = \frac{11}{6}.$ Hence, $\bar{z} = \frac{11/6}{4} = \frac{11}{24} \approx 0.4583.$

35. $2\int_0^a\int_0^{b\sqrt{1-\frac{x^2}{a^2}}}\int_0^{c\sqrt{1-\frac{x^2}{a^2}-\frac{y^2}{b^2}}}(xy + xz + yz)\,dzdydx = \frac{a^2b^2c}{15} + \frac{a^2bc^2}{15} + \frac{ab^2c^2}{15}$

Problem Set 16.8   Triple Integrals (Cylindrical and Spherical Coordinates)

1. $\int_0^{2\pi}\int_0^2\int_{r^2}^4 r\, dzdrd\theta = 8\pi \approx 25.1327.$

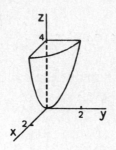

3. $\int_0^{2\pi}\int_0^2\int_{r^2/4}^{\sqrt{5-r^2}} r\, dzdrd\theta = \int_0^{2\pi}\int_0^2 \left[r(5-r^2)^{1/2} - \frac{r^3}{4}\right]drd\theta$

$$= \int_0^{2\pi}\frac{5^{3/2}-4}{3}\, d\theta = \frac{2\pi(5^{3/2}-4)}{3} \approx 15.0385.$$

5. Let $\delta(x,y,z) = 1$.   [See write-up of Problem 24, Section 16.7.]

$$m = \int_0^{2\pi}\int_0^2\int_{r^2}^{12-2r^2} r\, dzdrd\theta = 24\pi.$$

$$M_{xy} = \int_0^{2\pi}\int_0^2\int_{r^2}^{12-2r^2} zr\, dzdrd\theta = 128\pi.$$

$\bar{z} = 16/3.$ $\bar{x} = \bar{y} = 0$ (by symmetry).

7. $\delta(x,y,z) = kp$

$$m = \int_0^{\pi}\int_0^{2\pi}\int_a^b k_\rho\ \rho^2\sin\phi\ d\rho d\theta d\phi = k\pi(b^4-a^4).$$

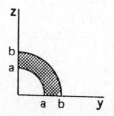

9. Let $\delta(x,y,z) = \rho$. [Letting k = 1 -- see comment at the beginning of the write-up of Problem 24 of the previous section.]

$$m = \int_0^{\pi/2}\int_0^{2\pi}\int_0^a \rho^3\sin\phi \; d\rho d\theta d\phi = \int_0^{\pi/2}\int_0^{2\pi} \frac{a^4\sin\phi}{4} \; d\theta d\phi$$

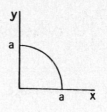

$$= \int_0^{\pi/2} \frac{\pi a^4\sin\phi}{2} \; d\phi = \frac{\pi a^4}{2}.$$

$$M_{xy} = \int_0^{\pi/2}\int_0^{2\pi}\int_0^a \rho^4\sin\phi\cos\phi \; d\rho d\theta d\phi \quad [z = \rho\cos\phi]$$

$$= \int_0^{\pi/2}\int_0^{2\pi} \frac{a^5\sin 2\phi}{10} \; d\theta d\phi = \int_0^{\pi/2} \frac{\pi a^5\sin 2\phi}{5} \; d\phi = (\pi/5)a^5.$$

$$\bar{z} = \frac{\pi a^5/5}{\pi a^4/2} = 0.4a; \quad \bar{x} = \bar{y} = 0 \quad \text{(by symmetry)}.$$

11. $$I_z = \iiint_S (x^2+y^2)k(x^2+y^2)^{1/2}dV = \int_0^{\pi/2}\int_0^{2\pi}\int_0^a k\rho^5\sin^4\phi \; d\rho d\theta d\phi = (k/16)\pi^2 a^6.$$

13. $$\int_0^\pi\int_0^{\pi/6}\int_0^1 \rho^2\sin\phi \; d\rho d\theta d\phi = \pi/9 \approx 0.3491.$$

15. Volume $$= \int_0^\pi\int_0^{\sin\theta}\int_{r^2}^{r\sin\theta} r \; dz dr d\theta$$

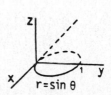

$$= \int_0^\pi\int_0^{\sin\theta} r(r\sin\theta - r^2)dr d\theta = \int_0^\pi \frac{\sin^4\theta}{12} \; d\theta$$

$$= \frac{1}{48}\int_0^\pi \left[1 - 2\cos 2\theta + \frac{1+\cos 4\theta}{2}\right]d\theta = \frac{\pi}{32} \approx 0.0982.$$

17. (a) Position the ball with its center at the origin. The distance of (x,y,z) from the origin is $(x^2+y^2+z^2)^{1/2} = \rho$.

$$\iiint_S (x^2+y^2+z^2)^{1/2}dV = 8\int_0^{\pi/2}\int_0^{\pi/2}\int_0^a (\rho^3\sin\phi)\,d\rho\,d\theta\,d\phi = \pi a^4.$$

Then the average distance from the center is $\pi a^4/[(4/3)\pi a^3] = 3a/4$.

(b) Position the ball with its center at the origin and consider the diameter along the z-axis. The distance of (x,y,z) from the z-axis is $(x^2+y^2)^{1/2} = \rho\sin\phi$.

$$\iiint_S (x^2+y^2)^{1/2}dV = 8\int_0^{\pi/2}\int_0^{\pi/2}\int_0^a (\rho\sin\phi)(\rho^2\sin\phi)\,d\rho\,d\theta\,d\phi = a^4\pi^2/4.$$

Then the average distance from a diameter is

$$[a^2\pi^4/4]/[(4/3)\pi a^3] = 3\pi a/16.$$

(c) Position the sphere above and tangent to the xy-plane at the origin and consider the point on the boundary to be the origin. The equation of the sphere is $\rho = 2a\cos\phi$, and the distance of (x,y,z) from the origin is ρ.

$$\iiint_S (x^2+y^2+z^2)^{1/2}dV = \int_0^{\pi/2}\int_0^{2\pi}\int_0^{2a\cos\phi} \rho(\rho^2\sin\phi)\,d\rho\,d\theta\,d\phi = 8\pi a^4/5.$$

Then the average distance from the origin is

$$[8\pi a^4/5]/[(4/3)\pi a^3] = 6a/5.$$

19. (a) $M_{yz} = \iiint_S kx\,dV = 4k\int_0^{\pi/2}\int_0^a\int_0^a (\rho\sin\phi\cos\theta)(\rho^2\sin\phi)\,d\rho\,d\theta\,d\phi$
$$= ka^4\pi(\sin a)/4.$$

$$m = \iiint_S k\,dV = 4k\int_0^{\pi/2}\int_0^a\int_0^a \rho^2\sin\phi\,d\rho\,d\theta\,d\phi = 4a^3ka/3.$$

Therefore, $\bar{x} = [ka^4\pi(\sin a)/4]/[4a^3ka/3] = 3a\pi(\sin a)/16a$.

(b) $3\pi a/16$ [See Problem 17(b).]

21. Let m_1 and m_2 be the masses of the left and right balls, respectively. Then $m_1 = \frac{4}{3} \pi a^3 k$ and $m_2 = \frac{4}{3} \pi a^3 (ck)$, so $m_2 = cm_1$.

$$\bar{y} = \frac{m_1(-a-b) + m_2(a+b)}{m_1+m_2} = \frac{m_1(-a-b) + cm_1(a+b)}{m_1+cm_1} = \frac{-a-b+c(a+b)}{1+c}$$

$$= \frac{(a+b)(-1+c)}{1+c} = \frac{c-1}{c+1}(a+b).$$

[Analogue] $\quad \bar{y} = \frac{m_1\bar{y}_1 + m_2\bar{y}_2}{m_1+m_2} = \bar{y}_1 \frac{m_1}{m_1+m_2} + \bar{y}_2 \frac{m_2}{m_1+m_2}.$

23. $x = \rho\sin\phi\cos\theta, \; y = \rho\sin\phi\sin\theta, \; z = \rho\cos\phi.$

$$J(\rho,\phi,\theta) = \begin{vmatrix} x_\rho & y_\phi & x_\theta \\ y_\rho & y_\phi & y_\theta \\ z_\rho & z_\phi & z_\theta \end{vmatrix} = \begin{vmatrix} \sin\phi\cos\theta & \rho\cos\phi\cos\theta & -\rho\sin\phi\sin\theta \\ \sin\phi\sin\theta & \rho\cos\phi\sin\theta & \rho\sin\phi\cos\theta \\ \cos\phi & -\rho\sin\phi & 0 \end{vmatrix}$$

$$= (\cos\phi)(\rho^2\cos\phi\sin\phi)(\cos^2\theta+\sin^2\theta) + (\rho\sin\phi)(\rho\sin^2\phi)(\cos^2\theta+\sin^2\theta)$$
$$= \rho^2\sin\phi(\cos^2\phi+\sin^2\phi) = \rho^2\sin\phi.$$

[Expansion was along the third row of the determinant.]

Problem Set 16.9   Chapter Review

True–False Quiz

1. True. Use result of Problem 25, Section 16.2, and then change dummy variable y to dummy variable x.

3. True. Inside integral is 0 since $\sin(x^3y^2)$ is an odd function in x.

5. True. It is less than or equal to $\int_1^2\int_0^2 1 \; dxdy$ which equals 2.

7. False. C^{ex}: Let $f(x,y) = x$, $g(x,y) = x^2$,
            $R = \{(x,y): x \text{ in } [0,2], y \text{ in } [0,1]\}$. The inequality holds
            for the integrals but $f(0.5,0) > g(0.5,0)$.

9. True.   See the write-up of Problem 24, Section 16.7.

11. True.   The integral is the volume between concentric spheres of radii 4
            and 1. That volume is 84π.

13. False. There are 6.

15. True. $|\nabla f|$ is the magnitude of the greatest increase in f.

$$|D_u f| = |\nabla f \cdot u| = |<f_x, f_y> \cdot u| = \sqrt{f_x^2 + f_y^2}\ (1)\cos\theta \leq \sqrt{4+4} = \sqrt{8}.$$

Therefore, $\text{Area}(G) \leq \text{Area}(R) \max\{\sec\gamma\} \leq (1)\sec(\tan^{-1}\sqrt{8}) = 3$.

Sample Test Problems

1. $\int_0^1 (1/2)(x^2 - x^3)dx = 1/24 \approx 0.0417.$

3. $\int_0^{\pi/2}\left[\dfrac{r^2\cos\theta}{2}\right]_{r=0}^{2\sin\theta} d\theta = \int_0^{\pi/2} 2\sin^2\theta\cos\theta\ d\theta = \left[\dfrac{2\sin^3\theta}{3}\right]_0^{\pi/2} = \dfrac{2}{3}.$

5. $\displaystyle\int_0^1\int_0^y f(x,y)dxdy.$ 7. $\displaystyle\int_0^{1/2}\int_0^{1-2y}\int_0^{1-2y-z} f(x,y,z)dxdydz.$

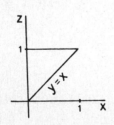

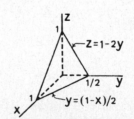

9. (a) $8\int_0^a \int_0^{\sqrt{a^2-z^2}} \int_0^{\sqrt{a^2-y^2-z^2}} dxdydz$.

(b) $8\int_0^{\pi/2} \int_0^a \int_0^{\sqrt{a^2-r^2}} r\ dzdrd\theta$.   (c) $8\int_0^{\pi/2} \int_0^{\pi/2} \int_0^a \rho^2\sin\phi\ d\rho d\theta d\phi$.

11. $8\int_0^1 \int_0^x \int_0^{1-y^2} z^2 dzdydx = 31/35 \approx 0.8857$.

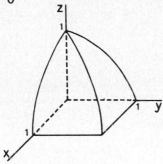

13. $m = \int_0^2 \int_1^3 xy^2 dxdy = 32/3$.

$M_x = \int_0^2 \int_1^3 xy^3 dxdy - 16$. $M_y = \int_0^2 \int_1^3 x^2 y^2 dxdy\quad 208/9$.

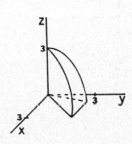

$(\bar{x}, \bar{y}) = (3/2, 13/6)$.

15. $z = f(x,y) = (9-y^2)^{1/2}$; $f_x(x,y) = 0$; $f_y(x,y) = -y(9-y^2)^{-1/2}$.

$\text{Area} = \int_0^3 \int_{y/3}^y \sqrt{y^2(9-y^2)^{-1}+1}\ dxdy$

$= \int_0^2 \int_{y/3}^y 3(9-y^2)^{-1/2} dxdy$

$= \int_0^3 (9-y^2)^{-1/2}(2y)dy = \left[-2(9-y^2)^{1/2}\right]_0^3 = 6$.

17. $\delta(x,y,z) = k\rho.$ $m = \int_0^{\pi}\int_0^{2\pi}\int_1^3 k\rho \ \rho^2 \sin\phi \ d\rho d\theta d\phi = 80\pi k.$

19. $m = \int_0^a\int_0^{(b/a)(a-x)}\int_0^{(c/ab)(ab-bx-ay)} kx \ dzdydx = (k/24)a^2bc.$

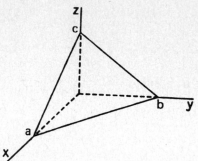

Problem Set 17.1   Vector Fields

1.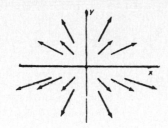

3.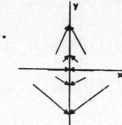

5.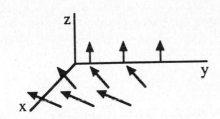

7. $\langle 2x-3y,-3x,2 \rangle$.

9. $f(x,y,z) = \ln|x| + \ln|y| + \ln|z|$; $\nabla f(x,y,z) = \langle x^{-1}, y^{-1}, z^{-1} \rangle$.

11. $e^y \langle \cos z, x\cos z, -x\sin z \rangle$.

13. div$\mathbf{F}$ = 2x-2x+2yz = 2yz.
    curl$\mathbf{F}$ = $\langle z^2, 0, -2y \rangle$.

15. div$\mathbf{F}$ = $\nabla \cdot \mathbf{F}$ = 0+0+0 = 0.
    curl$\mathbf{F}$ = $\nabla \times \mathbf{F}$ = $\langle x-x, y-y, z-z \rangle$ = 0. .

17. div$\mathbf{F}$ = $e^x\cos y + e^x\cos y + 1$.
    curl$\mathbf{F}$ = $\langle 0, 0, 2e^x\sin y \rangle$.

19. (a) meaningless.   (b) vector field.   (c) meaningless.
    (d) scalar field   (e) vector field.   (f) vector field.
    (g) vector field.   (h) meaningless.   (i) meaningless.
    (j) scalar field.   (k) meaningless.

21. Let $f(x,y,z) = -c|\mathbf{r}|^{-3}$, so $\text{grad}(f) = 3c|\mathbf{r}|^{-4}\frac{\mathbf{r}}{|\mathbf{r}|} = 3c|\mathbf{r}|^{-5}\mathbf{r}$. Then

$\text{curl}\mathbf{F} = \text{curl}[(-c|\mathbf{r}|^{-3})\mathbf{r}] = (-c|\mathbf{r}|^{-3})(\text{curl}\mathbf{r}) + (3c|\mathbf{r}|^{-5}\mathbf{r})\times\mathbf{r}$ [By 20(d)]

$= (-c|\mathbf{r}|^{-3})(0) + (3c|\mathbf{r}|^{-5})(\mathbf{r}\times\mathbf{r}) = \mathbf{0} + \mathbf{0} = \mathbf{0}.$

$\text{div}\mathbf{F} = \text{div}[(-c|\mathbf{r}|^{-3})\mathbf{r}] = (-c|\mathbf{r}|^{-3})(\text{div}\mathbf{r}) + (3c|\mathbf{r}|^{-5}\mathbf{r})\cdot\mathbf{r}$ [By 20(c)]

$= (-c|\mathbf{r}|^{-3})(1+1+1) + (3c|\mathbf{r}|^{-5})|\mathbf{r}|^2 = (-3c|\mathbf{r}|^{-3}) + 3c|\mathbf{r}|^{-3} = 0.$

23. $\text{grad } f = \langle f'(r)xr^{-1}, f'(r)yr^{-1}, f'(r)zr^{-1}$ [if $r \neq 0$]

$= f'(r)r^{-1}\langle x,y,z\rangle = f'(r)r^{-1}\mathbf{r}.$

$\text{curl}\mathbf{F} = [f(r)][\text{curl}\mathbf{r}] + [f'(r)r^{-1}\mathbf{r}]\times\mathbf{r}$

$= [f(r)][\text{curl}\mathbf{r}] + [f'(r)r^{-1}\mathbf{r}]\times\mathbf{r} = \mathbf{0} + \mathbf{0} = \mathbf{0}.$

25. (a) Let $P = (x_0, y_0)$.

div**F** = div**H** = 0 since there is no tendency toward P except along the line $x=x_0$, and along that line the tendencies toward and away from P are balanced; div**G** < 0 since there is no tendency toward P except along the line $x=x_0$, and along that line there is more tendencies toward than away from P; div**L** > 0 since the tendency away from P is greater than the tendency toward P.

(b) No rotation for **F**, **G**, **L**; clockwise rotation for **H** since the magnitudes of the forces to the right of P are less than those to the left.

(c) div**F** = div**H** = 0; curl**F** = curl**G** = **0**.

$\text{div}\mathbf{G} = -2ye^{-y^2} < 0$ since $y > 0$ at P; curl**L** = **0**.

$\text{div}\mathbf{L} = (x^2+y^2)^{-1/2} < 0$; $\text{curl}\mathbf{L} = \langle 0,0,-2xe^{-x^2}\rangle$ which points downward at P, so the rotation is clockwise in a right-hand system.

27. $\nabla f(x,y,z) = \frac{1}{2}m\omega^2\langle 2x,2y,2z\rangle = m\omega^2\langle x,y,z\rangle = \mathbf{F}(x,y,z).$

Problem Set 17.2   Line Integrals

1. $\int_0^1 (27t^3+t^3)(9+9t^4)^{1/2}dt = 14(2\sqrt{2}-1) \approx 25.5980.$

3. Let $x = t$, $y = 2t$, t in $[0,\pi]$.

   Then $\int_C (\sin x + \cos y)ds = \int_0^\pi (\sin t + \cos 2t)\sqrt{1+4}\, dt = 2\sqrt{5} \approx 4.4721.$

5. $\int_0^1 (2t+9t^3)(1+4t^2+9t^4)^{1/2}dt = (1/6)(14^{3/2}-1) \approx 8.5639.$

7. $\int_0^2 [(t^2-1)(2) + (4t^2)(2t)]dt = 100/3.$

9. $\int_C y^3dx+x^3dy = \int_{C_1} y^3dx+x^3dy + \int_{C_2} y^3dx+x^3dy$

   $= \int_1^{-2} (4)^3dy + \int_{-4}^2 (-2)^3dx = 192 + (-48) = 144.$

11. $y = -x+2.$ $\int_1^3 \left[[x+2(-x+2)](1) + [x-2(-x+2)](-1) \right] dx = 0.$

13. $\langle x,y,z \rangle = \langle 1,2,1 \rangle + t\langle 1,-1,0 \rangle.$

    $\int_0^1 [(4-t)(1) + (1+t)(-1) - (2-3t+t^2)(-1)]dt = 17/6 \approx 2.8333.$

15. On C_1: $y = z = dy = dz = 0$.
On C_2: $x = 2$, $z = dx = dz = 0$.
On C_3: $x = 2$, $y = 3$, $dx = dy = 0$.

$$\int_0^2 x\,dx + \int_0^3 (2-2y)\,dy + \int_0^4 (4+3-z)\,dz$$

$$= \left[\frac{x^2}{2}\right]_0^2 + \left[\,2y-y^2\,\right]_0^3 + \left[7z - \frac{z^2}{2}\right]_0^4 = 2 + (-3) + 20 = 19.$$

17. $m = \int_C k|x|\,ds = \int_{-2}^2 k|x|\,(1+4x^2)^{1/2}dx = (k/6)(17^{3/2}-1) \approx 11.5155k.$

19. $\int_C (x^3-y^3)dx+xy^2dy = \int_{-1}^0 [(t^6-t^9)(2t)+(t^2)(t^6)(3t^2)]dt = -7/44 \approx -0.1591.$

21. $W = \int_C \mathbf{F}\cdot d\mathbf{r} = \int_C (x+y)dx+(x-y)dy$

$$= \int_0^{\pi/2} [(a\cos t + b\sin t)(-a\sin t) + (a\cos t - b\sin t)(b\cos t)]dt$$

$$= \int_0^{\pi/2} [-(a^2+b^2)\sin t\cos t + ab(\cos^2 t - \sin^2 t)]dt$$

$$= \int_0^{\pi/2} \frac{-(a^2+b^2)\sin 2t}{2} +ab\cos 2t\ dt=\left[\frac{(a^2+b^2)\cos 2t}{4} + \frac{ab\sin 2t}{2}\right]_0^{\pi/2} = \frac{a^2+b^2}{-2}.$$

23. $\int_0^{\pi} [(\pi/2)\sin(\pi t/2)\cos(\pi t/2) + \pi t\cos(\pi t/2) + \sin(\pi t/2) - t]dt$

$$= 2 - 2/\pi \approx 1.3634.$$

25. $\int_C (1+y/3)ds = \int_0^{\pi/2}(1+10\sin^3 t)[(-90\cos^2 t\sin t)^2+(90\sin^2 t\cos t)^2]^{1/2}dt = 225.$

Karen needs $450/200 = 2.25$ gallons of paint.

27. C: $x+y = a$. Let $x = t$, $y = a-t$, t in $[0,a]$.

Cylinder: $x+y = a$; $(x+y)^2 = a^2$; $x^2+2xy+y^2 = a^2$.

Sphere: $x^2+y^2+z^2 = a^2$.

The curve of intersection satisfies:

$z^2 = 2xy$; $z = \sqrt{2xy}$.

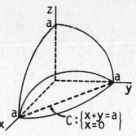

$$\text{Area} = 8\int_C \sqrt{2xy}\ ds = 8\int_0^a \sqrt{2t(a-t)}\ \sqrt{(1)^2+(-1)^2}\ dt = 16\int_0^a \sqrt{at-t^2}\ dt$$

$$= 16\left[\frac{t-\frac{a}{2}}{2}\sqrt{at-t^2} + \frac{\left[\frac{a}{2}\right]^2}{2}\sin^{-1}\left[\frac{t-\frac{a}{2}}{\frac{a}{2}}\right]\right]_0^a$$

$$= 16\left[\left[0 + \left[\frac{a^2}{8}\right]\left[\frac{\pi}{2}\right]\right] - \left[0 + \left[\frac{a^2}{8}\right]\left[\frac{-\pi}{2}\right]\right]\right] = 2a^2\pi.$$

Trivial way: Each side of the cylinder is part of a plane that intersects the sphere in a circle. The radius of each circle is the value of z in $z = \sqrt{2xy}$ when $x = y = a/2$. That is, the radius is $\sqrt{2(a/2)(a/2)} = a\sqrt{2}/2$. Therefore the total area of the part cut out is $4[\pi(a\sqrt{2}/2)^2] = 2a^2\pi$.

29. Note that $r = a\sin\theta$ along C.

Then $(a^2-x^2-y^2)^{1/2} = (a^2-r^2)^{1/2} = a\cos\theta$.

Let $\begin{Bmatrix} x = r\cos\theta = (a\sin\theta)\cos\theta \\ y = r\sin\theta = (a\sin\theta)\sin\theta \end{Bmatrix}$, θ in $[0,\pi/2]$.

Therefore, $x'(\theta) = a\cos2\theta$; $y'(\theta) = a\sin2\theta$.

Then Area $= 2\int_C (a^2-x^2-y^2)^{1/2}\ ds$

$$= 2\int_0^{\pi/2} (a\cos\theta)[(a\sin2\theta)^2 + (a\cos2\theta)^2]^{1/2}\ d\theta = 2a^2.$$

31. (a) $\int_C x^2 y\, ds = \int_0^{\pi/2}(3\sin t)^2(3\cos t)[(3\cos)^2 + (-3\sin t)^2]^{1/2}dt$

$$= 81\int_0^{\pi/2}\sin^2 t\,\cos t\,dt = 81\left[(1/3)\sin^3 t\right]_0^{\pi/2} = 27.$$

(b) $\int_{C_4} xy^2 dx + xy^2 dy = \int_0^3 (3-t)(5-t)^2(-1)dt + \int_0^3 (3-t)(5-t)^2(-1)dt$

$$= 2\int_0^3 (t^3 - 13t^2 + 55t - 75)dt$$

$$= -148.5$$

Problem Set 17.3 Independence of Path

1. $M_y = -7 = N_x$, so **F** is conservative. $f(x,y) = 5x^2 - 7xy + y^2 + C$.

3. $M_y = 90x^4 y - 36y^5 \neq N_x$ since $N_x = 90x^4 y - 12y^5$, so **F** is not conservative.

5. $M_y = (-12/5)x^2 y^{-3} = N_x$, so **F** is conservative. $f(x,y) = (2/5)x^3 y^{-2} + C$.

7. $M_y = 2e^y - e^x = N_x$ so **F** is conservative. $f(x,y) = 2xe^y - ye^x + C$.

9. $M_y = 0 = N_x$, $M_z = 0 = P_x$, and $N_z = 0 = P_y$, so **F** is conservative.

f satisfies $f_x(x,y,z) = 3x^2$, $f_y(x,y,z) = 6y^2$, and $f_z(x,y,z) = 9z^2$.

Therefore, f satisfies (1) $f(x,y,z) = x^3 + C_1(y,z)$,

(2) $f(x,y,z) = 2y^3 + C_2(x,z)$, and

(3) $f(x,y,z) = 3z^3 + C_3(x,y)$.

A function with an arbitrary constant that satisfies (1), (2), and (3) is $f(x,y,z) = x^3 + 2y^3 + 3z^3 + C$.

11. $M_y = 2y+2x = N_x$, so integral is path independent. $f(x,y) = xy^2+x^2y$.

$$\int_{(-1,2)}^{(3,1)} (y^2+2xy)dx + (x^2+2xy)dy = \left[xy^2+x^2y\right]_{(-1,2)}^{(3,1)} = 14. \quad \text{[Or use paths.]}$$

13. $M_y = 18xy^2 = N_x$, $M_z = 4x = P_x$, $N_z = 0 = P_y$. By paths $(0,0,0)$ to $(1,0,0)$; $(1,0,0)$ to $(1,1,0)$; $(1,1,0)$ to $(1,1,1)$.

$$\int_0^1 odx + \int_0^1 9y^2dy + \int_0^1 (4z+1)dz = 6. \quad \text{[Or use } f(x,y,z) = 3x^2y^3+2xz^2+z.\text{]}$$

15. $M_y = 1 = N_x$, $M_z = 1 = P_x$, $N_z = 1 = P_y$. (so path independent). From inspection observe that $f(x,y,z) = xy+xz+yz$ satisfies $\nabla f = \langle y+z, x+z, x+y\rangle$, so the integral equals $\left[xy+xz+yz\right]_{(0,0,0)}^{(-1,0,\pi)} = -\pi.$

[Or use line segments $(0,0,0)$ to $(-1,0,0)$, then $(-1,0,0)$ to $(-1,0,\pi)$.]

17. $f_x = M$, $f_y = N$, $f_z = P$. $f_{xy} = M_y$ and $f_{yx} = N_x$, so $M_y = N_x$.
$f_{xz} = M_z$ and $f_{zx} = P_x$, so $M_z = P_x$. $f_{yz} = N_z$ and $f_{zy} = P_y$, so $N_z = P_y$.

19. $\mathbf{F}(x,y,z) = k|\mathbf{r}|\frac{\mathbf{r}}{|\mathbf{r}|} = k\mathbf{r} = k\langle x,y,z\rangle$. $f(x,y,z) = (k/2)(x^2+y^2+z^2)$ works.

21. $$\int_C \mathbf{F}\cdot d\mathbf{r} = \int_a^b (m\mathbf{r}''\cdot\mathbf{r}')dt = m\int_a^b (x''x' + y''y' + z''z')dt$$

$$= m\left[\frac{(x')^2}{2} + \frac{(y')^2}{2} + \frac{(z')^2}{2}\right]_a^b = \frac{m}{2}\left[|\mathbf{r}'(t)|^2\right]_a^b = \frac{m}{2}[|\mathbf{r}'(b) - |\mathbf{r}'(a)|].$$

23. $f(x,y,z) = -gmz$ satisfies $\nabla f(x,y,z) = \langle 0,0,-gm\rangle = \mathbf{F}$. Then, assuming the path is piecewise smooth,

$$\text{Work} = \int_C \mathbf{F}\cdot d\mathbf{r} = \left[-gmz\right]_{(x_1,y_1,z_1)}^{(x_2,y_2,z_2)} = -gm(z_2-z_1) = gm(z_1-z_2).$$

25. (a) $M = y/(x^2+y^2)$; $M_y = (x^2-y^2)/(x^2+y^2)^2$.

$N = -x/(x^2+y^2)$; $N_x = (x^2-y^2)/(x^2+y^2)^2$.

(b) $M = y/(x^2+y^2) = (\sin t)/(\cos^2 t + \sin^2 t) = \sin t$

$N = -x/(x^2+y^2) = (-\cos t)/(\cos^2 t + \sin^2 t) = -\cos t$.

$$\int_C \mathbf{F} \cdot d\mathbf{r} = \int_C M dx + N dy = \int_0^{2\pi} [(\sin t)(-\sin t) + (-\cos t)(\cos t)] dt$$

$$= -\int_0^{2\pi} 1 \, dt = -2\pi \neq 0.$$

Problem Set 17.4   Green's Theorem in the Plane

1. $\oint_C 2xy\,dx + y^2 dy = \iint_S (0-2x)\,dA$

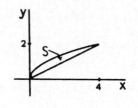

$$= \int_0^2 \int_{y^2}^{2y} -2x\,dx\,dy = -64/15 \approx -4.2667.$$

3. $\oint_C (2x+y^2)\,dx + (x^2+2y)\,dy = \iint_S (2x-2y)\,dA$

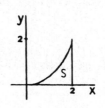

$$= \int_0^2 \int_0^{x^3/4} (2x-2y)\,dy\,dx = \int_0^2 \left[\frac{x^4}{2} - \frac{x^6}{16}\right] dx$$

$$= \frac{16}{5} - \frac{8}{7} = \frac{72}{35} \approx 2.0571.$$

5. $\oint_C (x^2+4xy)\,dx + (2x^2+3y)\,dy = \iint_S (4x-4x)\,dA = 0.$

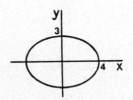

7. $A(S) = (1/2) \oint_C x\,dy - y\,dx$

$= (1/2)\int_0^2 [4x^2 - 2x^2]\,dx + (1/2)\int_2^0 [4x - 4x]\,dx = 8/3.$

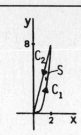
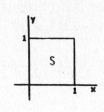

9. (a) $\iint_S \text{div}\mathbf{F}\,dA = \iint_S (M_x + N_y)\,dA = \iint_S (0+0)\,dA = 0.$

(b) $\iint_S (\text{curl}\mathbf{F})\cdot\mathbf{k}\,dA = \iint_S (N_x - M_y)\,dA = \iint_S (2x - 2y)\,dA$

$= \int_0^1\int_0^1 (2x - 2y)\,dx\,dy = \int_0^1 (1 - 2y)\,dy = 0.$

11. (a) $\iint_S (0+0)\,dA = 0.$ (b) $\iint_S (3x^2 - 3y^2)\,dA = 0,$ since for integrand,

$f(y,x) = -f(x,y).$

13. $\iint_S (\text{curl}\mathbf{F})\cdot\mathbf{k}\,dA$

$= \int_{C_1} \mathbf{F}\cdot\mathbf{T}\,ds - \int_{C_2} \mathbf{F}\cdot\mathbf{T}\,ds$

$= 30 - (-20) - 50.$

15. $W = \oint_C \mathbf{F}\cdot\mathbf{T}\,ds = \iint_S (N_x - M_y)\,dA = \iint_S (-2y - 2y)\,dA$

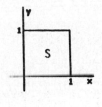

$= \int_0^1\int_0^1 -4y\,dx\,dy = \int_0^1 -4y\,dy = -2.$

17. $\mathbf{F}$ is a constant, so $N_x = M_y = 0.$ $\oint_C \mathbf{F}\cdot\mathbf{T}\,ds = \iint_S (N_x - M_y)\,dA = 0.$

19. (a) Each equals $(x^2 - y^2)(x^2 + y^2)^{-2}.$

(b) $\oint_C y(x^2 + y^2)^{-1}\,dx - x(x^2 + y^2)^{-1}\,dy = \int_0^{2\pi} (-\sin^2 t - \cos^2 t)\,dt = \int_0^{2\pi} -1\,dt.$

(c) M and N are discontinuous at $(0,0).$

21.   Use Green's Theorem with $M(x,y) = -y$ and $N(x,y) = 0$.

$$\oint_C (-y)\,dx = \iint_S [0-(-1)]\,dA = A(S).$$

Now use Green's Theorem with $M(x,y) = 0$ and $N(x,y) = x$.

$$\oint_C x\,dy = \iint_S (1-0)\,dA = A(S).$$

23. $A(S) = (1/2) \oint_C x\,dy - y\,dx$

$$= (1/2)\int_0^{2\pi} [(a\cos^3 t)(3a\sin^2 t)(\cos t) - (a\sin^3 t)(3a\cos^2 t)(-\sin t)]\,dt$$

$$= (3/8)a^2\pi.$$

25.   (a) $\mathbf{F}\cdot\mathbf{n} = \dfrac{x^2+y^2}{(x^2+y^2)^{3/2}} = \dfrac{1}{(x^2+y^2)^{1/2}} = \dfrac{1}{a}.$

Therefore, $\displaystyle\int_C \mathbf{F}\cdot\mathbf{n}\ ds = \dfrac{1}{a}\int_C 1\ ds = \dfrac{1}{a}(2\pi a) = 2\pi.$

(b) $\operatorname{div}\mathbf{F} = \dfrac{(x^2+y^2)(1) - (x)(2x)}{(x^2+y^2)^2} + \dfrac{(x^2+y^2)(1) - (y)(2y)}{(x^2+y^2)^2} = 0.$

(c) $M = x/(x^2+y^2)$ is not defined at $(0,0)$ which is inside C.

(d)   If origin is outside C, then $\displaystyle\oint_C \mathbf{F}\cdot\mathbf{n}\ ds = \iint_S \operatorname{div}\mathbf{F}\ dA = \iint_S 0\ dA = 0.$

If origin is inside C, let C' be a circle (centered at the origin) inside C and oriented clockwise. Let S be region between C and C'.

Then $0 = \displaystyle\iint_S \operatorname{div}\mathbf{F}\ dA$ [by "origin outside C" case]

$$= \int_C \mathbf{F}\cdot\mathbf{n}\ ds - \int_{C'} \mathbf{F}\cdot\mathbf{n}\ ds \quad \text{[by Green's Theorem]}$$

$$= \int_C \mathbf{F}\cdot\mathbf{n}\ ds - 2\pi \text{ [by part (a)], so } \int_C \mathbf{F}\cdot\mathbf{n}\ ds = 2\pi.$$

27. (a) divF = 4 x^2+y^2
     (b) 4(36) = 144

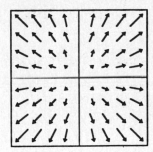

29. (a) divF = - 2sinx siny   sin(x) sin(y)
          divF < 0 in quadrants I and II
          divF > 0 in quadrants II and IV
     (b) Flux across boundary of S is 0

          Flux across boundary of T is $-2(1-\cos3)^2$

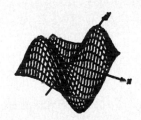

Problem Set 17.5   Surface Integrals

1. $\iint_R [x^2+y^2+(x+y+1)](1+1+1)^{1/2}dA = \int_0^1\int_0^1 \sqrt{3}(x^2+y^2+x+y+1)\,dx\,dy = \dfrac{8\sqrt{3}}{3}.$

3. $\iint_R (x+y)\sqrt{[-x(4-x^2)^{-1/2}]^2 + 0 + 1}\ dA = \int_0^{\sqrt{3}}\int_0^1 \dfrac{2(x+y)}{(4-x^2)^{1/2}}\ dy\,dx$

    $= \int_0^{\sqrt{3}} \dfrac{2x+1}{(4-x^2)^{1/2}}\ dx = \left[-2(4-x^2)^{1/2} + \sin^{-1}(x/2)\right]_0^{\sqrt{3}} = \dfrac{\pi+6}{3} \approx 3.0472.$

5. $\int_0^\pi \int_0^{\sin\theta} (4r^2+1) r\,dr\,d\theta = (5/8)\pi \approx 1.9635.$

7. $\iint_R (x+y)(0+0+1)^{1/2} dA$

Bottom: $\int_0^1 \int_0^1 (x+y)\,dx\,dy = 1.$ Top: Same Integral.
(z=0)   (z=1)

Left Side: $\int_0^1 \int_0^1 (x+0)\,dx\,dz = 1/2.$ Right Side: $\int_0^1 \int_0^1 (x+1)\,dx\,dz = 3/2.$
(y=0)   (y=1)

Back: $\int_0^1 \int_0^1 (0+y)\,dy\,dz = 1/2.$ Front: $\int_0^1 \int_0^1 (1+y)\,dy\,dz = 3/2.$
(x=0)   (x=1)

Therefore, the integral equals $1 + 1 + 1/2 + 3/2 + 1/2 + 3/2 = 6.$

9. $\iint_G \mathbf{F}\cdot\mathbf{n}\ dS = \iint_R (-Mf_x - Nf_y + P)\,dA = \int_0^1 \int_0^{1-y} (8y+4x+0)\,dx\,dy.$

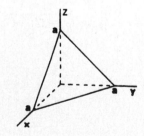

$\qquad = \int_0^1 [8(1-y)y + 2(1-y)^2]\,dy = \int_0^1 (-6y^2 + 4y + 2)\,dy = 2.$

11. $\int_0^5 \int_{-1}^1 [-xy(1-y^2)^{-1/2} + 2]\,dy\,dx = 20.$

[In the inside integral, note that the first term is odd in y.]

13. $m = \iint_G kx^2 dS = \iint_R kx^2 \sqrt{3}\ dA = \sqrt{3}k \int_0^a \int_0^{a-x} x^2 dy\,dx = (\sqrt{3}k/12)a^4.$

15. [Let $\delta = 1$.] $m = \iint_S 1\,dS = \iint_R (1+1+1)^{1/2} dA$

$\qquad = \sqrt{3} \int_0^a \int_0^{a-y} dx\,dy = \sqrt{3} \int_0^a (a-y)\,dy = \dfrac{a^2 \sqrt{3}}{2}.$

$$M_{xy} = \iint_S z\,dS = \iint_R (a-x-y)\,\sqrt{3}\,dA$$

$$= \sqrt{3}\int_0^a\int_0^{a-y}(a-x-y)\,dx\,dy = \sqrt{3}\int_0^a\left[a(a-y) - \frac{(a-y)^2}{2} - y(a-y)\right]dy$$

$$= \sqrt{3}\int_0^a\left[\frac{a^2}{2} - ay + \frac{y^2}{2}\right]dy = \frac{a^3\sqrt{3}}{6}.$$

$$\bar{z} = \frac{M_{xy}}{m} = \frac{a}{3};\ \text{then}\ \bar{x} = \bar{y} = \frac{a}{3}\ \text{[by symmetry]}.$$

17. (a) 0 [by symmetry, since $g(x,y,-z) = -g(x,y,z)$]

(b) 0 [by symmetry, since $g(x,y,-z) = -g(x,y,z)$]

(c) $\iint_G (x^2+y^2+z^2)\,dS = \iint_G a^2\,dS = a^2\text{Area}(G) = a^2(4\pi a^2) = 4\pi a^4.$

(d) Note: $\iint_G (x^2+y^2+z^2)\,dS = \iint_G x^2\,dS + \iint_G y^2\,dS = \iint_G z^2\,dS = 3\iint_G x^2\,dS$

[due to symmetry of the sphere with respect to the origin.]

Therefore, $\iint_G x^2\,dS = (1/3)\iint_G (x^2+y^2+z^2)\,dS = (1/3)4\pi a^4 = 4\pi a^4/3.$

(e) $\iint_G (x^2+y^2)\,dS = (2/3)4\pi a^4 = 8\pi a^4/3.$

19. (a) Place center of sphere at the origin.

$$F = \iint_G k(a-z)\,dS + ka\iint_G 1\,dS - k\iint_G z\,dS = ka(4\pi a^2) - 0 = 4\pi a^3 k.$$

(b) Place hemisphere above xy-plane with center at origin and circular base in xy-plane.

F = Force on hemisphere + Force on circular base

$$= \iint_G k(a-z)dS = ka(\pi a^2) = ka\iint_G 1 \ dS - k\iint_G z \ dS + \pi a^3 k$$

$$= ka(2\pi a^2) - k\iint_R z \sqrt{\frac{a^2}{a^2-x^2-y^2}} \ dA + \pi a^3 k = 3\pi a^3 k - k\iint_R z \frac{a}{z} \ dA$$

$$= 3\pi a^3 k - ka(\pi a^2) = 2\pi a^3 k.$$

(c) Place cylinder above xy-plane with circular base in xy-plane with center at origin.

F = Force on top + force on cylindrical side + Force on base

$$= 0 + \iint_G k(h-z)dS + kh(\pi a^2) = kh\iint_G 1 \ dS - k\iint_G z \ dS + \pi a^2 hk$$

$$= kh(2\pi ah) - 4k\iint_R z \sqrt{\frac{y^2}{a^2-y^2}} + 0 + 1 \ dA + \pi a^2 hk$$

[where R is region in yz-plane: $0 \le y \le a, \ 0 \le z \le h$]

$$= 2\pi ah^2 k + \pi a^2 hk - 4k\int_0^a\int_0^h \frac{az}{\sqrt{a^2-y^2}} \ dzdy$$

$$= 2\pi ah^2 k + \pi a^2 hk - \pi kah^2 = \pi ah^2 k + \pi a^2 hk = \pi ahk(h+a).$$

Problem Set 17.6   Gauss's Divergence Theorem

1. $\displaystyle\iiint_S (0+0+0)dV = 0.$

3. $\displaystyle\iint_{\partial S} \mathbf{F}\cdot\mathbf{r} \ dS = \iiint_S (M_x+N_y+P_z)dV = \int_0^c\int_0^b\int_0^a (2xyz+2xyz+2xyz)dxdydz$

$$= \int_0^c\int_0^b 3a^2yz \ dydz = \int_0^c \frac{3a^2b^2z}{2} \ dz = \frac{3a^2b^2c^2}{4}.$$

5. $2\iiint_S (x+y+z)dV = 2\int_0^{2\pi}\int_0^2\int_0^{4-r^2}(r\cos\theta + r\sin\theta + z)r\,dz\,dr\,d\theta = 64\pi/3 \approx 67.02.$

7. $\iiint_S (1+1+0)dV = 2(\text{volume of cylinder}) = 2\pi(1)^2(2) = 4\pi \approx 12.5664.$

9. $\iiint_S (M_x+N_y+P_z)dV = \iiint_S (2+3+4)dV = 9(\text{Volume of spherical shell})$

$$= 9(4\pi/3)(5^3-3^3) = 1176\pi \approx 3694.51.$$

11. $(1/3)\iiint_S (1+1+1)dV = V(S).$

13. Note: (1) $\iint_R (ax+by+cz)dS = \iint_R d\,dS = dD.$ [R is the slanted face]

   (2) $\mathbf{n} = \langle a,b,c\rangle/(a^2+b^2+c^2)^{1/2}.$ [for slanted face]

   (3) $\mathbf{F}\cdot\mathbf{n} = 0$ on each coordinate-plane face.

   Volume $= (1/3)\iint_S \mathbf{F}\cdot\mathbf{n}\,dS$ [where $\mathbf{F} = \langle x,y,z\rangle$]

   $= (1/3)\iint_R \mathbf{F}\cdot\mathbf{n}\,dS$ [by Note (3)]

   $= (1/3)\iint_R \dfrac{(ax+by+cz)}{\sqrt{a^2+b^2+c^2}}\,dS$

   $= \dfrac{dD}{3\sqrt{a^2+b^2+c^2}}.$

15. (a) $\text{div}\mathbf{F} = 2+3+2z = 5+2z$.

$$\iint_{\partial S} \mathbf{F} \cdot \mathbf{n}\, dS = \iiint_S (5+2z)\, dV = \iiint_S 5\, dV + 2\iiint_S z\, dV$$

$$= 5(\text{Volume of } S) + 2M_{xy} = 5\left[\frac{4\pi}{3}\right] + 2z(\text{Volume of } S)$$

$$= \frac{20\pi}{3} + 2(0)(\text{Volume}) = \frac{20\pi}{3}.$$

(b) $\mathbf{F} \cdot \mathbf{n} = (x^2+y^2+z^2)^{5/3}\langle x,y,z\rangle \cdot \dfrac{\langle x,y,z\rangle}{\sqrt{x^2+y^2+z^2}} = (x^2+y^2+z^2)^{13/6} = 1$ on ∂S.

$$\iint_{\partial S} \mathbf{F} \cdot \mathbf{n}\, dS = \iiint_S 1\, dV = 4\pi(1)^2 = 4\pi.$$

(c) $\text{div}\mathbf{F} = 2x+2y+2z$.

$$\iint_{\partial S} \mathbf{F} \cdot \mathbf{n}\, dS = \iiint_S 2(x+y+z)\, dV = 2\iiint_S x\, dV \quad [\text{Since } \bar{y} = \bar{z} = 0 \quad (\text{See (a)})]$$

$$= 2M_{yz} = 2(\bar{x})(\text{Volume of } S) = 2(2)\left[\frac{4\pi}{3}\right] = \frac{16\pi}{3}.$$

(d) $\mathbf{F} \cdot \mathbf{n} = 0$ on each face except the face R in the plane $x=1$.

$$\iint_{\partial S} \mathbf{F} \cdot \mathbf{n}\, dS = \iint_R \mathbf{F} \cdot \mathbf{n}\, dS = \iint_R \langle 1,0,0\rangle \cdot \langle 1,0,0\rangle\, dS = \iint_R 1\, dS = (1)^2 = 1.$$

(e) $\text{div}\mathbf{F} = 1+1+1 = 3$.

$$\iint_{\partial S} \mathbf{F} \cdot \mathbf{n}\, dS = \iiint_S 3\, dV = 3[\text{Volume of } S] = 3\left[\frac{1}{3}\left[\frac{1}{2}(4)(3)\right](6)\right] = 36.$$

(f) $\text{div}\mathbf{F} = 3x^2+3y^2+3z^2 = 3(x^2+y^2+z^2) = 3$ on ∂S.

$$\iint_{\partial S} \mathbf{F} \cdot \mathbf{n}\, dS = 3\iiint_S (x^2+y^2+z^2)\, dV = 3\left[\frac{3}{2}\frac{8\pi}{15}\right] = \frac{12\pi}{5}.$$

[That answer can be obtained by making use of symmetry and a change
to spherical coordinates. Or you could go to the solution for
Problem 24, Section 16.8, and realize that the value of the integral
in this problem is 3/2 (there are three terms instead of two) times
the answer obtained there for I_z, letting kabc = 1.]

(g) $\mathbf{F} \cdot \mathbf{n} = [\ln(x^2+y^2)]\langle x,y,0 \rangle \cdot \langle 0,0,1 \rangle = 0$ on top and bottom.

$$\mathbf{F} \cdot \mathbf{n} = (\ln 4)\langle x,y,0 \rangle \cdot \frac{\langle x,y,0 \rangle}{\sqrt{x^2+y^2}} = (\ln 4)\sqrt{x^2+y^2} = (\ln 4)\sqrt{4} = 2\ln 4 = 4\ln 2$$

on the side. $\displaystyle\iint_{\partial S} \mathbf{F} \cdot \mathbf{n}\ dS = \iint_R 4\ln 2\ dS = (4\ln 2)[2\pi(2)(2)] = 32\pi\ln 2.$

17. $\displaystyle\iint_{\partial S} D_{\mathbf{n}} f\ dS = \iint_{\partial S} \nabla f \cdot \mathbf{n}\ dS = \iiint_S \text{div}(\nabla f) dV = \iiint_S \nabla^2 f\ dV.$ [See next problem.]

19. $\displaystyle\iint_{\partial S} f D_{\mathbf{n}} g\ dS = \iint_{\partial S} f(\nabla g \cdot \mathbf{n}) dS = \iint_{\partial S} (f \nabla g) \cdot \mathbf{n}\ dS = \iiint_S \text{div}(f \nabla g)\ dV$ [Gauss]

$$= \iiint_S [f(\text{div}\nabla g) + (\nabla f) \cdot (\nabla g)] dV = \iiint_S (f \nabla^2 g + \nabla f \cdot \nabla g) dV.$$

[See Problem 20(c), Section 17.1]

Problem Set 17.7   Stokes' Theorem

1. $\displaystyle\oint_{\partial S} \mathbf{F} \cdot \mathbf{T} ds = \iint_R (N_x - M_y) dA = \iint_R 0\, dA = 0.$

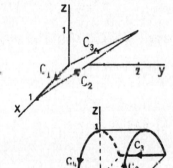

3. $\displaystyle\iint_S (\text{curl}\mathbf{F}) \cdot \mathbf{n}\ dS = \oint_{\partial S} \mathbf{F} \cdot \mathbf{T}\ ds$

$$= \oint_{\partial S} (y+z) dx + (x^2+z^2) dy + y\, dz$$

$$[*] = \int_0^1 1\, dt + \int_0^\pi [(1+\sin t)(-\sin t) + \cos t] dt + \int_0^1 -1\, dt + \int_0^\pi \sin^2 t\, dt$$

$$= \int_0^\pi (-\sin t + \cos t) dt = -2.$$

The result at [*] was obtained by integrating along S by doing so along C_1, C_2, C_3, C_4 in that order.

Along C_1: x=1, y=t, z=0, dx=dz=0, dy=dt, t in [0,1].
Along C_2: x=cost, y=1, z=sint, dx=-sindt, dy=0, dz=costdt, t in [0,π].
Along C_3: x=-1, y=1-t, z=0, dx=dz=0, dy=dt, t in [0,1].
Along C_4: x=-cost, y=0, z=sint, dx=sintdt, dy=0, dz=costdt, t in [0,π].

5. ∂S is the circle $x^2 + y^2 = 12$, z=2.

Parameterization of circle: $x = \sqrt{12}\cos t$, $y = -\sqrt{12}\sin t$, z=2, t in [0,2π].

$$\oint_{\partial S} \mathbf{F} \cdot \mathbf{T}\, ds = \oint_{\partial S} yz\,dx + 3xz\,dy + z^2\,dz = \int_0^{2\pi} (24\sin^2 t - 72\cos^2 t)\,dt = -48\pi$$

$$\approx -150.80.$$

7. $(\text{curl}\mathbf{F}) \cdot \mathbf{n} = \langle 3,2,1 \rangle \cdot (1/\sqrt{2})\langle 1,0,-1 \rangle = \sqrt{2}.$

$$\oint_{\partial S} \mathbf{F} \cdot \mathbf{T}\, ds = \iint_S \sqrt{2}\,dS = \sqrt{2}A(S) = \sqrt{2}[\sec(45°)](\text{Area of circle}) = 8\pi \approx 25.1327.$$

9. $(\text{curl}\mathbf{F}) = \langle -1+1, 0-1, 1-1 \rangle = \langle 0,-1,0 \rangle.$ The unit normal vector that is needed to apply Stokes's Theorem points downward, It is

$$\mathbf{n} = \frac{\langle -1,-2,-1 \rangle}{\sqrt{6}}.$$

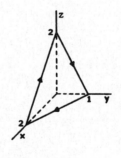

$$\oint_C \mathbf{F} \cdot \mathbf{T}ds = \iint_S (\text{curl}\mathbf{F}) \cdot \mathbf{n}\ dS = \iint_S (2/\sqrt{6})\,dS$$

$$= \iint_R (2/\sqrt{6})[1+4+1)]^{1/2}\,dA = \iint_R 2\,dA$$

$$= 2(\text{Area of triangle in xy-plane}) = 2(1) = 2.$$

11. $(\text{curl}\mathbf{F}) \cdot \mathbf{n} = \langle 0,0,1 \rangle \cdot \langle x,y,z \rangle = z.$

$$\iint_S z\,dS = \iint_R 1\,dA = \text{Area of R} = \pi(1/2)^2 = \pi/4 \approx 0.7854.$$

13. Let $H(x,y,z) = z - g(x,y) = 0$. Then $\mathbf{n} = \dfrac{\nabla H}{|\nabla H|} = \dfrac{<-g_x,-g_y,1>}{\sqrt{1+g_x^2+g_y^2}}$ points upward.

Thus, $\displaystyle\iint_S (\text{curl}\mathbf{F})\cdot\mathbf{n}\ dS = \iint_{S_{xy}} (\text{curl}\mathbf{F})\cdot\mathbf{n}\ \sec\gamma\ dA$

$\displaystyle = \iint_{S_{xy}} (\text{curl}\mathbf{F})\ \frac{<-g_x,-g_y,1>}{\sqrt{g_x^2+g_y^2+1}}\ \sqrt{g_x^2+g_y^2+1}\ dA$ [Theorem A, Section 17.5]

$\displaystyle = \iint_{S_{xy}} (\text{curl}\mathbf{F})\cdot<-g_x,-g_y,1>\ dA$

15. $\text{curl}\mathbf{F} = <0-x,0-0,z-0> = <-x,0,z>$.

$\displaystyle\oint_C \mathbf{F}\cdot\mathbf{T}\ ds = \iint_S (\text{curl}\mathbf{F})\cdot\mathbf{n}\ dS = \iint_{S_{xy}} (\text{curl}\mathbf{F})\cdot<-g_x,-g_y,1>\ dA$ [Problem 13]

where $z = g(x,y) = xy^2$.

$\displaystyle = \iint_{S_{xy}} <-x,0,z>\cdot<-y^2,-2xy,1>dA = \int_0^1\int_0^1 (xy^2+0+xy^2)dxdy$

$\displaystyle = \int_0^1\left[\left[x^2y^2\right]_{x=0}^1\right]dy = \int_0^1 y^2 dy = \frac{1}{3}$.

17. $\displaystyle\oint_{\partial S} \mathbf{F}\cdot\mathbf{T}\ ds = \iint_S (\text{curl}\mathbf{F})\cdot\mathbf{n}\ dS = \iint_{S_{xy}} (\text{curl}\mathbf{F})\cdot<g_x,g_y,1>\ dA$

$\displaystyle = \iint_{S_{xy}} <2,2,0>\cdot<x(a^2-x^2-y^2)^{-1/2},y(a^2-x^2-y^2)^{-1/2},1>$

$\displaystyle = 2\iint_{S_{xy}} (x+y)(a^2-x^2-y^2)^{-1/2}dA = 2\iint_{S_{xy}} y(a^2-x^2-y^2)^{-1/2}dA$

$\displaystyle = 4\int_0^{\pi/2}\int_0^{a\sin\theta} (r\sin\theta)(a^2-r^2)^{-1/2}r\ drd\theta = 4a^2/3\ \text{joules}.$

19. (a) Let C be any piecewise smooth simple closed oriented curve C that separates the "nice" surface into two "nice" surfaces, S_1 and S_2.

$$\iint_{\partial S} (\text{curl} \mathbf{F}) \cdot \mathbf{n} dS = \iint_{S_1} (\text{curl} \mathbf{F}) \cdot \mathbf{n} dS + \iint_{S_2} (\text{curl} \mathbf{F}) \cdot \mathbf{n} dS$$

$$= \oint_C \mathbf{F} \cdot \mathbf{T} ds + \oint_{-C} \mathbf{F} \cdot \mathbf{T} ds = 0. \quad [\text{-C is C with opposite orientation.}]$$

(b) $\text{div}(\text{curl} \mathbf{F}) = 0.$ [See Problem 20, Section 17.1.] Result follows.

Problem Set 17.8 Chapter Review

True-False Quiz

1. True. See Example 4, Section 17.1.

3. False. grad(curl$\mathbf{F}$) is not defined since curl$\mathbf{F}$ is not a scalar field.

5. True. See the three equivalent conditions in Section 17.3.

7. False. $N_z = 0 \neq z^2 = P_y$.

9. True. It is the case in which the surface is in a plane.

11. True. See discussion on text page 810.

Sample Test Problem

1.

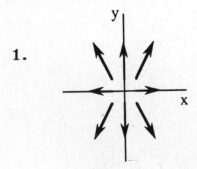

3. $\text{curl}(f \nabla f) = (f)(\text{curl} \nabla f) + (\nabla f \times \nabla f) = (f)(0) + 0 = 0.$

5. (a) Parametrization is x = sint, y = -cost, t in $[0, \pi/2]$.

$$\int_0^{\pi/2} (1-\cos^2 t)(\sin^2 t+\cos^2 t)^{1/2}dt = \pi/4 \approx 0.7854.$$

    (b) $\int_0^{\pi/2} [t\cos t - \sin^2 t \cos t + \sin t \cos t]dt = (3\pi-5)/6 \approx 0.7375.$

7. $\left[xy^2\right]_{(1,1)}^{(3,4)} = 47.$

9. (a) $\int_0^1 0dx + \int_0^1 (1+y^2)dy + \int_1^0 xdx + \int_1^0 y^2 dy$

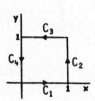

    $= 0 + \dfrac{4}{3} - \dfrac{1}{2} - \dfrac{1}{3} = \dfrac{1}{2}.$

    (b) A vector equation of C_3 is $\langle x,y \rangle$ = $\langle 2,1 \rangle$ + $t\langle -2, -1 \rangle$ for t in $[0,1]$, so let x = 2-2t, y = 1-t for t in $[0,1]$ be parametric equations of C_3.

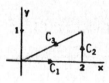

    $\int_0^2 0dx + \int_0^1 (4+y^2)dy + \int_0^1 [2(1-t)^2(-2) + 5(1-t)^2(-1)]dt$

    $- 0 + \dfrac{13}{3} - 3 = \dfrac{4}{3}.$

    (c) $\int_0^{2\pi} [(\cos t)(-\sin t)(-\sin t) + (\cos^2 t + \sin^2 t)(\cos t)]dt$

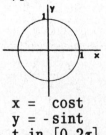

    $= \int_0^{2\pi} (1+\sin^2 t)\cos t \, dt$

    $= \left[\sin t + \dfrac{\sin^3 t}{3} \right]_0^{2\pi} = 0.$

    x = cost
    y = -sint
    t in $[0,2\pi]$.

11. By Gauss' Theory

$$\text{Flux} = \iiint_S \text{div}F \, dV$$

$$= \iiint_S 2 \, dV$$

$$= 2 \text{ (Vol. of Sphere)}$$

$$= \frac{8\pi}{3} \approx 8.3776$$

13. ∂S is the circle $x^2 + y^2 = 1$, $z = 1$.

A parametrization of circle is $x = \cos t$, $y = \sin t$, $z = 1$, t in $[0, 2\pi]$.

$$\oint_{\partial S} F \cdot T ds = \oint_{\partial S} [x^3 y dx + e^y dy + z\tan(xyz/4)dz] = \oint_{\partial S} (x^3 y + e^y dy)$$

$$= \int_0^{2\pi} [(\cos t)^3(\sin t)(-\sin t) + (e^{\sin t})(\cos t)]dt = 0.$$

15. $\text{curl}F = \langle 3-0, 0-0, -1-1 \rangle = \langle 3, 0, -2 \rangle$, $n = \dfrac{\langle a, b, 1 \rangle}{\sqrt{a^2+b^2+1}}$.

$$\oint_C F \cdot T \, ds = \iint_S (\text{curl}F) \cdot n \, dS$$

$$= \iint_S \frac{3a-2}{\sqrt{a^2+b^2+1}} \, dS = \frac{3a-2}{\sqrt{a^2+b^2+1}} [A(S)]$$

$$= \frac{3a-2}{\sqrt{a^2+b^2+1}} (9\pi) \qquad [\text{S is a circle of radius 3.}]$$

$$= \frac{9\pi(3a-2)}{\sqrt{a^2+b^2+1}}.$$

Problem Set 18.1 Linear First-Order Equations

Notes: (1) "I.F." denotes "Integrating Factor."

(2) After multiplying each side of $y + f(x)y = g(x)$ by an integrating factor, I.F., $D[y(I.F.)] = f(x)(I.F.)$

1. I.F. is e^x. $D(ye^x) = 1$. $y = e^{-x}(x+C)$.

3. $y' + \dfrac{x}{1-x^2}\, y = \dfrac{ax}{1-x^2}$. I.F.: $\exp 1\displaystyle\int \dfrac{x}{1-x^2}\, dx = \exp[\ln(1-x^2)^{-1/2}] = (1-x^2)^{-1/2}$.

$D[y(1-x^2)^{-1/2}] = ax(1-x^2)^{-3/2}$. [See note (2) above.]

Then $y(1-x^2)^{-1/2} = a(1-x^2)^{-1/2} + C$, so $y = a + C(1-x^2)^{1/2}$.

5. I.F. is $1/x$. $D[y/x] = e^x$. $y = xe^x + Cx$.

7. I.F is x. $D[yx] = 1$. $y = 1 + Cx^{-1}$.

9. $y' + f(x)y = f(x)$.   I.F.: $e^{\int f(x)dx}$.

$D[ye^{\int f(x)dx}] = f(x)\, e^{\int f(x)dx}$.

Then $ye^{\int f(x)dx} = e^{\int f(x)dx} + C$, so $y - 1 + Ce^{-\int f(x)dx}$.

11. I.F. is $1/x$. $D[y/x] = 3x^2$. $y = x^4 + Cx$. $y = x^4 + 2x$ goes through $(1,3)$.

13. I.F.: xe^x. $d[yxe^x] = 1$. $y = e^{-x}(1 + Cx^{-1})$. $y = e^{-x}(1 - x^{-1})$ through $(1,0)$.

15. Let y denote the number of pounds of chemical A after t minutes.
$\frac{dy}{dt}$ = (2 lbs/gal)(3gal/min) - (y lbs/20gal)(3gal/min) = $6 - \frac{3y}{20}$ lbs/min.

$y' + \frac{3}{20} y = 6.$ I.F.: $e^{\int (3/20)dx} = e^{3t/20}$.

$D[ye^{3t/20}] = 6e^{3t/20}$.

Then $ye^{3t/20} = 40e^{3t/20} + C.$ t=0, y=10, $\Rightarrow$ C = -30.
Therefore, $y(t) = 40 - 30e^{-3t/20}$, so $y(20) = 40 - 30e^{-3} \approx 38.506$ lbs.

17. dy/dt = 4 - [y/(120-2t)](6). $y' + [3/(60-t)]y = 4.$

I.F. is $(60-t)^{-3}$. $D[y(60-t)^{-3}] = 4(60-t)^{-3}$. $y(t) = 2(60-t)+C(60-t)^{3}$.

$y(t) = 2(60-t) - (1/1800)(60-t)^{3}$ goes through (0,0).

19. $I' + 10^{6}I = 1.$ I.F.= $\exp(10^{6}t)$. $D[I\exp(10^{6}t)] = \exp(10^{6}t)$.

$I(t) = 10^{-6}+C\exp(-10^{6}t)$. $I(t) = 10^{-6}[1-\exp(-10^{6}t)]$ goes through (0,0).

21. LI' + RI = E, so 3.5I' + 1000I = 120sin377t.

I' + 285.7143I = 34.2857sin377t.   I.F.: $e^{285.7143t}$.

$D[Ie^{285.7143t}] = 34.2857e^{285.7143t}\sin377t$.

Then, using intergration formula 67,

$$Ie^{285.7143t} \approx 34.2857\left[\frac{e^{285.7143t}}{223761.6612}(285.7143\sin377t - 377\cos377t)\right] + C$$
$$\approx 0.0001532e^{285.7143t}(285.7143\sin377t - 377\cos377t) + C.$$

t=0, I=0   C $\approx$ 0.05776.

(a) I $\approx$ 0.0001532(285.7143sin377t - 377cos377t) + $0.05776e^{-285.7143t}$

   $\approx$ 0.04377sin377t - 0.05776cos377t + $0.05776e^{-285.7t}$.

(b) As t → ∞, I → 0.04377sin377t - 0.05776cos377t.

23. Let y be the number of gallons of pure alcohol in the tank at time t.

(a) $y' = dy/dt = 5(0.25) - (5/100)y = 1.25 - 0.05y$ I.F. is $e^{0.05t}$.

$y(t) = 25 + Ce^{-0.05t}$; y=100, t=0 C=75.

$y(t) = 25 + 75e^{-0.05t}$; y=50, t=T T - $20(\ell n3) \approx 21.97$ min.

(b) Let A be the number of gallons of pure alcohol drained away.

$(100-A) + 0.25A = 50 \Rightarrow A = 200/3$.

It took $\dfrac{200/3}{5}$ minutes for the draining and the same amount of time

to refill, so $T = \dfrac{2(200/3)}{5} = \dfrac{80}{3} \approx 26.67$ min.

(c) C would need to satisfy $\dfrac{200/3}{5} + \dfrac{200/3}{C} < 20(\ln 3)$.

$C > 10/(3\ln 3 - 2) \approx 7.7170$.

(d) $y' = 4(0.25) - 0.05y = 1 - 0.05y$.

Solving for y, as in part (a), yields $y = 20 + 80e^{-0.05t}$.

We require that $(20+80e^{-0.05T}) + 0.25T = 50$, or $320e^{-0.05T}+T = 120$.

(e) Let $f(T) = 320e^{-0.05T}+T-120 = 0$; $f'(T) = -16e^{-0.05T}+1$.

Newton's Method: $T_{n+1} = \dfrac{16T_n + 320 - 120e^{0.05T_n}}{16 - e^{0.05T_n}}$

Beginning with $T_1 = 25$, one quickly obtains $T \approx 24.10$.

25 (a) $v_\infty = -32/0.05 = -640$.

$v(t) = [120-(-640)]e^{-0.05t} +(-640) = 0$ if $t = 20\ln(19/16)$.

$y(t) = 0 + (-640)t + (1/0.05)[120-(-640)](1-e^{-0.05t})$

$= -640t + 15200(1-e^{-0.05t})$. Therefore, the maximum altitude is

$y(20\ln(19/16)) = -12800\ln(19/16) + 45600/19 \approx 200.32$.

(b) $-640T + 15200(1-e^{-0.05T}) = 0$; $95 - 4T - 95e^{-0.05T} = 0$.

(c) Let $f(T) = 95 - 4T - 95e^{-0.05T}$; $f'(T) = -4 + 4.75e^{-0.05T}$.

$$T_n = [95+4.75T_n - 95\exp(0.05T_n)]/[4.75 - 4\exp(0.05T_n)]$$

Letting $T_1 = 7$, one quickly obtains that $T \approx 7.08$.

Problem Set 18.2 Second-Order Homogeneous Equations

1 Roots are 2 and 3. General solution is $y = C_1 e^{2x} + C_2 e^{3x}$.

3. Auxiliary equation: $r^2 + 6r - 7 = 0$, $(r+7)(r-1) = 0$ has roots $-7, 1$.
General solution: $y = C_1 e^{-7x} + C_2 e^x$.

$y' = -7C_1 e^{-7x} + C_2 e^x$.
If $x=0$, $y=0$, $y'=4$, then $0 = C_1 + C_2$ and $4 = -7C_1 + C_2$, so $C_1 = -1/2$ and $C_2 = 1/2$.

Therefore, $y = \dfrac{e^x - e^{-7x}}{2}$.

5. Repeated root 2. General solution is $y = (C_1 + C_2 x)e^{2x}$.

7. Roots are $2 \pm \sqrt{3}$. General solution is $y = e^{2x}(C_1 e^{\sqrt{3}x} + C_2 e^{-\sqrt{3}x})$.

9. Auxiliary Equation: $r^2 + 4 = 0$ has roots $\pm 2i$.
General Solution: $y = C_1 \cos 2x + C_2 \sin 2x$.
If $x=0$ and $y=2$, then $2 = C_1$; if $x = \pi/4$ and $y=3$, then $3 = C_2$.
Therefore, $y = 2\cos 2x + 3\sin 2x$.

11. Roots are $-1 \pm i$. General solution is $y = e^{-x}(C_1 \cos x + C_2 \sin x)$.

13. Roots are $0, 0, -4, 1$. General solution is $y = C_1 + C_2 x + C_3 e^{-4x} + C_4 e^x$.

15. Auxiliary Equation: $r^4 + 3r^2 - 4 = 0$, $(r+1)(r-1)(r^2+4) = 0$ has roots $-1, 1, \pm 2i$.
General Solution: $y = C_1 e^{-x} + C_2 e^x + C_3 \cos 2x + C_4 \sin 2x$.

17. Roots are $-2, 2$. General solution is $y = C_1 e^{-2x} + C_2 e^{2x}$.

$$y = C_1(\cosh 2x - \sinh 2x) + C_2(\sinh 2x + \cosh 2x)$$
$$= (-C_1 + C_2)\sinh 2x + (C_1 + C_2)\cosh 2x = D_1 \sinh 2x + D_2 \cosh 2x.$$

19. Repeated roots $(-1/2) \pm (\sqrt{3}/2)i$.

General solution is $y = e^{-x/2}[(C_1 + C_2 x)\cos(\sqrt{3}/2)x + (C_3 + C_4 x)\sin(\sqrt{3}/2)x]$.

21. (*) $x^2 y'' + 5xy' + 4y = 0$.

Let $x = e^z$. Then $z = \ln x$; $y' = \dfrac{dy}{dx} = \dfrac{dy}{dz}\dfrac{dz}{dx} = \dfrac{dy}{dz}\dfrac{1}{x}$;

$$y'' = \frac{dy'}{dx} = \frac{d}{dx}\left[\frac{dy}{dz}\frac{1}{x}\right] = \frac{dy}{dz}\left[\frac{-1}{x^2}\right] + \frac{1}{x}\frac{d^2y}{dz^2}\frac{dz}{dx} = \frac{dy}{dz}\left[\frac{-1}{x^2}\right] + \frac{1}{x}\frac{d^2y}{dz^2}\frac{1}{x}.$$

$$\left[-\frac{dy}{dz} + \frac{d^2y}{dz^2}\right] + \left[5\frac{dy}{dz}\right] + 4y = 0. \quad \text{[Substituting } y' \text{ and } y'' \text{ into (*).]}$$

$$\frac{d^2y}{dz^2} + 4\frac{dy}{dz} + 4y = 0.$$

Auxiliary Equation: $r^2 + 4r + 4 = 0$, $(r+2)^2 = 0$ has roots $-2, -2$.

General Solution: $y = (C_1 + C_2 z)e^{-2z}$, $y = (C_1 + C_2 \ln x)e^{-2\ln x}$.

$$y = (C_1 + C_2 \ln x)x^{-2}.$$

23. (a) $e^{bi} = 1 + (bi) + \dfrac{(bi)^2}{2!} + \dfrac{(bi)^3}{3!} + \dfrac{(bi)^4}{4!} + \dfrac{(bi)^5}{5!} + \cdots$

$$= \left[1 - \frac{b^2}{2!} + \frac{b^4}{4!} - \frac{b^6}{6!} + \right] \cdots + i\left[b - \frac{b^3}{3!} + \frac{b^5}{5!} - \frac{b^7}{7!} + \cdots\right]$$

$$= \cos(b) + i\sin(b).$$

(b) $e^{a+bi} = e^a e^{bi} = e^a[\cos(b) + i\sin(b)]$

(c) $D_x[e^{(a+\beta i)x} = D_x[e^{ax}(\cos\beta x + i \sin\beta x)]$

$$= ae^{ax}(\cos\beta x + i \sin\beta x) + e^{ax}(-\beta\sin\beta x + i\beta \cos\beta x)$$

$$= e^{ax}[(a+\beta i)\cos\beta x + (ai-\beta)\sin\beta x]$$

$$(a+\beta)e^{(a+\beta i)x} = (a+\beta i)[e^{ax}(\cos\beta x + i\sin\beta x)]$$

$$= e^{ax}[(a+\beta i)\cos\beta x + (ai-\beta)\sin\beta x]$$

Therefore, $D_x[e^{(a+\beta i)x} = (a+\beta)e^{(a+\beta i)x}]$

25. $y = 0.5e^{5.16228x} + 0.5e^{-5.16228x}$.

27. $y = 1.29099e^{-0.25x}\sin(0.968246x)$.

Problem Set 18.3   The Nonhomogeneous Equation

1. $y_h = C_1e^{-3x} + C_2e^{3x}$. $y_p = (-1/9)x+0$. $y = (-1/9)x + C_1e^{-3x} + C_2e^{3x}$.

3. Auxiliary Equation: $r^2-2r+1 = 0$ has roots 1,1.

$$y_h = (C_1+C_2x)e^x.$$

Let $y_p = Ax^2+Bx+C$; $y_p' = 2Ax+B$; $y_p'' = 2A$.

Then $(2A) -2(2Ax+B) + (Ax^2+Bx+C) = x^2+x$.
$Ax^2 + (-4A+B)x + (2A-2B+C) = x^2+x$.

Thus, $A=1$, $-4A+B = 1$, $2A-2B+C = 0$, so $A = 1$, $B = 5$, $C = 8$.
General Solution: $y = x^2 + 5x + 8 + (C_1+C_2x)e^x$.

5. $y_h = C_1e^{2x} + C_2e^{3x}$. $y_p = (1/2)e^x$. $y = (1/2)e^x + C_1e^{2x} + C_2e^{3x}$.

7. $y_h = C_1e^{-3x} + C_2e^{-x}$. $y_p = (-1/2)xe^{-3x}$. $y = (-1/2)e^{-3x} + C_1e^{-3x} + C_2e^{-x}$.

9. Auxiliary Equation: $r^2 - r - 2 = 0$, $(r+1)(r-2) = 0$ has roots $-1, 2$.

$y_h = C_1 e^{-x} + C_2 e^{2x}$.

Let $y_p = B\cos x + C\sin x$; $y_p' = -B\sin x + C\cos x$; $y_p'' = -B\cos x - C\sin x$.

Then $(-B\cos x - C\sin x) - (-B\sin x + C\cos x) - 2(B\cos x + C\sin x) = 2\sin x$.

$(-3B - C)\cos x + (B - 3C)\sin x = 2\sin x$, so $-3B - C = 0$ and $B - 3C = 2$; $B = \frac{1}{5}$; $C = \frac{-3}{5}$.

General Solution: $(1/5)\cos x - (3/5)\sin x + C_1 e^{2x} + C_2 e^{-x}$.

11. $y_h = C_1\cos 2x + C_2\sin 2x$. $y_p = (0)x\cos 2x + (1/2)x\sin 2x$.

$y = (1/2)x\sin 2x + C_1\cos 2x + C_2\sin 2x$.

13. $y_h = C_1\cos 3x + C_2\sin 3x$. $y_p = (0)\cos x + (1/8)\sin x + (1/13)e^{2x}$.

$y = (1/8)\sin x + (1/13)e^{2x} + C_1\cos 3x + C_2\sin 3x$.

15. Auxiliary Equation: $r^2 - 5r + 6 = 0$ has roots $2, 3$, so $y_h = C_1 e^{2x} + C_2 e^{3x}$.

Let $y_p = Be^x$; $y_p' = Be^x$; $y_p'' = Be^x$.

Then $(Be^x) - 5(Be^x) + 6(Be^x) = 2e^x$; $2Be^x = 2e^x$; $B = 1$.

General Solution: $y = e^x + C_1 e^{2x} + C_2 e^{3x}$.

$y' = e^x + 2C_1 e^{2x} + 3C_2 e^{3x}$.

If $x = 0$, $y = 1$, $y' = 0$, then $1 = 1 + C_1 + C_2$ and $0 = 1 + 2C_1 + 3C_2$; $C_1 = 1$, $C_2 = -1$.

Therefore, $y = e^x + e^{2x} - e^{3x}$.

17. $y_h = C_1 e^x + C_2 e^{2x}$. $y_p = (1/4)(10x+19)$.

$y = (1/4)(10x+19) + C_1 e^x + C_2 e^x$.

19. $y_h = C_1\cos x + C_2\sin x$. $y_p = -\cos x \ln|\sin x| - \cos x - x\sin x$.

$y = -\cos x \ln|\sin x| - x\sin x + C_3\cos x + C_2\sin x$. [Combined cos x terms].

---

21. Auxiliary Equation: $r^2 - 3r + 2 = 0$ has roots 1,2, so $y_h = C_1 e^x + C_2 e^{2x}$.

Let $y_p = v_1 e^x + v_2 e^{2x}$ subject to $v_1' e^x + v_2' e^{2x} = 0$,

$$\text{and } v_1'(e^x) + v_2'(2e^{2x}) = e^x(e^x+1)^{-1}.$$

Then $v_1' = \dfrac{-e^x}{e^x(e^x+1)}$ so $v_1 = \displaystyle\int \dfrac{-e^x}{e^x(e^x+1)}\,dx = \int \dfrac{-1}{u(u+1)}\,du = \int\left[\dfrac{-1}{u} + \dfrac{1}{u+1}\right]du$

$$= -\ln u + \ln(u+1) = \ln\left(\dfrac{u+1}{u}\right) = \ln \dfrac{e^x+1}{e^x} = \ln(1+e^{-x}).$$

$$v_2' = \dfrac{e^x}{e^{2x}(e^x+1)} \text{ so } v_2 = -e^{-x} + \ln(1+e^{-x}) \text{ [similar to finding } v_1\text{]}.$$

General Solution: $y = e^x \ln(1+e^{-x}) - e^x + e^{2x}\ln(1+e^{-x}) + C_1 e^x + C_2 e^{2x}$.

$$y = (e^x + e^{2x})\ln(1+e^{-x}) + D_1 e^x + D_2 e^{2x}.$$

23. $L(y_p) = (v_1 u_2 + v_2 u_2)'' + b(v_1 u_1 + v_2 u_2)' + c(v_1 u_1 + v_2 u_2)$

$= (v_1' u_1 + v_1 u_1' + v_2' u_2 + v_2 u_2')' + b(v_1' u_1 + v_1 u_1' + v_2' u_2 + v_2 u_2') + c(v_1 u_1 + v_2 u_2)$

$= (v_1'' u_1 + v_1' u_1' + v_1' u_1' + v_1 u_1'' + v_2'' u_2 + v_2' u_2' + v_2' u_2' + v_2 u_2'')$

$\qquad\qquad + b(v_1' u_1 + v_1 u_1' + v_2' u_2 + v_2 u_2') + c(v_1 u_1 + v_2 u_2)$

$= v_1(u_1'' + b u_1' + c u_1) + v_2(u_2'' + b u_2' + c u_2) + b(v_1' u_1 + v_2' u_2) +$

$\qquad\qquad + (v_1'' u_1 + v_1' u_1' + v_2'' u_2 + v_2' u_2) + (v_1' u_1' + v_2' u_2')$

$= v_1(u_1'' + b u_1' + c u_1) + v_2(u_2'' + b u_2' + c u_2) + b(v_1' u_1 + v_2' u_2) +$

$\qquad\qquad + (v_1' u_1 + v_2' u_2)' + (v_1' u_1' + v v_2' u_2')$

$= v_1(0) + v_2(0) + b(0) + (0) + k(x) = k(x).$

Problem Set 18.4 Applications of Second-Order Equations

1. $k = 20$ lb/ft, $w = 10$ lb, $g = 32$ ft/sec^2, $y_0 = -1$ ft, $B = 8$.

Then, using the result developed on text page 786, $y = -\cos 8t$.
Period is $\pi/4 \approx 0.7854$ sec.

3. Equilibrium position is where $y = 0$; $0 = -\cos 8t$; $8t = \pi/2, 3\pi/2, \cdots$.
$\quad t = \pi/16, 3\pi/16, \cdots$.
At each of these values of t, $|y'(t)| = |8\sin t| = 8$ ft/sec.

5. k=20 lb/ft; w = 10 lbs; y_o = 1 ft; q = 1/10 sec-lb/ft, B = 8, E = 0.32.

$E^2 - 4B^2 < 0$, so there is damped motion.

Roots of auxiliary equation are approximately $-0.16 \pm 8i$.

General solution is $y \approx e^{-0.16t}(C_1 \cos 8t + C_2 \sin 8t)$

$y \approx e^{-0.16t}(\cos 8t + 0.02 \sin 8t)$ satisfies the initial conditions.

7. Original amplitude is 1 ft. Considering the contribution of the sine term to be negligible due to the 0.02 coefficient, the amplitude is approximately $e^{-0.16t}$.

$e^{-0.16t} \approx 0.1$ if $t \approx 14.39$, so amplitude will be about one-tenth of original in about 14.4 seconds.

9. $LQ'' + RQ' + \dfrac{Q}{C} = E(t)$; $10^6 Q' + 10^6 Q = 1$; $Q' + Q = 10^{-6}$.   I.F.: e^t.

$D[Qe^t] = 10^{-6}e^t$; $Qe^t = 10^{-6}e^t + C$; $Q = 10^{-6} + Ce^{-t}$.
If t=0, Q=0, then $C = -10^{-6}$.
Therefore, $Q(t) = 10^{-6} - 10^{-6}e^{-t} = 10^{-6}(1 - e^{-t})$.

11. $Q/[2(10^{-6})] = 120 \sin 377t$.

(a) $Q(t) = 0.00024 \sin 377t$.   (b) $I(t) = Q'(t) = 0.09048 \cos 377t$.

13. $3.5Q'' + 1000Q + Q/[2(10^{-6})] = 120 \sin 377t$.

[Values are approximate (6 significant figures) the rest of the way.]

$Q'' + 285.714Q' + 142857Q = 34.2857 \sin 377t$.

Roots of the auxiliary equation are $-142.857 \pm 349.927i$.

$Q_h = e^{-142.857t}(C_1 \cos 349.927t + C_2 \sin 349.927t)$.

$Q_p = -3.18288(10^{-4})\cos 377t + 2.15119(10^{-6})\sin 377t$.

Then, $Q = -3.18288(10^{-4})\cos 377t + 2.15119(10^{-6})\sin 377t + Q_h$.

$I = Q' = 0.119995\sin 377t + 0.000811888\cos 377t + Q_h'$.

$0.000888\cos 377t$ is small and $Q_h' \to 0$ as $t \to \infty$, so the steady-state current is $I \approx 0.12\sin 377t$.

15. $C\sin(\beta t+\gamma) = C(\sin\beta t\cos\gamma + \cos\beta t\sin\gamma) = (C\cos\gamma)\sin\beta t + (C\sin\gamma)\cos\beta t$
$= C_1\sin\beta t + C_2\cos\beta t$, where $C_1 = C\cos\gamma$ and $C_2 = C\sin\gamma$.

[Note that $C_1^2 + C_2^2 = C^2\cos^2\gamma + C^2\sin^2\gamma = C^2$.]

17. The magnitudes of the tangential components of the forces acting on the pendulum bob must be equal.

Therefore, $-m\dfrac{d^2s}{dt^2} = mg\sin\theta$.

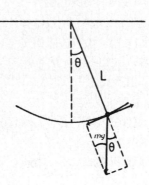

$s = L\theta$, so $\dfrac{d^2s}{dt^2} = L\dfrac{d^2\theta}{dt^2}$.

Therefore, $-mL\dfrac{d^2\theta}{dt^2} = mg\sin\theta$.

Hence, $\dfrac{d^2\theta}{dt^2} = -\dfrac{g}{L}\sin\theta$.

Problem Set 18.5 Chapter Review

True-False Quiz

1. False. y^2 is not linear in y.

3. True. $y' = \sec^2 x + \sec x\tan x$.
$2y'-y^2 = (2\sec^2 x + 2\sec x\tan x) - (\tan^2 x + 2\sec x\tan x + \sec^2 x)$

$= \sec^2 x - \tan^2 x = 1$.

5. True. $e^{\int(4/x)dx} = e^{4\ln x} = x^4$.

7. True.   -1 is a repeated root, with multiplicity 3, of the auxiliary equation.

9. False.   That is the form of y_h.

y_p should have the form Bxcos3x + Cxsin3x.

Sample Test Problems

1. I.F. is $|x|$. $D[y|x|] = 0$. $y = Cx^{-1}$.

3. (Linear first-order) $y' + 2xy = 2x$.   I.F.: $e^{\int 2xdx} = e^{x^2}$.

$D[ye^{x^2}] = 2xe^{x^2}$; $ye^{x^2} = e^{x^2} + C$; $y = 1 + Ce^{-x^2}$.

If x=0, y=3, then 3 = 1+C, so C = 2.

Therefore, $y = 1+2e^{-x^2}$.

5. I.F. is e^{-2x}. $D[ye^{-2x}] = e^{-x}$. $y = -e^x+Ce^{2x}$.

7. $u' + 3u = e^x$. I.F. is e^{3x}. $D[ue^{3x}] = e^{4x}$. $u = (1/4)e^x + C_1e^{-3x}$.

$y' = (1/4)e^x + C_1e^{-3x}$. $y - (1/4)e^x + C_3e^{-3x} + C_2$.

9. (Second-order homogeneous)
The auxiliary equation, $r^2-3r+2 = 0$, has roots 1,2.
The general solution is $y = C_1e^x + C_2e^{2x}$.

$y' = C_1e^x + 2C_2e^{2x}$.
If x=0, y=0, y'=3, then $0 = C_1+C_2$ and $3 = C_1+2C_2$, so $C_1=-3$, $C_2=3$.
Therefore, $y = -3e^x+3e^{2x}$.

11. $y_h = C_1e^{-x} + C_2e^x$ (Problem 8). $y_p = -1 + C_1e^{-x} + C_2e^x$.

13. $y_h = (C_1+C_2x)e^{-2x}$ (Problem 12). $y_p = (1/2)x^2e^{-2x}$.
$y = [(1/2)x^2 + C_1 + C_2x]e^{-2x}$.

15. (Second-order homogeneous)

     The auxiliary equation, $r^2+6r+25 = 0$, has roots $-3\pm4i$.

     General Solution: $y = e^{-3x}(C_1\cos4x + C_2\sin4x)$.

17. Roots are $-4,0,2$. $y = C_1e^{-4x} + C_2 + C_3e^{2x}$.

19. Repeated roots $\pm\sqrt{2}$. $y = (C_1+C_2x)e^{-\sqrt{2}x} + (C_3+C_4x)\ e^{\sqrt{2}x}$.

21. (Simple harmonic motion)

     $k = 5$; $w = 10$; $y_0 = -1$.

     $B = \dfrac{(5)(32)}{10} = 4$.

     Then the equation of motion is $y = -\cos4t$.

     The amplitude is $|-1| = 1$; the period is $2\pi/4 = \pi/2$.

23. $Q'' + 2Q' + 2Q = 1$. Roots are $-1\pm i$.

     $Q_h = e^{-t}(C_1\cos t + C_2\sin t)$ and $Q_p = 1/2$; $Q = e^{-t}(C_1\cos t + C_2\sin t) + 1/2$.

     $I(t) = Q'(t) = -e^{-t}[(C_1-C_2)\cos t + (C_1+C_2)\sin t]$.

     $I(t) = e^{-t}\sin t$ satisfies the initial conditions.

CONVERGENCE TESTS	Calculus with Analytic Geometry	Varberg & Purcell 6th Edition
Name	**Statement**	**Comment**
Geometric Series Page 492 Section 11.2 Ex. 1	The geometric series $a + ar + ar^2 + \cdots + ar^{n-1} + \cdots$ with $a = 0$ (i) converges and has sum $\frac{a}{1-r}$ if $\lvert r \rvert \epsilon \lvert$ (ii) diverges if $\lvert r \rvert \geq 1$	Note that $r = \frac{a_{n+1}}{a_n}$ This is one of the obvious series.
Divergence Test Page 493 Section 11.2 THM A	if $\lim\limits_{n \to \infty} a_n = 0$, then infinite Series $\sum a_n$ is divergent	If $\lim\limits_{n \to \infty} a_n = 0, \sum a_n$ may or may not converge.
Integral Test Page 501 Section 11.3 THM B	If a function f is positive valued, continuous, and decreasing for $x \geq 1$, then the infinite series $f(1) + f(2) + \cdots + f(n) + \cdots$ (i) converges if $\int_1^\infty f(x)dx$ converges (ii) diverges if $\int_1^\infty f(x)dx$ diverges	Use this test when $f(x)$ if easy to integrate. Choose f so that $a_k = f(k)$.
P-Series Page 502 Section 11.3 Ex. 2	The p-series $\sum\limits_{n=1}^{\infty} \frac{1}{n^p}$ (i) converges if $p > 1$ (ii) diverges if $p \leq 1$	This series is useful with comparison tests.
Ordinary Comparison Test Page 506 Section 11.4 THM A	Suppose $\sum a_n$ and $\sum b_n$ are positive term series (i) If $\sum b_n$ converges and $a_n \leq b_n$ for every positive integer n, then $\sum a_n$ converges. (ii) If $\sum b_n$ diverges and $a_n \geq b_n$ for every positive integer n, then $\sum a_n$ diverges	This test is to be used as a last resort. Other tests are often easier. Keep in mind the Harmonic series (Page 494 Section 11.2, Ex. 5).
Limit Comparison Test Page 508 Section 11.4 THM B	If $\sum a_n$ and $\sum b_n$ are positive term series and if: $\lim\limits_{n \to \infty} \frac{a_n}{b_n} = k > 0$ then either both series converge or both diverge. If $\sum b_n$ converges and $\lim\limits_{n \to \infty} \frac{a_n}{b_n} = 0$, then $\sum a_n$ converges. If $\sum b_n$ diverges and $\lim\limits_{n \to \infty} \frac{a_n}{b_n} = \infty$, then $\sum a_n$ diverges.	This is easier to apply than the comparison test, but still requires some ingenuity and skill in selecting the series $\sum b_k$ for comparison. Use b_n as the quotient of the leading terms from numerator and denominator if a_n is a rational expression in n.
Ratio Test Page 509 Section 11.4 THM C	Let $\sum a_n$ be a series of positive terms and suppose $\lim\limits_{n \to \infty} \frac{a_n + 1}{a_n} = \rho$ (i) if $\rho < 1$, the series converges. (ii) If $\rho > 1$, the series diverges. (iii) If $\rho = 1$, the test is inconclusive.	This test will always fail for a series whose n^{th} term is a rational expression in n, since in this case $\rho = 1$. Use for series whose n^{th} terms involves $n!$ or r^n or n^n.

Name	Statement	Comments
Alternating Series Test Page 514 Section 11.5 THM A	The series $a_1 - a_2 + a_3 - a_4 + \cdots$ and $-a_1 + a_2 - a_3 + a_4 - \cdots$ converge if (i) $a_1 \geq a_2 \geq a_3 \geq \cdots$ (ii) $\lim\limits_{k \to \infty} a_k = 0$	This test applies *only* to alternating series
Absolute Ratio Test Page 516 Section 11.5 THM C	Let Σa_n be an infinite series of non-zero terms such that $$\lim_{n \to \infty} \left| \frac{a_n + 1}{a_n} \right| = L$$ (i) The series converges absolutely if $L < 1$. (ii) The series diverges if $L > 1$ or $L = +\infty$. (iii) No conclusion if $L = 1$. (Series may converge or diverge.)	The series need not have positive terms and need not be alternating to use this test. An important test for absolute convergence. Use for interval of convergence.
Root Test Page 512 Section 11.4 Problem #41	Let Σa_n be an infinite series such that $$\lim_{n \to \infty} \sqrt[n]{|a_n|} = L$$ (i) The series converges absolutely if $L < 1$. (ii) The series diverges if $L > 1$ or $L = \infty$. (iii) No conclusion if $L = 1$ (Series may converge or diverge.)	Try this test when a_n involves only powers of n. Do not apply if a_n contains factorials.

In order to determine the convergence or divergence of a given series, the following procedure of testing is suggested.

Tests for Convergence

1. If $\lim a_n \neq 0$ as $n \to \infty$, then the series diverges.

2. If $\lim a_n = 0$ as $n \to \infty$, then try:

    (a) The alternating series test. If this test does not apply, try:

    (b) The ratio test. If this test fails, try:

    (c) The limit comparison test. If this is difficult, try:

    (d) The integral test.

    (e) You are on your own. Use your imagination.

Prepared by Professor Bruce Ransom, Clark College.